MODERN ASPECTS OF ELECTROCHEMISTRY

No. 27

LIST OF CONTRIBUTORS

A. R. DESPIĆ
Center of Electrochemistry
Institute of Chemistry, Technology and
Metallurgy
University of Belgrade
Belgrade, Yugoslavia

T. Z. FAHIDY
Department of Chemical Engineering
University of Waterloo
Waterloo, Ontario
Canada N2L 3G1

THOMAS F. FULLER
Department of Chemical Engineering
University of California
Berkeley, California 94720

JOSÉ R. GALVELE
Gerencia de Desarrollo
Comisión Nacional de Energía Atómica
1429-Buenos Aires, Argentina

Z. H. GU
Department of Chemical Engineering
University of Waterloo
Waterloo, Ontario
Canada N2L 3G1

V. D. JOVIĆ
Center for Interdisciplinary Studies
University of Belgrade
Belgrade, Yugoslavia

JOHN NEWMAN
Materials Sciences Division
Lawrence Berkeley Laboratory
Berkeley, California 94720

GERD SANDSTEDE
Battelle Europe
Battelle Institut e. V.
D-60486 Frankfurt, Germany

K. SCOTT
Department of Chemical and Process
Engineering
The University of Newcastle-upon-Tyne
Newcastle-upon-Tyne NE1 7RU,
United Kingdom

REINHOLD WURSTER
Ludwig-Bölkow-Systemtechnik GmbH
D-85521 Ottobrunn,
Germany

A Continuation Order Plan is available for this series. A continuation order will bring delivery of each new volume immediately upon publication. Volumes are billed only upon actual shipment. For further information please contact the publisher.

MODERN ASPECTS OF ELECTROCHEMISTRY

No. 27

Edited by

RALPH E. WHITE
Department of Chemical Engineering
University of South Carolina
Columbia, South Carolina

J. O'M. BOCKRIS
Department of Chemistry
Texas A&M University
College Station, Texas

and

B. E. CONWAY
Department of Chemistry
University of Ottawa
Ottawa, Ontario, Canada

PLENUM PRESS • NEW YORK AND LONDON

The Library of Congress cataloged the first volume of this title as follows:

Modern aspects of electrochemistry. no. [1]
　Washington Butterworths, 1954–
　v. illus., 23 cm.
　No. 1–2 issued as Modern aspects series of chemistry.
　Editors: no 1– J. Bockris (with B. E. Conway, No. 3–)
　Imprint varies: no. 1, New York, Academic Press.—No. 2, London, Butterworths.
　　1. Electrochemistry—Collected works. I. Bockris, John O'M.ed. II. Conway, B.E. ed.
(Series: Modern aspects series of chemistry)
QD552.M6   54-12732 rev

ISBN 0-306-44930-7

© 1995 Plenum Press, New York
A Division of Plenum Publishing Corporation
233 Spring Street, New York, N. Y. 10013

10 9 8 7 6 5 4 3 2 1

All rights reserved

No part of this book may be reproduced, stored in a retrieval system, or transmitted in any form
or by any means, electronic, mechanical, photocopying, microfilming, recording, or otherwise,
without written permission from the Publisher

Printed in the United States of America

Preface

This volume of *Modern Aspects of Electrochemistry* includes six chapters, ranging in their focus over a wide array of disciplines—from fundamental electrochemistry, as evidenced by electrode processes, to applied electrochemistry, represented by a discussion of water electrolysis in solar hydrogen demonstration projects.

In the first chapter, Kenneth Scott offers a comprehensive treatise delineating the trends and techniques of electrochemical analysis and investigation into a wide range of production and research applications.

Chapter 2, by Despić and Jović, describes developments in materials science that address the expanding industrial interest in alloy plating and its role in production in the corrosion protection electronics industries.

Galvele's third chapter discusses the electrochemical aspects of stress-corrosion-cracking and reviews significant contributions that have led to a better understanding of this interdisciplinary problem, as well as the limitations of the discipline of electrochemistry in offering solutions.

The fourth chapter, by Fuller and Newman, discusses the principles of operation of metal hydride electrodes and offers keys to improving their performance. In particular, the authors compare the NiOOH–metal hydride cell to competing technologies and present developmental technical challenges.

Chapter 5, by Fahidy and Gu, presents an overview of recent advances in the theoretical and experimental study of the dynamics of electrode processes, and discusses the oxidation of copper, iron, cobalt, nickel, silicon, hydrogen, organic compounds, and other systems. The authors refer particularly to the publications in the field that have appeared during the last five years.

The sixth chapter, co-authored by Sandstede and Wurster, is a presentation of current developments in water electrolysis and solar hydrogen demonstration projects. Characteristics of the coupling of water electrolysis with hydropower, wind energy, and photovoltaics are also discussed, and the authors present an overview of existing worldwide demonstration projects: hydropower/hydrogen, wind power/hydrogen, and solar hydrogen.

Ralph E. White
University of South Carolina
John O'M. Bockris
Texas A & M University
Brian E. Conway
University of Ottawa

Contents

Chapter 1

REACTION ENGINEERING AND DIGITAL SIMULATION IN ELECTROCHEMICAL PROCESSES

K. Scott

I. Introduction... 1
II. Electrode Kinetics and Reaction Rate Modeling............. 5
   1. Kinetic Equations for Electrochemical Reactions 7
   2. Mass Transfer and the Reaction Rate Model 9
   3. Categories of Electrochemical Rate Models............. 10
   4. The Role of Adsorption 13
   5. Conclusions 22
III. Reactor Modeling...................................... 22
   1. Ideal Reactors; Batch and Continuous................. 24
   2. Real and Nonideal Flow Models 42
IV. Electroanalytical Experiment and Reaction Modeling........ 49
   1. Introduction 49
   2. Steady-State Methods 51
   3. The Rotating Disk Electrode and Fast Reaction 58
   4. Digital Simulation and Electrode Processes 63
   5. Some Conclusions; Recent and Future Developments 73
V. Electrochemical Reactor Analysis 74
   1. Introduction 74
   2. A Review of Reactor Engineering Models 74
   3. The Requirements of a Reactor Model 80

4. An Approach to Mathematical Modeling of
   Electrochemical Reactions 81
5. Engineering Models of Complex Electrochemical
   and Homogeneous Chemical Reaction Schemes 96
6. Parallel-Plate Electrode Reactor Models 110
7. Fixed and Porous Bed Electrodes 113
8. Multiphase Electrochemical Reaction Systems 123
9. Conclusions 134
References 139

Chapter 2

ELECTROCHEMICAL DEPOSITION AND DISSOLUTION OF
ALLOYS AND METAL COMPOSITES—
FUNDAMENTAL ASPECTS

A. R. Despić and V. D. Jović

I. Introduction 143
II. Conditions for Codeposition of Metals
    by an Electrochemical Process 145
III. Equilibrium (Reversible) Potential of an Alloy in
     Contact with a Solution 147
   1. Effect of Gibbs Energy of Phase Formation 148
   2. Stability of an Alloy Phase in Solutions of the
      Corresponding Ions 151
   3. Effect of Gibbs Energies of Formation in
      View of the Type of Alloy 153
IV. Cathodic Processes in Electrodeposition of Alloys 159
   1. Equilibrium Codeposition 161
   2. Activation-Controlled (Irregular) Codeposition 166
   3. Effects of Adsorption on Irregular Deposition 173
   4. Transport-Controlled (Regular) Codeposition 176
   5. Anomalous Codeposition 179
   6. Induced Codeposition 182
V. Laminar Electrodeposition of Metals 183
   1. Spontaneous Formation of a Layered Deposit 183

2. Formation of Laminar Deposits by Pulsating Current 184
3. Dual-Current Pulse Laminar Deposition 190
VI. Deposition with Inclusion of Nonmetallic Materials 196
  1. Description of the Processes Involved in the
     Codeposition of Metals and Inert Particles 197
  2. The Mechanism of Codeposition of Inert
     Particles and Metals 201
VII. Characterization of Phase Structure of Alloys by
   Electrochemical Techniques 205
  1. Conditions for Identification of Phases in
     Thin Layers of Alloys 205
  2. Dissolution Characteristics of Different Alloys 206
  3. Quantitative Aspects 221
  4. Mechanism of Dissolution of Binary Alloys 222
  5. Kinetics of Dissolution of Binary Alloys. 225
  6. ALSV Investigation of Phase-Transformation Kinetics ... 226
References 229

Chapter 3

ELECTROCHEMICAL ASPECTS OF
STRESS CORROSION CRACKING

José R. Galvele

I. Introduction 233
  1. Definitions 233
  2. Range of the Phenomenon 245
  3. Experimental Techniques 245
  4. Historical Background 246
II. Crack Tip Chemistry 252
  1. Crack Chemistry Measurements 252
  2. Model Calculations for Pits, Crevices, and Cracks 253
III. Preexisting Localized Paths 261
  1. Aluminum Alloys 262
  2. Sensitized Stainless Steels 266
  3. Grain Boundaries 269

IV. Deleterious Anodic Films 270
V. Mechanisms Based upon Active Dissolution 272
   1. Introduction 272
   2. Elastic Strains 273
   3. Plastic Strains on Film-Free Metals 278
   4. Anodic Dissolution SCC Models 283
   5. Limitations of the Mechanisms 293
VI. Anodic Dissolution and Filmed Metals 297
   1. Introduction 297
   2. Plastic Strains on Filmed Metals 298
   3. Repassivation and Slip Step Dissolution Models 307
   4. Repassivation Rate Measurements 310
   5. Limitations of the Mechanisms 313
VII. Mechanisms Based upon Discontinuous Cleavage 315
   1. Continuous versus Discontinuous Cleavage 315
   2. The Two-Stage Model for Stress Corrosion Cracking .. 317
   3. Advantages and Limitations of the Mechanisms 318
VIII. Mechanisms Based upon Surface Mobility 321
   1. Surface-Mobility SCC Mechanism 322
   2. Surface Self-Diffusion Coefficients 326
   3. Effect of Hydrogen 330
   4. Experimental Observations 332
   5. Sources of Vacancies 342
   6. The Mechanism as a Predictive Tool 347
   7. Limitations of the Model 348
IX. General Conclusions 348
References 350

Chapter 4

METAL HYDRIDE ELECTRODES

Thomas F. Fuller and John Newman

I. Introduction 359
II. Fundamentals 362
   1. Thermodynamics 362

2. Electrode Kinetics 365
3. Transport Phenomena. 366
III. Analysis. 369
   1. The NiOOH–Metal Hydride Cell 369
   2. Optimization of Alloys. 370
   3. Stability 372
   4. Cell Design. 373
   5. Cell Pressure. 373
   6. Self-Discharge 375
   7. Competing Technologies 376
   8. Assessment of Research Needs 377
List of Symbols 379
References 380

Chapter 5

RECENT ADVANCES IN THE STUDY OF THE DYNAMICS OF ELECTRODE PROCESSES

T. Z. Fahidy and Z. H. Gu

I. Introduction. 383
II. Anodic Processes 384
   1. The Oxidation of Copper 384
   2. The Oxidation of Iron 387
   3. The Oxidation of Cobalt 389
   4. The Oxidation of Nickel 391
   5. The Oxidation of Silicon 391
   6. The Oxidation of Hydrogen. 393
   7. Organic Oxidations 394
   8. Other Oxidation Systems. 395
III. Cathodic Processes 396
   1. The Reduction of Zinc Ions 396
   2. The Reduction of Indium(III) Ions 397
   3. The Reduction of Hydrogen Peroxide 397
   4. Other Reduction Systems. 398

IV. Mathematical Techniques Applied to the Analysis of
Dynamic Phenomena: Data Interpretation and Modeling.... 399
   1. Geometrical Techniques—Construction of the
      Phase Portraits and Return Maps 399
   2. Fractional Brownian Motion (FBM) Theory 400
   3. The Time Differencing Technique:
      Box–Jenkins Approach 404
V. Summary and Conclusions 405
References 406

Chapter 6

WATER ELECTROLYSIS AND SOLAR HYDROGEN
DEMONSTRATION PROJECTS

Gerd Sandstede and Reinhold Wurster

I. Introduction 411
II. Technical Background for the Development of
    Water Electrolysis for the Energetic
    Application of Hydrogen 412
III. Historical Remarks about the Generation of
     Hydrogen, Water Electrolysis, and Solar Hydrogen 415
IV. Systematic Classification of Water Electrolysis Processes ... 420
V. Thermodynamics, Conductivity Problems, and
   Electrocatalysis of Water Electrolysis 422
VI. Material Technologies for Water Electrolysis,
    Including the Electrocatalysts 433
VII. Technical State of the Art of Water Electrolysis 440
VIII. Characteristics for Coupling of Water Electrolysis
      with Hydro Power, Wind Energy, and Photovoltaics 454
      1. Nonfluctuating Energy Sources 454
      2. Fluctuating Energy Sources 454
      3. Resulting Requirements for the Coupling of
         Solar Electric Sources with Electrolyzer
         Technology and Existing Coupling Concepts........... 455

4. Concepts of Selected Advanced Water Electrolysis
   Systems and Presently Ongoing Developments into
   Commercialization............................... 456
IX. Hydro Power/Hydrogen Demonstration Projects........... 463
   1. Rationale 463
   2. Euro-Québec Hydro-Hydrogen Pilot Project (EQHHPP).. 464
   3. Norwegian Hydro-Energy for Germany (NHEG)........ 469
   4. Hydrogen-Powered Automobiles Using Seasonal and
      Weekly Surplus of Electricity (HYPASSE)............. 471
   5. Concepts for Utilization/Export of Electricity
      Generated from Hydro Power/Geothermal
      Energy in Iceland and Greenland..................... 471
X. Wind Power/Hydrogen Demonstration Projects 474
   1. Rationale 474
   2. Hawaii Wind Energy Storage Test Facility 476
   3. Wind–Hydrogen Test Installation at the
      Fachhochschule Wiesbaden, Germany 477
   4. Photovoltaic/Wind–Hydrogen Test Installation at the
      University of Oldenburg, Germany................... 478
   5. Wind/Photovoltaic/Heat/Hot Water/Hydrogen
      Test System at the Technical University of
      Ilmenau, Germany 479
XI. Solar Hydrogen Test and Demonstration Projects 479
   1. Rationale 479
   2. Solar Hydrogen Project in Bavaria
      [Solar Wasserstoff Bayern GmbH (SWB)] 480
   3. Hysolar Project in Germany and Saudi Arabia......... 482
   4. Energy-Self-Sufficient Solar Houses in Europe 485
   5. Photovoltaics/Hydrogen Project at the Agricultural
      College of Triesdorf in Germany..................... 488
   6. Photovoltaics/Hydrogen Project at the Helsinki
      University of Technology in Finland................. 490
   7. Photovoltaics/Hydrogen Project at the Instituto
      Nacional Técnica Aerospacial (INTA) in Spain 490
   8. Autonomous Photovoltaic/Hydrogen/Electricity
      Supply System of the Library at KFA in
      Jülich, Germany................................... 490
   9. Decentralized Solar Hydrogen Energy Supply
      System at the Fachhochschule Wiesbaden, Germany 492

    10. Photovoltaic–Hydrogen Project at ENEA in Italy 492
XII. Outlook on Hydrogen Activities Worldwide 493
    1. Japanese Activities on International Clean
       Energy Network Using Hydrogen Conversion
       [World Energy Network (WE-NET)] and Others 493
    2. Other Hydrogen-Related Activities in the
       United States and Canada—Summary 495
    3. Other Hydrogen-Related Activities in
       Europe—Summary 497
    4. Other Hydrogen-Related Activities in the
       Rest of the World 498
    5. Arguments and Concepts for a Hydrogen Energy
       System and for International Hydrogen Trade 499
    6. Reasons for Application of Hydrogen in the
       Transportation Sector 502
XIII. Conclusions And Recommendations 504
References 505

Cumulative Author Index 515

Cumulative Title Index 529

Subject Index 539

1

Reaction Engineering and Digital Simulation in Electrochemical Processes

K. Scott

Department of Chemical and Process Engineering, The University of Newcastle-upon-Tyne, Newcastle-upon-Tyne NE1 7RU, United Kingdom

I. INTRODUCTION

Recent developments in electrochemistry have featured a number of techniques used as methods of investigation and analysis and in a wide range of production applications. The diversification of research applications is driven by an increasing need for selective, energy-efficient, nonpolluting, clean processes. Table 1 illustrates some of the trends in applied and fundamental electrochemical research. There is a strong emphasis on the generation of new electrocatalysts, the application of new solvent/electrolyte systems and the study of new reaction systems and appropriate kinetics. The study of kinetics, in which many established steady-state and dynamic methods of electrochemical analysis are applied, is becoming increasingly augmented by newer techniques which can give information about the electrode/electrolyte interface at the molecular level. Application of both classes of methods will enable the principles of electrochemical analysis and reactor design to be more firmly established, removing a large degree of empiricism often associated with this area.

The introduction and establishment of new electrochemical technologies requires early engineering evaluation of the cell behavior using robust procedures. This has seen an increasing use of mathematical tools in the

Modern Aspects of Electrochemistry, Number 27, edited by Ralph E. White *et al.* Plenum Press, New York, 1995.

Table 1
Trends in Applied and Fundamental Electrochemical Research

1. Precise and detailed kinetic and mechanistic studies of surface changes, electron transfer processes, and coupled homogeneous chemical reactions. The development of models for electron transfer, nucleation and growth of new phases, and the structure of interfaces. Characterization of electrode–solution interfaces at the molecular level
2. Intensification in mass transfer between phases and at electrode surfaces
3. Intensification of electrode area in novel cell designs and electrode structures
4. Reduction in energy requirements and cell voltages through new electrocatalysts
5. The development of new coatings: electrocatalysts, composite coatings, polymer films
6. Intensification of processes; reduction in energy consumption
7. Improvement in process strategies and the development of realistic synthetic reactions
8. New energy sources: solar energy, fuel cells, high energy and power density batteries
9. The application of electrochemistry to the manufacture of electronic devices
10. Electrochemical engineering development of novel cell designs and electrode structures
11. Developments in standard cell designs

simulation of electrochemical cells and processes both to determine kinetic and rate data and to evaluate cell and reactor design. Increasingly, these mathematical tools are being used in conjunction with experimental studies of fundamentals and bench/laboratory scale operation.

The mathematical modeling of multiple reaction sequences that combine heterogeneous and homogeneous processes, in which intermediate species undergo fast reaction and/or are at low concentration, equilibrium reactions and mass transport can present difficulties in solution. This is because many of the engineering model equation sets exhibit stiffness. As a result, extensive investigation and development of algorithms for mathematical solutions has produced a vast catalog of procedures.[1]

The evaluation of the distribution of current density in electrochemical cells is frequently an essential step in the modeling, design, or analysis of electrochemical systems. A substantial literature exists pertaining to the evaluation of current and potential fields although only a small proportion is relevant to the determination in electrochemical systems. Much of the early work was concerned with solving large classes of primary and secondary current distribution problems with the use of classical analytical procedures. However, the number of electrode geometries and cell configurations that can be treated in this way is relatively small. These analytical solution techniques include superposition, the method of images, series solutions, integral equations, and conformal mapping.

The typical complex geometries and nonlinear boundary conditions that characterize electrochemical systems are obviously suited to the application of numerical solution and digital computers to problems that are intractable by analytical techniques. There are several reviews on different aspects of current distribution problems covering areas such as rotating disk electrodes, parallel-plate systems, moving boundary problems, porous electrodes, and other geometries. The paper by Prentice and Tobias[2] on numerical methods and solutions for current distribution problems provides an excellent survey of the history of this area and the more significant developments in numerical techniques. What now follows is essentially a synopsis of this review that will give an indication of the methods of numerical solution required in the much broader area of electrochemical reaction systems.

The calculation of current distribution in an electrochemical system frequently requires the solution of the Laplace equation. In many practical cases, the presence of the mass transfer boundary layer allows the problem to be divided into two regions: the diffusion layer and the homogeneous electrolyte bulk. The equations that describe each region can then be solved separately, subject to suitable matching of properties at the defined interface between the regions. This usually requires an iterative scheme with one of a number of computer-based numerical techniques. The numerical methods that have been used in the solution of current distribution problems are:

- Green's functions
- Perturbation
- Straight lines
- Coordinate inversion
- Orthogonal collocation
- Variational Methods
- Finite elements
- Finite difference

For current distribution problems that deal with realistic practical situations, three of these methods seem to have the general applicability needed for efficient and effective solution. These methods are orthogonal collocation, finite difference, and finite elements.

Orthogonal collocation was first applied to a current distribution problem by Caban and Chapman,[3] who solved the current distribution in a flow channel previously solved by Parrish and Newman.[4] In this technique one of the variables is approximated by a finite series of orthogonal polynomials. The coefficients of the series are determined by satisfying the governing equations at a small number of collocation points. The number of points selected has, however, implications regarding the accuracy of solution.

In the finite-difference method, the differential equations are replaced by their approximate difference formulas. The solution domain is generally divided into rectilinear elements with some approximation of the field variable at a curved boundary. A general computer program for solving current distribution problems was developed by Fleck et al.[5] based on the finite-difference approximation. An algorithm was developed by Prentice and Tobias[2] to overcome the instabilities associated with nonlinear boundary conditions of certain current distribution problems. Generally, the square elements that are used in this method are well suited to computer manipulation.

In the finite-element method the domain of interest is divided into discrete polygons, and trial functions are used to approximate the variation in each polygon. The approximating functions must obey certain continuity conditions between the elements. The method has the advantage of offering a means of following an irregular boundary more accurately than other methods. A number of studies to delineate the advantages of the three methods have been made, although a final choice of technique depends on the accuracy required, the ease of programming, the size of machine available, and the cost of the simulation. The reader is advised to consult the paper of Prentice and Tobias[2] for more details of the methods as applied to current distribution problems.

The objective of this chapter is to bring together the necessary methods, principles, and procedures required for the reaction engineering of electrochemical processes. The title itself gives a clear indication of the part digital simulation must play in this task, although this subject does not dominate the contents but is introduced as a necessary tool for the solution of complex problems. The contents of this chapter can be conveniently separated into four parts:

- Reaction rate modeling
- Models for reactor systems

- Electrochemical analytical methods in reaction modeling
- The engineering of electrochemical reactors

The reaction rate modeling considers the principles and equations of kinetics and mass transport necessary to produce global expression for the overall rate in the vicinity of an electrode surface. Thus, electrochemical kinetics, adsorption, molecular and convective diffusion, and homogeneous kinetics are discussed for various categories of reaction types. For such models to be implemented, the relevant chemical and physical data must be obtained and the mechanism(s) by which species transformation occurs identified, if only in a semiempirical form. Here the need for integration of the scientific methods with the engineering models and strategy becomes an important feature. From this development of the reaction, the engineering of the reactor can, if required, be implemented by building in aspects of fluid mechanics, energy exchange, etc. This is the subject of the final part of this chapter, which considers a number of specific, but also by their nature quite general, examples, not all of which require complex digital simulation procedures or numerical methods. Notably, multiphase process operation is specifically addressed because of the importance of many systems where several phases are present. The phenomena of gas evolution and electrolysis with emulsions and dispersed gases are examples cited.

II. ELECTRODE KINETICS AND REACTION RATE MODELING

An analysis of an electrochemical system by digital simulation, or otherwise, relies on the availability of suitable kinetic or reaction rate data. The application of modeling is only as good as the data available, and a general qualitative or semiquantitative picture is often manufactured by exploring the effect of variables for nonspecific systems. The latter are often useful tools in design, especially where variables or parameters can be selected to approach required performance targets. However, electrochemical rate processes give only a limited scope to the manipulation of variables owing to restricting factors such as the range of operating temperatures, the specific electrode materials required, and the electrocatalysts and the range of overpotentials available.

The evaluation of rate data and the determination of the governing rate expressions are thus important components for successful simulation. This section provides an overview of the mathematical description of the rate processes that are frequently found to satisfactorily describe electrochemical processes on a macroscale. There are three general features to be considered:

(i) The rate of the electrochemical processes
(ii) The rate of chemical reactions
(iii) The rate of mass transport processes

The electrochemical processes are, to a greater or lesser extent, influenced by adsorption phenomena at the electrified interface between the electrolyte solution and the electrode. The charge transfer processes at the electrode are often accompanied by chemical reactions, which, on the time scale of the overall rate, are fast and are thus confined to this region. Their influence is therefore seen in the generated form of the electrode kinetic equation, which is classically devised on the basis of one step being slow and thus rate-determining. Overall, the electrochemical process is composed of coupled adsorption, electrochemical, chemical, and desorption steps, as depicted in Fig. 1, occurring either sequentially or not.

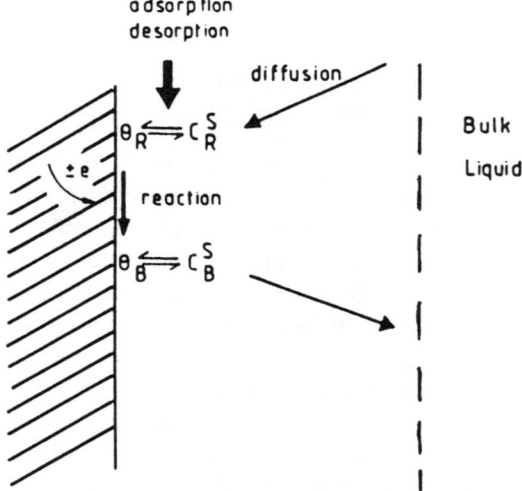

Figure 1. Rate processes in electrochemical reaction systems.

Multistep reactions leading to multiple product species give scope for greater interaction between constituent species. The chemical reactions that are introduced in the model of the system are considered as separate from those associated with processes at the electrode surface and are thus confined to the bulk electrolyte phase or to the boundary layer, or diffusion layer, regions. They are thus sequential to electrochemical reaction but may be fast (equilibrium) processes.

The mass transfer processes are taken as separate from surface adsorption phenomena, and their defined rate is dependent upon the hydrodynamic model adopted and whether or not transport is considered as a combined ionic, diffusional, and convective process. Chemical reaction may proceed during these transport processes.

1. Kinetic Equations for Electrochemical Reactions

In many cases electrode kinetics allow the current density i_j for reaction j to be expressed as an exponential function of the surface overpotential, η_{sj}, according to a Butler–Volmer relationship[6] (written for an anodic process):

$$i_j = i_{oj,\text{ref}} \left\{ \prod_i \left(\frac{C_{is}}{C_{i,\text{ref}}} \right)^{p_{ij}} \exp\left[\frac{\alpha_{aj} F}{RT}(Ea - \Phi_{oa} - U_{j,\text{ref}}) \right] \right.$$

$$\left. - \prod_i \left(\frac{C_{is}}{C_{i,\text{ref}}} \right)^{q_{ij}} \exp\left[\frac{-\alpha_{cj} F}{RT}(Ea - \Phi_{oa} - U_{j,\text{ref}}) \right] \right\} \quad (1)$$

where

$$U_{j,\text{ref}} = U^{\theta_j} - U^{\theta\text{ref}} - \frac{RT}{n_j F} \sum_1 s_{ij} \ln\left(\frac{C_{is}}{\rho_0} \right) + \frac{RT}{n_{\text{ref}} F} \sum_i s_{i,\text{ref}} \ln\left(\frac{C_{i,\text{ref}}}{\rho_0} \right) \quad (2)$$

and $i_{oj,\text{ref}}$ is the exchange current density evaluated at the reference concentration, $C_{i,\text{ref}}$; C_{is} is the surface concentration of species i; U^{θ_j} is the standard electrode potential of reaction j; Φ_{oa} is the potential of the solution just outside the diffuse double layer; and p_{ij} and q_{ij} are anodic and cathodic reaction orders, respectively, of species i in reaction j. These parameters and the characteristic parameters of the electrode reaction, α_{cj} and α_{oj}, are essentially the key components of the relationship describing the electrode kinetics. In view of this, a convenient, approximate expression of the rates involved in a redox reaction[7] is

$$A + e^- = B \tag{3}$$

$$i_1/F = k_{f1}(C_{As})^q - k_{b1}(C_{Bs})^p \tag{4}$$

The rate constants are usually defined as

$$k_{f1} = k_1 \exp(-\beta_1 E) \tag{5}$$

$$k_{b1} = k_1 \exp(\alpha_1 E) \tag{6}$$

where α_1 and β_1 are constants that describe the potential dependence of the rate constants and are related to the usual Tafel slopes (in millivolts), e.g.,

$$\alpha_1 = 2.303/a \tag{7}$$

where a is the Tafel slope. The rate constants can thus be defined as

$$k_{f1} = i_o/(C_{Ab})^q \exp[-\alpha_1(E - E_{e1})] \tag{8}$$

$$k_{b1} = i_o/(C_{Bb})^p \exp[\beta_1(E - E_{e1})] \tag{9}$$

where C_{Ab} and C_{Bb} are reference (bulk) concentrations.

Clearly, the transition from Eq. (1) to Eq. (4) involves a number of assumptions that may not be generally valid. Species A and B may be ionic, neutral, or solid, in which case the activity of the component can be used.

If the reaction is of the general form

$$s_{ij}M_i^{z_i} \rightarrow n_j e^- \tag{10}$$

then an equivalent form to Eq. (4) becomes

$$i_j/n_j F = k_{fj}\prod_i (C_i)^{q_i} - k_{bj}\prod_i (C_i)^{p_i} \tag{11}$$

Overall, the information required to use Eq. (11) is of two general types:

(i) Electrochemical polarization parameters in the rate constants, k_{fj} and k_{bj}
(ii) Reaction orders for all components influencing the rate of reaction

Experimental methods for determining such parameters are discussed later.

2. Mass Transfer and the Reaction Rate Model

The aspect of mass transport is of considerable importance in electrochemical systems and is directly related to the transport laws of solutions and to fluid mechanics. These phenomena are, in principle, the basis for formulating the reactor models into which the required aspects of rate processes are introduced. This, in design, is a rigorous approach requiring the application of sophisticated methods and is not always the most appropriate, especially when an approximate, less rigorous procedure would suffice. Such an approximation is based on the concept of the diffusion layer, which can enable a reaction model to be formulated incorporating both electrochemical surface phenomena and mass transport.

The concept of the diffusion layer is depicted in Fig. 2. This layer offers the sole resistance to mass transfer and is taken, in the first instance, to have a constant thickness over the electrode surface. From Fick's first law of diffusion, the concentration gradient of a reacting species at the electrode surface is proportional to the sum of the current densities associated with that reacting species:

$$s_{ij}\frac{i_j}{n_j F} = D_j \frac{dC_j}{dx}\bigg|_{x=0} \tag{12}$$

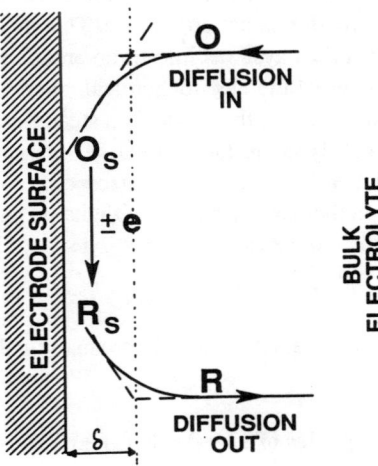

Figure 2. Model of the diffusion layer.

By replacing the actual concentration profile with a linear approximation, the current density is written in terms of a mass transfer coefficient $D_j/\delta = k_{Lj}$ as

$$s_{ij}i_j/n_j F = k_{Lj}(C_j - C_{js}) \tag{13}$$

where δ is the diffusion layer thickness, and C_j is the bulk concentration of species j. The model therefore depicts the system in terms of a single mass transfer parameter, which increases in value with an increase in convection owing to a thinning of the stagnant diffusion layer. This model is widely adopted in engineering, and significant efforts have been made by researchers to produce correlations for the prediction of mass transport coefficients for many practical systems.[8,9] The model has limitations when, for example, mass transport of ionic species is more appropriately described by the convective diffusion equation (see Section III). Here, however, because electrochemical systems tend to have large Schmidt numbers, $Sc = \mu/\rho D_j$ (ratio of kinematic viscosity to diffusion coefficient of around 1000), then diffusion layers are of the order of 100 μm and appropriate simplifications can be adopted.

It is worth noting that Eq. (13) states that the total supply of species by mass transport is equal to the total consumption of species j at the electrode by all relevant reactions, which will also include all nonelectrochemical processes. When reactions are not confined to the electrode surface, then this situation is not applicable. This then leads to a classification of electrochemical systems into three broad categories depending upon the relative rates of the electrochemical, chemical, and mass transport processes. Additionally, the model is a steady-state interpretation of the behavior, in that dynamic (accumulative) response in the electrode region is ignored, which for systems with fast response could lead to erroneous behavior. Perhaps, in the developing age of digital simulation and computational techniques, the more exact representation should be used:

$$k_{Lj}(C_j - C_{js}) = (s_{ij}i_j/n_j F) + dC_{js}/dt \tag{14}$$

This, of course, is the case with many electroanalytical methods discussed in later sections.

3. Categories of Electrochemical Rate Models

In a sense, Section II.2 has focused on the first category of rate models, namely, "fast reactions"—those in which the surface reactions of intermediates, both electrochemical and chemical, are fast in comparison to the mass transport of the intermediates. Here, putting to one side the single-electron-transfer redox process, this category treats intermediate reaction steps involved in the overall reaction(s) as rapid processes in which the quasi-steady-state approximation is applied. By drawing analogies to the similar analyses of two-phase chemical and catalytic reaction systems,[10,11] this category can be defined in terms of a Hatta number,

$$\mathrm{Ha} = \sqrt{kD/k_L^2} > 100 \tag{15}$$

representing the ratio of the kinetic rate to the mass transport rate.

The second category of reaction systems apply when homogeneous reactions occur at a comparable rate to diffusion of the associated species and such reactions occur mainly in the diffusion layer and thus simultaneously with the mass transport process. The reaction model then becomes one of electrochemical surface behavior combined with that of the concurrent diffusion and reaction, the latter being described by a Fickian second law of mass transport with reaction, i.e.,

$$D_i(d^2C_i/dx^2) - \sum r_j = 0 \tag{16}$$

where Σr_i is the sum of the reaction rates of species i. The dimensional form of this equation shows the significance of the Hatta number in these systems:

$$d^2C_i/dz^2 = \mathrm{Ha}^2 C_i \tag{17}$$

where $C_i = C_i/C_{ir}$ with C_{ir} a reference concentration and $z = x/\delta$.

This category of reaction is defined by Hatta numbers in the range

$$0.01 < \mathrm{Ha} < 100 \tag{18}$$

At large values of Hatta number, chemical reaction is mainly at regions near the electrode surface, whereas at lower values the reaction is located more toward the bulk electrolyte side of the diffusion region. This is not to say that there is no reaction in the bulk electrolyte region but rather that this contribution to the overall reaction is relatively small.

This category of reaction was discussed above for relatively simple processes involving single species or with other reacting species in excess and must be considered in a wider context when complex chemical interactions occur. For example when a chemically reactive species, generated electrochemically, undergoes reaction with, say, an electro-

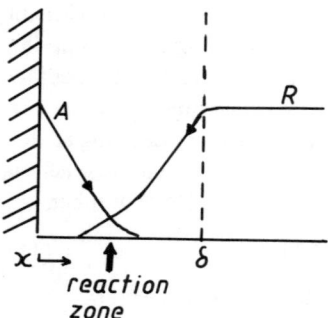

Figure 3. Reaction zone model for fast reaction between electrochemically generated species and inert reagent.

chemically inert reagent and this reaction is fast, or near instantaneous, then reaction is confined to within a zone (or plane) in the diffusion layer (see Fig. 3). Thus, the reaction is determined by the relative diffusion rates of the species and, in the limit of infinitely fast reaction, the process occurs at a reaction plane. When the reagent is of a sufficiently high concentration, then the reaction zone can be considered to be at the electrode surface and therefore this becomes a category 1 process.

The third category of reaction is when the chemical reactions are slow in comparison to mass transport and thus the majority of reaction takes place in the bulk electrolyte. This system is then considered as a succession of steps:

(i) Mass transport to the electrode surface
(ii) Electrochemical transformation
(iii) Mass transport to the bulk electrolyte
(iv) Chemical reaction in the bulk electrolyte

This category of reaction is described approximately by a Hatta number of < 0.01.

Overall, the three categories of reaction enable a suitable rate model to be adopted which is a good picture of the processes occurring in practice. Thus, the truly general picture in which all electrochemical, chemical, and mass transfer steps occur simultaneously is avoided. In truth, rarely is this situation considered; the Butler–Volmer equation of electrode kinetics is testimony. However, the closest situation to this is the category 2 system where reaction can occur in both the bulk electrolyte and the diffusion layer. The total chemical reaction is then determined by the sum of, for example, the rate of product species diffusing across the artificial diffusion layer barrier ($D_i dC_i/dx|_{x=d}$) and the product formed by homogeneous bulk reaction.

4. The Role of Adsorption

Adsorption of ionic or neutral species onto electrode surfaces can have a significant impact on the progression and direction of electrochemical processes.These species may be electroreactive reagents, intermediates, or solvent molecules that form some type of bond with the surface and thereby accelerate, decelerate, or alter the pathway the reaction(s) takes to form products. The adsorption bond may be covalent or electrostatic, or the molecule may have a preferred affinity for the surface. The role of adsorption in electrochemical processes will generally result in:

- A modification of the kinetic pathway by avoiding a slow step in the process, i.e., electrocatalysis
- A change in the interfacial environment at the electrode, which may induce depletion of particular species, thereby altering reaction pathways

These factors are influenced by the degree of surface coverage, which itself depends on the type and concentration of adsorbate, the solution composition, temperature, the electrode material, and the electrode potential. The degree of surface coverage by particular species can be measured in the absence of faradaic processes and provides useful information for the understanding of the role of adsorption when such processes occur. However, the interaction of the faradaic reactions with the surface coverage prevents the measurement of the latter, and the behavior is usually inferred from kinetic studies. This will now form the focus of this section; the reader is referred to a number of texts that deal with theoretical and experimental aspects of adsorption in electrochemistry.[12-14]

(i) Adsorption on Electrodes

The surface coverage of adsorbate is described by an adsorption isotherm which relates the degree of coverage to the concentration of adsorbate in the electrolyte and the free energy of adsorption. There are a number of isotherm models, among which that derived from the classical theory of Langmuir is frequently adopted. This theory is based on a number of hypotheses:

- Monolayer coverage
- Uniformly energetic sites of adsorption
- No interaction between adsorbed molecules

Consider the adsorption of a single component A,

$$A + l = Al \qquad (19)$$

where l represents the adsorption site, and Al the adsorbed species. Each surface is said to have a maximum available area for adsorption, which is defined in terms of the total concentration of available sites C_T (kmol/m² surface). These sites either contain adsorbed species or are vacant, and therefore the sum of the vacant site and occupied site concentrations, C_l and C_{Al}, respectively, is equal to the total concentration of available sites:

$$C_T = C_l + C_{Al} \qquad (20)$$

The net forward rate of reaction (20) is

$$r = k_a C_A C_l - k_{-a} C_{Al} \qquad (21)$$

where k_a and k_{-a} are adsorption and desorption rate coefficients.

At equilibrium ($r = 0$) the adsorption isotherm is found from Eqs. (20) and (21) as

$$C_{Al} = C_T K_A C_A / (1 + K_A C_A) \qquad (22)$$

where $K_A = k_a/k_{-a}$, the adsorption equilibrium constant. It is convenient to normalize this equation to obtain the isotherm in terms of fractional coverage θ:

$$\theta_A = C_{Al}/C_T = K_A C_A / (1 + K_A C_A) \qquad (23)$$

The above treatment of adsorption can be extended to adsorption of more than one species, which results in the following expression:

$$\theta_i = K_i C_i / (1 + \sum K_i C_i) \qquad (24)$$

These species may be neutral, and their presence may limit an electron transfer process by blocking the sites of adsorption of species that are electroactive. Application of a potential field at an electrode may be viewed as intensifying the adsorption process of electroactive species, which displace the neutral molecules and thus enable electron transfer to occur at increased rates.

The adsorption of certain molecules can occur by the breaking of bonds or by dissociation:

$$A_2 + 2l = 2Al \tag{25}$$

or

$$HA + 2l = Hl + Al \tag{26}$$

If the molecule's constituents show no preference for different sites, then the isotherm can be obtained from the equilibrium of reaction (25):

$$\theta_{Al} = \sqrt{K_A C_{A_2}}/(1 + \sqrt{K_A C_{A_2}}) \tag{27}$$

When the molecule's fragments show preference for adsorption at different sites and when several bond-breaking steps are involved, the surface coverage characteristics are more complicated and are determined by kinetic processes and, to a lesser extent, the equilibrium thermodynamic factors. Thus, the application of more sophisticated isotherm models, where the free energy of adsorption varies linearly (Frumkin, Temkin) or logarithmically (Freundlich) with surface coverage, can introduce an unnecessary degree of complexity to the formulation of engineering reaction rate models.

(ii) Reaction Rate Models for Adsorption

For adsorption to be adequately incorporated into a reaction rate model, it is important to establish a basis for correlating observed kinetic and rate behavior. In electrochemical systems it is possible to distinguish between two types of behavior, one in which adsorption and electron transfer occur simultaneously in one step, and one in which electrochemical reaction occurs after adsorption. In both cases, and generally in the formulation of reaction rate models at interfaces, a rigorous treatment of the dynamic rate processes is not needed and a steady state of intermediates is assumed.

(a) Simultaneous adsorption and electron transfer

Consider the overall two-electron-transfer process

$$R - 2e^- \rightarrow O \tag{28}$$

which proceeds in two consecutive steps:

$$R + l \underset{k_{b1}}{\overset{k_{f1}}{\rightleftharpoons}} Rl + e^- \qquad (29)$$

$$Rl = O + l + e^- \qquad (30)$$

where k_{f1} and k_{b1} are electrochemical rate constants.
The net rates of reactions (29) and (30) are

$$r_R = k_{f1}(1 - \theta)C_R - k_{b1}\theta \qquad (31)$$

$$r_{Rl} = k_{f2}\theta - k_{b2}C_O(1 - \theta) \qquad (32)$$

The overall rate of this process is derived by applying the stationary-state approximation to Rl and gives

$$i/nF = (k_{f1}C_R - k_{b1}Y)(1 + Y)^{-1} \qquad (33)$$

where $Y = (k_{f1}C_R + k_{b2}C_O)/(k_{b1} + k_{f2})$. For irreversible processes, this reduces to

$$i/2F = k_{f1}k_{f2}C_R/(k_{f2} + k_{f1}C_R) \qquad (34)$$

Equation (32) gives the reaction rate in terms of the electrolyte composition and appropriate rate constants of the various steps. The expression for an irreversible process is a familiar rate form for heterogeneous systems but is limited to moderate conversions of reagent O. It is often found that one of the steps is much slower than the other in these reactions, and it is then termed the rate-determining step (rds). However, this assumption is often adopted as a matter of convenience when there is every possibility that more than one of the steps is slow. For example, one step may be controlling in one region of operation, whereas another step is rate-controlling in another region, and then a middle region of operation is likely where both steps are of approximate equal importance in determining the overall rate. An example of this is in the electrochemical reduction of vanillin,[15] where the effect is the observation of a shifting reaction order from two to zero.

(b) Adsorption rate models

The system is

$$R + l \underset{k_{-a}}{\overset{k_a}{\rightleftharpoons}} Rl \qquad (35)$$

$$Rl \underset{k_{b1}}{\overset{k_{f1}}{\rightleftharpoons}} O'l + e^- \qquad (36)$$

$$O'l \underset{k_{b2}}{\overset{k_{f2}}{\rightleftharpoons}} Ol + e^- \qquad (37)$$

$$Ol \underset{k_{-d}}{\overset{k_d}{\rightleftharpoons}} O + l \qquad (38)$$

or the steps given by Eqs. (37) and (38) combine to give

$$O'l \underset{k_{b2}}{\overset{k_{f2}}{\rightleftharpoons}} O + l + e^- \qquad (39)$$

If the step given by Eq. (38) is rate-determining, then the assumption is that the other steps rapidly approach equilibrium, and then the concentrations of adsorbed intermediates are related through appropriate equilibrium constants, i.e.,

$$Rl = K_a C_R \theta; \quad O'l = K_1 Rl; \quad Ol = K_2 O'l \qquad (40)$$

where $K_a = k_a/k_{-a}$, $K_1 = k_{f1}/k_{b1}$, and $K_2 = k_{f2}/k_{b2}$. The surface coverage is therefore given by

$$1 = \theta(1 + K_a C_R + K_1 K_a C_R + K_1 K_2 K_a C_R) \qquad (41)$$

The rate of the reaction in Eq. (38) is thus

$$r = \frac{i}{2F} = k_d Ol - k_{-d} C_O \theta \qquad (42)$$

which becomes

$$\frac{i}{2F} = \frac{(k_d K_2 K_1 K_a C_R - k_{-d} C_O)}{K_a C_R (1 + K_1 + K_1 K_2) + 1} \qquad (43)$$

This expression is one of many rate equations that can be developed based on the concept of the rds, and the use of this concept generally results in a catalog of possible rate expressions. The evaluation and use of these will depend on whether or not a knowledge base exists on the chemistry of the scheme. For example, the rapid desorption of species Ol may occur, making the step in Eq. (35) [or Eq. (39)] irreversible; then if the step in Eq. (37) or (39) is rate-determining and with the preceding steps in equilibrium, the rate is given by

$$\frac{i}{2F} = \frac{k_{f2}K_1K_aC_R}{1 + K_aC_R(K_1 + 1)} \qquad (44)$$

If adsorption of R is rate-controlling, then with both electrochemical steps considered in equilibrium, the rate of reaction is derived as

$$\frac{i}{2F} = \frac{k_a[C_R - (C_OK_2/K_1)]}{1 + C_OK_2 + (C_OK_2/K_1)} \qquad (45)$$

Alternatively, with the step in Eq. (36) rate-determining, then

$$\frac{i}{2F} = \frac{k_{f1}K_a[C_R - (C_OK_2/K_1)]}{1 + K_aC_R + K_2C_O} \qquad (46)$$

With preceding adsorption, we see that there are similarities to the theoretical rate expressions in the previous section. The lumped rate coefficients associated with the respective species concentration (e.g., $k_{f1}K_a$) are not significantly different and will exhibit a different current versus potential dependency; that is, the slope of polarization plots of log i versus E will be different. In the case of Eq. (45) above, in the absence of product species, O, the rate is governed by the adsorption rate coefficient.

With the reaction sequence having irreversible steps, the rate equation is

$$\frac{i}{2F} = \frac{k_aC_R}{1 + C_R[(k_a/k_{f1}) + (k_a/k_{f2}) + (k_a/k_d)]} \qquad (47)$$

Hence, if $k_{f2} \gg k_a$ and $k_d \gg k_a$, then at low potentials (small k_{f1}) the process is kinetically controlled, whereas at high potentials it is adsorption-controlled.

It is instructive to see how these expressions compare with the general rate equation for the reaction mechanism (Eqs. 35–38) that is obtained by applying the stationary-state approximation to Rl, O′l, and Ol. This gives the rate as

$$\frac{i}{2F} = \frac{k_{f1}k_a C_R \theta}{k_{f1} + k_{-a}} - \frac{k_{-a}k_{b1}\theta_0'}{k_{f1} + k_{-a}} \quad (48)$$

where the fraction of sites unoccupied is given by

$$\theta = \frac{1}{1 + P_1\left(1 + \dfrac{k_{b1}}{k_{f1} + k_{-a}} + \dfrac{k_{f2}}{k_d + k_{b2}}\right) + \dfrac{k_a C_R}{k_{f1} + k_{-a}} + \dfrac{k_{-d}C_O}{k_d + k_{b2}}} \quad (49)$$

and $\theta_0' = P_1 \theta$ with

$$P_1 = \frac{k_{f1}k_a C_R + k_{-d}P_2 C_O}{[(k_{f1} + k_{-a}) - (k_{f2} \cdot P_2)](k_{b1} + k_{f2})}$$

and

$$P_2 = \frac{k_{b2}(k_{f1} + k_{-a})}{k_d + k_{b2}}$$

This is a complicated equation which represents the rate for, in principle, a simple two-electron-transfer reaction. Clearly, the rds limiting forms are more convenient, both for engineering use and in experimental data analysis.

The interpretation of the experimental data will aid in identifying the range of variables in which one, or more, of the limiting forms is applicable. If this exercise leads to the conclusion that there is more than one rds, then the model must be based on this criterion and the appropriate equation derived.

The characteristics of the rate equations derived above can be summarized in the general form:

$$\text{Rate} = \frac{\text{(Kinetic factor) (Driving force)}}{\text{(Adsorption group)}} \quad (50)$$

The potential catalog of rate forms for nonunimolecular transformations, bimolecular reactions, dissociations, etc., is immense, and consultation with publications in the area of heterogeneous catalysis[16,17] can provide a valuable aid. These publications give tabulated expressions for the three key groups in the general rate from Eq. (50). A logical extension of such an approach is the inclusion of models in which reaction is between adsorbed and unadsorbed species,[18] which leads to rate models that are different in detail but similar in formulation.

The kinetic factor in Eq. (50) is generally a lumped parameter term that is assigned a single parameter constant.

Overall, the occurrence of adsorption of species in electrochemical rate models can be written as

$$\frac{i_j}{n_j F} = \frac{k_{fj} \prod_i (C_i)^{q_i}}{(1 + \sum_{j=1} C_i^q K_j)^m} - \frac{k_{bj} \prod_i (C_i)^{P_i}}{(1 + \sum_{j=1} C_i^P K_j)^w} \quad (51)$$

The exponents m and w on the adsorption group will generally have values of 0, 1, or 2. Within this group, the exponents on the concentrations, p and q, may be positive or negative, and the terms K_j are lumped parameters.

The lumped electrochemical kinetic terms k_{fj} and k_{bj} are based on the use of a working electrode potential versus a suitable reference potential (e.g., NHE), and their preexponential rate coefficients are the values of the forward and reverse rate constants at zero potential. This convention is preferred, as often the reversible potentials of multireaction paths are not known, and thus extrapolation from relatively simple polarization measurements is then used as the basis for parameter determination.

A brief digression from the thrust of this section is perhaps appropriate and relates to the structure of the electrode. It is assumed that the electrode surface, on the macroscale, is smooth and flat and that electroactivity is uniform over all of the surface. With porous electrocatalysts, then the rate will be influenced by slow pore diffusion and a distribution of potential in the pores. The general treatment of these aspects is complicated and beyond the scope of this work, although readers will find a good introduction to the subject in the text by Newman.[19] However, for thin porous electrodes the distribution in electrode potential is small and can be ignored. Thus, the distribution of the concentration of species in the pores is obtained from the diffusion equation[17]

$$\frac{D_j d^2 C_j}{dx^2} - r_j = 0 \quad (52)$$

The solution of this equation leads to a simplified analysis based on an electrochemical effectiveness factor,[20] ε, for all independent reactions. This is defined as the ratio of the actual rate to the rate based on the surface concentration,

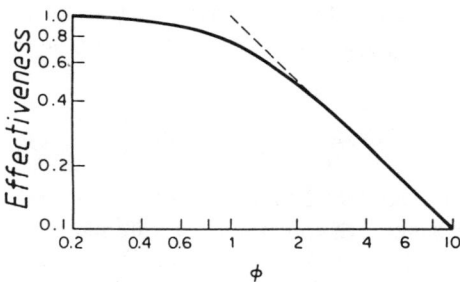

Figure 4. Effectiveness factors as a function of a Thiele modulus, ϕ, for electrochemical reactions.

$$\varepsilon = \frac{l}{L_p} \int_0^1 r_j \, dx / (r_j)_s \tag{53}$$

where L_p is the equivalent length of a pore. The observed rate is then written as

$$(r_j)_{obs} = \varepsilon (r_j)_s \tag{54}$$

Equations and correlations for effectiveness factors are readily available in publications dealing with the analogous area of heterogeneous catalysis. Effectiveness factors are published (see Fig. 4) as a function of a Thiele modulus, ϕ, for electrochemical reactions, which is given generally by

$$\phi = L_p (r_j)_s / \sqrt{2} \left(\int_0^{C_{js}} D_j r_j \, dC_j \right)^{1/2} \tag{55}$$

where D_j is an effective diffusivity in the porous electrocatalyst, and L_p is its thickness. This expression has been derived on the basis of an asymptotic solution to Eq. (52) for all reaction rate forms and gives remarkably good agreement with values from exact solutions.

Overall, therefore, inclusion of pore diffusion in the reaction rate model is accommodated by a single parameter, which is evaluated through the integration of the reaction rate equation in the absence of diffusion.

In the case of multiple dependent reactions, the replacement of the rate models for each step by equations of the form of Eq. (54) when pore diffusion occurs is not valid, and recourse to solution of the coupled diffusion equations (52) is necessary.

5. Conclusions

The reaction rate models described in this section form the basis of an overall reactor modeling and simulation exercise that can be carried out in two stages. The formulation of models for the surface phenomena of charge transfer and adsorption gives a semimicroscopic view of the kinetic processes. The integration of the kinetic and mass transport processes over the electrode/electrolyte interface to bulk regions provides the basis for the formulation of the reaction model. It is abundantly clear that there is need for access to a reliable data base of thermodynamic, charge transfer, kinetic, and transport properties. For classical systems involving electron transfer confined to the double layer, there is generally adequate data. Moving beyond such systems can often be impeded by the lack of data, and it is in this area that electrochemists can start to unlock the model by applying both the classical methods of potential scanning, AC impedance, etc., and the more recent, *in situ* spectroscopic techniques that allow identification of stable intermediates at electrode surfaces. These methods will enable the engineering of complex reaction systems, in which charge transfer and homogeneous processes are interspersed and in which diffusion of intermediates occurs. This subject area is dealt with in Section V of this chapter. The formulation of the engineering problem must be augmented by the acquisition of experimental product distribution data using, for example, classical batch reactors. In this context, it is imperative to explore the influence of key variables, such as temperature and pH, on the rate processes, product distributions, and physical properties.

III. REACTOR MODELING

The third stage in the exercise of reactor modeling is the integration of the reaction rate models over the confines of the reactor with the appropriate hydrodynamic characteristics of flow and mixing. An integral feature of the latter can be localized distributions of mass transfer arising from changes in boundary layer thickness and the inducement of turbulence by suitable promoters or by gas evolution.[21] At some stage a strategic decision is required regarding the influence of the counter electrode reaction(s) on the reactor model or their relevance to the model. These processes may be seen as passive, in, for example, the generation of oxygen at an anode where protons are liberated into an acidic electrolyte, causing only minor changes in pH. Alternatively, they may be seen as imperative to the progression of a reaction while also being detrimental to selectivity. This would be the case in the epoxidation of olefin gases[22] by electrogenerated bromine, where hydroxide ions generated by the cathodic evolution of hydrogen promote both the saponification of the alkene bromohydrin to the epoxide and also the latter's saponification to the glycol. The implications of the counter electrode reactions must also be considered in the context of the use of a cell separator, or membrane, which is generally introduced to prevent certain species from contacting electrodes or electrolyte regions where derogatory reactions might occur. The transport processes through membranes, and even reactions within them, as realized in chlor-alkali diaphragm electrolyzers,[23] yet further extend the domain of reactor modeling.

Several reactors of various configurations have been used in electrochemical processes. These range from the simpler parallel-plate systems to fluidized beds and two- or three-phase units. There is little scope for mixing on a macroscale in some of these reactors, for example, thin-gap, or narrow-channel, electrolyzers and filter press type cells.[24] With these, one wall or surface is the working electrode, and plug flow is assumed with complete lateral mixing and no axial mixing. As a reasonable approximation, this assumption is valid for turbulent flow with cells having aspect ratios (the ratio of the characteristic diameter, e.g., the channel hydraulic mean diameter or, in the case of packed beds, the particle diameter, to the reactor length) of less than 0.02. With relatively low velocities, turbulence may not be achieved, and the flow will be laminar. The reactor models for this type of flow are generally more complicated, requiring the solution of the convective diffusion equation. These are dealt with in more detail in Section V. A turbulent flow analog of this has also been considered.[25]

Departure from the ideal plug flow behavior occurs in practice due to localized mixing and can be modeled using a number of methods that rely on the measurement of residence time distributions of fluid in the reactor. Two popular single-parameter models are the dispersed plug flow model and the tanks in series model.[26] These models, apart from the hydrodynamic exact model of laminar flow, give the best engineering interpretation of real reactor behavior. The laminar flow reactor can also be modeled in this way.[17] The use of these models does extend the required data base requirements for simulation.

At the extreme of mixing are both the batch reactor (BR) and the continuous stirred tank flow reactor (CSTR). Mixing in these units is assumed to be so good that, even on a microscopic level, there is a uniform composition of each particular species and a uniform temperature throughout the reaction medium. Electrochemical reactors that approach this ideal of mixing include rotating electrolyzers with relatively small fluid flow, fluidized beds,[27] and gas-sparged,[28] slurry,[29] sieve plate,[30] and other similar multiphase systems as well as channel flow reactors with large recycle.[31]

This section will lay the foundations of the basic design procedures for electrochemical reactors, illustrating behavior with simple reaction rate models.

1. Ideal Reactors; Batch and Continuous

The three types of ideal reactors—batch, plug flow, and continuous flow stirred tank—are depicted in Fig. 5. These reactors can, for a number

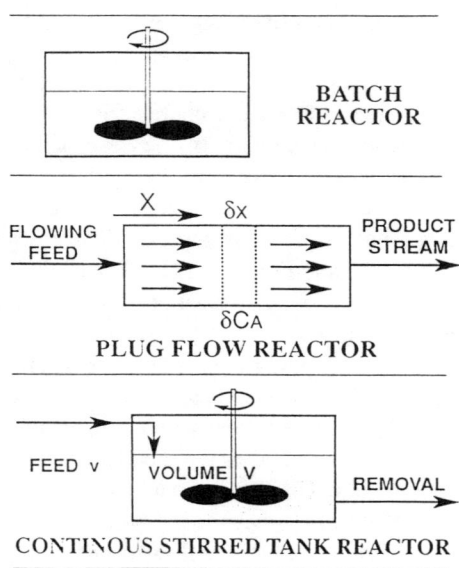

Figure 5. Ideal reactor configurations.

of reasons, operate either with or without recycle, which leads to different performance behavior in many situations. A different performance will also be realized by applying current or power to the reactor in different ways, for example, by adopting a galvanostatic or potentiostatic control. This is particularly true of the batch reactor.

(i) The Batch Reactor

Batch electrochemical reactors are frequently adopted for laboratory investigations as they enable performance data to be obtained as a function of reactant concentrations or conversion. Typically, reactors are designed to have a "uniform current distribution" over the electrode surfaces, and it is assumed that this condition applies. In industry, batch reactors are frequently the most economical option when production rates are small. In its simplest form, a batch reactor is any closed vessel in which a unit volume of feed V of uniform composition is discharged from the vessel, which is then made ready to receive the next charge of reactant.

The analysis of a batch reactor is based on the material balance, which for solely electrochemical reaction is

$$\frac{dC_i}{dt} = \frac{ai_j}{nF} \tag{56}$$

where a is the specific electrode area, and i_j is the current density associated with the formation of j. For a single reaction of the type A $\rightarrow$ B under galvanostatic conditions at almost 100% current efficiency, Eq. (56) gives

$$C_B - C_{B0} = \frac{ai_jt}{nF} \tag{57}$$

where C_{B0} is the initial concentration of B; that is, product concentration increases linearly with time. This is a rather global expression giving little insight into the reactor behavior, and generally we need to consider that more than one reaction will occur and introduce the appropriate reaction model. At this juncture we adopt the following kinetic expression (from Section II.1) exhibiting a first-order dependence on reactant concentration:

$$i_j = nFk_jC_{is}\exp(\beta E) = nFk_{fj}C_{is} \tag{58}$$

where β is a potential coefficient for an anodic process. For a cathodic process, where current density increases with a more negative rise in potential, Eq. (58) is still applied with the condition $-E = E$. Introducing the mass transport rate equation (13) as

$$i_j/n_jF = k_{ij}(C_i - C_{is}) \qquad (59)$$

and applying the steady-state approximation at the electrode surface to eliminate the surface concentration C_{js} gives

$$i_j = \frac{nFk_{fj}C_i}{1 + (k_{fj}/k_{Lj})} \qquad (60)$$

This is a convenient form of rate equation in which the influence of mass transport is defined through a dimensionless group $k_{fj}/k_{Lj} = \text{Da}_j$, a modified Damköhler number. If k_{Lj} is large (Da_j small), then mass transport is not rate limiting. Conversely, if k_{Lj} is small compared to k_{fj}, then limiting current conditions are approached.

The mode of current supply that minimizes secondary or side reaction is potentiostatic control. The reactor model is obtained by combining Eqs. (56) and (60) in the form

$$\frac{-dC_A}{dt} = \frac{ak_{fA}C_A}{1 + (k_{fA}/k_{LA})} \qquad (61)$$

This is integrated (noting that k_{fA} = constant) with the initial condition $C_A = C_{A0}$, $C_B = 0$ at $t = 0$ to give

$$\ln\left(\frac{C_A}{C_{A0}}\right) = \frac{-ak_{fA}t}{1 + (k_{fA}/k_{LA})} \qquad (62)$$

This is the classic exponential decay relationship of a first-order reaction, which with operation at the limiting current truncates to

$$\frac{C_A}{C_{A0}} = \exp(-ak_{LA}t) \qquad (63)$$

If a side reaction $P \rightarrow Q$ occurs to a small extent with the main reaction, then an expression identical to Eq. (62) is written for the associated species. The major influence of this reaction is with regard to the current efficiency, $\emptyset$, given by

$$\emptyset = \frac{\int_0^t i_A dt}{\int_0^t (i_A + i_P)\, dt} \qquad (64)$$

or, alternatively,

$$\emptyset = \frac{C_B}{C_B + a\int_0^t \frac{i_p}{nF} dt} \qquad (65)$$

A common side reaction in electrochemical reactors is solvent decomposition, which will usually occur under kinetic control with negligible change in reactant concentration. Thus, the associated reaction rate is constant, and estimation of current efficiency is greatly simplified.

In the above it is assumed that current densities are truly additive and "independent." Moreover, a commonly applied assumption is that when the two current densities are measured "separately" (when only one reaction occurs), the total current density is the sum of the constituents. This is not always the case, and then the current balance requires current efficiency data to split the relative contributions. This clearly results in the need for more data in formulating the reaction model.

An alternative, often preferred mode of supplying current to the reactor is under galvanostatic control. One advantage of galvanostatic operation over potentiostatic operation is that often much shorter reaction times are required to process a unit mass of feed. Thus, for a given production target, capital costs of reactors operating galvanostatically are generally much lower. A penalty for this mode of operation can be a higher energy consumption ensuing from the rising electrode potential.

Rising electrode potential can also contribute to higher energy consumptions during galvanostatic operation by causing a reduction of current efficiency due to the occurrence of side reactions. If this reaction is solvent decomposition, then the additional problem of handling potentially explosive gases arises. Ideally, during galvanostatic operation the limiting current should not be exceeded, although a compromise between overall production rate and energy consumption will have to be made.

Galvanostatic analysis of Tafel reaction with solvent decomposition

An important category of reactions is those that occur in regions where independent side reactions occur. Such reactions can often proceed with current efficiencies near 100% if the reaction rate is low enough. However, this can lead to economically unacceptable reactor sizes and hence costs for industrial operation. A reduction in reactor cost is therefore often made at the expense of current efficiency and energy consumption.

Consider the following reaction scheme:

$$A + n_1 e^- \xrightarrow{k_{f1}} B$$

$$C + n_2 e^- \xrightarrow{k_{f2}} D \tag{66}$$

With current densities for both reactions written according to Eq. (4) for galvanostatic operation, we have the current balance

$$i_T = i_1 + i_2 \tag{67}$$

The batch reactor material balances are

$$\frac{-dC_A}{dt} = \frac{ai_1}{n_1 F} \tag{68}$$

$$\frac{-dC_C}{dt} = \frac{ai_2}{n_2 F} \tag{69}$$

A useful expression in the analysis of such schemes is the charge balance given by

$$C_C - C_{C0} + \frac{n_1}{n_2}(C_A - C_{A0}) = \frac{-ai_T t}{n_2 F} \tag{70}$$

This simply states that the current supplied to an electrode is used to convert A and C and is merely a combination of Eqs. (67)–(69). Equations (68) and (69) are both functions of concentration and electrode potential, and the solution of this problem in general requires numerical procedures.

The numerical procedure uses the current balance to determine the electrode potential (by numerical integration) at any time interval, enabling Eqs. (68) and (69) to be integrated successively over selected incremental times.

A useful limiting solution of Eqs. (68) and (69) is obtained when mass transport is not rate-limiting and the coefficients β_1 and β_2 are equal:

$$\frac{C_A}{C_{A0}} = \left(\frac{C_C}{C_{C0}}\right)^{k_1/k_2} \tag{71}$$

When coupled with Eq. (70), this gives the following relationship between C_C and t:

$$\frac{C_C - C_{C0}}{C_{A0}} + \frac{n_1}{n_2}[(\frac{C_C}{C_{C0}}) - 1]^{k_1/k_2} = \frac{-ai_T t}{n_2 F C_{A0}} \tag{72}$$

Thus, the ratio of rate constants k_1/k_2, as might be expected, has a significant bearing on performance and particularly the current efficiency, which is given (for component A) by

$$\text{C.E.} = \frac{(C_{A0} - C_A)}{i_T at/n_1 F} \times 100\% \tag{73}$$

Frequently, side reactions involve the decomposition of water to either oxygen or hydrogen with little change in the corresponding concentration. Typically, this will proceed under kinetic control, and the current balance becomes

$$i_T = \frac{n_1 F k_{f1} C_A}{1 + (k_{f1}/k_{LA})} + n_2 F k_{f2} \tag{74}$$

Combining Eqs. (68) and (74) we obtain

$$\frac{dC_A}{dt} = \frac{-a(i_T - n_2 F k_{f2})}{n_1 F} \tag{75}$$

Thus, the problem reduces to the coupled expressions of the form

$$\frac{dC_A}{dt} = f_1(E) \tag{76}$$

and

$$C_A = f_2(E)$$

This problem is readily solved numerically using standard numerical integrations employing Simpson's rule or Runge–Kutta methods.

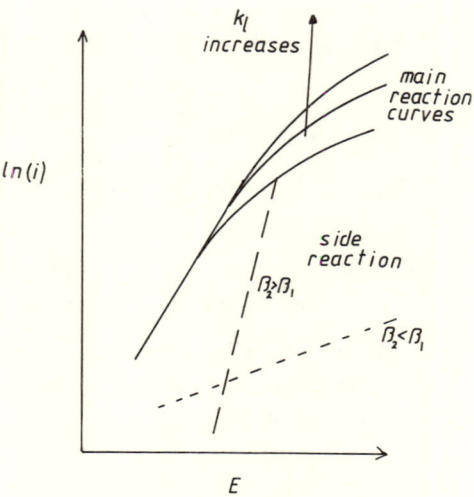

Figure 6. The effect of reaction rate model parameters on two independent reactions. k_l is mass transfer coefficient.

Frequently, electrochemical kinetics exhibit characteristics such that Tafel slopes are approximate integer ratios, i.e., $\beta_1/\beta_2 = 1, 2$, etc. This enables analytical solutions to be obtained[32] relating the variation of electrode potential with time. The following is typical of these when $\beta_1 = \beta_2$:

$$at = \frac{1}{k_{f10}} - \frac{1}{k_{f1}} + \frac{1}{k_{LA}}\left[\frac{i_T - n_2Fk_{f20}}{i_T - n_2Fk_{f2}}\right] + \frac{n_2Fk_2}{k_1i_T}\ln\left[\frac{k_{f2}(i_T - n_2Fk_{f20})}{k_{f20}(i_T - n_2Fk_{f2})}\right] \quad (77)$$

The variation of reactant concentration with time is hence obtained from Eq. (74). Qualitatively, with reference to Fig. 6, we expect the following behavior. The efficiency of conversion of A is favored by a high ratio k_1/k_2, a high rate of mass transport, and a low total current density when $\beta_2 < \beta_1$. When $\beta_2 > \beta_1$, then efficiency will be highest at some intermediate current density. This behavior is primarily due to the current density of the desired reaction approaching its mass transport limiting value.

An important point to note with galvanostatic operation is that the total current density should never exceed that defined by the limiting current at the initial concentration ($n_1Fk_{LA}C_{A0}$). Exceeding this value merely wastes electrical energy and causes excessive gas evolution. In fact, operation at a current density well below this value may be advised if current efficiencies are not to be low when high conversions are required. Actual operation may well approach a situation in which mass transport is not rate-limiting and thus, as seen in Fig. 6, high current efficiencies can be achieved.

(ii) The Batch Reactor with Recirculation

The traditional concept of a batch reactor often presents design problems when heterogeneous reaction occurs. This is due to the fact that the reaction is not distributed uniformly in the fluid but located at one small region. Requirements of moderate electrode areas per unit volume, good mass transfer rates, and small interelectrode gaps conflict with simple batch vessel design. To overcome this, the recirculating or recycle batch flow system shown in Fig. 7 is used.

In its ideal form, this system rapidly recirculates electrolyte from a well-mixed vessel of volume V_m through the reactor of volume V. In this way, high mass transfer rates can be attained in reactors with low interelectrode space. The reactor itself may be of a plug flow or continuous stirred tank configuration. Although it is feasible to design a reactor with internal electrolyte recirculation, this generally results in a more complex and costly design.

The type of configuration is now analyzed to see how its performance compares with that of a simple batch reactor. In doing this, design aspects of both the continuous stirred tank reactor and plug flow reactor are introduced.

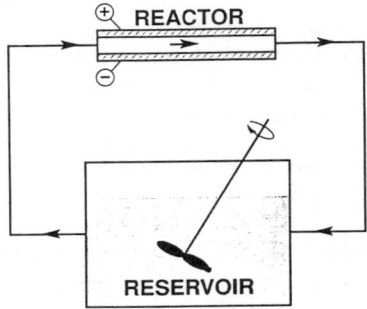

Figure 7. The Batch recirculation reactor unit.

(a) Batch recirculation with plug flow reactor

In a plug flow reactor (PFR), reactant flows, at a flow rate v, through the channel formed between the electrodes and separator. There is no mixing or dispersion of the species along the reactor, only a gradual decrease in reactant concentration due to electrochemical reaction. At any place a distance x along the reactor, the composition is assumed uniform, except at regions close to the electrodes where mass transfer takes place. A steady-state material balance on a differential volume of reaction, δV, relates the change in concentration to the local current density in that section, i_x, according to

$$-v\, dC_A = \frac{i_x}{nF} a\, dV \tag{78}$$

Equation (78) is written in terms of residence time $\tau = V/v$ as

$$\frac{-dC_A}{d\tau} = \frac{a i_x}{nF} \tag{79}$$

This expression is analogous to the design equation for a batch reactor. Evidently, an element of electrolyte passing from inlet to outlet reacts for an equivalent batch reaction time. This expression is considered in detail in Section III.1(v). In the context of the recirculating batch reactor, analysis of limiting current operation (i.e., at every position x the current density $i_x = nFk_{LA}C_A$) gives

$$\ln\left(\frac{C_A}{C_{A1}}\right) = -ak_{LA}\tau \tag{80}$$

where C_{A1} is the inlet concentration, representing an exponential decay of concentration with "length."

The above analysis applied to a steady-state operation of a PFR, whereas, in the recirculating batch system, inlet and exit concentrations vary with time. The material balance for the PFR is therefore

$$\frac{\partial C_A}{\partial t} + \frac{\partial C_A}{\partial \tau} = \frac{-ai_x}{nF} = r_A \tag{81}$$

Solution of this partial differential equation requires quite detailed numerical routines. Often this need not be treated rigorously, but rather a pseudo-steady-state approximation can be adopted[33] in which $\partial C_A/\partial t = 0$ in the reactor. With batch recirculation this approximation is

Table 2
Expressions for Fractional Conversion under Mass-Transport-Controlled Conditions

Model and reactor operation[a]	Expression for fractional conversion, X_A
Batch or PFR	$1 - \exp[-(k_L a t)]$
PFR with batch recycle	$1 - \exp\{-t/\tau_m[1 - \exp(-(k_L a \tau))]\}$
CSTR with batch recycle	$1 - \exp\{-t/\tau_m[1 - 1/(1 + k_L a \tau)]\}$
CSTR single pass	$1 - 1/(1 + k_L a \tau)$
CSTR cascade of n tanks	$1 - 1/(1 + k_L a\, \tau_n)^n$
Dispersed plug flow	$1 - \dfrac{4a\,\exp(\mathrm{Pe}/2)}{(1+a)^2 \exp(\mathrm{Pe}\,a/2) - (1-a)^2 \exp(-a\cdot\mathrm{Pe}_a/2)}$
	$a = (1 + 4\tau k_L/\mathrm{Pe})^{1/2}$, $\mathrm{Pe} = uL/\mathrm{De}$

[a]PFR, Plug flow reactor; CSTR, continuous stirred tank reactor.

more accurate, the smaller the volume of the reactor in comparison to the volume of the mixing tank.

For the batch recycle unit, an instantaneous material balance over the mixer with holding time τ_m can be written,

$$C_{A2} = C_{A1} + \tau_m \frac{dC_{A1}}{dt} \tag{82}$$

relating the outlet and inlet concentrations from the reactor. Eliminating C_{A2} (the reactor exit concentration) between Eqs. (80) and (82) gives

$$\tau_m \frac{dC_{A1}}{dt} + C_{A1} = C_{A1} \exp(-ak_{LA}\tau) \tag{83}$$

The solution of this equation is given in Table 2, together with equivalent expressions for other ideal reactor operations.

The analysis assumes that the interconnecting pipework is of negligible volume or the fluid therein is in plug flow. An allowance for mixing in this pipework can be included in the volume V_m.

(b) Batch recirculation with a continuous stirred tank reactor

A single-pass continuous stirred tank reactor is illustrated in Fig. 5. In its ideal form, all conditions within the electrolyte are uniform, and reactant flows into the reactor at a concentration C_{A1} and immediately mixes with the electrolyte within. The material balance for the reactor is

$$C_{A1} = C_{A2} + \tau \frac{dC_{A2}}{dt} + \frac{ia\tau}{nF} \qquad (84)$$

Equations (82) and (84) can be combined to give

$$\tau\tau_m \frac{d^2 C_{A2}}{dt^2} + (\tau + \tau_m)\frac{dC_{A2}}{dt} + \tau\tau_m \frac{dr_A}{dt} + \tau r_A = 0 \qquad (85)$$

When the electrochemical reaction rate model is a set of linear functions of concentration, Eq. (85) may be solved analytically, if the value of the electrochemical rate constant is fixed. Otherwise, a numerical procedure is required. The adoption of a pseudo-steady-state approximation to the reactor, as in the case of a PFR, can simplify this procedure. For the case of a mass-transport-limited reaction, the solution is given in Table 2.

A large number of electrochemical reactors, especially experimental ones, adopt batch recycle operation, and the solution for mass-transport-limited operation of a single reaction is of limited use. Although the solution of Eq. (83) or (85) will generally require numerical procedures, if one is embarking on an exercise that has an objective of determining an appropriate reaction rate model, certain simplifications can be beneficial. One such approximation is to consider that the reactor is so small in comparison to the holding tank that the unit behaves as a closed batch reactor. Inspection of Table 2 shows this to be a reasonable approximation, at least for limiting current operation.

(iii) The Continuous Flow Stirred Tank Reactor

For the continuous stirred tank reactor, the material balance for any component, A, flowing in at a concentration C_{A0} and out at a concentration C_A is, in the steady state,

$$C_{A0} = C_A + \tau r_A \qquad (86)$$

where r_A represents the rate of conversion of A by electrochemical-reaction and chemical reaction.

The significant feature of the CSTR is that operating parameters are time and space independent, with concentration and current density ideally uniform over the electrode surface.

As an illustration of the construction of the reactor model, consider a Tafel reaction

$$A + n_1 e^- \rightarrow B$$

accompanied by solvent decomposition. Combination of the current density of the main reaction, given by Eq. (60), with the material balance gives

$$C_{A0} = C_A + \frac{a\tau k_{f1} C_A}{1 + (k_{f1}/k_{LA})} \qquad (87)$$

This rearranges to give the exit product concentration as

$$\frac{C_A}{C_{A0}} = \frac{1}{1 + \dfrac{a\tau k_{f1}}{1 + Da_1}} \qquad (88)$$

The current balance for this system is

$$i_T = \frac{n_1 F k_{f1} C_A}{1 + (k_{f1}/k_{LA})} + n_2 F k_{fd} \qquad (89)$$

In essence, the use of the CSTR models requires the solution of simultaneous equations. For example for a known value of i_T, k_{f1} is obtained from Eq. (89) and hence X_A is obtained from Eq. (88).

A major limitation of-the CSTR is that a large residence time is generally needed when a high degree of conversion is required. Owing to the mechanical design of some electrochemical reactors, high residence times can mean that electrolyte velocities are low. A conflict may therefore result between the requirements of large residence time and good mass transport rate. Two means exist by which this can be resolved:

(i) Operate the CSTR with recycle of the product stream to the reactor inlet. This has the effect of giving a high flow rate through the cell, and hence good mass transport, while overall flow from inlet to exit is low and compatible with a large residence time. Reactor designs based on this principle are identical to that described above except the mass transport rate is defined by the overall flow in the cell.

(ii) Operate with more than one CSTR connected in series as a cascade. In such a cascade, only the last unit operates at the final reactant concentration, the other units operating at successively higher concentrations upstream. The overall residence time is therefore reduced. The reactor design is implemented by applying the design equation successively to each unit, where exit concentration from one unit is the inlet concentration to the next.

(iv) The Plug Flow Reactor

The plug flow reactor was introduced in Section III.1(ii) on the batch recirculation reactor. In practice, the plug flow reactor operates at a constant cell voltage (assuming negligible electrode ohmic losses), with local current densities determined by the voltage balance and current balance if more than one electrode reaction occurs. The voltage balance for a PFR is written as

$$E_{cell} = E_a + E_c + I_T \sum R_e \qquad (90)$$

where a and c refer to the anode and cathode, respectively, and the last term represents the sum of the ohmic voltage losses, in electrolytes, separators, and electrodes. Both electrode potentials are functions of reactant concentration and current density.

For high-field Tafel type kinetics, the voltage balance becomes (for single electrode reactions)

$$E_{cell} = E_D + \frac{1}{\beta_a}\ln(\frac{i_x}{n_a F k_a C_a}) + \frac{1}{\beta_c}(\frac{i_x}{n_c F k_c C_c}) + I_T \sum R_e \qquad (91)$$

where E_D is the cell equilibrium potential.

The PFR is thus modeled on the basis of three sets of coupled equations:

1. Material balances for all species
2. Current balance
3. Voltage balance

A procedure for solution would be to solve the current and voltage balances simultaneously, at known or specified concentrations of species, to determine the local electrode potential. The potential is then used to facilitate integration of the material balances over prescribed incremental reactor lengths or residence times.

The PFR is, like the CSTR, often limited in application as a single-pass unit because of conflicts between the requirements of large residence time and good mass transport and the mechanical engineering. To meet the conflicting requirements, a number of smaller reactors can be connected in series. Each reactor is then analyzed as an individual unit where a relatively small (incremental) conversion occurs. The mode of operation can then be approximated as either galvanostatic or potentiostatic, depending on a number of factors prevailing in the unit.[34]

An alternative means of meeting the requirement of large residence times is to operate the reactor with "recycle" of product stream back to the reactor inlet. In this case, overall flow through the unit is small, while the flow through the cell is high to give high mass transfer rates. Conversions per pass in the cell are low, and the mode of operation can again be approximated as either galvanostatic or potentiostatic. However, the effect of recycle can influence the reactor model, depending upon the specific reaction system.

Clearly, either of these approximations ignores the aspect of both the current and potential varying along the reactor, which for some circumstances must be considered. In industrial electrochemical reactors, the preferred operating mode is generally specified by supplying a known overall current at a prescribed voltage. Thus, it is often assumed that galvanostatic operation exists in that there is no local variation of current density. However, if there is a significant decrease in electrolyte conductivity due to, say, gas evolution, then it can be more appropriate to assume that the reactor is operating potentiostatically (i.e., electrode potential is uniform along the reactor).

(v) The Recycle Plug Flow Reactor

The model of the recycle plug flow reactor will be developed using an example of the classic ECE mechanism to stress the important difference between a reactor with and without recycle. This difference is essentially due to the influence of chemical reaction.

The ECE mechanism, in which a reactant A undergoes successively charge transfer, chemical transformation, and charge transfer, is represented as

$$A + n_1 e^- \xrightarrow{k_{f_1}} B$$

$$B \xrightarrow{k_c} C$$

$$C + n_2 e^- \xrightarrow{k_{f_2}} D \tag{92}$$

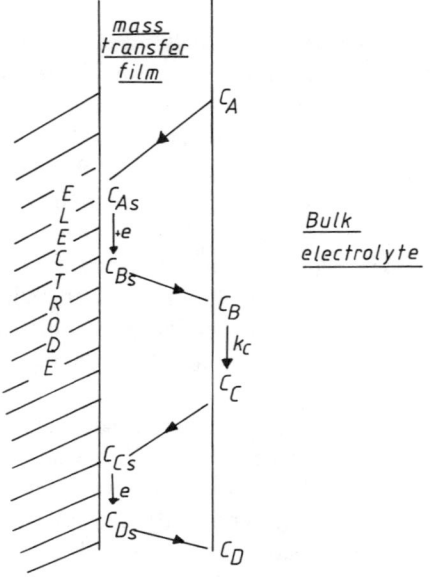

Figure 8. Schematic reaction model for the ECE sequence.

It is assumed that the chemical transformation is not too fast and that reaction in the diffusion layer is negligible. The reaction and mass transfer paths for this sequence are depicted in Fig. 8. With both steps following Tafel kinetics, the current densities are

$$\frac{i_1}{n_1 F} = \frac{k_{f1} C_A}{1 + Da_1} = C_A Y_1$$

$$\frac{i_2}{n_2 F} = \frac{k_{f2} C_C}{1 + Da_2} = C_C Y_2 \qquad (93)$$

where $Da_2 = k_{f2}/k_{LC}$.

Material balances for the recycle PFR in Fig. 5 are as follows, where the recycle ratio r defines the fraction of total flow through the cell that is returned to the inlet:

$$\frac{dC_A}{d\tau_r} = \frac{-ai_1}{n_1 F} \qquad (94)$$

where τ_r is the residence time based on the overall flow rate $(r + 1)$;

$$\frac{dC_B}{d\tau_r} = \frac{ai_1}{n_1 F} - k_c C_B \tag{95}$$

$$\frac{dC_C}{d\tau_r} = k_c C_B - \frac{ai_2}{n_2 F} \tag{96}$$

$$\frac{dC_D}{d\tau_r} = \frac{ai_2}{n_2 F} \tag{97}$$

For the solution of this set of equations, the concentrations of species at the reactor inlet C_{ji} are required. This is found from a mass balance at the junction of the feed and recycle streams, as

$$C_{Ai} = \frac{C_{A0} + rC_{Ae}}{r+1} \tag{98}$$

where C_{Ae} is the concentration out of the reactor.

For component B we must account for chemical reaction in the recycle loop. This degree of reaction will depend on the characteristics of flow pertaining to this loop, that is, either plug flow or well mixed.

(a) For plug flow with first-order reaction, the concentration of B at the recycle junction C_{Be} is given by

$$\frac{C_{Be}}{C_{Bp}} = \exp(-k_c \tau_p) \tag{99}$$

where τ_p is the residence time of fluid in the recycle loop. The mass balance for B at the recycle loop is given by

$$C_{Bi} = \frac{rC_{Be} + C_{B0}}{r+1} \tag{100}$$

(b) For a CSTR the concentration C_{Be} is given by

$$\frac{C_{Be}}{C_{Bp}} = \frac{1}{1 + k\tau_p} \tag{101}$$

For component C, the transformation of B to C in the recycle loop is also required. This is given by a material balance as

$$C_{Ce} = C_{Cp} + C_{Bp} - C_{Be} \tag{102}$$

The concentration C_{Ci} is therefore given by

$$C_{Ci} = \frac{C_{C0} + rC_{Ce}}{r + 1} \quad (103)$$

Equations (94)–(96) can now be solved subject to the mode of control. For potentiostatic operation, the solution follows standard procedures for first-order linear differential equations and gives

$$\frac{C_A}{C_{Ai}} = \exp(-aY_1\tau_r) \quad (104)$$

$$\frac{C_B}{C_{Ai}} = \frac{Y_1 a}{(k_c - Y_1 a)}[\exp(-aY_1\tau_r) - \exp(-k_c\tau_r)] + \frac{C_{Bi}}{C_{Ai}}\exp(-k_c\tau_r) \quad (105)$$

$$\frac{C_C}{C_{Ai}} = \frac{k_c}{(k_c/Y_1 a) - 1}\left[\frac{\exp(-aY_1\tau_r) - \exp(-aY_2\tau_r)}{(aY_2 - aY_1)}\right]$$

$$+ \frac{\exp(-aY_2\tau_r) - \exp(-k_c\tau_r)}{(aY_2 - k_c)}$$

$$+ \frac{k_c C_{Bi}}{(aY_2 - k_c)C_{Ai}}[\exp(-k_c\tau_r) - \exp(-aY_2\tau_r)] + \frac{C_{Ci}\exp(-aY_2\tau_r)}{C_{Ai}} \quad (106)$$

The concentration C_D is found from an overall material balance.

For the purpose of this model, we will consider that intermediate C is the desired product. If B were desired, then it would be necessary to reconvert C to B by perhaps reoxidation if C is the reduced state. With the ECE mechanism, slow chemical reaction will tend to produce high concentrations of the first intermediate B in the reactor outlet. This maintains the concentration of final product D at low values, and if the chemical reaction of B to C proceeds in the holding tank downstream from the reactor, high yields of C will result.

A key factor that determines the yield of C is the recycle ratio and its interaction with mass transport, due to the dependence of mass transfer coefficient on flow rate.

To illustrate the reactor model, Fig. 9 gives typical variations of intermediate product C for recycle operation. Under conditions of kinetic control ($k_L > k_{f1}$), the effect of recycle decreases yields, which typically

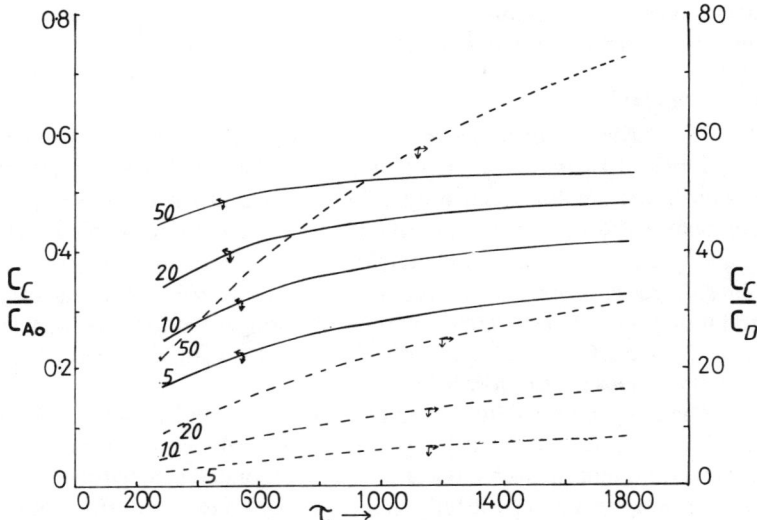

Figure 9. Variation of intermediate product of an ECE reaction in a recycle plug flow reactor. Values of recycle ratio on Figure $\sigma L/Q = 10^5$; C_C/C_{Ao} (—), solid lines; C_C/C_D (- - -), dashed lines (a) $k_L \to \infty$; (b) $k_L = 10^{-7}$.

exhibit maximum values characteristic of series reactions. With large recycle ratios ($r > 100$), the recycle reactor behaves similarly to a CSTR.

Under mixed kinetic and mass transport control, the fall of yield with increasing recycle is not as severe. As mass transport control is approached, increasing recycle serves to increase the yield of C.

In commercial continuous electrolytic processes, recycle operation of the reactor appears to be favored. Its attraction lies in the high rates of mass transport coupled with large overall conversion. Its performance characteristics lie between those of the PFR ($r = 0$) and those of the CSTR ($r \to \infty$). The reactor model is similar to that for the PFR except that the inlet stream contains product species. The solution of the fundamental model equations are generally of a similar form to that for the PFR, although a major difference arises with coupled chemical reaction.

Application of galvanostatic operation of the recycle reactor invariably requires numerical solution of model equations, except under limiting conditions previously identified or when single charge transfer reactions occur. Numerical solutions are ideally performed using simulation languages or "library" programs of packaged software.

(vi) Conclusions

The derivation of a reactor model always relies on some assumed conditions of operation or approximation to ensure that solutions are within the realms of practicality. In these days of rapidly developing computer software, models are becoming more and more sophisticated, enabling exploration of the interaction between large numbers of variables. Regardless of this, there is still a use and indeed a need for the tried and tested procedures, which are the usual springboard for more sophisticated models. Concepts such as ideal flow are still in many cases realistic approximations to real flow behavior.

The analysis of real flow, often referred to as "nonideal flow," leading to multiparameter flow models is a fairly well developed area in chemical engineering. Indeed, such models can be incorporated in electrochemical reactor design, and occasionally their use may be justified. Electrolytic reactors will, under certain conditions, exhibit nonideal flow of significant proportions to warrant consideration in design.

In addition, our models of reaction kinetics and mass transport may prove to be unsatisfactory for reactor design. The phenomenon of adsorption may play a leading role, and its effect may be to modify surface structures or lead to deactivation of electrode surfaces. This latter phenomenon was considered for an electroorganic reduction,[35] and a batch reactor model was proposed that incorporated "activity coefficiencies" to empirically quantify the effect.

2. Real and Nonideal Flow Models

A reactor model that accounts for some degree of nonideal flow will require a greater degree of sophistication in solution than is needed for the case of ideal flow models. The model, by its very nature, will be more rigorous and may involve aspects of reactor geometry, hydrodynamics, ionic transport, and simultaneous current and potential distributions as well as modeling of the electrochemical and chemical processes. The role of digital simulation in the application of these models is of great importance owing to the complexities of the coupled equations describing the physicochemical interactions. The requirements of numerical routines in simulation will clearly depend upon the structure of the model. Such routines are often generally available and have been applied to the simulation of many electrochemical systems of industrial significance. These are, however, open to further development in areas in which, for the moment, methodologies based partly on empirical engineering concepts are utilized. These include technologically important areas where several phases are involved, such as gas evolution systems, phase transformations, batteries, films, corrosion, and multiphase reactions.

This section outlines the equations required to model electrochemical reactors with real flow. There are two broad categories of models in this area, referred to as thin film models[36] or thin gap channel models.[37] The former isolates boundary layers from the remainder of the electrolyte bulk, whereas the latter accommodates direct interactions between adjacent electrode regions. In the latter there is inevitably laminar flow in the small interelectrode gap used to give low ohmic voltage losses.

Reaction models based on residence time distributions are based on a knowledge of the local history of reacting fluid elements within the reactor. By identifying the localities of fluids and quantifying those localities, and the time spent there, the extent of reaction of fluid elements and thus the overall reaction can, in principle, be determined. The measurement and modeling of the residence time distribution then becomes the means of formulating the basic reactor model. Extensive treatments of this subject are available[38-40] in the chemical engineering literature, and only a brief synopsis is given here.

In the treatment of residence time distribution, there are two commonly used single-parameter models, the dispersed plug flow and the tanks in series. The dispersed plug flow model superimposes a form of diffusion mixing on the plug flow behavior and takes the form of a one-dimensional diffusion equation with a convection term:

$$\frac{\partial C}{\partial t} + u\frac{\partial C}{\partial x} = \frac{\partial}{\partial x}\left(D_e\frac{\partial C}{\partial x}\right) + r(C) \qquad (107)$$

where D_e is the mixing/dispersion coefficient, determined experimentally, and u is taken to be the mean velocity through the reactor. The tanks in series model pictures the reactor as a series of n well-mixed regions of equal size, and the reactor is analyzed on the basis of the CSTR model applied successively, n times. The parameter n is found by statistically matching experimental residence time distribution data, obtained generally from tracer studies, with the theoretical dynamic model of the unit.

There are a number of similarities in the residence time distribution behavior of these two models, although there are subtle differences between them which can direct the user to the most appropriate one for a particular application. The tanks in series model has no provision for backmixing, or movement of fluid upstream. Thus, it has been suggested that the tanks in series model can be modified to include backflow streams[41]; i.e.,

$$\frac{\tau}{n}\frac{dC_i}{dt} = (1 + \beta_r)C_{i-1} - (1 + \beta_r)C_i + \beta_r C_{i+1} \tag{108}$$

where i represents the tank number ($i = 1, \ldots, n$), and β_r is the ratio of backflow to through flow between stages. This therefore introduces a second parameter into the model, making it in principle more sophisticated. There are a number of examples where models based on dispersion effects are not appropriate, particularly dealing with multiphase systems, and then multiparameter and multizone models are adopted. A typical example in electrosynthesis is the behavior of cells generating streams of small bubbles moving through the continuous electrolyte phase.

The convenience in using the dispersed plug flow model is that mathematical routines and solutions are well developed. For the case of mass-transport-controlled reactions, the steady-state fractional conversion can be expressed in terms of an axial Peclet number, $\mathrm{Pe}_a = uL/D_e$, as given in Table 2. The case of the laminar flow reactor has been theoretically considered.[42]

The solution of the convective diffusion equation has been compared with the one-dimensional dispersion equation (with a large length-to-diameter ratio) to give the relationship

$$D_e = D + \frac{u^2 R^2}{48D} \tag{109}$$

which is the dispersion coefficient for laminar flow. The laminar flow model is thus considered in much the same way as the dispersed plug flow model.

When applied to electrochemical reactors, the dispersion model approach gives a means of applying the thin film treatment in nonideal flow. The behavior of the bulk electrolyte is thus that of dispersed plug flow, with concentrations perpendicular to the flow considered uniform. This bulk interacts with the thin boundary layers formed at the electrodes (or

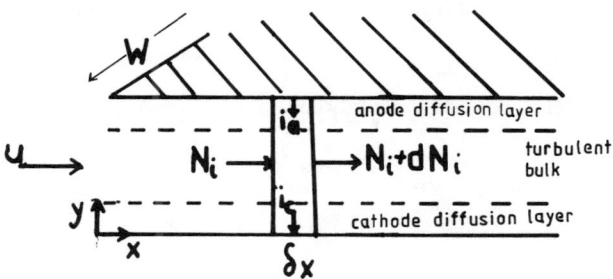

Figure 10. Model of the thin film electrochemical reactor.

at the separator), in which species diffuse or migrate to and from the surface (see Fig. 10). In other words, the model is analogous to the Nernst diffusion layer concept, which is, however, extended to enable aspects of migration and convective transport to be included.

(a) The thin film model

The thin film model applies to systems with forced convection in which the hydrodynamic velocity distributions are known. The basic equations defining the model are considered here; for detailed discussions of the model, the reader is referred to the works of Newman.[19,43]

The model defines the concentration and potential distributions by virtue of four basic equations:

1. The flux N_i of a species due to the combined action of migration in an electric field, diffusion through the existence of a concentration gradient, and convection from the velocity of the fluid is given by

$$N_i = -z_i u_i F C_i \nabla \phi - D_i \nabla C_i + u C_i \tag{110}$$

where u_i is the ionic mobility, and u is the local velocity vector. Ionic diffusion coefficients are related, approximately, to ionic mobility by the Nernst–Einstein equation:

$$D_i = RT u_i \tag{111}$$

2. A material balance for species i in the steady state is

$$N_i - r_i = 0 \tag{112}$$

In the absence of homogeneous reaction, this is

$$N_i = 0 \tag{113}$$

This is subject to boundary conditions at the electrode ($y = 0$),

$$\frac{dC_i}{dy} = \frac{i_j}{nFD_i}$$

and in the bulk ($y \to \infty$),

$$C_i = C_{i\infty}$$

3. The current density, i, is due to the movement of charged species

$$i = F \sum_i z_i N_i \tag{114}$$

4. Electrical neutrality of the electrolyte solution

$$\sum_i z_i C_i = 0 \tag{115}$$

This is a good approximation up to a distance of approximately a hundred angstroms from the interface.

Equation (110) is valid for dilute solutions, and extensions to concentrated solutions have been discussed.[44] For reaction of ionic species in a solution of an excess supporting electrolyte, Eq. (110) is written as

$$N_i = -D_i \nabla C_i + u C_i \tag{116}$$

Substitution in Eq. (113) leads to the equation of convective diffusion

$$u \nabla C_i = D_i \nabla^2 C_i \tag{117}$$

It is usually appropriate to consider only a two-dimensional problem (electrode axial dimension x and the perpendicular distance y) and, in electrolytic systems, the case of high Schmidt numbers (>1000), which means that the diffusion layer is thin even compared to the hydrodynamic boundary layer. The velocity terms can then be represented by

$$u_x = yB(x) \quad u_y = -0.5y^2 B'(x)$$

where $B(x)$ is the velocity derivative at the interface. Equation (117) is therefore written as

$$y\beta \frac{\partial C_i}{\partial x} - \frac{1}{2}y^2\beta' \frac{\partial C_i}{\partial y} = D_i \frac{\partial^2 C_i}{\partial y^2} \tag{118}$$

This equation has formed the basis for the analysis of current distribution and mass transfer problems at several geometries under conditions of an axial invariable concentration. From this, several expressions describing mass transfer distributions at electrode surfaces have been developed.[45] This is an important area of electrochemical engineering as it often forms an integral part of a reaction model. The measurement and interpretation of these mass transport distributions is still a developing field.

The equations outlined above form the basis of certain electrochemical reactor models analyzed in Section V, which by their nature require digital simulation methods, particularly when multiple reactions occur. A major simplification is possible if the convection component of ionic transport is negligible in Eq. (110), especially if a one-dimensional model of the diffusion film is adopted.[46] The flux equation for active species is then

$$N_i - z_i u_i F C_i \frac{d\phi}{dy} - \frac{D_i dC_i}{dy} \tag{119}$$

and for inactive species,

$$z_i F u_i C_i \frac{d\phi}{dy} + \frac{D_i dC_i}{dy} = 0 \tag{120}$$

The local current density is defined by Eq. (114) and for the case of an electrochemically active component is given by

$$\frac{i_x}{zFD} = \frac{dC}{dy} + \left(\frac{zFC}{RT}\right)\frac{d\phi}{dy} \tag{121}$$

Within the limits of the approximations adopted, this approach can provide a useful model of a plug flow thin film system, in which the bulk solution is turbulent. The solution procedure is one of numerically integrating the plug flow design equation, in which the local current density is determined from the integration of Eqs. (120) and (121) in conjunction with the condition of electroneutrality, the expressions for the electrolyte potential drop in the three regions (two films, one bulk), and the reaction electrode kinetics.

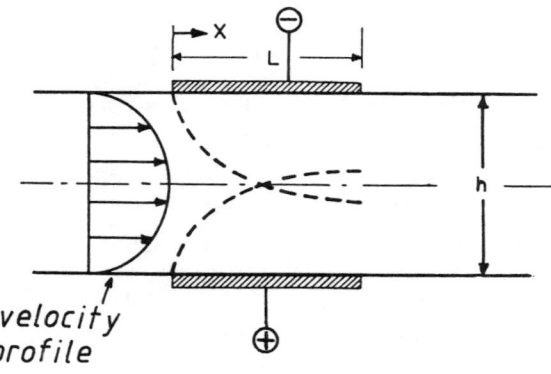

Figure 11. Model of the thin gap electrochemical cell.

(b) Thin gap models

A thin gap model can be distinguished from a thin layer model in that, in the former, specific interactions can occur between opposing boundary layers formed at the electrodes (Fig. 11). The gap between the electrodes is so thin that species can diffuse from one electrode region to the other, across overlapping boundary layers and undergo corresponding reactions, for example, metal dissolution and metal deposition. Generally, in these systems axial diffusion is negligible. The equations that form the basis of the model are, in principle, identical to those of the thin layer model. A recent model of thin gap cells[37] has been used to predict the current, potential, and concentration distributions of a metal dissolution–deposition system. The model, although ignoring the effect of ionic migration (i.e., excess supporting electrolyte is assumed to be present), showed that when the product of the aspect ratio of the channel and the Peclet number, (d_e/L)Pe, is less than 100, the boundary layers interact. Now, as the Peclet number is the product of the Reynolds number, Re, and the Schmidt number, Sc, and with Sc values typically around 1000, the condition for a thin gap cell is approximately

$$\text{Re}(d_e/L) < 0.1$$

With a 0.1-mm gap and an aqueous electrolyte, this gives the condition that Re < 500L. Thus, even with industrial-sized units (L = 2 m), the condition is met by laminar flow, an undesirable flow regime in terms of mass transport requirements in synthesis. This does not mean that in general the model is limited; for example, redox flow batteries can adopt laminar flow to keep pressure losses low. However, there is an acute danger in using this type of model, if one does not allow for the limitations in electrochemical technology, of exploring the behavior of reactors that are of little practical significance.

In summary, there is now available a powerful range of models and mathematical methods for electrochemical systems which can be used for the purpose of engineering design and analysis. These models are, more and more, being used in the implementation of experimental investigations so that a better understanding of the reaction mechanism and a better formulation of the model can be made. This aspect and the development of an adequate data base are of vital importance to the implementation of electrochemical technology.

IV. ELECTROANALYTICAL EXPERIMENT AND REACTION MODELING

This section will focus on a range of established electroanalytical methods used in determining reaction mechanisms and associated rate data for electrochemical-based reactions. The importance of the integration of these typically dynamic techniques with steady-state and product distribution data is stressed. The role of numerical methods and digital simulation in this task is an important factor, and their use is now widespread, with several computer packages available for ready implementation.

1. Introduction

Early work on the modeling of electrochemical reactions was concerned with the field of electroanalytical chemistry. Electrochemists sought methods of differentiating between various reaction schemes with a view to gaining insights into the fundamental aspects of electrode processes, including the effects of double-layer structure, adsorption, and reversibility. At first, techniques such as polarography at planar electrodes with the inclusion of a Luggin capillary to measure electrode potential were used. Steady-state currents were measured at a number of discrete potentials to determine the current–potential characteristics. Mass transport in such systems was purely diffusional over a short period of time (over longer periods, density changes led to natural convection) and was modeled by Fick's laws.

Uncertainties about the thickness of the diffusion layer at a planar electrode led to the adoption of a dropping mercury electrode, which provided better defined diffusion layer equations. With the introduction of modern, fast-acting potentiostats and pulse and sweep generators, new methods were developed, and these have been well summarized by Thirsk and Harrison.[47] Diffusion control is often accomplished by the use of rotating disk or ring-disk electrodes.

To overcome the difficulties of obtaining accurate measurements close to the equilibrium potential, unsteady-state techniques are often used. These are perturbation methods that depend upon the fact that the effect of a change in applied potential or current will be seen over a short time span as an almost instantaneous, but transient, change in other variables. If a potential step is applied, then current can immediately rise without the normal diffusion limitations, but after a short time, mass transport will reassert itself and a steady state will again be reached. Most perturbation methods use a fast, continuous potential sweep, or potential or current pulses, so that no steady state is attained during the time required to take the necessary measurements. It is the transient behavior of current or potential that is of importance in such experiments, and use of the unsteady-state techniques can reveal many details that are of interest.

There is other work which has dealt with steady-state experiments on a longer time scale which is intended mainly as an electroanalytical tool. Gelb and Meites[48] derived analytical solutions for a number of systems involving series of consecutive electrochemical and chemical reactions. Their work included the calculation of concentration–time profiles and so could be adapted to the purpose of obtaining expected product distributions in industrial reactors. Only first-order reactions in batch or plug flow reactors and mass-transfer-controlled situations were considered, and it was assumed that chemical reactions would occur only in the bulk of the electrolyte.

Karp and Meites[49] recognized that fast chemical reactions could occur in the diffusion layer as well as in the bulk and investigated the effects of this on current–time characteristics in a number of reaction schemes. They concluded that if a chemical intermediate was detectable in the bulk electrolyte, then its reaction within the diffusion layer could be neglected. In the case of a reaction so fast that the intermediate wholly reacted before entering the bulk, being confined within a reaction layer with a thickness less than that of the diffusion layer, then significant effects were noted. Again, the work was presented as applying only to mass-transfer-controlled cases, but it is possible to expand their equations to obtain concentration–time characteristics of the reaction schemes under kinetic and mixed control as well. The only remaining constraints are that operation must be at constant potential in a batch or plug flow reactor and the reactions must be first-order.

Amatore and Saveant,[50] and others, have done much work on the modeling of electrochemical reactions, and of particular interest is a series of articles[51] dealing with the product distributions in preparative scale electrolyses. A number of complex electrochemical/chemical reaction schemes were considered, mainly with a view to experimental classification of reactions: however, the equations derived may also be used to examine concentration–time characteristics. It was assumed that chemical reactions were either so fast that the intermediate existed only within a thin reaction layer or so slow that the extent of reaction within the diffusion layer could be neglected. Karp and Meites's work indicates that this assumption will not lead to great errors.

For each reaction scheme, the expected product distributions were investigated under each of three electrolysis regimes. These were exhaustive potentiostatic operation and both potentiostatic and galvanostatic batch operation under conditions where additional reactant was added to maintain the concentration of the initial reactant constant throughout the experiment. Exhaustive potentiostatic operation, in which a batch of solution is electrolyzed at constant potential until all electroactive species have been consumed, is sometimes used industrially. Batch electrolysis with constant concentration of one species (the concentration of intermediates and products are free to change) is not a common mode of industrial operation.

2. Steady-State Methods

A simple model of an electrochemical surface reaction presented in Section II was one of coupled electron transfer and mass transfer. The response of an experimental current density–electrode potential measurement will show a transition from one region of rate control to the other (Fig. 12). The precise shape of the i–E response will depend on whether the reaction is reversible or not. As a guide, a standard rate constant greater than 2×10^{-4} m/s leads to a reversible system.

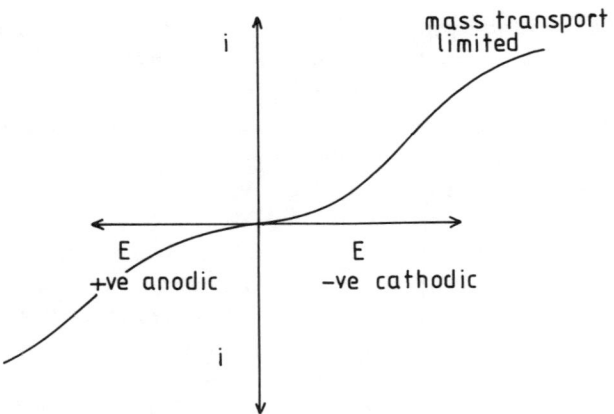

Figure 12. Current density–electrode potential response for an electron transfer reaction.

Although electrodes of a variety of shapes are used in electrochemical experiments (e.g., plate, rod, wire, mercury drop), the rotating disk electrode (r.d.e.) is a valuable and frequently used device in electroanalysis owing to its well-defined and controllable hydrodynamics.

(i) The Rotating Disk Electrode

The r.d.e. consists of a planar disk of the electrode material rotating at a constant speed about an axis perpendicular to the disk and passing through its center.[52] The theoretical result for the thickness of the diffusion layer under laminar flow conditions

$$\delta = 1.61 D^{1/3} v^{1/6} w^{-1/2} \quad (122)$$

is valid when the thickness of the Prandtl layer is much less than the radius of the disk and when edge effects are minimized, by insetting the disk electrode within an insulating holder with which it forms a continuous plane. The units of diffusion coefficient D, viscosity v, angular velocity w, and diffusion layer thickness δ are centimeters and seconds.

The corresponding mass transfer limiting current density, i_d, is

$$i_d = 0.62 n F D^{2/3} v^{-1/6} w^{1/2} C \quad (123)$$

where C is the bulk concentration of reactant in moles per cubic centimers. The range of rotating speeds of the r.d.e. is between 1 rad/s and 2000 rad/s for a disk of 1.0-cm diameter. The upper limit is imposed by the requirement of laminar flow; that is, the Reynolds number has the value

$$\text{Re} = r^2 w/v < 50{,}000 \tag{124}$$

(ii) Applications of the r.d.e.

A study of limiting currents permits the use of Eq. (122) to determine values for n, δ, and C if any pair are already known. Usually, a plot of i_d versus $w^{1/2}$ is used for this purpose.

In electrode kinetics, mass transport limitations can in some cases be rendered negligible by the use of high rotation speeds. More commonly, the use of high values of w, which give rise to low values of δ, cause those reactions that would be fully reversible to become quasi-reversible. Thus, values for the electrochemical rate constants can be obtained by standard means in which mass transport and Butler–Volmer type kinetic processes are combined[52] to give suitable correlations for linear regression; for example, for reactions of type 3,

$$\ln\left[\frac{(1 - \frac{i}{i_{df}})(1 - \frac{i}{i_{dr}})\exp(nF\eta)}{i}\right] = \alpha nF\eta - \ln(i_o) \tag{125}$$

where i_{df} and i_{dr} are diffusion limiting current densities for forward and reverse reactions, respectively.

For the r.d.e., Eqs. (58), (5), and (123) can be combined to give a convenient method for the determination of kinetic parameters. This leads to the following expression:

$$\frac{nF}{i} = (k_{f1}C_A - k_{b1}C_B)^{-1} + \frac{\left(\dfrac{\delta_A}{B_A} + \dfrac{\delta_R}{B_R}\right)}{(k_{f1}C_A - k_{f2}C_B)w^{1/2}} \tag{126}$$

where $B = 0.62 nFD^{2/3}v^{-1/6}$.

For a linear plot of $1/i$ against $1/w^{1/2}$, the intercept gives the kinetic limited current, $nF(k_{f1}C_0 - k_{b1}C_R)$, and the slope yields the value of $(\delta_A/B_A + \delta_R/B_R)$. The electrode kinetics can be treated in the usual way, that is, by employing Tafel plots to obtain exchange current densities and transfer coefficients.

The rotating disk can be used to obtain the order of an irreversible electrode reaction without the need to change the bulk concentration of electrolyte. When

$$i = k_f (C_{As})^p \tag{127}$$

where p is the order of the reaction, we obtain

$$\log i = \log k_{f1} + p \log[(i_d - i)/w^{1/2}] - p \log B_A \tag{128}$$

Thus, one can obtain p from a plot of $\log i$ versus $\log[(i_d - i)/w^{1/2}]$ at constant potential without even knowing the value of C_A.

The r.d.e. is also useful in the study of coupled chemical reactions as the variation of rotation speed enables the relative rates of mass transfer and chemical reaction to be changed, and thus the two processes distinguished and the rate constants determined. For example, in an ECE reaction

$$O_1 + n_1 e^- = R_1$$

$$R_1 \overset{k}{=} O_2 \tag{129}$$

$$O_2 + n_2 e^- = R_2$$

sufficiently rapid rotation enables R_1 to diffuse to the bulk solution more rapidly than it is transferred to O_2. Thus, with increasing rotation speed, the system passes from an $(n_1 + n_2)$-electron reaction to an n_1-electron reaction.

The evaluation of the chemical rate constant, k, will in practice require a comparison of the experimental and the theoretical number of electrons with variation in rotation speed. The theoretical variations of the system are obtained by digital simulation. It is useful at this stage to consider an example of the application of an r.d.e. in the formulation of a reaction model.

(ii) The Reduction of Nitrobenzene

A recent example of the use of steady-state techniques in the determination of a reaction model for a multiple-reaction system is in the reduction of nitrobenzene on a copper electrode in a sulfuric acid/1-propanol electrolyte.[53] The simplified mechanism of this reduction, as analyzed previously[54,55] using both analytical and numerical simulation of product distribution data, is shown in Fig. 13. Two major products are

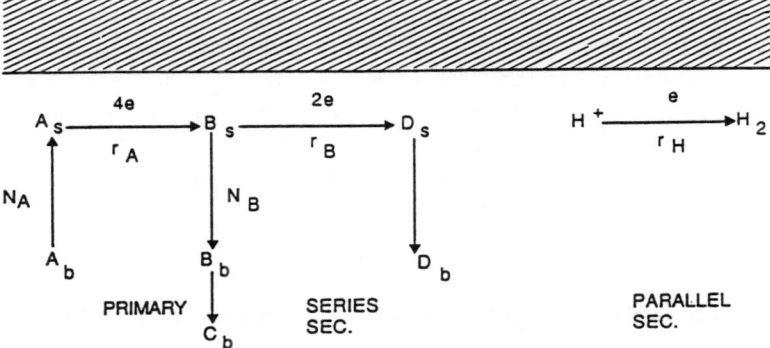

Figure 13. Schematic mechanism for the reduction of nitrobenzene.

formed in this scheme, aniline and p-aminophenol, by the electrochemical reduction and chemical rearrangement, respectively, of the first reduction product, phenylhydroxylamine. This scheme is an engineering model of the system, not a detailed mechanistic picture, and thus a reaction engineering approach is appropriate. Therefore, as well as investigation of the electrochemical kinetics, the integration of concentration–time data and their analysis is an important factor in determining the overall reaction model. In nitrobenzene reduction, clearly three processes interact: mass transport, electrochemical kinetics, and chemical reaction. In this system the rate of chemical rearrangement is slow compared to mass transport, and thus a category 3 system exists, where chemical reaction is mainly confined to the bulk electrolyte.

The formulation of the reaction model requires studies of the individual processes, which are:

- Electroreduction of nitrobenzene (NB) to phenylhydroxylamine (PH)
- Chemical reaction of PH to p-aminophenol
- Electroreduction of PH to aniline
- Mass transport behavior
- Side reactions

In this study the mass transfer behavior is that of the rotating disk electrode, and thus the diffusion coefficient for nitrobenzene was determined from the i_d versus $w^{1/2}$ plot discussed earlier.

With the reduction of phenylhydroxylamine, the hydrogen evolution reaction masked the limiting current behavior, and so the electrooxidation of this compound to nitrosobenzene on platinum was treated in an analogous manner to determine its diffusion coefficient. The values obtained can be checked from data generated in the plots of $1/i$ versus $1/w^{1/2}$.

The electroreduction of nitrobenzene was studied at an r.d.e., and values of the kinetic current density,

$$i_k = nFk_{f1}C_{NB}^p \tag{130}$$

were obtained from the usual inverse current versus $w^{1/2}$ plots. Before the normal Tafel type plots (Fig. 14) can be obtained for this system, it is necessary to:

(a) Determine the order of the reaction. This was done by showing that the variation of i_k with C_{NB} was linear.

(b) Establish that the kinetic current is almost entirely due to the electrochemical reaction of interest. Several methods may be adopted here including the measurement of the background current in the absence of the reacting species, nitrobenzene. This will indicate the extent of the hydrogen evolution side reaction. This procedure relies on the currents being truly additive.

The chemical reaction of PH was studied using classical batch reactor conversion–time experiments at several temperatures. In this reaction the order with respect to PH was 1.0, as indicated by a linear plot of $\ln(1 - X)$ versus time, where X is the fractional conversion of PH. The variation of

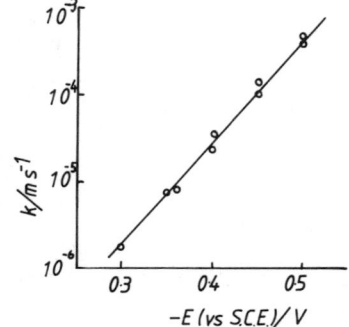

Figure 14. Tafel plot for nitrobenzene reduction (from Ref. 53).

the rate constant closely followed the Arrhenius equation as demonstrated by a linear plot of $\ln k$ versus $1/T$.

Investigating the effect of electroreduction on PH required that two factors be accounted for: the interference of the hydrogen evolution side reaction and the parallel chemical rearrangement. The latter was not a significant factor over the time scale of the experiments, although for other systems with faster chemical reactions, dynamic methods may be required. The rate of hydrogen evolution was extrapolated from the kinetic measurements by determining the current density in the absence of PH at different potentials. As mentioned earlier, this method should be used with caution as the reacting species may well influence the kinetics of the reduction, a typical example being the influence of adsorbed hydrogen on a number of electroreductions. The method does, however, provide a means of correlating hydrogen evolution kinetics for the system. The kinetics of the reduction of PH were determined from the recommended plot of Eq. (128) (see Fig. 15) and the subsequent plot of the Tafel equation, k_{f2} versus potential, from the data generated. The order of reaction with respect to PH was determined from the recommended plot of Eq. (128), the value being one.

An intriguing part of the work was the observed dynamic behavior of the first-order electrochemical rate constant for the reduction of PH. The value of this constant decreased with increasing exposure time to PH, to limiting values that depended on the value of the electrode potential. This feature is particularly relevant to electrochemical engineering, in which processing may be for many hours or on a continuous basis.

The final strategy in setting up the reaction model is to perform integral-time or preparative experiments in which product distribution

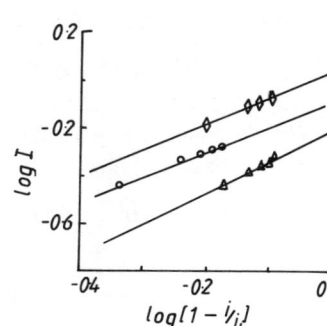

Figure 15. Plot according to Eq. (128) to determine the reaction order of phenlhydroxylamine electroreduction (Ref. 53).

data are compared to the results from analytical or numerical simulation of the reaction model. At this stage it is acceptable to adjust certain parameters, whose measurement may have been subject to significant experimental error, to improve the agreement between model and experiment. This procedure was adopted by Scott[54] and by Haines[55] to confirm the model of the reaction scheme in the absence of a detailed experimental study. An extension of this work is the simulation of reactor performance, for both batch and continuous reactors, which is considered in more detail in Section V.

Overall, this example illustrates well the methodology required in setting up a reaction model for a multiple-reaction system based on steady-state r.d.e. experiments. The importance of combining reactant conversion and product yield data with the experimental electrochemical data is again stressed. The further comparison of experiment with model simulation is another important part of the methodology. In terms of reactor design or modeling, the next stage is the integration of this model with the mass transport and hydrodynamic characteristics of the reactor. The limitation of the above treatment is with regard to one category of reaction, namely, type 3, slow chemical reaction. When reactions are fast with chemical reaction in diffusion layers, then a greater reliance on digital or computer simulation is necessary, as was demonstrated in the work of Haines.[55] The r.d.e. can still be of value in cases of coupled fast chemical reaction and electrochemical reaction as demonstrated in the following section, where digital simulation methods are discussed.

3. The Rotating Disk Electrode and Fast Reaction

With an r.d.e., when simultaneous reaction and diffusion take place, the following assumptions lead to a system of steady-state mass transfer equations that can be solved numerically:

(i) Migrational transport is insignificant in a well-supported electrolyte.
(ii) Physical properties are constant.
(iii) The electrode potential is sufficiently high to cause limiting current behavior of electroactive species.

The equation of continuity for any species in the vicinity of a rotating electrode is

$$u_x \frac{dC_i}{dx} = D_i \frac{d^2 C_i}{dx^2} - r_i \tag{131}$$

where

$$u_x = 0.51023w(w/v)^{1/2}x^2 \quad (132)$$

The solution of Eq. (131) with appropriate boundary conditions can generate suitable current response curves from which comparison with experiment can be made to reveal kinetic parameters.

Consider as an example the catalytic reaction scheme

$$R = aO + ne^- \quad (133)$$

$$O + bS = cR + \text{Products} \quad (134)$$

where species O performs an indirect oxidation of species S to products. The overall reaction can be written as

$$(1 - ac)R + abS = a(\text{Products}) + ne^- \quad (135)$$

and three cases can be distinguished:

(a) $ac = 1$; R is completely regenerated.
(b) $ac < 1$; R is partially consumed and partially regenerated.
(c) $ac > 1$; R is generated as a net product.

This system represents an important category of reactions whereby chemical products are formed by indirect electrochemical oxidations (or reductions). A typical example of this of some industrial interest is the bromohydrinaton of olefins, which leads to the production of epoxides. In studies in which the electroreactive component is in excess, pseudo-first-order reaction behavior is often applied, and thus the conservation equation is linear and, accordingly, can be solved analytically. However, with low concentrations of redox agent, this is not possible, and a numerical approach is required to simulate behavior. In the bromohydrination of olefins, for example, with high concentrations of bromide, brominated organic compounds are formed as well as the epoxide. The reaction may be written as

$$2Br^- = Br_2 + 2e^- \quad (136)$$

$$Br_2 + -C{=}C{-} \rightarrow -C(Br)-C- + Br^- + H^+ \quad (137)$$

Thus, it is apparent that rapid bromination of the olefin can replenish the local concentration of bromide near the electrode and give enhanced bromine oxidation currents. This increased current compared to that without the olefin is related theoretically to the rate constant k.

The model for the r.d.e. consists of three conservation balances for the bromine (O), bromide (R), and olefin (S), which, when cast in dimensionless form, become

$$\alpha_O \frac{d^2\theta_O}{dz^2} + 3z^2 \frac{d\theta_O}{dz} - \phi^2 \theta_O \theta_S = 0 \tag{138}$$

$$\frac{d^2\theta_R}{dz^2} + 3z^2 \frac{d\theta_R}{dz} - c\phi^2 \theta_O \theta_S = 0 \tag{139}$$

$$\alpha_S \frac{d^2\theta_S}{dz^2} + 3z^2 \frac{d\theta_S}{dz} - b\phi^2 \theta_O \theta_S = 0 \tag{140}$$

where

$$\theta_i = \frac{C_i}{C_R}, \quad \alpha_O = \frac{D_O}{D_R}, \quad \alpha_S = \frac{D_S}{D_R}$$

and $z = x/\delta$ and $\phi = \delta(kC_R/D_R)^{1/2}$.

$$x = 0; \quad \theta_R = 0, \quad \alpha_O \frac{d\theta_O}{dx} + a \frac{d\theta_R}{dx} = 0, \quad \frac{d\theta_S}{dx} = 0$$

$$x \to \infty; \quad \theta_O = O, \quad \theta_R = 1, \quad \theta_S = \theta_{S\infty}$$

The system of equations has been studied by Machado and Chapman.[56] For an infinitely fast reaction, the system reduces to a "reaction plane problem" for which an analytical solution is obtained. For the case of a finite chemical rate, the solution was obtained by the numerical technique of orthogonal collocation, convergence to steady numerical results being obtained generally with 10 interior collocation points. These results are presented in terms of a mass transfer enhancement factor (MTEF), which is defined as the ratio of current with chemical reaction to that without chemical reaction. The MTEF is a function of the reaction modulus Θ, as shown in Fig. 16. Experimental MTEF values for allyl alcohol oxidation were compared with the theoretical plots of Machado and Chapman,[56] from which the rate constants were estimated at between 180 and 520 m^2 mol^{-1} s^{-1}, which agreed favorably with previously reported values.

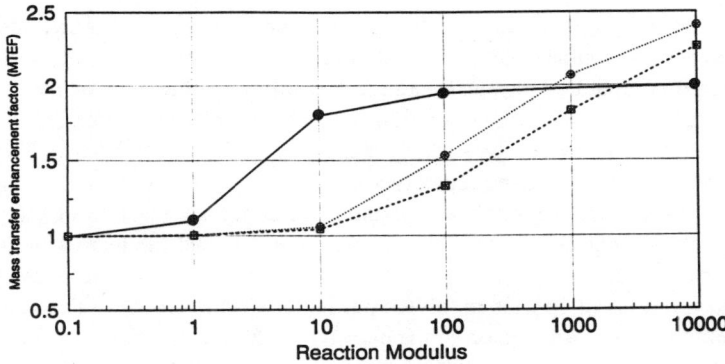

Figure 16. Plot of MTEF for olefin bromohydrination as a function of reaction modulus.

The general method of analyzing the r.d.e. with catalytic reaction has recently been applied to a more complex system of chemical reactions involving the bromide/bromine couple.[57] The numerical simulation of this system was performed by using a false transient finite-difference method to approach the steady-state solution. This application allowed for the effect of pH on the bromine equilibrium reactions to the tribromide, hypobromous acid, and hypobromite:

$$Br_2 + Br^- = Br_3^-$$
$$Br_2 + H_2O = HOBr + Br^- + H^+ \qquad (141)$$
$$HOBr = OBr^- + H^+$$

Other Applications of the Rotating Disk Electrode

For several multiple reactions involving chemical and electrochemical steps, the steady-state current/electrode potential characteristics of the r.d.e. have been derived either from analytical solution or digital simulation of the governing conservation equations. In practice, what is generally required is appropriate qualitative tests as to the validity of the reaction mechanism under study. Several techniques are available for this purpose and include the use of the rotating ring-disk electrode,[58] potential scanning methods, and potential switching methods.[59]

The rotating-ring disk electrode was developed from the r.d.e. and consists of a coplanar ring electrode that circumscribes the disk, from which it is separated physically and electrically by a thin annular gap of an insulating material. In operation, when this bielectrode rotates, the electrolyte moves over the surface of the disk to the ring electrode. The ring and disk are polarized separately, and thus the ring can function as an *in situ* electroanalytical device to monitor the composition of the electrolyte solution adjacent to the disk. Thus, it can be used in electrokinetic experiments to test mechanisms of electrode processes by determining concentrations of species involved, particularly intermediates. The ring potential can be varied in order to characterize the species that react there in terms of their half-wave potentials. For quantitative analysis, the ring potential is normally selected to give mass-transfer-controlled conditions. Theoretical development of the system has tended to concentrate on the derivation and use of the collection efficiency of the ring in relation to the disk generation of species. Although, in principle, model mechanisms can be tested by using an r.d.e., or alternatively by using non-steady-state methods, the ring-disk electrode offers definite advantages in many cases, especially with complex systems involving more than one reaction path. The advantages of the ring-disk electrode arise from the following features:

(i) The ring is usually operated under a diffusion limiting steady-state current at constant potential and therefore free from adsorption and other capacitance effects.
(ii) The use of different materials for the fabrication of the ring and disk enables less severe (or more severe) conditions to be applied on the disk. Thus, intermediates can be detected by their reduction or oxidation while the disk reactant is unreactive at the ring owing to the higher overpotential required at the different surface.
(iii) Very low concentrations of intermediates can be detected.
(iv) It is suitable for the study of reactions in which mixed kinetics pertain.
(v) It is suitable for the elucidation of systems in which two reactions occur simultaneously on the disk.

In linear potential sweep (and cyclic) voltammetry, the electrode potential is varied linearly with time between fixed potential limits. If the direction of the potential scan is reversed, then the method is known as cyclic voltammetry. Many variations of the method have been used,[58] but only symmetric potential waveforms will be discussed here.

The method in principle embraces several different techniques. When stirred solutions are employed, then in the limit of a slow scan rate it is simply an automatic means of obtaining steady-state polarization curves. At higher scan rates, in stirred solutions, hysteresis effects develop for various reasons. One inevitable contribution to hysteresis is the capacitance current density required to charge the electrode double layer. In practice, some systems never show true steady-state behavior.

When unstirred solutions are used and on scanning from a region of electroinactivity to one in which a charge transfer process is possible, the current due to a simple redox process rises with potential, initially with the same form as in stirred solutions, but soon the effects of the progressive depletion of reactant at the electrode surface with time become important and the current commences to fall. The method can be applied to several systems, such as metal deposition, the formation of insulating films, and in organic electrochemistry. Here it is possible to obtain several peaks on the curve due to progressive reductions or oxidations, which help characterize the electrode processes. The lifetimes of intermediates can be investigated when their charge transfer formation reactions are reversible, by reversing the direction of the potential scan and applying variable sweep rates.

Combined with the rotating disk electrode, the method provides a powerful tool to locate phenomena and to identify those which involve mass transfer control. Theoretical treatments have been developed for many model reacting systems with the use of numerical simulation and analytical models.[59] For the acquisition of precise quantitative data, other methods are often preferred, such as impedance and potential switching. These methods are, however, not considered here as the intention of this brief treatment is to merely give a flavor of the electroanalytical methods available and the significance of simulation in their use.

4. Digital Simulation and Electrode Processes

The application of electrochemical techniques to the evaluation of kinetic data and to the determination of a reaction mechanism is based on determining the response of the system to a change in an imposed variable. The set of partial differential equations that typically result when dealing with coupled chemical and electrochemical reactions can be solved generally in one of three ways. Firstly, analytical solutions may be obtained, for which there is a large catalog available. Secondly, the partial differential equations can be solved by converting them into integral equations and solving these numerically. Thirdly, solution can be made by digital simulation of the electrode process. A technique described by Feldberg[59] has frequently been used and has been adapted[60] when necessary to increase the speed of solution. Valuable digital simulation methods based on these finite-difference methods have been described by Newman[61,19] and modified and improved over the years.

Consider now the simulation of the convective diffusion equation for an r.d.e. given by an equation of the form

$$\frac{\partial C_k}{\partial t} = D_k \frac{\partial^2 C_k}{\partial x^2} - k C_k - u_x \frac{\partial C_k}{dx} \qquad (142)$$

With reference to Fig. 17, the solution space is divided into mesh points of distance h. The central difference approximation of derivatives is used, i.e.,

$$\frac{d^2 C_k}{dx^2} = \frac{C_k(x_j + h) + C_k(x_j - h) - 2C_k(x_j)}{h^2} + O(h^2) \qquad (143)$$

$$\frac{dC_k}{dx} = \frac{C_k(x_j + h) - C_k(x_j - h)}{2h} + O(h^2) \qquad (144)$$

With this approximation, the following difference equation is obtained:

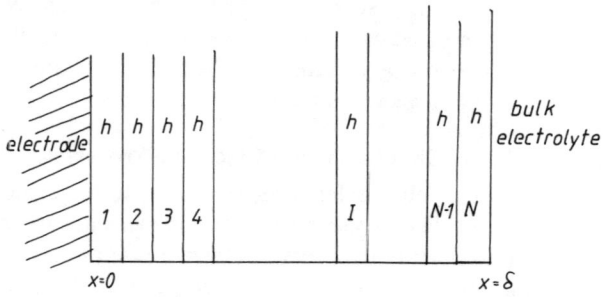

Figure 17. Division of space for digital simulation.

$$\sum_{k=1}^{n} A_{i,k}(j)C_k(j-1) + B_{i,k}(j)C_k(j) + D_{i,k}(j)C_k(j+1) = \frac{\Delta C_k(j)}{\Delta t} \quad (145)$$

where

$$C_k(j) = C_k(x_j)$$

$$A_{i,k}(j) = D_k - h/2(-u_x)$$

$$D_{i,k}(j) = D_k + h/2(-u_x)$$

$$B_{i,k}(j) = -2D_k + h^2(-k)$$

with u_x given by Eq. (132). A stable solution based on this explicit method is obtained if

$$\Delta t\, D/h^2 < 0.5$$

In addition, the solution method must ensure that the chemical reaction takes place in several space elements and that the thickness of the kinetic layer $[(D/k)^{1/2}]$ must be larger than h. The basic time interval must be chosen so that it has a value considerably smaller than the chemical reaction half-life ($dt < 1/k$).

In electrochemical systems where intermediates are removed by fast chemical reactions, a finely spaced grid will be required, and thus the number of iterations required to satisfy stability will be large. Joslin and Pletcher[60] have discussed a method that reduces the overall number of space elements by incorporating a variable grid size. The grid size is smaller at the electrode and increases smoothly out into the electrolyte. This is achieved by transforming the equation using the substitution $y = f(x)$, which has the effect of converting the grid space into one of a uniform dimension in y. This procedure transforms the derivatives in Eq. (142) to

$$\frac{\partial C}{\partial x} = \frac{\partial C}{\partial y} \frac{\partial f(x)}{\partial x} \quad (146a)$$

$$\frac{\partial^2 C}{\partial x^2} = \frac{\partial f(x)}{\partial x} \frac{\partial}{\partial y} \left[\frac{\partial C}{\partial y} \frac{\partial f(x)}{\partial x} \right] \quad (146b)$$

Equation (146b) can now be written in finite-difference form as

$$\frac{\partial^2 C}{\partial x^2} = \frac{1}{(\Delta y)^2} \left(\frac{\partial f(x)}{\partial x}\right)_j \left[\left(\frac{\partial f(x)}{\partial x}\right)\right]_{j+1/2}$$

$$\times [C_k(j+1) - C_k(j)] - \left(\frac{\partial f(x)}{\partial x}\right)_{j-1/2} [C_k(j) - C_k(j-i)] \quad (147)$$

The function $f(x)$ is chosen so that the grid size increases with distance from the electrode surface; for example, $y = [a/b - a/(b+x)]$, where a and b are constants. In the solution procedure the values of a, b, and dt are chosen and used to calculate the grid size by using the defined stability criteria. Joslin and Pletcher[60] also discussed the above techniques using an implicit solution in which the finite-difference equations employed are those described by Crank and Nicolson.[62] In the context of digital simulation, Newman[19] pointed out that partial differential equations can often be reduced to ordinary differential equations either because one independent variable alone is pertinent or through the application of similarity transformation or the method of separation of variables. The latter method uses the central finite-difference approximation applied to a set of coupled linear second-order differential equations. If the boundary conditions are complex and involve derivatives, then an image point at $x = -h$, outside the domain of interest, is introduced such that the electrode is at the position $j = 2$, $x = 0$.

For cases in which the coupled differential equations are nonlinear, Newman[19] proposed a method based on linearizing the equations on the basis of a trial solution. From successive linear approximations based on the trial solution, convergence to the solution of the nonlinear equations is achieved.

(i) Collocation Methods

The application of orthogonal collocation and related methods to the simulation of second-order differential equations has quite recently been demonstrated for a variety of problems in electrochemistry.[63] These methods provide a powerful, fast, and easily applied means to simulate kinetic diffusion problems and are based on approximating the solution of the differential equations over discrete time intervals by a set of polynomial approximations.[64]

The orthogonal collocation technique can be summarized as follows. The second-order partial differential equations are first made suitably dimensionless. This in itself is common practice in the numerical solution of differential equations in general. The resulting equations are then discretized at the roots, x_i, of an orthogonal polynomial; e.g.,

$$\frac{dC(x,t)}{dx}\bigg|_{x_i} = \sum_{j=1}^{N+2} A_{i,j} C(x_j,t) \tag{148a}$$

$$\frac{d^2C(x,t)}{dx^2}\bigg|_{x_i} = \sum_{j=1}^{N+2} B_{i,j} C(x_j,t) \tag{148b}$$

$A_{i,j}$ and $B_{i,j}$ are matrix elements dependent only on the kind and on the degree, N, of the polynomial chosen. The resulting set of equations are now first-order ordinary differential equations. The coefficients of the polynomial in $C(x_j,t)$ are fitted such that the differential equations are fulfilled exactly at certain points. Integration of the set of simultaneous differential equations to obtain the set of concentration profiles with respect to time and distance can be carried out by appropriate third-order Runge–Kutta techniques.

In applying the method of collocation, it is advisable and expedient to use the wide range of computer packages available that readily handle numerical integration, matrix inversion, etc. At the moment, the vast majority of published works on the application of this method in electrochemistry have analyzed established problems to which other simulation techniques, such as finite-difference methods, were previously applied. This, among other things, has demonstrated the utility of the method, which now awaits greater application.

(ii) Numerical Simulation and Product Distribution Data

It was shown earlier that the use of preparative electrolysis and or batch reaction experiments, when integrated with electrochemical kinetic measurements, can be a powerful tool for the formulation of a reaction rate model. This was done for the case of a slow chemical reaction in an electrochemical system, whereas here the case of relatively fast chemical reaction, which can occur within the diffusion/reaction layer near the electrode, is considered. Work which typifies this approach is that of Saveant and Amatore,[50,51,65,66] who analyzed the relationships between the distribution of products in preparative electrolysis and the characteristic rates, or ratio of rates, and other operating parameters for a number of reaction schemes. Detailed knowledge of these relationships will help to reduce empiricism in the design of electrochemical processes The analysis of product distribution data as a function of the various operational parameters may, in many practical circumstances, be the only way of tracing the reaction mechanism. The use of standard electrochemical techniques in the investigation of mechanisms can indeed suffer several limitations:

(i) On the time scales that are accessible by means of these techniques, intermediates having lifetimes shorter then 10 microseconds will practically escape detection. Beyond this limit, information regarding reaction orders of the rate-determining steps may be derived from the voltammetric curves. However, this may be obscured by the initial electron transfer becoming rate-determining as the follow-up reactions are faster and faster.

(ii) Detection and rate characterization of the unstable intermediates generally concerns the immediate product of the first electron transfer, because its chemical transformation is the rate-determining step. Detection and determination of its lifetime can be obtained through the observation of its reoxidation (or rereduction) pattern. This does not provide information about competition between other reaction steps in the reaction sequence. Note that, in principle, it is possible to observe and characterize the intermediates of such follow-up reactions by electrical or spectral means.

The analysis is based on the Nernst approximation with a diffusion layer of thickness δ, which has typical values of the order of 0.005 to 0.01 cm. The concentrations of the unstable intermediates are assumed to be nonzero in only a small portion (the reaction layer) of the diffusion layer next to the electrode surface. The thickness of the reaction layer is given approximately by

$$\mu_r = (D/k)^{1/2}$$

for a first-order rate-determining step. For $D = 0.00001$ cm/s, as soon as $k = 40 \text{ s}^{-1}$, the reaction layer is only 0.1δ, that is, small enough for the pure kinetic conditions to be met. In these conditions both the concentrations and the concentration gradients are tending toward zero outside the reaction layer.

The transport and the reactivity of reactants are assumed not to be significantly affected by adsorption at the electrode surface. The influence of the double layer on the transport processes of charged species is, in the first instance, neglected. Double-layer effects have been addressed as regards the kinetic competition between electron transfer and follow-up chemical reaction.[50] It was found that the double layer significantly affects the competition for rate constants greater than 10^7 s^{-1}.

The reaction system is divided into three successive regions, each corresponding to a particular type of variation of the reactant concentration with space and/or with time. Starting from the electrode, these are:

1. The reaction layer, in which all reactants are considered and their concentrations are given by time-independent diffusion conservation equations with kinetic terms corresponding to the various chemical reactions:

$$D\frac{d^2C}{dx^2} + \text{kinetic term} = 0 \qquad (16)$$

2. The remainder of the diffusion layer, in which the concentrations of the unstable intermediates are zero. The concentrations of stable reactants and products are described by time-independent diffusion equations without a kinetic term; that is, the profiles are linear.

3. The bulk solution, where again the concentrations of the unstable intermediates are zero. The concentrations of the other stable species are space independent but may depend upon time.

Several types of electrolysis regimes are conceived,[50] depending upon whether the electrode potential or the current is controlled and whether continuous or discontinuous operation is carried out:

(i) Exhaustive potentiostatic electrolysis (EPE). The electrode potential is controlled and set up at the plateau of the reactant wave. Electrolysis is then performed until virtually all the reactant has been consumed.

(ii) Continuous operation where the reactant concentration is held constant by continuous replenishment. In this context two types of electrical control may be employed: in constant-concentration potentiostatic electrolysis (CCPE), the potential is held constant at the plateau of the reactant wave and whereas in constant-concentration galvanostatic electrolysis (CCGE), the current is held constant at a value that is less than the diffusion limiting value of the current of the reactant so that the electrolysis does not shift to an undesired reduction (or oxidation) wave.

The basic reaction building block considered by Amatore and Saveant[66] is characteristic of a number of electrochemical processes.

Starting from A, the transfer of an electron leads to a species B, which undergoes an irreversible chemical reaction to species C. This species can undergo several competing reactions, one of which is the further addition of one electron, which is often thermodynamically easier than the addition of the first electron. This second electron transfer can occur either at the electrode surface or in the solution:

$$A \pm 1\,e^- \rightleftharpoons B$$

$$B \xrightarrow{k} C$$

$$C \pm e^- \rightleftharpoons D^E$$

$$C + B \xrightarrow{k_d} D^D + A$$

This so-called "ECE–Disp" process may therefore be a competing reaction pathway for a number of other reactions involving B or C. In such conditions, the characteristics of the competition will differ depending on whether the second electron is exchanged predominantly between B and C or at the electrode. This scheme is thus a relevant preliminary problem to address before more elaborate systems are analyzed. The ECE–Disp competition has been previously analyzed using relaxation techniques such as cyclic voltammetry and potential step chronoamperometry.[67–70]

The system is described by the following set of differential equations, for the convenience of the reader, the notation of Amatore and Saveant[66] is retained, cast in a dimensionless form.

(a) $0 < y < \mu'(\mu' = \mu/\delta)$:

$$d^2a/dy^2 = -\lambda_d bc$$

$$d^2b/dy^2 = \lambda b + \lambda_d bc$$

$$d^2c/dy^2 = -\lambda b + \lambda_d bc$$

$$d^2 d^E/dy^2 = 0$$

$$d^2 d^D / dy^2 = -\lambda_d bc$$

$$\mu' < y < 1:$$
(150)

$$d^2 a / dy^2 = 0$$

$$d^2 d^E / dy^2 = 0$$

$$d^2 d^D / dy^2 = 0$$

(b) $y = 0$:

$$da/dy + db/dy = 0$$

$$dc/dy + dd^F/dy = 0$$

$$dd^D/dy = 0$$

$$c = 0 \qquad (151)$$

and with the boundary conditions

$$a = 0 \qquad \text{for potentiostatic conditions}$$
$$da/dy + dc/dy = \Psi^0 = i^0 \delta / FSc_0 D \qquad \text{for galvanostatic conditions}$$

$$y \to \mu': \quad b \to 0, \, dc/dy \to 0 \qquad (152)$$

where the concentrations of all reactants, intermediates, and products have been ratioed against c_0, the starting concentration of the substrate (i.e., $a = c_A/c_0$, $b = c_B/c_0$, etc.); $y = x/\delta$; time $\tau = t\,(DS/dV)$, where V/S is the volume-to-surface ratio of the cell; $\lambda = k\delta^2/D$ and $\lambda_d = kc_0\delta^2/D$ are the first-order and second-rate constants, respectively; and current $\Psi^0 = i\delta/FSDc_0$.

(c) $y = 1$. For exhaustive electrolysis, the bulk concentrations are related to the concentration gradients at the edge of the diffusion layer:

$$da^b/d\tau = -(da/dy)_1$$

$$dd^{E,b}/d\tau = -(dd^E/dy)_1$$

$$dd^{D,b}/d\tau = -(dd^D/dy)_1$$

$$\tau = 0, \quad a^b = 1, \quad d^{Eb} = d^{Db} = 0 \qquad (153)$$

These sets of equations are readily manipulated and simplified to enable solutions to be obtained pertinent to the electrolysis regime. For the case of CCPE, the bulk concentration of A is constant, i.e., $a_b = 1$. The concentrations of D^E and D^D increase in proportion with time, and the yields R are independent of time:

$$R^E_{CCPE} = d^{Eb}/(d^{Eb} + d^{Db}) = \frac{dc}{dy}\Big|_0 \tag{154}$$

$$R^D_{CCPE} = 1 - R^E_{CCPE} \tag{155}$$

$$C^b_{D^E} = C^0 R^E_{CCPE}(\frac{DS}{\delta V})t \tag{156}$$

$$C^b_{D^D} = C^0 R^D_{CCPE}(\frac{DS}{\delta V})t \tag{157}$$

The yield of E as obtained in this situation is expressed in Fig. 18 as a function of a parameter $p^* = \pi^{1/2}\lambda_d\lambda^{-3/2}$.

Similar curves have been obtained for the case of EPE and CCGE electrolysis. In the case of extended potentiostatic electrolysis, the yield is a function of time or conversion of A.

The ECE–Disp competition has very similar features for all three electrolysis regimes. The Disp situation is reached for large values of the parameter p^*, and, conversely, the ECE behavior at small values of p^*. As expected, an increase in the homogeneous reaction electron transfer rate constant favors the Disp pathway. An increase in the chemical reaction rate constant has the opposite effect. Large values of k involve the

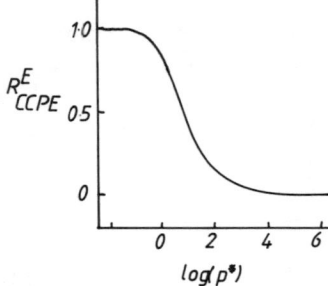

Figure 18. Yield of product E during constant-concentration potentiostatic electrolysis for the ECE Disp system.

formation of C at a short distance from the electrode surface. Thus, C can diffuse back to the electrode and be reduced (or oxidized) there before reacting significantly with B, leading thus to the predominance of the ECE reaction path. Conversely, the Disp route will be favored by low values of the chemical rate constant k, which is the major intrinsic parameter influencing the competition because D does not vary greatly from one system to the other and because k_d is likely to be close to the diffusion limit for a large number of organic systems. For typical values of other parameters and with $c_0 = 100$ mol/m^3, more than 95% of the final product D will be formed through the ECE pathway when $k > 7 \times 10^5$ s^{-1} under CCPE conditions. Conversely, more than 95% of D will be formed along the Disp route when $k < 2000$ s^{-1}.

The treatment of fast chemical reaction exemplifies the approach to be adopted in reaction modeling of electrochemical systems. The above example is sufficient at the moment to outline the method of analysis, and it is clear that numerical methods and digital simulation are crucial in obtaining the required solution to the problem. There are a number of other studies along the lines of that of Amatore and Saveant, including the work of Wendt and Plzak [71,72] on the addition reaction of anodically generated radicals to olefins and the work of Clifton and Savall[73] on the use of Fenton's reagent for hydroxylating aromatic compounds. In both of these studies the simulated model behavior was compared with experimental data.

5. Some Conclusions; Recent and Future Developments

A feature of recent developments in electrochemistry has been the use of electrochemical techniques as methods of investigation. Some current trends in research are:

- Precise and detailed kinetic and mechanistic studies of surface changes, electron transfer processes, and coupled homogeneous reactions
- Development of models for electron transfer, nucleation, and growth of new phases and the structure of the interfaces
- The characterization of the electrode/solution interface at the molecular level and the comparisons with the metal/gas and the metal/ultrahigh vacuum (UHV) systems

Successful exploitation of ideas based on electrochemistry requires a detailed understanding of the kinetics of the processes and of the structure and properties of the electrode/electrolyte interface. Research in electrochemistry has been characterized by the application of a much wider range of nontraditional techniques. Research is turning increasingly to the simultaneous examination of electrode processes by conventional means combined with appropriate *ex situ* structural probes[74,75] or with the direct application of *in situ* spectroscopic or structural methods.[76–80] The interest in the study of electrode processes at microelectrodes is to investigate the kinetics of fast electrochemical reactions in the steady state. As it is only possible to measure the rate constant of an electrode process when this process is slow compared to mass transport, the use of an electrode geometry that makes the mass transfer coefficient solely dependent on the electrode geometry is advantageous. When the characteristic dimension is, say, of the order of 10 nm, then the mass transfer coefficient will be approximately 0.1 m/s.[81] This therefore allows the study of very fast electrochemical reactions.

V. ELECTROCHEMICAL REACTOR ANALYSIS

1. Introduction

This section will consider in more detail the design methods and solution techniques required and used in the analysis of the performance of electrochemical reactors. In order for this exercise to be effective, it is necessary and indeed relevant to consider a number of specific example systems. These will draw on many areas of work and will include simpler first-order systems in which simulation is achieved by numerical integration of first-order ordinary differential equations constituted as initial-value problems. Reaction systems where diffusion and chemical reaction take place in the vicinity of an electrode and which involve two phase flows are also broached. The solution techniques here can involve methods of finite differences, finite elements, or collocation to solve appropriate boundary value problems.

2. A Review of Reactor Engineering Models

This section is not intended to provide a comprehensive review of all electrochemical reactor models and simulations but to indicate the range of models considered—from simple first-order analytical models to complex multiple-reaction, multidimensional dynamic models. Thus, several of the more relevant models are considered which later form the greater part of the section discussing digital simulation methods in reactor modeling.

(i) Analytical Engineering Models

Work on the development of engineering or process models of electrochemical reactions has followed two courses, one of which is the search for analytical solutions. This approach requires limiting assumptions about certain aspects of the system, which means that much of the work is not generally applicable.

Many engineering models concentrate on only one electrode region, where the reactions of interest occur, and assume that the counter electrode reaction has no effect on the reactor performance. In many examples of electroorganic synthesis, this is the case since, in aqueous solution, the most likely counter electrode reactions are the evolution of oxygen at the anode (accompanying organic reductions at the cathode) or hydrogen at the cathode. In the former case, protons are also formed in solution at the anode but are usually consumed in equal quantities at the cathode, preserving the overall concentration at the original value. Hence, the only effect of the anode reaction is a low consumption of water, which can be neglected in many cases. Hydrogen evolution accompanying an anodic oxidation can similarly be neglected insofar as its effects on the working electrode reaction are concerned.

Pickett[8] deals with the design of batch, plug flow, recycle, and continuous stirred tank reactors. Much of this work considers the case in which a single electrode reaction is operating at a mass-transfer-controlled limiting current, and with this assumption the possibility of a simultaneous side reaction, such as solvent decomposition, is also included. It should be noted here that operation at limiting current generally precludes the possibility of galvanostatic operation except in the presence of a non-mass-transfer-controlled secondary reaction. Consideration is also given to situations where the electrode reaction is operating at sub-limiting current.

Pickett's work is, in general, not applicable to electroorganic syntheses, because no situations involving competing or consecutive reactions are included. Most electroorganic processes involve complex reaction sequences which cannot be adequately represented by the simpler schemes that he considered.

Sakellaropoulos and Francis[82] analyzed a scheme of multiple competing (parallel) reactions, and Sakellaropoulos[83] also considered a three-component series reaction mechanism. In both cases, reactions at solid electrodes were considered to occur under wholly kinetic control, with mass transport much faster than electrochemical reaction, and under potentiostatic conditions. Reactions of other than first order were included in the analysis; first- and zero-order reactions were considered in one paper,[83] and first-, half-, and zero-order reactions in the other.[82] Two ideal reactor types were examined, a stirred tank and a plug flow cell. The emphasis of this work was on the determination of selectivities of the reaction schemes with regard to the desired product. The effects of varying potential and varying residence time or space/time were examined, and the analysis was recommended as an aid to electrochemical reactor design.

It was stated that use of Sakellaropoulos's techniques would indicate the reactor volume required to yield the maximum concentration of any desired product in the reaction scheme. Another use of this approach is the possibility of gaining mechanistic information from experimental investigations of product distributions and concentration–time characteristics. This arises from the presentation of the results of the model in the form of design charts showing predicted behavior for a variety of kinetic parameters. Comparison of experimental results with model predictions can indicate values of transfer coefficients and rate constants not accessible through other methods.

Fleischmann *et al.*[84] derived analytical solutions for a three-component series reaction scheme in a batch or plug flow reactor and compared the theoretical results with those obtained experimentally. The assumptions required in this work were that mass transfer was not rate-controlling and that there was no simultaneous decomposition of the solvent, though the authors noted that the experimental results indicated that solvent decomposition did occur to a marked extent. The aim of the work was to predict the yield of a reaction intermediate in terms of either time or the conversion of the initial reactant. Constant-potential electrolysis was considered, and also, with the assumption of equal Tafel slopes of the two electrochemical reactions, constant-current operation. The ratio of the rate constants of the two reactions is fixed under these conditions, and the concentration–conversion and concentration–time graphs produced were plotted for various values of this parameter.

Scott[85] has produced reactor engineering models of a number of complex reaction schemes. Starting with a single reversible electrode reaction, the analysis is extended to include two electrochemical reactions in series or in parallel, the reactions being carried out in a batch, recycle, plug flow, or stirred tank reactor under potentiostatic control. The work includes the effects of reversibility of one or both reactions, the possibility of kinetic, mass transfer, or mixed rate control, and the simultaneous decomposition of the electrolyte. Mass transport is modeled by the use of a mass transfer coefficient k_L. The results of the analysis are expressed in terms of concentration–time and current–time characteristics, and from these are derived expressions for the maximum yield of one of the components and the residence time required to achieve this yield.

Other papers by Scott deal with the further complication of a homogeneous chemical reaction along with two electrochemical steps. Two different schemes are analyzed to determine their behavior in a batch reactor[54] or a recycle reactor.[86] Much of the work is concerned with potentiostatic operation, and equations are derived to account for the behavior of concentrations with time in this mode. From this, it is illustrated that the product distributions obtained experimentally can be used to derive approximate kinetic data. The case of galvanostatic operation has also been considered by Scott.[32]

(ii) Numerical Engineering Models

The limitations that must be imposed in order to derive analytical solutions have led to solutions of the governing differential equations by numerical methods. In this way, particularly if solutions in dimensionless forms are generated, the results may be more generally useful.

Scott et al.[87] analyzed concurrent and consecutive electrochemical reaction schemes operating under galvanostatic control in which both mass transport and solvent decomposition reactions occur.

A detailed review by Feldberg[88] covers methods of solving the governing equations for electrochemical/chemical reaction schemes by finite-difference techniques. Six categories of equations are required—those relating to diffusional flux (e.g., Fick's laws), electroneutrality, current, species conservation, equations of motion, and overall continuity. Feldberg attempted to write a program that would be able to model many different reaction schemes, reactor types, and electrolysis regimes by variation of a few parameters. For instance, the shape of the electrode was described as being part of a sphere, but by setting the sphere radius to infinity, a planar electrode could be modeled.

The method of solution involves rewriting the differential equations in finite-difference form and linearizing them. These form a set of matrices, which can be inverted with a standard subroutine, and, with a set of boundary conditions, it is possible to numerically integrate the equations to model the system at discrete intervals

Ruzic and Feldberg[89] presented an extension to these techniques for use when very fast homogeneous chemical reactions occur in a reaction layer close to the electrode. This situation results in "stiffness" of the equations and would require a very fine grid (i.e., many volume elements and long computation times) if solved in the normal way. Ruzic and Feldberg's method involves the replacement in the model of a homogeneous chemical reaction in the reaction layer by a "heterogeneous equivalent" reaction that can be assumed to occur at the electrode surface and is treated satisfactorily by the numerical techniques.

Bennion[90] has published a more recent review of these techniques, but his paper illustrates no substantial advances over Feldberg's work. Results have not been published in a form that would allow them to be easily used in reactor design.

Alkire and Mirarefi[91,92] developed a different computational method to calculate current distribution along the length of a tubular electrode under fully developed laminar flow conditions. The counter electrode in this system was assumed to be remote from the tubular working electrode. These authors were able to incorporate the effects of potential distribution within the cell and mixed control of reaction rate into a model that was described by three algebraic equations, an integral equation, and a second-order differential equation, all of which had to be solved simultaneously. Solution was achieved by an iterative method, and results were presented in the form of potential, current, and surface concentration distributions along the reactor length as a function of various parameters. The original work dealt with only a single electrode reaction, but a later paper[93] extended the model to the case of two independent reactions proceeding in parallel. Chemical reactions were not included.

Alkire and Gould[94] used numerical techniques similar to those reviewed by Feldberg[88] to analyze the performance of multiple reaction sequences in a porous electrode. Two differential equations for each component, plus one algebraic equation, were solved simultaneously for each of several different types of reaction sequence. The solution of the equations was achieved by first using simple models to obtain a trial solution and then linearizing the equations about this solution to reduce all the equations to first order. The equations were then converted to finite-difference form and solved by Newman's method.[61]

White and co-workers have developed models to predict behavior at rotating disk electrodes[95] and in parallel-plate cells.[96,97] The rotating disk model includes the effect of multiple electrode reactions occurring under potentiostatic control but does not include the effects of chemical reactions. The parallel-plate model considers the case of the electrowinning of copper, in which the counter electrode reaction has a significant effect on the overall reactor performance and so must be modeled along with the working electrode reaction. An undesirable side reaction is also included in this analysis, which aims to predict current efficiencies and product selectivities throughout the reactor using Newman's numerical methods. Again, there is no provision for any chemical reaction.

White's first paper on parallel-plate cells[96] dealt with the case in which the electrodes are close together and took no account of the effects of axial diffusion or migration. A second paper[97] extended the model to the situation where the electrode separation is comparable to the length of the electrodes, in which case axial dispersion effects would be expected to become more significant.

Fedkiw and Scott[98] examined the effects of modulated potential or current control on complex electrochemical/chemical reaction sequences. The influence of frequency of modulation on yields and product selectivities was investigated using numerical techniques based on polynomial approximation to solve the time dependent differential equations.

Scott and co-workers[99-101] have presented models for the simulation of adiponitrile production in undivided batch and plug flow reactors under galvanostatic control. For a postulated reaction scheme, the application of a pseudo-steady-state theory to all intermediates allowed the derivation of reaction rate equations dependent only on the surface concentration of adiponitrile. Chemical reactions were considered to be so fast that they occurred only at the electrode surface, and all reactions were first order. Mass transport was modeled using a mass transfer coefficient related to the Reynolds number; current-potential behavior was assumed to be described by a Tafel type equation; and kinetic rate constants were input to the model in terms of dimensionless ratios of rate constants. The model was solved numerically using an improved Euler technique, with an iteration routine included to reevaluate the potential at each step to keep the current constant. The model also included the effects of hydrogen evolution as a side reaction at the cathode, again described by a Tafel equation.

The second of these models[101] considered the adiponitrile synthesis reaction operating with an emulsion of acrylonitrile. The complex reaction network, consisting of six coupled reactions, was influenced by mass transport processes between phases. The influence of the rate of interphase mass transport on reaction selectivity was determined.

Alkire and Lisius[22] developed a detailed model of propylene oxide production in a parallel-plate reactor. This reaction utilizes both anodically and cathodically generated intermediates, and hence both electrode regions must be modeled. It was assumed that only low conversions per pass were achieved in the reactor, so that concentration changes along its length could be neglected. Second-order differential equations for each component were solved by dynamic simulation to find the steady-state solution. The spatially dependent ordinary differential equations were converted to partial differential equations in both time and distance perpendicular to the electrodes. With boundary conditions at the electrode surface and at the interface between the diffusion layer and the bulk of the electrolyte, the equations were solved using a numerical code (Forsim VI) until no further changes with time were found. The results of the simulation examined the effects of pH, reactant concentration, mass transfer, space/time, and current density on yield, selectivity, current efficiency, and cell voltage.

Yung and Alkire[102] have published a paper illustrating the use of experimentally derived product distribution and current–potential characteristics in estimating kinetic reaction parameters. Often it is not possible to determine absolute values by this method alone, but ratios of rate constants can be derived. Since many engineering models are formulated in dimensionless terms, rate constant ratios are commonly used. This work demonstrated that experimental measurements of product distributions can, in some cases, provide sufficient information for predictions of performance to be made over a much wider range of conditions, if used in conjunction with engineering models of the process.

3. The Requirements of a Reactor Model

A model to assess an electrochemical reaction of industrial significance must predict the values of the variables that determine economic viability of the proposed process. It should require only a small amount of experimentally derived kinetic and physical data, since if extensive experimental studies are undertaken, the need for modeling of behavior is reduced.

Data to be input to the model will be:

(a) The reaction pathway, including undesirable side reactions and solvent decomposition
(b) Kinetic and thermodynamic data relating to electrochemical and chemical rate constants
(c) Correlations of mass transport
(d) Physical data relating to mass transport
(e) The reactor type and mode of electrical supply
(f) Concentrations of species input to the reactor

Once a model is formulated, the simulation of its performance must give predictions of a range of design factors, which include:

(a) Product distributions
(b) Current efficiency
(c) Total current density
(d) Total cell voltage
(e) Production rates per unit electrode area or per unit volume of the reactor

The change of these design factors with variations in a range of parameters can enable the reactor performance to be optimized in relation to plant production. This is a function of a reactor model but will not be considered here.

4. An Approach to Mathematical Modeling of Electrochemical Reactions

The simulation of electrochemical reactors covers a broad spectrum of methods and numerical techniques. The model must provide a good representation of the system of interest, and thus assumptions adopted in deriving the model must not be unrealistic. With coupled chemical and electrochemical reactions, it is reasonable to suppose that reactions occur both at the surface of the electrode and throughout the electrolyte solution, including the diffusion layer. An approach based on this condition is thus relevant and is considered here.

The modeling approach described in this section has been developed to model sequences of electrochemical and chemical reactions. Particular emphasis is given to reaction schemes under galvanostatic control and in the presence of solvent decomposition. Only one electrode region is considered in this modeling approach.

In electrochemical reactions, concentrations of participating species vary both with time and with distance from the electrode within the diffusion layer and so are described by the solutions of partial differential equations. At any instant in time, these reduce to spatially dependent, ordinary differential equations which may give analytical solutions describing the concentration profile across the diffusion layer in terms of various parameters. Thus, eliminating the dimension normal to the electrode from the original partial differential equations gives time-dependent, initial-value, ordinary differential equations describing the variation of concentrations within the reactor, which are solved numerically.

The differential equations are different for each reaction pathway, and the concentration profiles within the diffusion layer must be calculated for each scheme used. The data required for each model are:

(a) Expressions for reaction rates, given in terms of bulk concentrations, and specific to the reaction under consideration. Solvent decomposition can be included in these reaction rates.

(b) Rate constants, or appropriate dimensionless ratios of rate constants.

(c) Current–potential behavior, as modeled by the Butler–Volmer equation with appropriate dimensionless ratios of Tafel slopes.

(d) Mass transport, as represented by mass transfer coefficients, and input as the dimensionless ratio k_L/k_{fl}.

(e) Flow or mixing models; here ideal batch, plug flow, continuous stirred tank, and recycle reactors can be modeled under potentiostatic or galvanostatic control.

(f) Dimensionless ratios of initial concentrations.

The model results describe the behavior of a number of parameters, given below. For batch reactor models, the variation with reaction time is modeled. Plug flow reactor models describe the variation with space/time, but because galvanostatic or potentiostatic operation along the whole length of a plug flow reactor is not possible, the results for the simplest of such models are only approximate. The extension of plug flow reactor models to include operation at constant voltage, rather than constant-electrode-potential or constant-current operation, is also included. For continuous stirred tank or recycle reactor models, dynamic simulation models the transition from the initial conditions to steady-state operation.

Dimensionless parameters accessible from the models are:

(a) Dimensionless product distributions

(b) Current efficiencies for the production of each species, including the current efficiency of accompanying solvent decomposition

(c) Dimensionless partial and total current densities

(d) A dimensionless overpotential of the working electrode, relative to the initial value

(e) Dimensionless production rates

In order to generate results in other than dimensionless terms, absolute values are required for some input data. Thus, in the case of (a) above, if actual input concentrations are used, then absolute output concentrations result. In the case of (c) and (e), if the kinetic data are known as absolute values, then actual current densities and production rates are calculated. In the case of (d), if the value of a kinetic parameter β is known, then the working electrode overpotential is calculated. With the assumptions that the counter electrode overpotential and the electrolyte resistivity are known, this can be used to determine total cell voltage. In electroorganic applications, the counter electrode reaction will often be either oxygen or hydrogen evolution, and the overpotential will have a known, constant value for any current density. The resistivity of electrolytes with high conductive salt concentrations will not be significantly altered by the electrochemical reactions in the cell and can often be considered constant.

In Section V.4(i) the method adopted in this approach to model general reaction sequences is outlined. Section V.4(ii) deals with the application of this method to the modeling of one example reaction scheme in a batch or plug flow reactor.

(i) Generalized Description of the Method

In this section the steps involved in the modeling of a general reaction sequence of electrochemical and chemical steps are briefly described. Figure 19 is a block diagram of the steps required to produce a computer program describing the mathematical model of such a process. The five steps shown in the diagram are discussed in the following subsections.

(a) Reaction sequence

The reaction sequence to be modeled must be chosen and the rate-controlling steps specified. In some cases, the relevant reaction pathways may be indicated by previous publications. In others, doubts over the

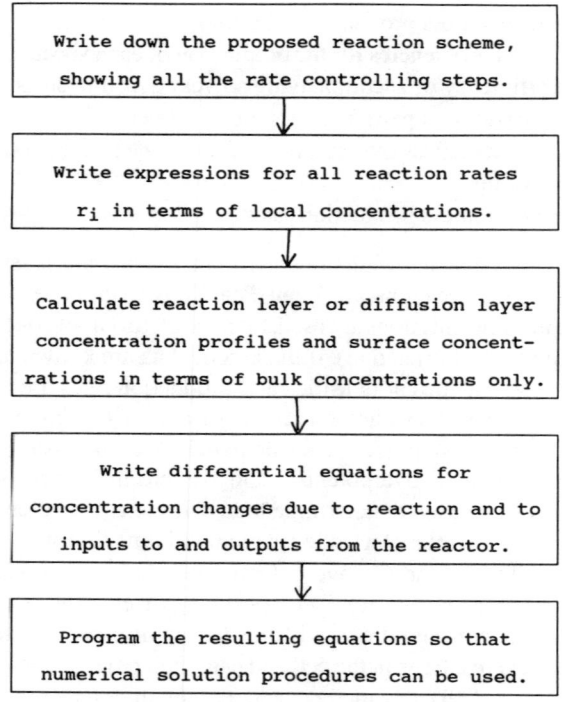

Figure 19. Block diagram of steps involved in the formulation of a model.

pathways exist, and it may be necessary to model several different pathways. Comparison of the model results with experimental results may then determine which model pathway best described the reaction.

In certain reaction schemes, not all steps in the sequence are important in determining reaction rates. For instance, cathodic reactions are often followed by a protonation that is very much faster than the electron transfer step. The overall reaction rate is controlled by this latter step, and so it may be permissible to neglect the protonation in formulating the reaction rate equations. Likewise, in cases in which a reaction is operating at its limiting current density, the electrode kinetics can be neglected in the determination of reaction rates, since mass transfer is rate-controlling.

(b) Controlling rate equations

The second step is to write down expressions that define the controlling electrochemical and chemical reaction rates or mass transfer rates in terms of kinetic or mass transport parameters, specific electrode area of the reactor, and the concentrations of reactant species. Reactant concentrations may vary within the reactor, and at this stage, the local concentrations should be used. These are the concentrations in the bulk of the electrolyte, at the electrode surface, or in the electrolyte reaction layer or the diffusion layer close to the electrode surface.

(c) Reduction in bulk concentrations

The third step involves the elimination of all concentration terms except those relating to the bulk electrolyte. By assuming steady-state behavior, it is often possible to derive analytical equations expressing the relationship between concentrations in the bulk electrolyte and those in other regions of the reactor. This allows all the controlling rate equations to be rewritten in terms of bulk concentrations.

(d) Formulation of differential equations

A material balance over the reactor for each of the species involved in the reaction sequence will give differential equations describing the variation of their concentrations with time. Reaction rates relating to each species have been calculated in step 3 above, and the rates of input and output of each species are governed by the mode of operation of the reactor. A number of different ideal reactor types can be modeled using the same reaction rate equations, with the differential equations altered to account for the different inputs and outputs.

In a batch reactor, there are no input and output streams, and the variation of concentrations with batch reaction time is modeled. A plug flow reactor model assumes that each element of fluid passes through the reactor without mixing, and so again there are no input and output streams for the volume under consideration Plug flow reactor models approximately describe the variation of concentrations with space/time.

A continuous stirred tank reactor model assumes that the fluid leaving the reactor in the output stream has the same composition as the fluid in the well-mixed bulk of the electrolyte within the reactor. The input stream has a composition independent of time. The aim of a CSTR model is to determine the steady-state outlet composition from the reactor for a given space/time. This is achieved by dynamic simulation—the concentration–

time behavior of the reactor is examined until no further variations of concentration with time are found, and the reactor has then attained its steady-state performance.

A recycle reactor is similar to a plug flow reactor in operation, except that part of the outlet stream is mixed with fresh feed to form the inlet stream, the composition of which is not independent of time until steady state has been reached. The recycle stream may pass through a length of pipe equivalent to a plug flow reactor before being mixed with the fresh feed, or it may be passed to a reservoir equivalent to a CSTR. The extent of chemical reaction in the recycle loop may be significant and should be included in the reactor model. Dynamic simulation can again be used to determine the steady-state behavior of a recycle reactor.

(e) Simulation program

The mathematical models developed using the methods described can be solved numerically using a computer simulation language (ACSL) designed specifically "for modeling and evaluating the performance of continuous systems described by time-dependent, non-linear, differential equations."[103] ASCL is a FORTRAN-based, preprocessor language.

(ii) Example Reaction Model

An example of the way in which electrochemical/chemical reaction schemes can be modeled using this approach is now considered. The following series reaction involving competitive electrochemical and chemical reaction is analyzed.

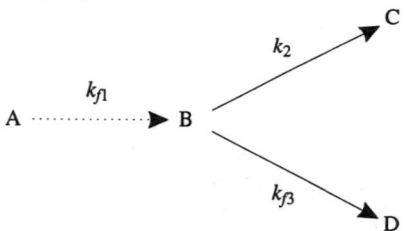

A schematic view of the model is given in Fig. 20.

The following assumptions are made in the analysis:

(i) There is no accumulation of any species within the diffusion layer or at the electrode surface.

(ii) Mass transfer coefficients for each species are mutually exclusive (i.e., noninteractive).

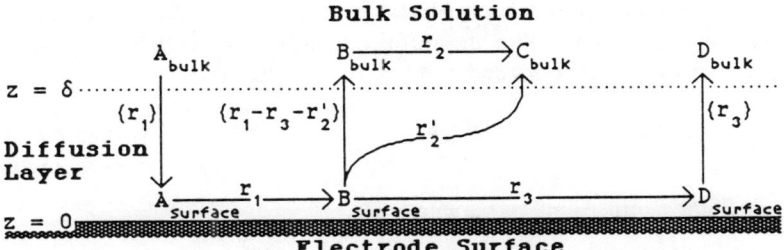

Figure 20. Schematic representation of a model for a series reaction involving competitive electrochemical and chemical reaction. See text for details.

(iii) Isochoric conditions prevail.

The reaction rates are:

$$r_1 = ak_{f1}C_{As} \tag{158}$$

$$r_2 = k_2 C_B \tag{159}$$

$$r'_2 = ak_2 \int_0^\delta C_B \, dy \tag{160}$$

$$r_3 = ak_{f3}C_{Bs} \tag{161}$$

Material transfer may be written in the form:

$$N_t = k_L a(C_{ib} - C_{is}) \tag{162}$$

To solve the differential equations describing the concentration–time behavior of the reacting species, all the surface concentrations and concentration profiles within the reaction layer in the equations above must be written in terms of bulk concentrations.

For species A, a linear concentration profile within the diffusion layer is assumed, leading to

$$C_{As} = \left(\frac{k_L}{k_{f1} + k_L}\right) C_A \tag{163}$$

and, from Eq. (158),

$$r_1 = a\left(\frac{k_{f1} k_L}{k_{f1} + k_L}\right) C_A \tag{164}$$

Species B will have a nonlinear concentration profile if the rate of chemical reaction within the diffusion layer is appreciable. A molar balance over thickness y and unit area of the diffusion layer gives

$$\frac{d^2 C_B}{dy^2} - \frac{k_2}{D_B} C_B = 0 \tag{165}$$

Defining a dimensionless parameter Ha,

$$\text{Ha} = \delta \left(\frac{k_2}{D_B}\right)^{1/2} = \left(\frac{k_2 D_B}{k_L^2}\right)^{1/2}$$

the solution of Eq. (165) is

$$C_B^z = A' \exp(-\text{Ha} \cdot z) + B' \exp(\text{Ha} \cdot z) \tag{166}$$

where $z = y/\delta$, and A' and B' are constants of integration evaluated through two boundary conditions:

(i) At the electrode surface

$$r_1 - r_3 - a(N_B)|_{z=0} = 0 \tag{167}$$

$$r_1 - a k_{\beta}(C_B)|_{z=0} + a D_B \left.\frac{dC_B}{dz}\right|_{z=0} = 0 \tag{168}$$

(ii) At the interface between the diffusion layer and the bulk electrolyte:

$$(C_B)|_{z=1} = C_B \tag{169}$$

Solving for A' and B' using the boundary conditions yields

$$A' = \frac{C_B(\text{Ha} \cdot k_L - k_{\beta}) + (r_1/a) \exp \text{Ha}}{\text{Ha } k_L[\exp \text{Ha} + \exp(-\text{Ha})] + k_{\beta}[\exp \text{Ha} - \exp(-\text{Ha})]} \tag{170}$$

$$B' = \frac{C_B(\text{Ha} \cdot k_L + k_{\beta}) - (r_1/a) \exp(-\text{Ha})}{\text{Ha } k_L[\exp \text{Ha} + \exp(-\text{Ha})] + k_{\beta}[\exp \text{Ha} - \exp(-\text{Ha})]} \tag{171}$$

$$C_B = \frac{C_B[\text{Ha} \cdot k_L \cosh(\text{Ha} \cdot z) + k_{\beta} \sinh(\text{Ha} \cdot z)] + (r_1/a) \sinh[\text{Ha}(1-z)]}{\text{Ha } k_L \cosh \text{Ha} + k_{\beta} \sinh \text{Ha}} \tag{172}$$

$$r_2' = \frac{a \cdot \text{Ha}\{k_L C_B \cdot \text{Ha} \cdot k_L \sinh \text{Ha} + [C_B k_\beta + (r_1/a)][(\cosh \text{Ha}) - 1]\}}{\text{Ha} \cdot k_L \cosh \text{Ha} + k_\beta \sinh \text{Ha}} \tag{173}$$

and, from Eq. (161),

$$r_3 = \frac{a \cdot \text{Ha} \cdot k_L k_\beta C_B + k_\beta r_1 \sinh \text{Ha}}{\text{Ha} \cdot k_L \cosh \text{Ha} + k_\beta \sinh \text{Ha}} \tag{174}$$

(a) Batch reaction system

The above expressions for the reaction rates in the diffusion layers are substituted into reactor balances for components A, B, C, and D, represented by the following equations:

$$\frac{dC_A}{dt} = -r_1 \tag{175}$$

$$\frac{dC_D}{dt} = r_3 \tag{176}$$

$$\frac{dC_B}{dt} = r_1 - r_2 - r_2' - r_3 \tag{177}$$

$$\frac{dC_C}{dt} = r_2 + r_2' \tag{178}$$

These equations were programmed using the ACSL language.

A typical synthesis used to demonstrate the applicability of the model is the manufacture of *p*-aminophenol from nitrobenzene with aniline as the other product (Fig. 21). In determining reaction parameters, attempts were made to reproduce the experimental data of Marquez and Pletcher[104] for an electrolyte of 50% acetone/50% water containing $3M$ sulfuric acid. Initially, the values determined by Scott[54] were used in the program:

$$k_{f1} = 1.2 \times 10^{-7} \exp(-13.55E) \text{ m/s} \quad k_2 = 3.5 \times 10^{-4} \text{s}^{-1}$$

$$k_\beta = 1.6 \times 10^{-8} \exp(-13.55E) \text{ m/s} \quad k_L = 10^{-5} \text{ m/s}$$

The results compared well with Scott's analytical solutions (a check on the accuracy of the numerical integration) and reproduced the general trends of the experimental data. For a close fit, the parameters were varied

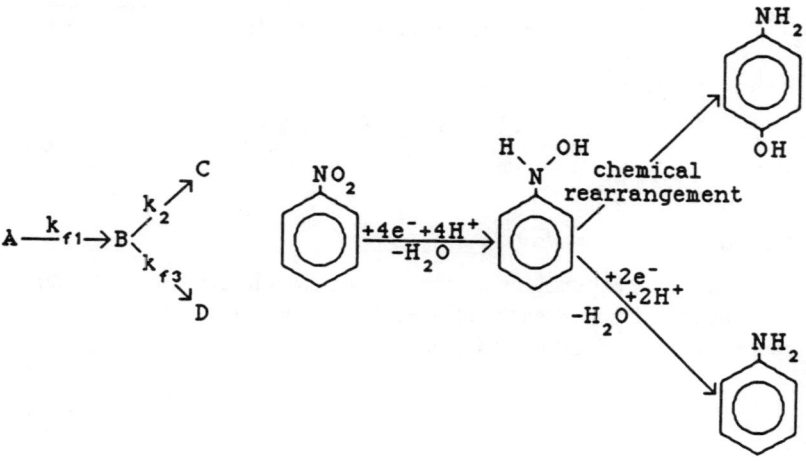

Figure 21. A model for the electrochemical formation of *p*-aminophenol.

from these bases to fit first the organic yield versus potential curve and then the initial current density versus potential plot.

A consistent set of values were observed for k_{f1} and k_L while a ratio of k_2/k_{f3} in the range of 0.41 to 0.59 was required. The typical results obtained from numerical solutions using these values are shown in Fig. 22. The ACSL program was run until 99% of the initial nitrobenzene had reacted, whereas the experimental electrolyses were continued until "the current had decayed to a low value."

Figure 23 shows the effect of variation of electrode potential on the concentration–time curves for potentiostatic operation.

(b) Galvanostatic operation

Galvanostatic operation can also be simulated by calculating a new value of electrode potential E (and hence k_{f1} and k_{f3}) after each time interval during the numerical integration. If both electrochemical steps are considered to have the same potential dependence, then an analytical solution for E is possible. The current, which must be kept at its initial value, is given by

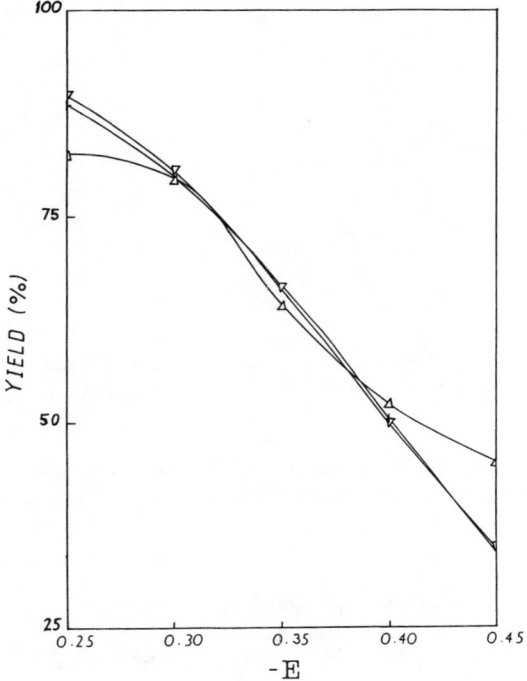

Figure 22. Molar yield (%) of p-aminophenol versus potential for complete conversion in a potentiostatic batch reactor. Data are from Marquez and Pletcher.[104]

$$i_T = \frac{F}{a}(n_1 r_1 + n_3 r_3) \qquad (179)$$

Substituting from Eqs. (164) and (174) and putting $x = k_{f3}/k_{f1}$ yields a quadratic equation in k_{f1}:

$$a k_{f1}^2 + b k_{f1} + c = 0 \qquad (180)$$

where

$$a = x\{k_L[n_1 + n_3](\sinh \text{Ha})C_A + n_3 \cdot \text{Ha} \cdot C_B] - (i_T/F) \sinh \text{Ha}\} \qquad (181\text{a})$$

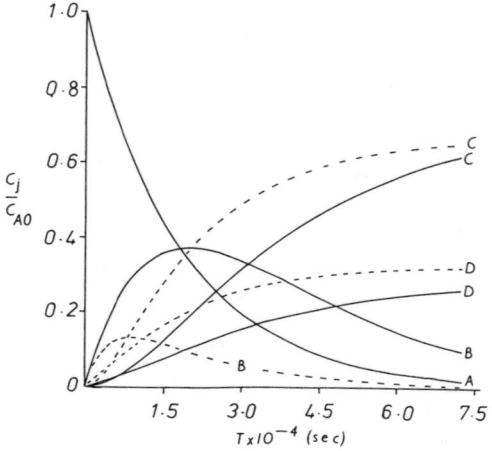

Figure 23. Concentration versus time curves for potentiostatic operation in a batch reactor.

$$b = k_L\{\text{Ha} \cdot k_L[n_1(\cosh \text{Ha})C_A + n_3 x C_B] \\ - (i_T/F) \text{Ha} \cosh \text{Ha} + x \sinh \text{Ha})\} \quad (181b)$$

$$c = -(i_T/F)\text{Ha} \cdot k_L^2 \cosh Ha \quad (181c)$$

For any values of C_A and C_B, k_{f1} is the positive solution of the quadratic. When the coefficient a becomes negative, this indicates that mass transfer limitations will prevent the maintenance of constant current in the absence of another electrochemical reaction such as solvent decomposition, as shown below.

For any values of C_A and C_B, the maximum rates of the electrochemical reactions will be

$$r_1^{\max} = a k_L C_A \quad (182)$$

$$r_3^{\max} = \frac{a \cdot \text{Ha} \cdot k_L C_B}{\sinh \text{Ha}} + r_1^{\max} \quad (183)$$

and the mass-transfer-limited maximum current is

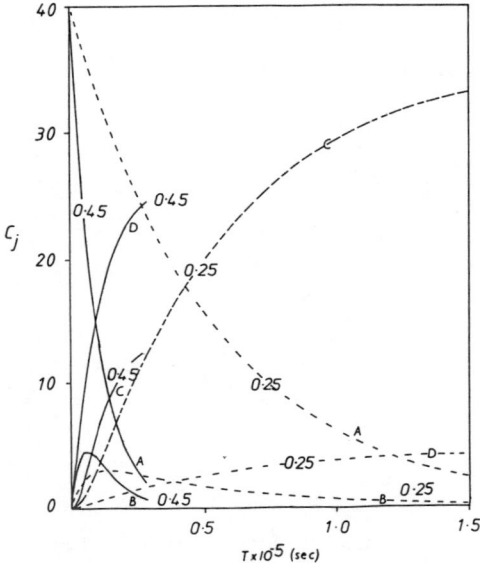

Figure 24. Concentration versus time curves for galvanostatic operation in a batch reactor. ---, $E = -0.25$ V; —, $E = -0.45$ V.

$$i_L = \frac{F}{a}(n_1 r_1^{max} + n_3 r_3^{max}) = F[(n_1 + n_3)k_L C_A + \frac{n_3 \cdot \text{Ha} \cdot k_L C_B}{\sinh \text{Ha}}] \quad (184)$$

Galvanostatic operation cannot continue when $i_L < i_T$ i.e., when

$$k_L\left((n_1 + n_3)C_A + \frac{n_3 \cdot \text{Ha} \cdot C_B}{\sinh \text{Ha}}\right) < \frac{i_T}{F} \quad (185)$$

or $a < 0$.

After k_{f1} has been calculated, k_{f3} is given by $k_{f3} = xk_{f1}$, and E is calculated from the expression for k_{f1}.

The remaining figures produced from numerical simulations will concentrate on only one set of parameters:

$$k_{f1} = 6.0 \times 10^{-8} \exp(-13.55E) \text{ m/s} \quad k_L = 1.7 \times 10^{-4} \text{ s}^{-1}$$

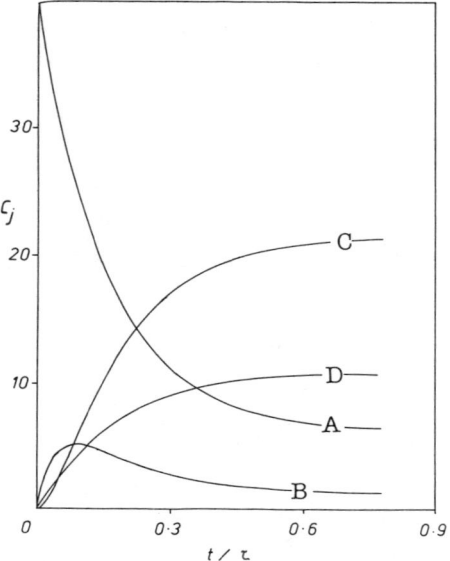

Figure 25. Concentration versus time curves for potentiostatic operation in a CSTR with $\tau = 10^5$ s and $E = -0.35$ V.

$$k_{f3} = 2.9 \times 10^{-8} \exp(-13.55E) \text{ m/s} \quad k_L = 1.3 \times 10^{-5} \text{ m/s}$$

and the initial (nitrobenzene) concentration $C_{A0} = 40$ mol/m^3.

Figure 24 shows the effect of variation of electrode potential on the concentration–time curves for galvanostatic operation; the electrode potential is now the initial potential and becomes more cathodic during the course of the reaction to maintain constant current.

(c) Other reactor types

Finally, the effect of a different reactor type can be studied by altering the differential equations (175) to (178). For a CSTR of volume V m^3 and flow rate v m^3/s, the outlet composition being identical to the composition within the well-stirred tank and the inlet containing nitrobenzene at a concentration of C_{A0} mol/m^3 and none of the other organic species, then the differential equations become

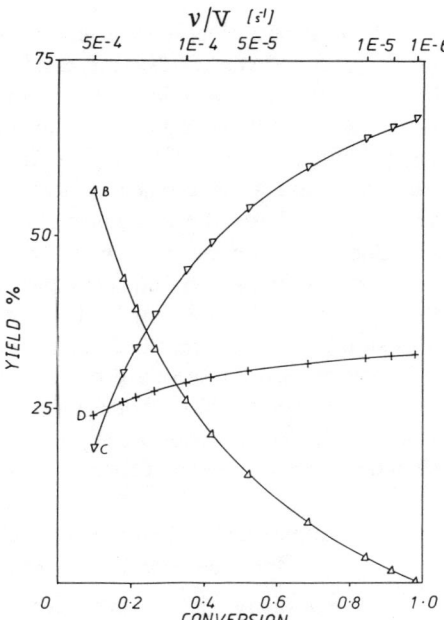

Figure 26. Steady-state molar yield versus nitrobenzene conversion for potentiostatic operation in a CSTR. $E = -0.35$ V.

$$\frac{dC_A}{dt} = \frac{v}{V}(C_{A0} - C_A) - r_1 \qquad (186)$$

$$\frac{dC_B}{dt} = r_1 - r_2 - r_2' - r_3 - \frac{v}{V}C_B \qquad (187)$$

$$\frac{dC_C}{dt} = r_2 + r_2' - \frac{v}{V}C_C \qquad (188)$$

$$\frac{dC_D}{dt} = r_3 - \frac{v}{V}C_D \qquad (189)$$

These equations can be included in the ACSL program and will dynamically model the behavior of a CSTR. The simulation will proceed from the starting conditions to reach steady state, and yields of organic products can then be calculated. Figure 25 shows how steady state is approached for one particular set of conditions. In Fig. 26, steady-state product yields are plotted against conversion in the reactor, conversion being varied by altering the ratio v/V.

Overall, a parallel chemical–electrochemical reaction scheme allowing for reaction in the diffusion layer has been considered. The analysis is limited to first-order reaction steps, but the effects of operation both under potentiostatic and under galvanostatic conditions were investigated. Using a limited amount of published data, it was demonstrated that the parallel chemical–electrochemical reaction scheme can be used to describe the synthesis of p-aminophenol from nitrobenzene.

The simulation method applied here can be used generally for ideal reactors operating at a constant potential, current, or cell voltage. Although the example considered involved analytical solutions for reaction and diffusion in the mass transfer film, the method can be extended to introduce approximate collocation procedures in this reaction. On a wider note, the principle behind the method is valid for systems in which, for example, reaction and film diffusion are approximated by a finite-difference representation. This application features in the following section.

5. Engineering Models of Complex Electrochemical and Homogeneous Chemical Reaction Schemes

Some of the restrictions introduced in the previous section, particularly with regard to the lack of involvement of counter electrode reactions in the main reaction processes, will sometimes not be a good representation of the actual reactor behavior. This will be the case in, for example, paired synthesis, where by necessity products from both electrode regions interact in a complex way. This situation has been modeled by Alkire and Lisius[22] for the electrochemical synthesis of propylene oxide in an undivided continuous flow stirred tank electrochemical reactor, and this is a good system to consider further.

(i) The Paired Synthesis of Propylene Oxide

The synthesis of propylene oxide from propylene is an indirect oxidation involving the anodic generation of bromine from bromide:

$$2\,Br^- \rightarrow Br_2 + 2e^- \tag{190}$$

The pH plays a crucial role in the selectivity of the reaction.

The system chemistry is summarized in Fig. 27 for the case of the electrolysis of an aqueous solution of sodium bromide saturated with dissolved propylene. The rate of addition of bromine to propylene is reported to be 200 times faster than the hydrolysis of bromine. The

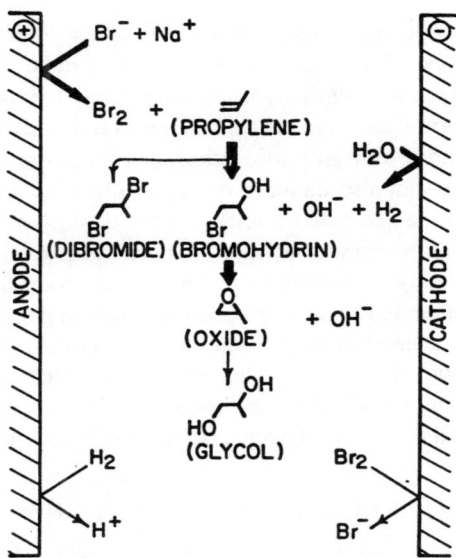

Figure 27. The reaction chemistry of the epoxidation of propylene (from Ref. 22 with permission).

reaction of dissolved propylene was assumed to occur exclusively with molecular bromine to form propylene bromohydrin:

$$Br_2 + CH_3CH=CH_2 + OH^- \rightarrow CH_3CH(OH)-CH_2Br + Br^- \quad (191)$$

Propylene oxide is formed by base-promoted dehydrohalogenation of the bromohydrin:

$$CH_3CH(OH)-CH_2Br + OH^- \rightarrow CH_3CH\overset{O}{-\!\!\!-\!\!\!-}CH_2 + H_2O + Br^- \quad (192)$$

The electrolysis of water occurs at the cathode:

$$2H_2O + 2e^- \rightarrow H_2 + 2OH^- \quad (193)$$

The overall cell reaction is

$$CH_3CH=CH_2 + H_2O \rightarrow CH_3CH\overset{O}{-\!\!\!-\!\!\!-}CH_2 + H_2 \quad (194)$$

As products from both the anodic and the cathodic reactions are involved in the overall reaction, the process is termed a paired synthesis; the halogen species does not appear in the overall process.

A number of potential side reactions can occur with this system. These include reactions of bromine species and of organic propylene compounds. Bromine can form polybromides, notably tribromide, and may hydrolyze to form hypobromous acid. Hypobromous acid can hydrolyze to the anion and, on disproportionation, can yield the bromate. These bromine species can also undergo further electrochemical reaction, but notably bromine can be reduced back to bromide.

Although all bromine species of valence states of 0 and above are capable of oxidizing propylene, these reactions are not taken to be significant in this system. However, side reactions involving organic propylene compounds do occur since the carbonium ion intermediate is available to nucleophiles other than water. The major by-product is propylene dibromide, formed by the nucleophilic attack of Br^- on the carbonium ion intermediate resulting from the reaction of propylene with bromine:

$$CH_3\overset{+}{C}HCH_2Br + Br^- \rightarrow CH_3\overset{Br}{\overset{|}{C}H}-CH_2Br \qquad (195)$$

Propylene oxide is susceptible to water attack, which in basic solution gives the glycol:

$$CH_3\overset{O}{\overset{/\backslash}{CH}}\!\!-\!\!CH_2 + H_2O \rightarrow CH_3\overset{OH}{\overset{|}{C}H}-CH_2OH \qquad (196)$$

To an insignificant extent, the bromohydrin can react to form dibromoisopropyl ether and may also hydrolyze to form the glycol. Overall, the reactions outlined above were considered in this model system.

(a) Electrolytic cell model

The model of the electrolytic reactor is that of an undivided parallel-plate cell having a well-mixed central core region with a diffusion layer at either electrode. The convective mass transfer processes that define the thickness of each of the diffusion layers are caused by the continuous flow of the reactant stream into the unit. The cell operates with a constant current I, has a holding time of τ, and operates under isothermal conditions at constant pressure. The model is therefore a good approximation to turbulent flow conditions in a parallel-plate cell. The feed to the cell is sodium bromide saturated with propylene. The reactor model is illustrated schematically in Fig. 28. Overall, 11 reaction species were considered in the reaction sequence.

The equations used to describe the overall model were as follows:

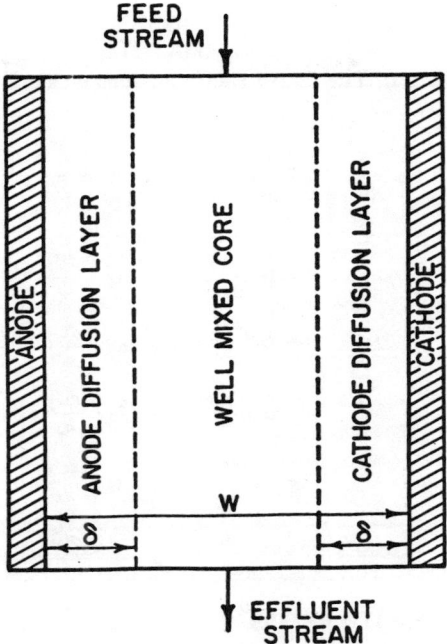

Figure 28. Schematic diagram of the well-mixed reactor (from Ref. 22 with permission).

1. The electrochemical reaction of the Br_2/Br^- couple was described by a Butler–Volmer rate law, in which the back reduction of bromine to bromide was included. The chemical reactions that produce the epoxide, the glycol, and the dibromide derivatives of propylene were expressed by second-degree polynomial expressions.

2. A set of conservation of species equations were applied to the anode and the cathode diffusion films and represent transport by diffusion with homogeneous chemical reactions. These were based on dilute solution theory and ignored the influence of migration to minimize the complexity of the model. Both are reasonable assumptions. These one-dimensional conservation balances are shown in Table 3 along with the boundary conditions for the model. The boundary conditions satisfy several criteria:

(i) Nonelectroactive species have zero flux at the electrodes.

Table
The Conversion Equations for the Epoxidation of

Species	Number	Conversion equation	Boundary Anode
Bromide ion	1	$0 = D_1 \dfrac{d^2C_1}{dx^3} + R_{3,4} + R_{4,6}$	$\dfrac{dC_1}{dx} = \dfrac{1000(I_{app} - j_2)}{FD_1}$
Bromine	2	$0 = D_2 \dfrac{d^2C_2}{dx^3} - R_{3,4} - R_{3,5}$	$\dfrac{dC_2}{dx} = \dfrac{500(I_{app} - j_2)}{FD_2}$
Propylene	3	$0 = D_3 \dfrac{d^2C_3}{dx^3} + R_{3,4} - R_{3,5}$	$\dfrac{dC_3}{dx} = 0$
Propylene bromohydrin	4	$0 = D_4 \dfrac{d^2C_4}{dx^3} + R_{3,4} - R_{4,5}$	$\dfrac{dC_4}{dx} = 0$
Propylene dibromide	5	$0 = D_5 \dfrac{d^2C_5}{dx^3} + R_{3,5}$	$\dfrac{dC_5}{dx} = 0$
Propylene oxide	6	$0 = D_6 \dfrac{d^2C_6}{dx^3} + R_{4,6} - R_{6,7}$	$\dfrac{dC_6}{dx} = 0$
Propylene glycol	7	$0 = D_7 \dfrac{d^2C_7}{dx^3} + R_{6,7}$	$\dfrac{dC_7}{dx} = 0$
Hydroxide ion	8	$0 = D_8 \dfrac{d^2C_8}{dx^3} - R_{3,4} - R_{4,6}$	$\dfrac{dC_8}{dx} = 2\dfrac{D_9}{D_8}\dfrac{dC_9}{dx}\Big|_{anode}$

[a]After Alkire and Lisius (Ref. 22).

(ii) Electroactive species flux at the electrodes is governed by the sum of the two electrode reactions. The anolyte boundary conditions satisfy the mass balance for each species that may undergo reaction in the central well-mixed core. This therefore defines the conditions in this region, where the molar flow out of species is balanced by the sum of the molar flow in plus the molar rate of production by reaction ($R_{1,2}$, etc.) and the molar rates of diffusion from both diffusion films.

3. A voltage balance across the cell was expressed by Ohm's law based on ionic mobilities and concentrations in solution. The potential field in the central well-mixed region is given by

3
Propylene in a Well-Mixed Electrochemical Reactor[a]

Conditions
Anolyte

$$0 = \frac{D_1}{W}\left[\frac{dC_1}{dx}\bigg|_{ab1} - \frac{dC_1}{dx}\bigg|_{cb1}\right] + \frac{C_1^{in} - C_1}{\tau} + R_{3,4} + R_{4,5}$$

$$0 = \frac{D_2}{W}\left[\frac{dC_2}{dx}\bigg|_{ab1} - \frac{dC_2}{dx}\bigg|_{cb1}\right] + \frac{C_2}{\tau} - R_{3,4} - R_{3,5}$$

$$0 = \frac{D_3}{W}\left[\frac{dC_3}{dx}\bigg|_{ab1} - \frac{dC_3}{dx}\bigg|_{cb1}\right] + \frac{C_3^{in} - C_3}{\tau} - R_{3,4} + R_{3,5}$$

$$0 = \frac{D_4}{W}\left[\frac{dC_4}{dx}\bigg|_{ab1} - \frac{dC_4}{dx}\bigg|_{cb1}\right] - \frac{C_4}{\tau} + R_{3,4} + R_{4,6}$$

$$0 = \frac{D_5}{W}\left[\frac{dC_5}{dx}\bigg|_{ab1} - \frac{dC_5}{dx}\bigg|_{cb1}\right] - \frac{C_5}{\tau} + R_{3,5}$$

$$0 = \frac{D_6}{W}\left[\frac{dC_6}{dx}\bigg|_{ab1} - \frac{dC_6}{dx}\bigg|_{cb1}\right] - \frac{C_6}{\tau} + R_{4,6} + R_{6,7}$$

$$0 = \frac{D_7}{W}\left[\frac{dC_7}{dx}\bigg|_{ab1} - \frac{dC_7}{dx}\bigg|_{cb1}\right] - \frac{C_7}{\tau} + R_{6,7}$$

$$0 = \frac{D_8}{W}\left[\frac{dC_8}{dx}\bigg|_{ab1} - \frac{dC_8}{dx}\bigg|_{cb1}\right] - \frac{C_8^{in} - C_8}{\tau} - R_{3,4} - R_{4,6}$$

(continued)

$$dE/dx = -I/F2(u_1 C_1 + u_8 C_8 + u_{10} C_{10} + u_{11} C_{11}) \qquad (197)$$

(b) Solution method for simulation

The system of equations of this model was stiff, and the method of solution was dynamic simulation. This involved introducing the time derivatives of the variables and converting the ordinary differential equations to partial differential equations. These equations are integrated in time until the time derivatives become negligible, that is, until a steady state is reached. The equations were first scaled before solution to reduce the range of magnitude of the variables. Solution was achieved using a

Table 3
(continued)

	Boundary Conditions	
Species	Catholyte	Cathode
Bromide ion	$C_1\|_{cb1} = C_1\|_{ab1}$	$\dfrac{dC_1}{dx} = -2\dfrac{D_2}{D_1}\dfrac{dC_2}{dx}\|_{\text{cathode}}$
Bromine	$C_2\|_{cb1} = C_2\|_{ab1}$	$C_2 = 0$
Propylene	$C_3\|_{cb1} = C_3\|_{ab1}$	$\dfrac{dC_3}{dx} = 0$
Propylene bromohydrin	$C_4\|_{cb1} = C_4\|_{ab1}$	$\dfrac{dC_4}{dx} = 0$
Propylene dibromide	$C_5\|_{cb1} = C_5\|_{ab1}$	$\dfrac{dC_5}{dx} = 0$
Propylene oxide	$C_6\|_{cb1} = C_6\|_{ab1}$	$\dfrac{dC_6}{dx} = 0$
Propylene glycol	$C_7\|_{cb1} = C_7\|_{ab1}$	$\dfrac{dC_7}{dx} = 0$
Hydroxide ion	$C_8\|_{cb1} = C_8\|_{ab1}$	$\dfrac{dC_8}{dx} = \dfrac{1000(I_{\text{app}} - j_1)}{FD_8}$

numerical package (Forsim VI). The potential profile was calculated by solving the finite-difference form of Eq. (197) across the diffusion layer.

Overall, the model was used to calculate yields, selectivities, potential fields, and concentration profiles across the reactor and to explore the response of the system to changes in variables. The typical concentration profiles obtained with this model are shown in Fig. 29 for a reactor with a holding time of 10 s operating with an electrolyte of NaBr (500 mol/m^3) at a pH of 14. In this figure, it can be seen that on entering the cell the propylene concentration in the well-mixed region has fallen to approximately 0.37 of its original value owing to the reaction with bromine, which takes place virtually exclusively in the anode diffusion layer. Note that the concentrations are adimensional and referenced against the saturation concentration of propylene. In the anode diffusion layer, it is apparent that

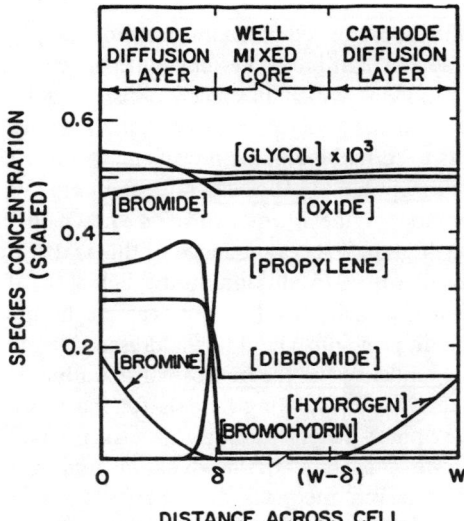

Figure 29. Variation of species concentration in the reactor during propylene epoxidation (from Ref. 22 with permission).

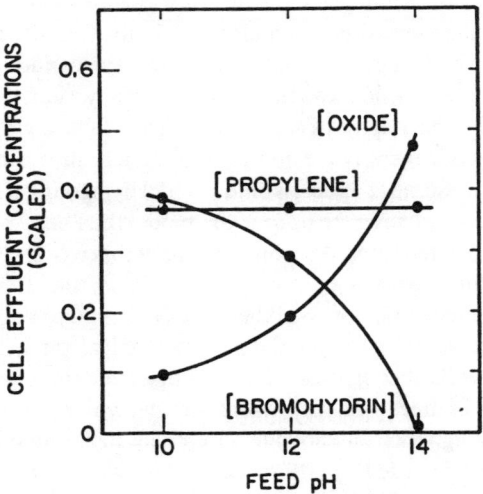

Figure 30. The effect of pH on product distributions in the epoxidation of propylene (from Ref. 22 with permission).

only over a small part of this region do the bromine and propylene species react, and as an approximation the system could have been modeled as a reaction plane problem, which has been considered by Scott.[34] This reaction produces an instantaneous concentration of the bromohydrin, which then reacts to form the organic propylene species. These reactions are particularly susceptible to pH as can be seen in Fig. 30, where almost complete conversion of the bromohydrin is shown to be induced by operating at very high pH. A consequence of this is that high concentrations of the glycol result. Additionally, the behavior of the chemical reactions of bromine at such pH values, for example, hydrolysis, may now be important and, in principle, should be addressed.

Overall, this model serves the purpose of initially demonstrating the procedures required in formulating models for reactions in electrolytic cells. The data required in performing this task included specific rate equations and rate constants, thermodynamic equilibrium constants, knowledge of reaction intermediates, transport properties, and mass transport coefficients. The availability of such data will generally dictate the degree of sophistication that can be introduced into a reactor model.

(ii) Other Studies

The model of Alkire and Lisius[22] has demonstrated a number of features associated with digital simulation. Firstly, formulating the model can introduce a high degree of stiffness, due to large variations in parameters and variables, which can tax severely many numerical routines. However, the current degree of sophistication available coupled with large computing power generally means that a solution routine can be obtained. This may require some adaptation of the model by making it less rigorous, although this should never be done if the model then becomes unrealistic or becomes subject to error. A second feature of this work was the knitting together of three aspects of numerical analysis, the finite-difference method, dynamic simulation, and the use of a computer package.

On the subject of modeling the electrochemical synthesis of propylene oxide, several other studies have been made. Goodridge *et al.*[105] also considered the indirect oxidation with bromine, which was again a paired synthesis, although there are notable differences in the chemistry between their model and that used by Alkire and Lisius, which is not at issue here. Their model did introduce aspects of transport between phases in terms of the absorption of propylene, although it was a steady-state model for a CSTR with bulk reaction only. This therefore required only a numerical

Table 4
The Reaction System for the Cathodic Reduction of N-Nitroso-2-methylindolene

Homogeneous chemical reactions	Kinetic laws
$A + H^+ \underset{r_{-1}}{\overset{r_1}{\rightleftharpoons}} C + NO^+$	$r_1 = k'_{C1} \cdot C_A$; $k'_{C1} = k_{C1} \cdot [H^+]$
	$r_{-1} = k_{C-1} \cdot C_C \cdot C_{NO^+}$
$NO^+ + EtOH \overset{r_2}{\rightarrow} EtONO + H^+$	$r_2 = k'_{C2} \cdot C_{NO^+}$; $k'_{C2} = k_{C2} \cdot [EtOH]$
$B + NO^+ \overset{r_3}{\rightarrow} C + N_2O + H^+$	$r_3 = k_{C3} \cdot C_B \cdot C_{NO^+}$
$B + H^+ \underset{r_{-4}}{\overset{r_4}{\rightleftharpoons}} BH^+$	$r_4 = k'_{C4} \cdot C_B$; $k_{C4} = k_{C4} \cdot [H^+]$
	$r_{-4} = k_{C-4} \cdot C_{BH^+}$

iteration procedure for its solution. Neither group considered the influence of the *in situ* generation of gases, such as hydrogen, which, among other things, would tend to volatilize species such as propylene oxide. This aspect has recently been considered for the direct oxidation of propylene[106] and also for the indirect oxidation.[107] In both cases the reactor was modeled as a dynamically operated CSTR, and simulation of the nonlinear ordinary differential equations was achieved by the use of a numerical package for initial-value problems (ACSL).

Yu *et al.*[108] have developed models for other paired syntheses, involving butanone and also glucose, that utilized three-dimensional electrodes. This work will be considered later in this chapter.

Weise *et al.*[109] have presented an engineering model of a batch reactor and a CSTR in which homogeneous chemical reactions and heterogeneous electrochemical reactions occur. The example considered was the cathodic reduction of N-nitroso-2-methylindolene (A) to N-amino-2-methylindolene (B). A set of homogeneous reactions involving species A and B occur in which a major by-product is 2-methylindolene (C) (see Table 4). The

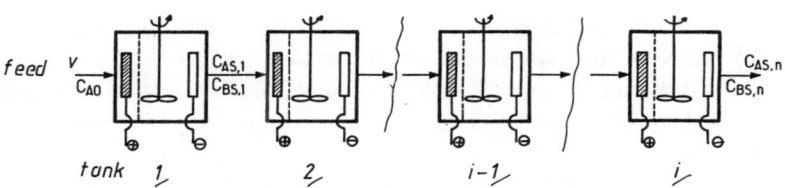

Figure 31. A cascade of CSTRs to approximate plug flow behavior.

model considers only the cathode region and consists of conservation equations for a well-mixed reaction core ($y > \delta$) and for a diffusion layer ($0 < y < \delta$) as shown in Table 5. Simulation of the reactor was achieved by solving the second-order differential equations for the diffusion layers using a fifth-order Runge–Kutta method. The boundary conditions were estimated by a numerical optimization method (Simplex). The authors pointed out that the chemistry of the system can be simplified, for engineering purposes, to

$$A + 4e^- \to B$$
$$A + B \to 2C$$
(198)

This reaction model system was then considered in some detail in a batch (or plug flow) reactor, which was constructed by setting up a cascade arrangement of CSTRs (Fig. 31). It was noted that, in the limit of an infinite number of series-connected CSTRs, the behavior approaches that of a batch system; thus, the latter may be modeled by moving along successive tanks and solving the CSTR model repeatedly. The simulations of the model showed that the selectivity of the product B is better in a batch reactor than in a CSTR, as the latter gives the less favorable, well-mixed conditions for the reaction between A and B.

Clifton and Savall[73] have modeled a batch reactor in which Fenton's reagent (a mixture of hydrogen peroxide and ferrous ions) is used for the hydroxylation of benzene to phenol. Fenton's reagent is produced by the cathodic reduction of oxygen to peroxide and by the cathodic reduction of ferric ions. These species are, overall, involved in 11 homogeneous reactions with benzene and other species. The mathematical model is different from those previously developed in this section and is based on a type of surface renewal, or penetration, approach to simulate the behavior of a mercury pool electrode stirred by a magnetic bar on its surface. In the model it is assumed that:

(i) A diffusion layer forms at the surface of the electrode and spreads progressively into a stagnant electrolyte phase.

(ii) Mass transfer by diffusion is confined to this mass transfer layer; in the bulk, concentrations change only as a result of homogeneous reaction.

(iii) This diffusion layer is periodically remixed with the bulk electrolyte. This renewal rate is taken to be half the rotation rate of the stirrer.

Table 5
Reactor Conservation Equations for the Cathodic Reduction of *N*-Nitroso-2-methylindolene: Mass Balance Equations for the *i*th Reactor Cascade in a CSTR

Species	Mass balances for: $y > \delta$	$0 < y < \delta$	Boundary conditions for: $y = 0$	$y = \delta$
A	$\nu C_{AS,i-1} = V(r_1 - r_{-1})_{y=\delta} + D_A a \left(\dfrac{dC_A}{dy}\right)_{y=\delta} + \nu C_{AS,i}$	$D_A \cdot \left(\dfrac{d^2 C_A}{dy^2}\right) = r_1 - r_2$	$D_A \cdot \left(\dfrac{dC_A}{dy}\right)_{y=0} = \dfrac{i}{nF}; \ (C_A)_{y=0} = 0$	$(C_A)_{y=\delta} = C_{AS,i}$
B	$\nu C_{BS,i-1} = V(r_3 + r_4 + r_{-4})_{y=\delta} + D_B a \left(\dfrac{dC_B}{dy}\right)_{y=\delta} + \nu C_{BS,i}$	$D_B \cdot \left(\dfrac{d^2 C_B}{dy^2}\right) = r_3 + r_4 - r_{-4}$	$D_B \cdot \left(\dfrac{dC_B}{dy}\right)_{y=0} = 0$	$(C_B)_{y=\delta} = C_{BS,i}$
BH$^+$	$\nu C_{BH^+S,i-1} = V(-r_4 + r_{-4})_{y=\delta} + D_{BH^+} a \left(\dfrac{dC_{BH^+}}{dy}\right)_{y=\delta} + \nu C_{BH^+S,i}$	$D_{BH^+} \cdot \left(\dfrac{d^2 C_{BH^+}}{dy^2}\right) = -r_4 + r_{-4}$	$D_{BH^+} \cdot \left(\dfrac{dC_{BH^+}}{dy}\right)_{y=0} = -\dfrac{i}{nF}$	$(C_{BH^+})_{y=\delta} = C_{BH^+S,i}$
C	$\nu C_{CS,i-1} = V(-r_1 + r_{-1} - r_3)_{y=\delta} + D_C a \left(\dfrac{dC_C}{dy}\right)_{y=\delta} + \nu C_{CS,i}$	$D_C \cdot \left(\dfrac{d^2 C_C}{dy^2}\right) = -r_1 + r_{-1} - r_3$	$D_C \cdot \left(\dfrac{dC_C}{dy}\right)_{y=0} = 0$	$(C_C)_{y=\delta} = C_{CS,i}$
NO$^+$	$\nu C_{NO^+S,i-1} = V(-r_1 + r_{-1} + r_2 + r_3)_{y=\delta} + D_{NO^+} a \left(\dfrac{dC_{NO^+}}{dy}\right)_{y=\delta} + \nu C_{NO^+S,i}$	$D_{NO^+} \cdot \left(\dfrac{d^2 C_{NO^+}}{dy^2}\right) = -r_1 + r_{-1} + r_2 + r_3$	$D_{NO^+} \left(\dfrac{dC_{NO^+}}{dy}\right)_{y=0} = 0$	$(C_{NO^+})_{y=\delta} = C_{NO^+S,i}$

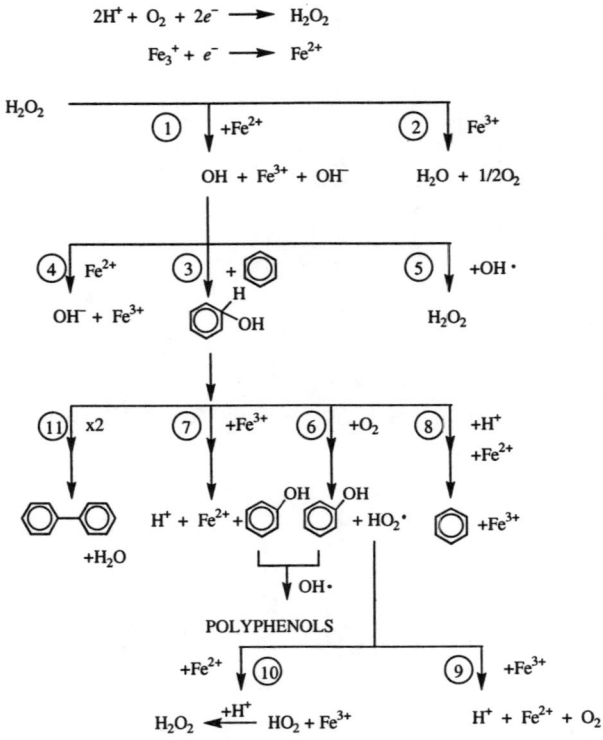

Figure 32. Reaction scheme for benzene oxidation using Fenton's reagent (after Clifton and Savall[73]).

The formation of the diffusion layer is described mathematically by the dynamic form of the diffusion reaction equation

$$\frac{\partial C_i}{\partial t} = D_i \frac{\partial^2 C_i}{\partial x^2} + r_i(C_1, C_2, C_3...)$$

(199)

with the following boundary conditions:

(a) At the beginning of each stirrer cycle, the concentrations are uniform.

(b) Electroinactive species have a zero flux at the electrode surface.

(c) Electrode reactions are mass transfer limited, corresponding to zero values of surface concentrations.

(d) Diffusion fluxes of oxidized species toward the electrode are equal to diffusion fluxes away from the surface.

(e) Outside the diffusion film, concentrations change only as a result of chemical reactions:

$$dC_i/dt = r_i(C_1, C_2, ...) \qquad (200)$$

The system of equations for the reaction system depicted in Fig. 32 was solved using a finite-difference technique, a modified Crank–Nicolson method. The reaction term was calculated explicitly and incorporated into the discretized form of Eq. (199), owing to the great complexity of the reaction scheme. At the end of an integration cycle (i.e., one period of the mixing), concentration profiles of the species were obtained. These concentration profiles were then integrated and new bulk concentrations calculated after remixing was performed. The difference between the concentrations at each successive cycle can be used to calculate the overall rate of change of concentration, dC_i/dt.

The simulation of this model enabled the concentration distributions of species to be determined for the batch reactor, which also enabled the current efficiencies to be evaluated; typical current yield data are shown in Fig. 33. The initial phase of the simulation, in which current yields are low and rise with time, reflects the time required for the concentrations of Fe^{2+} and H_2O_2 to reach sufficiently high values for the bulk to become productive.

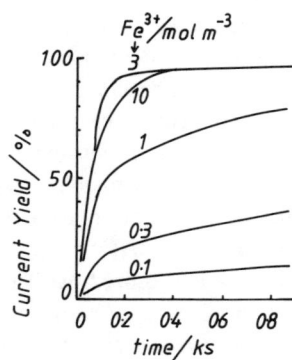

Figure 33. Current yields for the oxidation of benzene using Fenton's reagent. Initial Fe^{3+} concentrations are marked on the curves.

6. Parallel-Plate Electrode Reactor Models

Parallel-plate electrochemical reactors are of obvious industrial importance because of their many applications, and, as a result, numerous studies of such units have been conducted. Rudimentary models for these systems were discussed briefly in Section V.2. These models, in which several restrictive assumptions are made, are frequently applied because of the unnecessary complexity introduced in attempting to solve a full set of model equations, describing actual hydrodynamics and charge, mass, and energy balances coupled with electrochemical and chemical kinetics. However, situations exist in which this more rigorous approach must be applied, and these will now be discussed further. Several of these were introduced in the brief literature review in Section V.2.

Newman[19] proposed a model of a cell in the laminar flow region using an approximation of thin diffusion layers near the electrode surfaces. Lapicque and Storck[110] have applied this model approach, more realistically, to the case of turbulent flow and analyzed the case of copper recovery from a sulfuric acid-based electrolyte. In this work, three zones were defined in the reactor (see Fig. 10): a turbulent core in which concentration and velocity profiles are flat, and two thin boundary layers at the electrodes in which species move by diffusion and migration. The reactor operates at a constant cell voltage everywhere along its length. This potential is composed of the potential losses in the three regions plus the surface potentials at both electrodes. Simulations of the model were achieved by relatively straightforward numerical integration.

The thin layer model is often not considered to be appropriate when the electrolyte flow is laminar. Nguyen *et al.*[97] have described a rigorous two-dimensional model of a parallel-plate unit (Fig. 34) in which the electrodes are not close together. The assumptions adopted in this model include well-developed laminar flow, Butler–Volmer kinetics, no gas generation, constant physical and transport properties, and the use of dilute solution theory. The model consists of a set of time-dependent, nonlinear, coupled, multidimensional differential equations. The influence of dynamics, axial diffusion, and axial migration can be determined from this model. The solution to the model was obtained using an implicit alternating direction (IAD) algorithm with Newman's technique and, on implementation, required two finite-difference equations for each differential equation. A major result of the model was the conclusion that when the aspect ratio of the reactor (ratio of electrode separation to electrode length)

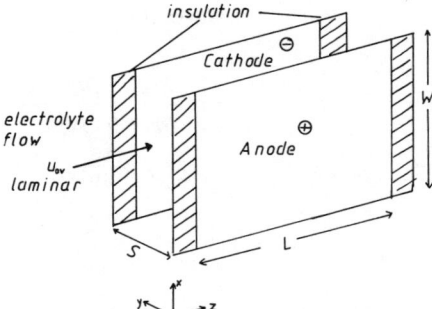

Figure 34. Schematic model of a two-dimensional parallel-plate reactor.

is less than 0.5, the influence of axial migration and axial diffusion is not significant. Thus, for virtually all practical systems, the model is of little benefit except if dynamic simulation is of interest, in which case the numerical method is quite appropriate. The earlier model of White et al.[96] is then a suitable method for the simulation of this type of reactor and is thus outlined here.

In the absence of homogeneous reaction, the steady-state material balance for a species i is

$$\Delta N_i = 0 \tag{201}$$

where

$$N_i = -D_i C_i - z_i u_i C_i F\theta + u C_i \tag{202}$$

The velocity distribution for well-developed laminar flow is

$$u_x = 6u_{\text{avg}}[y/S - (y/S)^2] \tag{203}$$

and

$$u_y = 0 \tag{204}$$

Substituting the Nernst equation,

$$u_i = D_i/RT \tag{111}$$

and Eqs. (202), (203), and (204) in Eq. (201) gives

$$6u_{avg}\left(\frac{y}{s} - \frac{y^2}{s^2}\right)\frac{\partial C_i}{\partial x} = D_i\left(\frac{\partial^2 C_i}{\partial x^2} + \frac{\partial^2 C_i}{\partial y^2}\right)$$

$$+ \frac{z_i D_i F}{RT}\left(C_i\frac{\partial^2 \Phi}{\partial x^2} + \frac{\partial C_i}{\partial x}\frac{\partial \Phi}{\partial x} + C_i\frac{\partial^2 \Phi}{\partial y^2} + \frac{\partial C_i}{\partial y}\frac{\partial \Phi}{\partial y}\right) \quad (205)$$

The condition for electroneutrality

$$\sum_i z_i C_i = 0 \quad (115)$$

completes the set of $i + 1$ equations needed to determine the unknowns C_i and the potential θ.

Equation (205) can be simplified for small aspect ratios, α. This is because terms associated with this parameter in Eq. (205), when cast in an dimensional form, have a negligible influence on the system. These terms are derivatives of potential and concentration in the axial direction. The resultant equation for the model is

$$3\frac{D_R}{D_i}\text{Pe}\cdot\alpha(\eta - \eta^2)\frac{\partial \theta_i}{\partial \zeta} = \frac{\partial^2 \theta_i}{\partial \eta^2} + \frac{z_i F}{RT}\left(\theta_i\frac{\partial^2 \Phi}{\partial \eta^2} + \frac{\partial \theta_i}{\partial \eta}\frac{\partial \Phi}{\partial \eta}\right) \quad (206)$$

where $\zeta = x/L$, $\eta = y/S$, $\theta_i = C_i/C_{i,\text{ref}}$, and Pe is the Peclet number, equal to $2Su_{av}/D_R$, and S is the interelectrode gap.

This model is solved subject to initial conditions at the reactor inlet:

$$\sum_i z_i C_{i,\text{ref}}\theta_i = 0, \quad \text{at } \zeta = 0; \theta_i = 1 \quad (207)$$

and the boundary conditions are at $\eta = 0$ (anode)

$$\sum_j \frac{s_{ij} i_{nj}}{n_j F} = -N_{ni}$$

at $\eta = 1$ (cathode)

$$\sum_j \frac{s_{ij} i_{nj}}{n_j F} = N_{ni} \quad (208)$$

and at both $\eta = 0$ and $\eta = 1$

$$\sum_i z_i C_i = 0$$

An implicit stepping technique in the axial direction and Newman's technique in the normal direction were used by White *et al.* to simulate the reactor behavior. This enabled suitable predictions of the changes in current, concentration, and selectivity along the electrode. The model was applied to the deposition of copper from a chloride electrolyte solution in an undivided cell.

More recently, Edwards and Newman[37] presented a model for the prediction of current, potential, and concentration distributions in a thin gap parallel-plate cell. The model, like that of White *et al.*, can predict the effect of interacting boundary layers and also allows for the two-dimensional nature of the current flow. The model is restricted to systems with excess supporting electrolyte (i.e., no migration). Simulations of the model demonstrated the errors that can result if axial current flow is neglected for thick cells (i.e., aspect ratio > 0.1). Again, for a reasonable scale of practical operation, this would suggest that with thin cells, of the order of a few millimeters or so, axial current flow is not important. This also concurs with the conclusions of the thin film model of Weise *et al.*[109] discussed earlier.

Overall, it would seem that with parallel-plate electrochemical reactors of a practical scale, by which one generally means benchtop synthesis size and upward, the simulations can be based on two general models:

(i) For narrow channel electrolyzers, where energy consumption is of concern with, say, dilute electrolytes, axial diffusion, migration, and flow of current can be neglected, although the thin film model may not be appropriate.

(ii) For cells with larger gaps, a thin film model can be used, especially when pumping costs do not restrict flows to the laminar region.

7. Fixed and Porous Bed Electrodes

The mathematical modeling of three-dimensional electrodes based on fixed or porous beds has been described by numerous authors and has been the subject of several reviews.[111–113] A great deal of the activity in this area was geared toward cells for effluent treatment, particularly metal recovery, and for batteries. A typical model is that of Alkire and Ng,[114] who considered the two-dimensional distributions in a packed bed electrochemical flow reactor. The feature of this model that distinguishes it from models for parallel-plate (and concentric cylinder) units is that both the electrolyte phase and the porous electrode matrix behave as a continuum that obeys Ohm's law. The electrical conductivity of each phase is corrected for the presence of the other phase fraction by assigning effective conductivities for both phases. The relative current flow in each phase thus determines the distribution of the local current density in the structure of the electrode. In the model of Alkire and Ng, the frequently adopted assumption that the electrical conductivity of the electrode phase is much greater than that of the electrolyte is used. For many electrode materials, this assumption is quite reasonable and leads to simplifications in the model formulation and the subsequent numerical solution. In addition, it is usually assumed that the hydrodynamic conditions approximate to those of plug flow on which can be superimposed axial or radial dispersion. Simulation of the potential and concentration distribution was achieved by a numerical finite-difference procedure. The model differential equation is parabolic, and integration over the two-dimensional volume can proceed stepwise, using the Crank–Nicolson symmetric form.

An engineering model for a complex electroorganic synthesis in a continuous flow-through porous electrode was developed by Alkire and Gould.[94] The reaction considered was the oxidation of 9,10-diphenylanthracene (DPA), which proceeds via an EEC reaction mechanism. The model gives a one-dimensional representation of the concentration and potential distribution in the porous matrix. The model of the porous structure is shown in Fig. 35. Due to the excess supporting electrolyte, the migration contribution to the current flow of active species is neglected.

In this model, a one-dimensional conservation equation is written for each species:

$$D_i \frac{d^2 C_i(y)}{dy^2} - u \frac{dC_i(y)}{dy} = \sum v_{ij} r_{ij} \qquad (209)$$

where the reaction terms r_{ij} are either for homogeneous processes occurring in the bulk of the electrolyte or for heterogeneous processes occurring at the surface of the porous electrode structure. In the latter case, these surface reaction rates are related to the local concentration difference between the bulk and surface through the usual film mass transport equation. A one-dimensional charge balance equation is also needed:

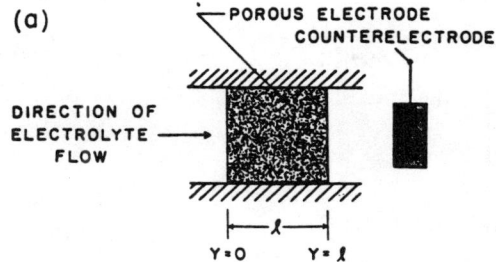

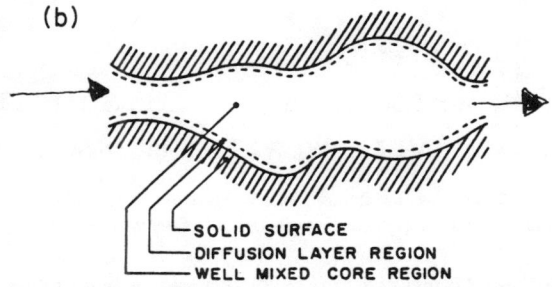

Figure 35. Schematic model of a pore in a three-dimensional structure. (a): the electrode (b): the pore region. From Ref. 99 with permission.

$$\frac{d^2\phi}{dy^2} = \frac{ai_j}{K} \qquad (210)$$

The boundary conditions for a counter electrode downstream are

$$y = 0: \quad C = C_i^0, \frac{d\phi}{dy} = 0$$

$$y = 1: \quad \frac{dC_i}{dy} = 0, \phi = \phi^0 \qquad (211)$$

The reaction system investigated by this model is as follows

$$DPA = DPA^+$$

$$DPA^+ = DPA^{2+} + e^-$$

$$DPA^{2+} + Nu = \text{Product} \qquad (212)$$

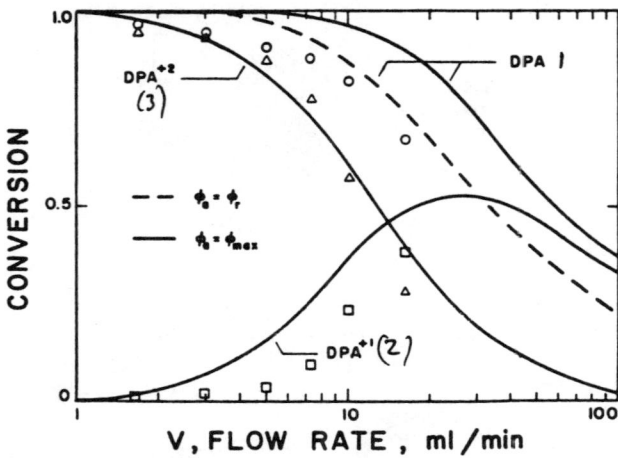

Figure 36. Product distributions in the oxidation of 9,10-diphenylanthracene (DPA) in a porous electrode (from Ref. 99 with permission).

Overall, the model consisted of five conservation equations for the reacting species, the one charge balance equation, and Butler–Volmer electrode kinetics. As is usual, the six differential equations were put into a dimensionless form prior to numerical solution. This procedure involved first the linearization of the equations about a trial solution and then recasting into finite-difference form by employing central difference operators. The resulting set of tridiagonal matrices was inverted by a Gauss–Jordan elimination method. This work was supported by a detailed experimental study, and the simulations of the model showed good agreement with the data obtained, as can be seen in Fig. 36. This figure shows the influence of electrolyte flow on the conversion in the porous electrode. At the lower flow rates, the residence time is sufficiently long to permit near quantitative formation of the dication species. Increasing the flow rate, that is, decreasing the residence time, improves the relative conversion to the radical intermediate. Readers who are interested in further details of this system should consult the original referenced work.

Oloman and Reilly[115] modeled a fixed bed reactor used for the oxidation of benzene to p-benzoquinone. The reactor was actually a trickle flow cell with a liquid/liquid/gas dispersion flowing down over a bed of lead dioxide as the anode (Fig. 37a). The complexity of the reactions, the

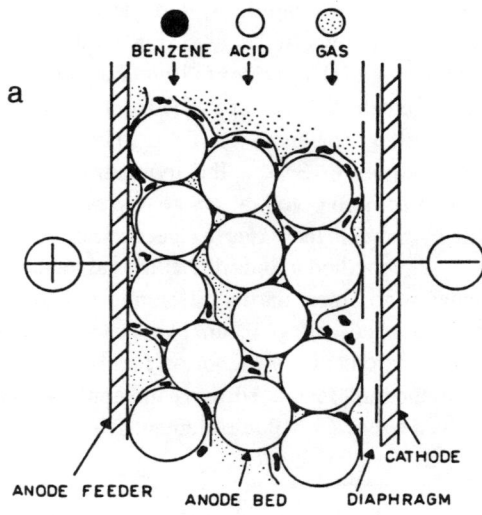
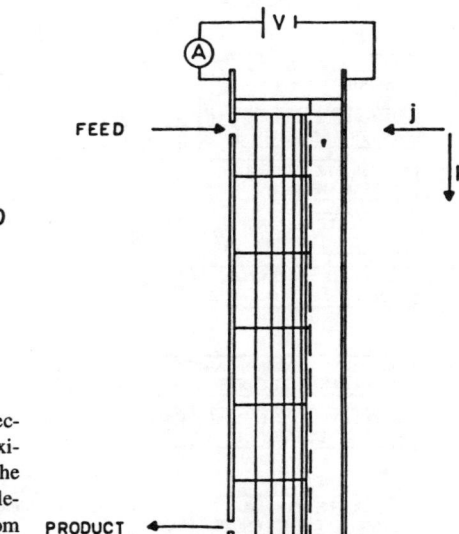

Figure 37. Model of three-phase electrochemical reactor for benzene oxidation: (a) three-phase flow in the fixed bed electrode; (b) reactor elements in numerical analysis (from Ref. 115 with permission).

hydrodynamic conditions, and the mass transfer behavior directed the simulation toward the use of a modified finite-difference procedure, employing trial and error to find boundary conditions in the three-dimensional electrode structure. The reactor was modeled using iterative mass and voltage balances on finite increments of the reactor volume, as depicted in Fig. 37b. In the application of this solution procedure, in the region of high potential gradients near the anode feeder, the thickness of the increments increases in a geometric progression away from the feeder.

The method avoids difficulties associated with the solution of simultaneous nonlinear partial differential equations. With reference to Fig. 38, the mass and voltage balances are carried out with the reactor divided into finite increments of thickness j and of length k. The calculation begins with the inlet feed conditions and applied voltage. The mass balances are solved over each volume element at a common position j for each species.

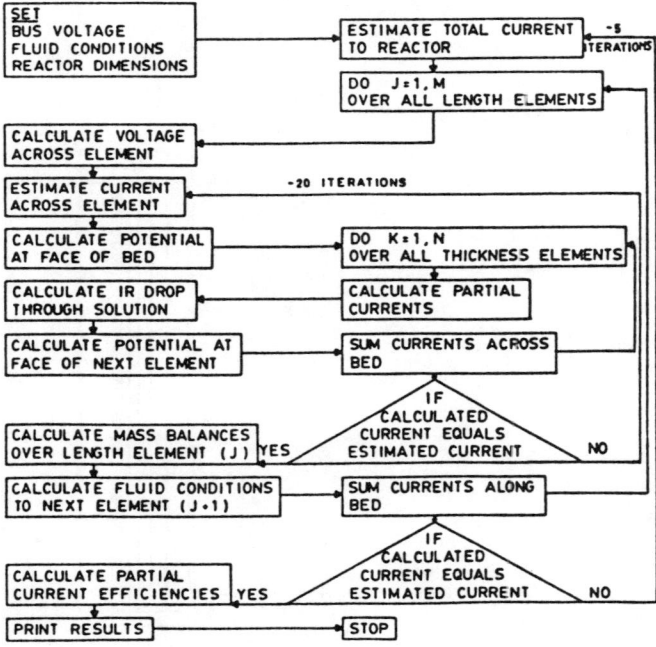

Figure 38. Logic diagram for numerical analysis of a fixed bed reactor (from Ref. 115 with permission).

After each lengthwise step, the fluids are mixed to simulate infinite dispersion. The benzoquinone produced at each stage is assumed to be distributed between the organic and the aqueous phase according to the phase equilibrium condition. Thus, the mass balances on benzoquinone over each length increment are, for each volume increment,

$$v\Delta C_{BQk} = \left(\sum r_{BQ,jk}\right) V_k \tag{213}$$

and over a complete length increment,

$$v\Delta C_{BQk} = v(C_{BQk+1} - C_{BQk}) + v_B(C_{B,k+1} - C_{Bk}) \tag{214}$$

where v and v_B are acid flow and benzene flows, respectively, and B and BQ refer to benzene and benzoquinone, respectively.

The generation of gases is allowed for by calculating the new gas flow rates, the gas loadings, pressure gradient, liquid holdup, effective conductivity, and mass transfer coefficients from appropriate correlations.

The superficial current density, the potential distribution, and the partial current densities are found from the iterative solution of the voltage balance at each increment. The complete procedure is described by a computer logic diagram given in Fig. 38.

The modeling of this trickle flow reactor gave fair agreement with experimental data for the synthesis. The simulation accuracy suffers from the common problem of not having sufficient, accurate data, which is accentuated by the very nature of the reactor and the three-phase flow conditions.

Yu et al.[108] have considered the mathematical modeling of packed bed flow reactors using two examples of paired electroorganic synthesis. The first considers the paired synthesis of 2-butanone from 2,3-butanediol, the reaction chemistry of which is summarized in Table 6. Electrogenerated bromine oxidizes the 2,3-butanediol to acetoin in the bulk electrolyte. Acetoin is then reduced to 2-butanone in the cathode, where hydrogen evolution is a side reaction. Other side reactions include the oxidation of bromide to bromate and the cathodic reductions of bromine and bromate to bromide.

The set of reactions take place in a batch recycle reactor consisting of an electrochemical packed bed tubular reactor with an amalgamated zinc shot cathode and a graphite anode. As the reactions are considered to be slow, the reactor is considered to operate at a constant concentration at any given time (i.e., negligible conversion per pass). This then satisfies

Table 6
Reaction Chemistry for the Synthesis of 2-Butanone from 2,3-Butanediol

Cathode

$$CH_3COCH(OH)CH_3 + 2H^+ + 2e^- \rightarrow CH_3COC_2H_5 + H_2O \quad (C1)$$
$$\text{Acetoin} \qquad\qquad\qquad\qquad \text{2–Butanone}$$

$$2H_2O + 2e^- \rightarrow H_2 + 2OH^- \quad (C2)$$

$$Br_2 + 2e^- \rightarrow 2Br^- \quad (C3)$$

$$BrO_3^- + 6H^+ + 6e^- \rightarrow Br^- + 3H_2O \quad (C4)$$

Anode

$$2Br^- \rightarrow Br_2 + 2e^- \quad (A1)$$

$$Br^- + 3H_2O \rightarrow BrO_3^- + 6H^+ + 6e^- \quad (A2)$$

Bulk solution

$$CH_3CH(OH)CH(OH)CH_3 + Br_2 \rightarrow CH_3COCH(OH)CH_3 + 2HBr$$
$$\text{2,3–Butanediol} \qquad\qquad\qquad \text{Acetoin}$$

differential reactor operation. A number of assumptions are used in setting up the model:

- The porous electrode is considered to be one-dimensional (in length).
- The electrical conductivity of the electrode is much greater than that of the electrolyte.
- The reactor operates in plug flow, and the reservoir is well mixed.

The overall model consists of equation sets for the cathode and for the anode, and each set is comprised of two groups as illustrated in Table 7 for the cathode. One group consists of Butler–Volmer equations, mass transfer correlations, Ohm's law, and conservation of charge. These determine the current density, the overpotential and the reaction rate distributions. The equations in the second group essentially define the conservation of species for all components, expressed in a time-independent form. The set of equations for the anode is similar to that for the cathode.

Table 7
Model Equations for the Cathode in the Synthesis of 2-Butanone[a]

Group I

Kinematic rate equation (reactions C1–C4)[b]

$$i_{cl} = \frac{i_{oi}}{n_{ci}F}\left[\exp\left(\frac{\alpha_{ai}F(\eta_{c1} - \eta^e_{ci})}{RT}\right) - \exp\left(\frac{-\alpha_{ci}F(\eta_{c1} - \eta^e_{ci})}{RT}\right)\right]$$

$i = 1, 2, 3, 4$

Mass transfer equation

$$i_{ci} = k_k(C^b_k - C^b_k)$$

$i = 1, 3, 4; k = $ acetoin, Br_2, BrO_3^-

Ohm's law

$$\frac{d\eta_{ci}}{dx} = \frac{i_c}{\kappa}$$

Boundary condition: $i_c = 0$ at $x = 0$

Conservation of charges

$$\frac{di_c}{dx} = A_c(n_{c1}Fi_{c1} + n_{c2}Fi_{c2} + n_{c3}Fi_{c3} + n_cFi_{c4})$$

Boundary condition: $i_c = $ constant at $x = 4$

Group II

Conservation of species (acetoin, 2-butanone, OH^-, Br_2, BrO_3^-)

$$i_{ci}A_c = D_k\frac{d^2C^b_k}{dx^2} - u\frac{dC^b_k}{dx}$$

$i = 1, 2, 3, 4; k = $ acetoin, 2-butanone, OH^-, Br_2, BrO_3^-
Boundary conditions: $C_k = C^b_{k,CSTR}$ at $x = 0$; $dC^b_k/dx = 0$ at $x = l_c$

Surface concentration equation (acetoin, 2-butanone, OH^-, Br_2, BrO_3^-)

$$C^s_k = \frac{i_{ci}}{k_4} - C^b_k$$

$i = 1, 2, 3, 4; k = $ acetoin, 2-butanone, OH^-, Br_2, BrO_3^-

[a] After Yu et al. (Ref. 108).
[b] See Table 6 for the reaction chemistry.

Table 8
Finite-Difference Equations for the CSTR in the Synthesis of 2-Butanone

The change of species concentration due to the cathode reaction (acetoin, 2-butanone, Br_2, BrO_3^-)

$$\Delta C_{k,CSTR} = \sum_{\substack{\text{all} \\ \text{mesh} \\ \text{points}}} \left[\frac{i_{ci}}{4 \sum_{1} (n_{ci}F_{ci})} \Delta i_c \right] \frac{A}{V} \Delta t$$

$i = 1, 3, 4; k =$ acetoin, 2-butanone, Br_2, BrO_3^-

The change of species concentration due to anode reaction (Br_2, BrO_3^-)

$$\Delta C_{k,CSTR} = \sum_{\substack{\text{all} \\ \text{mesh} \\ \text{points}}} \left[\frac{i_a}{4 \sum_{1} (n_{ai}F_{ai})} \Delta i_c \right] \frac{A}{V} \Delta t$$

$i = 1, 2; k = Br_2, BrO_3$

The CSTR mass balance

$$C_{k,\text{CSTR}} - C_{k,\text{in}} = r_k \tau$$

$k = $ 2,3-butanediol, acetoin, Br_2

The above electrochemical reactor equations are then combined with the conservation of mass equations for the CSTR, which can be conveniently expressed in a more practical finite-difference form as shown in Table 8. The conservation of mass equation is a statement that the amount of species consumed or produced at the cathode per unit time is equal to the rate of change in the amount of that species in the entire system. Similar equations apply to the anode. The rate equation for the reaction of bromine with 2,3-butanediol is expressed as

$$-dC_{Br_2,\text{CSTR}}/dt = k_2 C_{Br_2,\text{CSTR}} C_{2,3-\text{butanediol,CSTR}} \tag{215}$$

The method of solution is based on a modified fourth order Runge–Kutta algorithm using a shooting method and a finite-difference form for the second-order conservation equation in the electrode. The procedure is defined in a computer flow chart and is based on dividing the computation into a small number (40) of incremental time intervals. In each time interval, the group I equations are first solved for the cathode to give the current density and potential profiles along the cathode. Then the group II equations are solved to give the concentration profiles along the cathode. The solution then turns to the anode, where similar computations are performed. Following this, the solution procedure then turns to the CSTR to determine the new inlet concentrations for the next time increment of the procedure. The whole procedure is then repeated for the prescribed number of time intervals until the desired final reaction time is reached. Thus, the overall variation in species concentrations is predicted. Agreement between the simulated results and the experimental data for acetoin and 2-butanone current efficiencies is generally good.

The second example of Yu et al.[108] considers the paired synthesis of gluconate and sorbitol from glucose. There are similarities between this synthesis and that involving butanone, discussed above, in that both involve the use of electrogenerated bromine as an oxidant. The reactor configuration is the same in this work, as is the mode of operation. Furthermore, the methodology and solution method for simulation of the model are similar. Simulated overall current efficiencies of the model agree well with experimental data. On a more practical note, it is worth looking at the implications of a dual synthesis of this nature where two products are generated in a common electrolyte, notably the potential problems associated with product separation.

A purpose of this work is to consider digital simulation in the context of reactor design, and the procedures adopted for three-dimensional electrodes are in principle no different from those used for two-dimensional electrodes. The examples referred to have illustrated the models and solution techniques typically used in this task. The example of the work by Oloman and Reilly[115] highlights an area where further research is required, not just with regard to a need for more data, but also in the treatment of multiphase systems. This area of work will be considered further in the following section.

8. Multiphase Electrochemical Reaction Systems

Multiphase reaction systems involving gas/liquid and liquid/liquid mixtures are common in electrochemistry. This is particularly true where gas evolution reactions are involved, and the phenomena associated with bubble formation and growth and motion in electrolyte media require much greater study. The analysis of mass transport behavior and current distribution has undergone considerable investigation. As this area of research is large, only a limited treatment is given here, again focusing mainly on work relevant to simulation of electrochemical reactors. Similarly, selected examples of liquid/liquid electrochemical processes are considered, where the focus is generally the reactions of organic species, which have low electrolyte solubilities, either by direct or indirect electrochemical synthesis.

(i) Liquid/Liquid Reactions

The work of Oloman and Reilly discussed earlier is an example of a liquid/liquid reaction system, although, because one product is gaseous, it is in fact a three-phase system. The aspects of the three-phase flow were dealt with by utilizing existing correlations in the literature for pressure drop in the gas and in the liquid phases and for mass transport rates between the two liquids. The latter data are particularly important and for many systems are reported as liquid-phase mass transfer capacity coefficients, sometimes referred to as specific mass transfer coefficients (units of reciprocal seconds). These are essentially terms combining the mass transfer coefficient and the interfacial area per unit volume of the system ($k_{La} = k_L a$) and are often used because for many systems the interfacial area cannot be accurately estimated.

The transport of species from, say, the organic phase to the aqueous phase is then defined by an equation of the form

$$r = k_{La}(C_e - C) \qquad (216)$$

where C_e is the concentration of solute species in the aqueous phase in equilibrium with the organic phase. This can be related to the organic phase concentration through the use of a distribution coefficient.

The treatment of the model can then follow two general paths, depending on whether the reactions of the organics occur in the bulk electrolyte or are electrochemical surface processes (see Fig. 39). In the latter case, a stationary-state situation can be considered, where the rates of interphase mass transport, intraphase mass transport, and reaction are in series and are equal; i.e.,

$$k_{La}(C_e - C) = ak_L(C - C_s) = ak_{fj}C_{sj}^m \qquad (217)$$

This condition was adopted by Oloman and Reilly for the simulation of benzene oxidation in the trickle bed reactor. This equation should be applied to all species that show phase distribution, although in Oloman and Reilly's model the benzoquinone distribution was assumed to be at

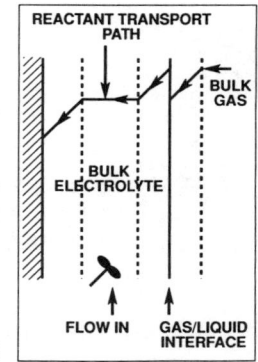

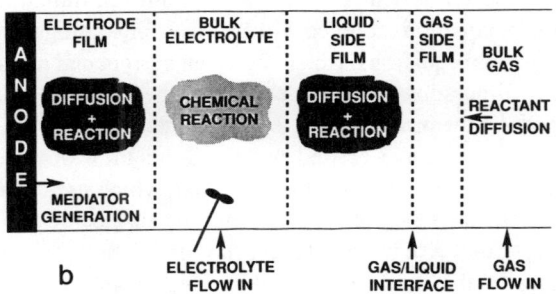

Figure 39. Schematic reactor models for two-phase electrochemical processes: (a) direct electrochemical reaction; (b) indirect electrochemical reaction.

the equilibrium condition; that is, rapid mass transport between phases prevailed.

The aspect of interphase mass transport with surface electrochemical reaction was considered by Scott and Hayati for the industrial synthesis of adiponitrile from acrylonitrile.[101] This process involves a set of coupled surface reactions, with interphase mass transport between the dispersed organic phase, containing acrylonitrile, adiponitrile, and several other products, and the aqueous electrolyte. The electrochemical reaction takes place in a bank of undivided narrow parallel-plate electrodes in which oxygen gas is simultaneously evolved. The three-phase mixture then passes to a gas/liquid separator and then to a liquid/liquid separator (Fig. 40), from which the organic product is decanted and the aqueous electro-

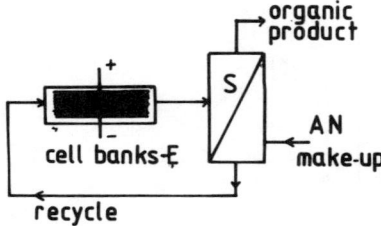

Figure 40. Schematic model of two-phase reaction loop for the synthesis of adiponitrile from acrylonitrile (AN).

lyte recycled to the reactor. Simulation of the reactor for a batch laboratory process was achieved using a computer package (NAG Fortran) based on Merson's method for solving simultaneous nonlinear, initial-value ordinary differential equations coupled with routines for solving simultaneous nonlinear algebraic equations, which represent current and mass balances in the system. Simulation of the model enabled the influence of interphase mass transport on the product yield to be obtained.

Shih and Lee[116] have carried out the simulation of an extractive electrochemical reaction in a parallel-plate cell, which used an extractive liquid to separate an adherent product from the surface of the electrode, as this product otherwise would prevent further reaction. This model was representative of the oxidation of sulfide ion to sulfur at a platinum electrode in basic media.[117] The equations derived are not complex, and simulation was achieved using a simple iterative procedure. There is, however, scope to develop this model further.

There is a substantial body of literature (e.g., Refs. 118 and 119) on the subject of organic synthesis reactions mediated by electrochemically generated species, typically metal-ion oxidants such as Co(III), Mn(III), and Ce(III). A typical example is the oxidation of toluene or substituted toluenes for preparation of the corresponding aldehydes. Several examples of this technique are reported in the literature,[120-122] and it seems to be an attractive method owing to the low solubility of the organic in the aqueous electrolyte. However, if the electrolyte is saturated with the substituted aromatic compound, then severe passivation of the electrode can occur. Consequently, in order to prevent these effects, the oxidation is performed in a separate reactor. This procedure has proved to be efficient in the oxidation of several toluenes, with the metallic mediator being regenerated anodically in the electrochemical reactor after the aqueous electrolyte has been cleared of the organic. Figure 41 shows a schematic

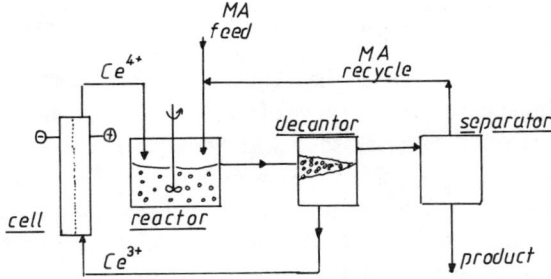

Figure 41. Schematic diagram of the reactor unit for two-phase mediated oxidation of toluenes. MA = methylanisol.

diagram of this indirect oxidation process, in which the organic species is oxidized in an emulsion reactor and the two phases are separated to achieve electrolyte recycle, and oxidant regeneration, and subsequent product recovery.

Wendt and Schneider[123] carried out the mediated oxidation of substituted toluenes and suggested as a modification to the generally proposed simple process the incorporation of a countercurrent reactor extractor (Fig. 42). This additional unit was devised to prevent any carryover of organic

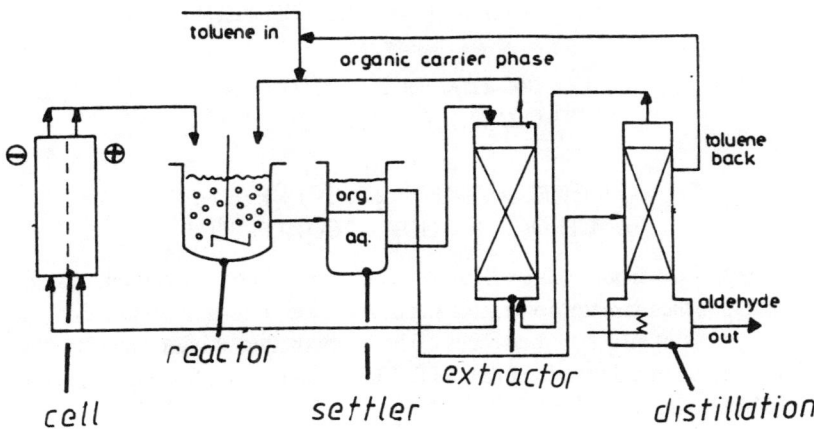

Figure 42. Schematic diagram for two-phase reactor with a countercurrent reactor extractor stage.

species to the cell. Their preliminary simulations of the process were based on a steady-state operation and did not require digital procedures. Kreysa and Medin[121] also considered the mediated oxidation of substituted toluenes and simulated the performance of batch reactor operation. The model involved interphase mass transport with homogeneous reaction between the mediator and the organic species in the aqueous solution. The reaction was slow enough (low Hatta number) to be considered to occur solely in the bulk solution. Accordingly, the model equations were not complex, and the solution for this initial-value problem was achieved by standard Runge–Kutta procedures. Simulation of the model using mass transfer and rate constant values determined from experimental data by a Simplex procedure gave close agreement to experimental performance.

Tzedakis and Savall[124] considered in detail the simulation of the indirect electrooxidation of methylanisole using a liquid/liquid CSTR. In their approach, reaction occurs simultaneously with diffusion of reactant from the liquid/liquid interface into the bulk electrolyte. The solubility of methylanisole in the aqueous electrolyte is relatively low, and, to compensate for this, the organic substrate is emulsified to try to maintain a near saturation concentration. The mediated oxidation occurs in several steps:

$$CH_3O(C_6H_4)CH_3 + 2Ce(IV) + H_2O \xrightarrow{k_4}$$
$$CH_3O(C_6H_4)CH_2OH + 2Ce(III) + 2H^+$$

$$CH_3O(C_6H_4)CH_2OH + 2Ce(IV) \xrightarrow{k_5}$$
$$CH_3O(C_6H_4)CHO + 2Ce(III) + 2H^+$$

$$CH_3O(C_6H_4)CHO + 2Ce(IV) + H_2O \xrightarrow{k_6}$$
$$CH_3O(C_6H_4)COOH + 2Ce(III) + 2H^+ \qquad (218)$$

The product from each oxidation step is extracted by the organic phase, and thus the mass transport rates and the phase equilibrium of species have a significant effect on the process. The mathematical model considers the case of chemical reaction in a CSTR. The molar balance for methylanisole (MA) in the bulk electrolyte is

$$-V_a a D_2 \left(\frac{dC_2}{dx} \right)_{x=\delta} = k_4 C_1^\alpha C_2^\beta V_a (1 - a\delta) + v_a C_2 \qquad (219)$$

The material balances for the other species did not consider the aspect of the distribution of concentration and thus of rate in the liquid/liquid diffusion film. The inorganic species were confined to the aqueous phase, and the organic species, other than MA, reached rapid equilibrium between the phases; thus, the material balances become

$$r_i V_{aq} = v_{aq} C_{i,aq} + v_{org} K_{Pi} C_{i,aq} \quad (220)$$

where K_{Pi} is the distribution coefficient.

In order to solve the model, the transfer term of MA from the interface must be evaluated, which was achieved by the use of an isothermal liquid film model of diffusion and reaction for MA (species 2) in this case; i.e.,

$$\frac{D_2 d^2 C_2}{dx^2} = k C_1 C_2. \quad (221)$$

The model assumed the concentration of Ce(IV), species 1, in this case to be constant in the film and equal to its bulk value. Simulation of the reactor behavior was achieved by numerically integrating the film model using a fourth order Runge–Kutta procedure with application of a Newton procedure to achieve convergence. From this, the value of the flux of MA at the edge of the diffusion film was obtained. The set of equations representing the liquid film model was solved by an iterative method in which a mass transfer coefficient (k_t as defined by Tzedakis and Savall) was used as an adjustable parameter, which required the solution of the following equation:

$$D_2 \left(\frac{dC_2}{dx} \right)_{x=d} = -k_t (C_{2sat} - C_2) \quad (222)$$

Overall, the model was used for the purposes of simulating the reaction system with a view to scale-up and predicting the influence of holding time, mixing rate, and mediator concentration. Application to scale-up is one of the major uses of digital simulation.

(ii) Gas/Liquid Systems

The work of Yung and Alkire[102] is a suitable starting point from which to introduce aspects of importance in the simulation of reaction systems involving gas/liquid mixtures. This work was involved in a study of the production of 1,2-dichloroethane and ethylene chlorohydrin by the electrolysis of ethylene-containing hydrochloric acid solutions in a continuous flow, undivided parallel-plate cell. The objective of the work was the development of a reactor model of the system that could be used in a scale-up and optimization study of the process. These aspects were incorporated into an economic analysis of the system to study the influence of design and operating parameters. A successive quadratic programming method was used to identify optimum operating conditions. Although this analysis is of obvious importance in process engineering, what is of interest in the context of this chapter is the treatment of the reactor system.

The significant feature of the reactor is the generation of gases at both of the electrodes (i.e., chlorine and hydrogen). This generation of gases essentially produced three regions in the reactor:

(i) A thin layer next to the anode containing a chlorine bubble swarm. Bubbles in this region were observed to move with similar velocities and were approximately 100 μm in diameter. The thickness of this chlorine layer was estimated to be 0.2 mm.

(ii) A similar bubble layer next to the cathode containing hydrogen. The diameter of the hydrogen bubbles was estimated to be 200 μm, and the thickness of the bubble layer was approximately 0.5 mm.

(iii) The central portion of the flow channel free of bubbles.

The thicknesses of the bubble layers were inversely proportional to the square root of the electrolyte velocity. The rise velocity of the bubbles was in agreement with predictions from Stokes' law and was thus proportional to the cell current. A schematic representation of the bubble flow pattern is shown in Fig. 43, which is based on the work of Yung and Alkire and other investigations of bubble patterns in cells. In this diagram, there is a thin layer of small stationary bubbles below the moving bubble layers that is mainly unaffected by flow. This bubble flow pattern is of great significance in setting up the engineering model of the system.

In terms of the reaction system, chlorine produced at the anode reacts homogeneously with ethylene dissolved in solution. The concentration profiles near the electrode were investigated using a transport model based on the following assumptions and behavior:

- Well-developed laminar flow with a linear velocity field in the thin reaction layer
- 100% chlorine evolution efficiency
- First-order homogeneous reaction between chlorine and ethylene

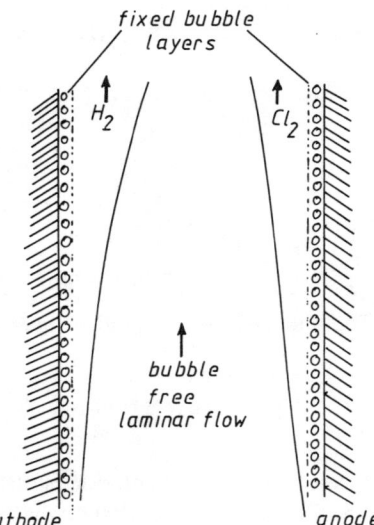

Figure 43. Schematic diagram of the flow pattern in the reactor for chlorination of ethylene, showing fixed bubble layers developing gas bubble flow region.

- Transport by convection along the electrode and by diffusion normal to the electrode

The last assumption above clearly does not allow for the bubble flow behavior in practical situations, which would be likely to induce a mixing process. Nevertheless, this assumption is adopted because it is a limiting case of operation.

The differential mass balances for chlorine and ethylene over the reaction region of thickness δ_R are

$$D_{Cl_2}\frac{\partial^2 C_{Cl_2}}{\partial y^2} - u_x \frac{\partial C_{Cl_2}}{\partial x} = kC_{Cl_2}C_{C_2H_4} \tag{223}$$

$$D_{C_2H_4}\frac{\partial^2 C_{C_2H_4}}{\partial y^2} - u_x \frac{\partial C_{C_2H_4}}{\partial x} = kC_{Cl_2}C_{C_2H_4} \tag{224}$$

where the linear velocity field is given by

$$u_x = 6uy/h$$

There are six boundary conditions as follows:

$$y = \delta: \quad C_{Cl_2} = 0, \; C_{C_2H_4} = C_0 \text{ for all } x$$

$$y = 0: \quad -D_{Cl_2}\frac{\partial C_{Cl_2}}{\partial y} = \frac{i}{2F} \quad \text{for all } x$$

$$\frac{\partial C_{C_2H_4}}{\partial y} = 0 \tag{225}$$

$$x = 0; \quad C_{C_2H_4} = C_0, \; C_{Cl_2} = 0 \quad \text{for all } y$$

The thickness of the reaction layer was determined from a mass balance

$$\frac{iL}{2F} = \int_0^\partial C_{Cl_2}(y)_{x=L}u_x(y)dy + \int_0^L \int_0^\partial r(x,y)dy\,dx \tag{226}$$

A mass balance for chloride is also needed because of the influence of chloride-ion concentration on the reaction selectivity, given by

$$D_{HCl}\frac{\partial^2 C_{Cl^-}}{\partial y^2} - u(y)\frac{\partial C_{Cl^-}}{\partial x} + \frac{R}{R+1}r = 0 \tag{227}$$

where R is the product ratio of ethylene chlorohydrin to dichloroethane.

To assist in the interpretation and the application of the model, both chemical kinetic measurements and electrochemical measurements on a rotating disk system were used. The latter produced appropriate Tafel equations for the anodic and the cathodic processes.

The influence of the bubble layer phenomenon was accounted for in the model in terms of the influence on the cell voltage. Effective conductivities were assigned to the stationary layers, and the conductivities of the moving bubble layer were described by the Bruggeman equation, which corrects the actual electrolyte conductivity using an empirical correlation that is a function of the gas bubble voltage. A correlation for mass transfer that accounted for the microconvection due to bubble growth was also used.

The concentration profiles were formulated on the assumption of dilute solution theory as well as with linear concentration profiles in the mass transfer boundary layer. With the current balance and the voltage balances, overall there were 18 independent equations to be considered at any location in the cell. The solution to the model was obtained by implementing a finite-difference procedure, using an in-house subroutine, to calculate the concentration profiles in the reaction layer. In the solution it was convenient to arrange the mass balance integral equation for HCl in the following approximate form:

$$C_b(x) = C^* - \frac{W}{2FQ}\left(\frac{R_0+2}{R_0+1}\right)\int_0^x i_a(x')\,dx'$$

$$\int_0^x i_a(x')\,dx' \cong \sum_{j=1}^{n-1} i_a(j)\frac{L}{N} + \left(x - \frac{L}{N}(n-1)\right)i_a(n) \cong \sum_{j=1}^{n-1} i_a(j)\frac{L}{N} \quad (228)$$

where

$$(n-1)\frac{L}{N} < x \leq n\frac{L}{N}$$

No iterations were required in the solution of the equations, and so integration was carried out by dividing the cell into N sections along the flow path. The set of equations were combined into two nonlinear algebraic equations, which were solved numerically.

In this study the most sophisticated numerical procedures were those used to solve the model of the reaction layer at the anode. Although this study generated two dozen or so equations, these were readily solved by using commercial software. This aspect is of growing significance, as the development of a reactor or any model should not be limited by the time-consuming efforts required to develop software, if at all possible.

A feature of the system of Yung and Alkire[102] was the effect of the generation of gas on volatile components in the electrolyte. This will cause a stripping of such components into the gas, which will also contain water vapor. The mass transfer of this behavior could well play an important role in the model as well as in practical aspects of reactor operation. This situation has recently been broached[107] in a reactor system for the direct oxidation of propylene in a sieve plate reactor (Fig. 44), comprising a bank of bipolar undivided parallel-plate electrodes. This reactor provides very good mass transport conditions for the intimate contacting of gas and liquid electrolyte. The model of this system, consisting of a set of dynamic material balances describing electrochemical and chemical reaction and gas absorption and desorption, was solved quite effectively with the use of a computer simulation package.

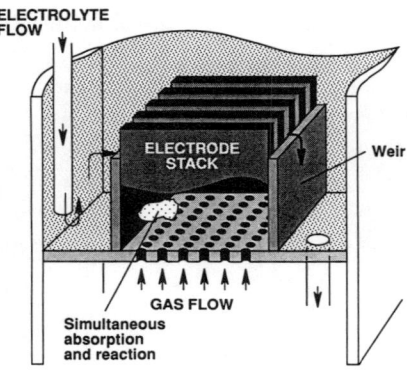

Figure 44. Schematic diagram of the sieve plate electrochemical reactor.

9. Conclusions

In this section, an attempt has been made to demonstrate the range of problems to which digital simulation procedures have been applied in the area of electrochemical reaction engineering. Both the setting up of a reaction model based on electroanalytical techniques and the integration of such models into the reactor environment have been discussed. The area is extremely broad and covers almost all sectors of electrochemical technology, and hence the focus has been intentionally limited to the application of reaction models in synthesis. It was not intended that this section be a review of this area of work but rather that it serve more as a source of information from which the reader can acquire knowledge on the building of mathematical models for electrochemical reactors and the methods available for their solution and resultant simulation.

The simulation methods are varied and range from procedures for linear ordinary differential equations, and even algebraic equations, to sets of dynamic two-dimensional, nonlinear partial differential equations. Thus, at one extreme, solutions may be analytical, whereas, at the other extreme, they are heavily machine-based and use sophisticated iterative finite-difference methods, finite elements, or collocation.

This area of activity has at the present time seen the fruits of the many research programs involving digital simulation appear as a range of software packages, in addition to hard-copy forms of suitable programs. This activity has helped to encourage the use of more sophisticated models of electrochemical systems, with the obvious advantages of enabling more rigorous and accurate simulation. In addition, the computer package has now taken on a role of replacing analytical integration and solution of, for example, first-order ordinary differential equations. This is not to say that there is no longer a role for analytical techniques but rather points to a greater use of numerical routines and digital simulation, sometimes coupled with analytical techniques, as seen in Section V.4.

It may not have gone unnoticed that the reaction systems considered as examples in this work were, for all intents and purposes, isothermal. This is a quite satisfactory approximation for many electrochemical systems, where only a few degrees' change in temperature is realized. However, very few electrochemical reactors have integral heat transfer facilities within the reaction environment, and, because of the ohmic heating effect induced by the passage of current through the cell, temperature increases of 10 or more degrees can easily be realized. Apart from the influence of such a temperature change on electrode kinetics, a major effect will be with regard to chemical reactions, this level of temperature rise can cause an order of magnitude increase in the rate of reaction. In principle, the analysis of electrochemical reactors should mimic that of their chemical and catalytic counterparts and incorporate within the overall model the required energy balances for the reactor system. Thus, the effect of a variation in temperature on both physical and chemical properties can be satisfactorily introduced into the digital simulation methods. This has been done for relatively simple reaction systems,[125] where the required data were available to implement the method. On this subject, I would like to conclude by referring to a recent simulation of an electrochemical system in this area, that is, the work by Wright[126,127] on the simulation of the operation of an aluminum cell. As is common knowledge, such cells operate with a molten salt electrolyte at very high temperatures, and thus simulation of the system must require the application of rigorous energy balance with all other aspects of electrochemical cell modeling. In addition to this, the work of Wright[126,127] considered the practical implications of anode positioning during continuous operation and the dynamic behavior of the melt through current surging and other phenomena. At this point, it is not possible to consider this system in further detail, but again the reader is recommended to consult with the original reference if interested.

ACKNOWLEDGMENT

I would like to acknowledge the contributions of the many research workers in this field, from which a number of examples used here have been abstracted.

NOTATION

a	Area per unit volume of electrode
a, b	Tafel coefficients
a_j	Activity of component j
C_e	Equilibrium concentration between two liquid phases
C_i	Concentration of species i
C_{ii}	Inlet concentration of species i
C_{iref}	Reference concentration of species i
C_{is}	Concentration of species i at the surface
C_{i0}	Initial concentration of species i
C	Total concentration
d_e	Equivalent diameter of channel
D_e	Dispersion coefficient
D_i	Diffusivity of species i
e^-	Electron
E	Electrode potential
E_{cell}	Cell voltage
E_e	Equilibrium potential
E_o	Open-circuit voltage
E_{IR}	Ohmic voltage loss
F	Faraday
h	Mesh distance
i	Current density
i_k	Kinetic current density
i_L	Limiting current density
i_T	Total current density
i_0	Exchange current density
I	Current
I_j	Partial current for reaction j
k	Homogeneous reaction rate constant
k_a, k_b	Adsorption rate constants
k_{-a}, k_{-b}	Desorption rate constants

k_{b1}	Reverse electrochemical rate constant
k_{bi}	Reverse electrochemical rate parameter
k_{f1}	Forward electrochemical rate constant
k_{fi}	Forward electrochemical rate parameter
k_j	Electrochemical rate constant in forward direction for step j
k_{La}	Specific mass transfer coefficient
k_{Li}	Mass transport coefficient of species i
K	Equilibrium constant
K_e	Electrochemical equilibrium parameter $= k_{f1}/k_{b1}$
K_j	Electrochemical equilibrium constant for step j
K_m	Distribution coefficient
K_A	Equilibrium constant for adsorption of species A
l	Adsorption site
L	Electrode length
L_p	Thickness of porous electrode
m	Order of electrochemical reaction
m	Distribution coefficient
n	Number of increments
n_j	Number of electrons in reaction j
N	Number of sections
N	Number of tanks
N_j	Flux of species
p, q	Order of reaction
P	Pressure
P_i	Partial pressure of component j
r	Recycle ratio
r	Radial dimension or radius of disk
r_j	Rate of reaction of step j
R	Gas constant
R	Radius of pipe
R_e	Internal electrical resistance
s_{ij}	Stoichiometric coefficient of species i in reaction j
S	Electrode gap
t	Time
T	Temperature
u	Velocity
u_i	Mobility of species i
v	Volumetric flow rate of electrode
V	Volume

w	Angular velocity
W	Electrode width or cell width
x	Dimension of reactor
X_j	Fractional conversion of species j
y	Coordinate direction
z	Electronic charge

Dimensionless Numbers

Da	Damköhler (modified) number
ha	Hatta number
Pe	Peclet number
Re	Reynolds number
Sc	Schmidt number

Greek Symbols

α	Constant describing the potential dependency of the reaction rate
α_i	Transfer coefficient of species i
β	Constant describing the potential dependency of the reaction rate
β_r	Ratio of backflow to throughflow
δ	Diffusion film thickness
ε	Effectiveness factor
η	Overpotential
θ	Dimensionless concentration
θ_i	Fractional coverage of species i
θ_v	Number or concentration of vacant sites
θ_T	Total number or concentration of active sites
ρ	Fluid density
μ	Kinematic fluid viscosity
μ_r	Reaction layer thickness
v	Dynamic fluid viscosity
v	Potential sweep rate
τ	Residence time or holding time
τ_g	Dimensionless residence time = $\sigma i_T \tau / n_1 F C_{Ai}$
τ_r	Residence time under recycle conditions
ϕ	Potential of solution
ϕ	Current efficiency
ϕ	Dimensionless group = $\sqrt{(k D_L / k_L^2)}$
Φ_{oa}	Potential of solution outside double layer

Subscripts

ads	Adsorbed species
a	Anode
c	Cathode
e	Equilibrium
i	Inlet condition
i	Species
j	Step j
m	Mixing unit
n	Number of the reactor
0	Initial amount or condition
o	Overall
PFR	Plug flow reaction
r	Reactor
ref	reference concentration or condition
s	Surface
sat	Saturated
T	Total
x	Local position along electrode

Superscripts

o	Standard state
max	Maximum value

REFERENCES

[1] R. A. Willoughby, ed., *Stiff Differential Systems*, Plenum, New York, 1974.
[2] G. A. Prentice and C. W. Tobias, *J. Electrochem. Soc.* **129** (1982) 72.
[3] R. Caban and T. W. Chapman, *J. Electrochem. Soc.* **123** (1976) 1036.
[4] W. R. Parrish and J. Newman, *J. Electrochem. Soc.* **116** (1969) 169.
[5] R. N. Fleck, D. N. Hanson, and C. W. Tobias, Lawrence Radiation Laboratory, Berkeley, California, UCRL 11612, September 1964.
[6] L. N. Klatt and W. J. Blaedel, *Anal. Chem.* **40** (1968) 512.
[7] J. A. Harrison, *Electrochim. Acta* **27**(8) (1982) 1113.
[8] D. J. Pickett, *Electrochemical Reactor Design*, Elsevier, Amsterdam, 1979.
[9] A. A. Wragg, *The Chemical Engineer* **1977** (January) 39.
[10] K. R. Westerterp, W. P. M. von Swaaji, and A. C. M. Beenackers, *Chemical Reactor Design and Operation*, John Wiley & Sons, New York, 1984.
[11] L. K. Doraiswamy and M. M. Sharma, in *Heterogeneous Reactions, Analysis and Design*, Vol. 12, *Fluid–Fluid–Solid Reactions*, John Wiley & Sons, New York, 1984.
[12] B. Damaskin, O. Petrii, and V. Batrakov, *Adsorption of Organic Compounds on Electrodes*, Plenum, New York, 1971.
[13] W. J. J. Albery, *Electrode Kinetics*, Clarendon Press, Oxford, 1975.
[14] P. Delahay, *Double Layer and Electrode Kinetics*, Interscience, New York, 1965.

[15] K. Scott, *J. Electroanal Chem.* **325** (1992) 1.
[16] K. H. Yang and O. A. Hougen, *Chem. Eng. Prog.* **46** (1950) 146.
[17] G. F. Froment and K. B. Bischoff, *Chemical Reactor Analysis and Design*, 2nd ed., John Wiley & Sons, New York, 1990.
[18] J. A. Harrison and Z. A. Khan, *J. Electronanal. Chem.* **26** (1970) 1.
[19] J. S. Newman, *Electrochemical Systems*, Prentice-Hall, Englewoods Cliffs, New Jersey, 1973.
[20] P. Stonehart and P. N. Ross, *Electrochim. Acta* **21** (1976) 441.
[21] D. Economou, M.Sc. Thesis, University of Illinois, 1983.
[22] R. C. Alkire and J. D. Lisius, *J. Electrochem. Soc.* **132** (1985) 1879.
[23] J. van Zee, R. E. White, and A. T. Watson, *J. Electrochem. Soc.* **133** (1986) 508.
[24] F. Goodridge and C. J. H. King, in *Technique of Electro-organic Synthesis*, Part I, Ed. by N. L. Weinberg, *Techniques of Chemistry Series*, Vol. V, John Wiley & Sons, New York, 1974.
[25] R. Barz, C. Bernstein, and W. Vielstich, in *Advances in Electrochemistry and Electrochemical Engineering*, Vol. 13, Ed. by H. Gerischer and C. W. Tobias, John Wiley & Sons, New York, 1984.
[26] O. Levenspiel, *Chemical Reaction Engineering*, John Wiley & Sons, New York, 1972.
[27] M. Fleischmann and J. W. Oldfield, *J. Electroanal. Chem. Interfacial Electrochem.* **29** (1971) 211.
[28] D. Pletcher and F. C. Walsh, *Industrial Electrochemistry*, 2nd ed., Chapman and Hall, London, 1990.
[29] H. Gerischer, *Ber. Bunsenges. Phys. Chem.* **67** (1967) 164.
[30] C. F. Odouza and K. Scott, *Chem. Eng. Symp. Ser. Inst.* No. 127 (1992), 37.
[31] G. P. Sakellaropoulos, *AIChE J.* **15** (1979) 781.
[32] K. Scott, *J. Appl. Electrochem.* **15** (1985) 837.
[33] A. T. S. Walker and A. A. Wragg, *Electrochim. Acta* **22** (1977) 1129.
[34] K. Scott, *Electrochemical Reaction Engineering,* Academic Press, London, 1991.
[35] K. Scott, *Chem. Eng. Res. Des.* **64** (1986) 226.
[36] W. R. Parrish and J. Newman, *J. Electrochem. Soc.* **117** (1970) 43.
[37] V. Edwards and J. Newman, *J. Electrochem. Soc.* **134** (1987) 1181.
[38] D. M. Himmelbau and K. B. Bischoff, *Process Analysis and Simulation*, John Wiley & Sons, New York, 1968.
[39] E. B. Nouman, in *Scale Up of Chemical Processes*, Ed. by A. Bisio and R. L. Kabel, John Wiley & Sons, New York, 1985.
[40] C. Y. Wen and L. T. Fan, *Models for Flow Systems and Chemical Reactors*, Marcel Dekker, New York, 1975.
[41] A. Elinkenberg, *Chem. Eng. Sci.* **26** (1971) 1133.
[42] B. Hunt, *Int. J. Heat Mass Transfer* **20** (1977) 393.
[43] J. Newman, *Int. J. Heat Mass Transfer* **10** (1967) 983.
[44] E. N. Lightfoot, E. L. Cussler, and R. L. Rettig, *AIChE J.* **8** (1962) 708.
[45] J. Newman, *Ind. Eng. Chem.* **60** (1968) 12.
[46] F. Lapicque and A. Storck, *J. Appl. Electrochem.* **15** (1985) 925.
[47] H. R. Thirsk and J. A. Harrison, *A Guide to the Study of Electrode Kinetics,* Academic Press, London, 1972.
[48] R. I. Gelb and L. Meites, *J. Phys. Chem.* **68** (1964) 636.
[49] S. Karp and L. Meites, *J. Electroanal. Chem.* **17** (1968) 253.
[50] C. Amatore and J. M. Saveant, *J. Electroanal. Chem.* **123** (1981) 198–201, 203–217, 219–229, 231–242.
[51] C. Amatore and J. M. Saveant, *J. Electroanal. Chem.* **125** (1981) 1.
[52] Southampton Electrochemistry Group, *Instrumental Methods in Electrochemistry*, Ellis Horwood, Chichester, England, 1985.

[53] T. R Nolen and P. S. Fedkiw, *J. Appl. Electrochem.* **20** (1990) 370.
[54] K. Scott, *Electrochim. Acta* **30** (1983) 245.
[55] A. N. Haines, Ph.D Thesis, Teesside Polytechnic, 1988.
[56] R. M. Machado and T. W. Chapman, *J. Electrochem. Soc.* **134** (1987) 385.
[57] K. Scott and W. Hui, in press.
[58] W. J. Albery and M. L. Hitchman, *Ring-Disc Electrodes*, Clarendon Press, Oxford, 1971.
[59] S. Feldberg, in *Electroanalytical Chemistry*, Vol. 3, Ed. by A. J. Bard, Marcel Dekker, New York, 1969, p. 199.
[60] T. Joslin and D. Pletcher, *Electroanal. Chem. Interfacial Electrochem.* **49** (1974) 171.
[61] J. Newman, *Ind. Eng. Chem. Fundam.* **7** (1968) 514.
[62] G. D. Smith, *Numerical Solution of Partial Differential Equations*, Oxford University Press, Oxford, 1969.
[63] B. S. Pons, *Can. J. Chem.* **59** (1981) 1538.
[64] J. Villadsen and M. L. Michelsen, *Solution of Differential Equation Models by Polynomial Approximation*, Prentice-Hall, Englewood Cliffs, New Jersey, 1978.
[65] J. M. Saveant, *J. Electroanal. Chem.* **125** (1981) 23.
[66] C. Amatore and J. M. Saveant, *J. Electroanal. Chem.* **126** (1981) 1.
[67] C. Amatore and J. M. Saveant, *J. Electroanal. Chem.* **102** (1979) 25.
[68] C. Amatore and J. M. Saveant, *J. Electroanal. Chem.* **86** (1977) 27.
[69] C. Amatore and J. M. Saveant, *J. Electroanal. Chem.* **88** (1978) 127.
[70] C. Amatore and J. M. Saveant, *J. Electroanal. Chem.* **107** (1980) 353.
[71] H. Wendt and V. Plzak, *J. Electroanal. Chem.* **154** (1983) 13.
[72] H. Wendt and V. Plzak, *J. Electroanal. Chem.* **154** (1983) 29.
[73] M. J. Clifton and A. Savall, *J. Appl. Electrochem.* **16** (1986) 812.
[74] A. T. Hubbard, *Acc. Chem. Res.* **13** (1980) 177.
[75] P. M. A. Sherwood, *Chem. Soc. Rev.* **1** (1985).
[76] P. N. Ross and F. T. Wagner, in *Advances in Electrochemistry and Electrochemical Engineering*, Vol. 13, Ed. by H. Gerischer and C. Tobias, John Wiley & Sons, New York, 1985, p. 69.
[77] A. Bewick and S. Pons, in *Advances in Infrared and Raman Spectroscopy*, Vol. 12, Ed. by R. J. H. Clark and R. E. Hester, Heyden, London, 1985, p. 1.
[78] J. Foley, C. Korzeniewski, J. Daschbach, and S. Pons, in *Electroanalytical Chemistry*, Vol. 14, Ed. by A. Bard, Marcel Dekker, New York, 1986, p. 309.
[79] T. E. Furtak and R. E. Chang, eds., *Surface Enhanced Raman Scattering*, Plenum Press, New York, 1982.
[80] M. Fleischmann and I. R. Hill, in *Comprehensive Treatise of Electrochemistry*, Vol. 12, Ed. by R. E. White, J. O'M Bockris, B. E. Conway, and E. Yeager, Plenum Press, New York, 1984.
[81] M. Moskovits, *Rev. Mod. Phys.* **57** (1985) 783.
[82] G. P. Sakellaropoulos and G. A. Francis, *J. Electrochem. Soc.* **126** (1979) 1928.
[83] G. P. Sakellaropoulos, *AIChE J.* **25** (1979) 781.
[84] M. Fleischmann, C. L. K. Tennakooon, P. Gough, J. H. Steven, and S. R. Korn, *J. Appl. Electrochem.* **13**, (1983) 603.
[85] K. Scott, *Electrochim. Acta* **30**, (1985) 235.
[86] K. Scott, *J. Appl. Electrochem.* **15** (1985) 659.
[87] K. Scott, A. N. Haines, and I. F. McConvey, *J. Appl. Electrochem.* **17** (1987) 925.
[88] S. W. Feldberg, *Electroanalytical Chemistry*, Vol. III, Ed. by A. J. Bard, Marcel Dekker, New York, 1969.
[89] I. Ruzic and S. W. Feldberg, *J. Electroanal. Chem.* **50** (1974) 153.
[90] D. N. Bennion, *AIChE Symp. Ser.* **79** (1983) 25.
[91] R. Alkire and A. A. Mirarefi, *J. Electrochem. Soc.* **120** (1973) 1507.
[92] R. Alkire and A. A. Mirarefi, *J. Electrochem. Soc.* **124** (1977) 1043.

[93] R. Alkire and A. A. Mirarefi, *J. Electrochem. Soc.* **124** (1977) 1214.
[94] R. Alkire and R. Gould, *J. Electrochem. Soc.* **123** (1976) 1842.
[95] R. E. White, S. E. Lorimer, and R. Darby, *J. Electrochem. Soc.* **130** (1983) 1123.
[96] R. E White, M. Bain, and M. Raible, *J. Electrochem. Soc.* **130** (1983) 1037.
[97] T. V. Nguyen, C. W. Walton, R. White, and J. Zee, *J. Electrochem. Soc.* **133** (1986) 81.
[98] P. S. Fedkiw and W. D. Scott, *J. Electrochem. Soc.* **131** (1984) 1304.
[99] A. N. Haines, I. F. McConvey, and K. Scott, *Electrochim. Acta* **30** (1985) 291.
[100] K. Scott, I. F. McConvey, and B. Hayati, *Inst. Chem. Eng. Symp. Ser.* **112** (1989) 263.
[101] K. Scott and B. Hayati, *Chem. Eng. Sci.* **45** (1990) 2341.
[102] E. K. Yung and R. C. Alkire, *J. Electrochem. Soc.* **132** (1985) 2341.
[103] ACSL User Guide/Reference Manual, Mitchell and Gauthier Associates, USA 1981.
[104] J. Marquez and D. Pletcher, *J. Appl. Electrochem.* **10** (1980) 567.
[105] F. Goodridge, S. Harrison, and R. E. Plimley, *J. Electroanal. Chem.* **214** (1986) 283.
[106] C. F. Odouza and K. Scott, *I. Chem. E. Research Event* (Jan 1991) Cambridge.
[107] K. Scott, C. F. Odouza, and W. Hui, paper presented at International Society Chemical Reaction Engineering 12, Torino, Italy, June 1992.
[108] J. C. Yu, M. M. Baizer, and K. Nobe, *J. Electrochem. Soc.* **135** (1988) 1392, 1400.
[109] L. Weise, M. Giron, G. Valentin, and A. Storck, *Inst. Chem. Eng. Symp. Ser.* **98** (1986).
[110] F. Lapicque and A. Storck, *J. Appl. Electrochem.* **15** (1985) 925.
[111] N. Ibl, in *Comprehensive Treatise of Electrochemistry*, Vol. 6, Ed. by E. Yeager, J. O'M Bockris, B. E. Conway, and S. Sarangapani, Plenum Press, New York, 1983.
[112] F. Goodridge and A. R. Wright, in *Comprehensive Treatise of Electrochemistry*, Vol. 6, Ed. by E. Yeager, J. O'M Bockris, B. E. Conway, and S. Sarangapani, Plenum Press, New York, 1983, Chapter 6.
[113] J. Newman and W. Tiedemann, *AIChE J.* **21** (1975) 21.
[114] R. Alkire and P. K. Ng, *J. Electrochem. Soc.* **121** (1974) 95.
[115] C. Oloman and P. Reilly, *J. Electrochem. Soc.* **134** (1987) 859.
[116] Y. S. Shih and J. L. Lee, *J. Appl. Electrochem.* **17** (1987) 480.
[117] H. Binder, A. Kohling, and G. Sandatede, *Angew Chem. Int. Ed. Engl.* **6** (1967) 884.
[118] H. Fess and H. Wendt, in *Technique of Electroorganic Synthesis*, Part 3, *Performance of Two Phase Electrolyte Electrolysis*, Ed. by N. L. Weinberg and B. V. J. Tilak, John Wiley & Sons, New York, 1982, Chapter 2.
[119] E. Steckhan, in *Volume 142 Topics in Current Chemistry, Electrochemistry*, Springer-Verlag, New York, 1987, Chapter 1.
[120] N. Ibl, K. Kramer, L. Ponto, and P. Robertson, *AIChE Symp. Ser.* **75** (1979) 45.
[121] G. Kreysa and H. Medin, *J. Appl. Electrochem.* **16** (1986) 753.
[122] C. Comninellis, E. Plattner, and P. Javet, *J. Appl. Electrochem.* **9** (1979) 753.
[123] H. Wendt and H. Schneider, *J. Appl. Electrochem.* **16** (1986) 134.
[124] T. Tzedakis and A. Savall, *Chem. Eng. Sci.* **9** (1991) 2269.
[125] R. P. Chandran and D. T. Chin, *Electrochim. Acta* **31** (1986) 39.
[126] A. R. Wright, Ph.D Thesis, The University of Newcastle-upon-Tyne, 1989.
[127] A. R. Wright and A. W. Wright, *Inst. Chem. Eng. Symp. Ser.* **105** (1987) 101.

2

Electrochemical Deposition and Dissolution of Alloys and Metal Composites—Fundamental Aspects

A. R. Despić and V. D. Jović

Center of Electrochemistry, Institute of Chemistry, Technology and Metallurgy, 11001 Belgrade, and Center for Multidisciplinary Studies, University of Belgrade, 11001 Belgrade, Yugoslavia

I. INTRODUCTION

It is general experience in materials science that alloys can exhibit qualities that are unobtainable with the parent metals. This is true of electroplated deposits as well. Thus, such properties as hardness, tensile strength, ductility, Young's modulus, density, corrosion resistance, solderability, wear resistance, and antifriction service may be enhanced. Also, special properties not exhibited by the parent metals can be obtained, such as high magnetic permeability or other desired magnetic and electrical properties, amorphous structure, etc. Alloy plates may be more suitable than the parent metals for subsequent electroplate overlays and conversion chemical treatments.

Some alloys are more easily obtained by electrodeposition than by metallurgical processes; this may be the case for instance, when the metals to be alloyed have large differences in melting temperatures or are difficult to mix in the liquid state. Electroplating is capable of codepositing metals into an alloy even when the alloy cannot be obtained metallurgically because the components do not mix at all with each other even in the molten state (e.g., Ag–Ni or Ag–Co).

Modern Aspects of Electrochemistry, Number 27, edited by Ralph E. White *et al.* Plenum Press, New York, 1995.

The fast-growing requirements of modern industry for materials with special qualities have given rise to increasing interest in alloy plating. Hence, all the above-mentioned characteristics of alloy plating make it an important element of industrial production, particularly in corrosion protection and in the modern electronics industry.

Electroplating of at least two alloys (brass and bronze) is as old as electroplating of single metals. It followed fairly soon after the discovery of the first relatively stable direct current sources (Daniel element in 1836). De Ruolz,[1] in 1842, was the first to deposit both alloys, shortly after the preparation of the first cyanide baths, which were essentially similar to the ones used to the present day in that they contained complexes of Cu, Zn, and Sn.

In the century and a half of electroplating practice, over 180 alloys involving 40 elements have been electrodeposited in the combinations shown in Fig. 1.[2] An excellent review of the achievements up to 1962 is found in the two-volume work of Brenner.[3] For practical work, a more recent (1980) book by Bondar, Grimina, and Pavlov,[4] containing recipes and references for about 1100 baths, is recommended.

The first attempts at a scientific approach to alloy plating and an understanding of the processes involved came rather late, with the work in 1905 of Spitzer,[5] who discussed the role of cathode potential in the deposition of brass. A more comprehensive attempt was that of Schlötter[6] in 1914. Yet, the field of alloy electroplating had to await developments through the better part of this century, in the understanding of electrochemical thermodynamics and kinetics, as well as in complexometry and some other fields, in order to obtain a sound scientific basis. Some useful attempts at applying modern theory to alloy deposition processes were made by Gorbunova and Polukarov,[7] Fedoteev *et al.*,[3] and Faust.[9] They remained, however, at a rather elementary level, obviously oriented to help practical electroplaters.

Inasmuch as the development of new plating processes has remained empirical up to the present, the ever increasing requirements for technical and economic optimization as well as for the design of high technologies yielding products of narrowly defined qualities, warrant a strict scientific approach to problems. It is possible to maintain that the present state of development of fundamental (theoretical) electrochemistry offers a sound (for the most part, mathematical) background for a detailed understanding of the alloy plating processes. It is the aim of this chapter to justify this statement.

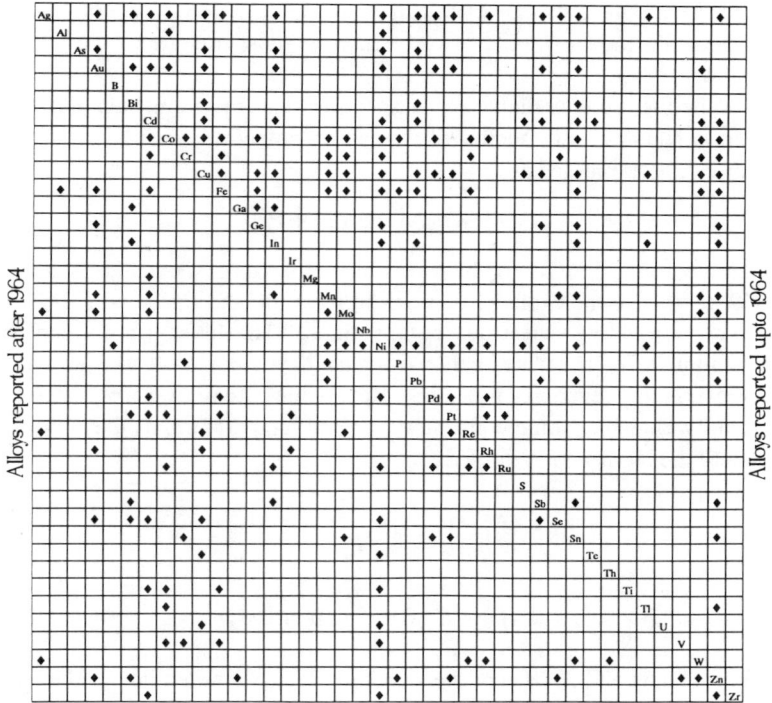

Figure 1. Binary electrodeposited alloys from aqueous solutions (represented by "diamonds") reported in the literature[2] up to 1974.

II. CONDITIONS FOR CODEPOSITION OF METALS BY AN ELECTROCHEMICAL PROCESS

Metals immersed in solutions of their simple salts are known to establish reversible electrode potentials over a wide range of values covering a span of about 3 V. The deposition of a metal can take place only when the electrode potential becomes negative (cathodic) with respect to its reversible potential. In a mixed-salt solution, a potential at which a metal on the positive side of the scale (e.g., Cu) normally deposits at a high rate ("deposition potential") is usually far too positive for a metal on the negative side of the scale (e.g., Zn) to deposit at any notable rate.

Deposition potentials, however, can be manipulated, and thus conditions created, for codeposition of two or more metals, in spite of the above statement.

The literature available so far specifies that for two metals to codeposit, a condition of equality of deposition potentials must be satisfied. This can be expressed as

$$E_r(A) + P(A) = E_r(B) + P(B) \tag{1}$$

where $E_r(A)$ and $E_r(B)$ are the reversible potentials of the metals A and B with respect to their ions present in the solution, and $P(A)$ and $P(B)$ are the corresponding "electrode polarizations." The latter are defined as potential differences between the actual deposition potential and the corresponding reversible potentials. Reversible potentials can be manipulated by changing ionic concentrations and temperature, as they should follow the Nernst equation; i.e.,

$$E_r(A) = E^{\ominus}(A) + \frac{RT}{pF} \ln a(A^{p+}) \tag{2}$$

$$E_r(B) = E^{\ominus}(B) + \frac{RT}{qF} \ln a(B^{q+}) \tag{3}$$

where $E^{\ominus}(A)$ and $E^{\ominus}(B)$ are the standard electrode potentials of the two metals, and p and q are the number of electrons needed for reduction of the two metal ions to the corresponding metals and are equal to the number of positive charges on the ions.

Obviously, the closer the reversible potentials of the two metals are made, the more similar the polarizations can be.

The condition given by Eq. (1), however, can serve only for a rough assessment of whether or not codeposition can occur, as the deposition potential, as such, is an undefined quantity unless the current density of deposition is specified. It could be taken to imply the requirement that the two partial currents of deposition j_A and j_B be equal, in which case an equimolar alloy of the two metals would be formed if $p = q$. Moreover, the standard potentials of the components in the alloy are, as a matter of principle, different from those of the pure metals.

A better definition of the condition of codeposition would be that the partial current densities of deposition are in the same range. More precisely, if, for example, a binary alloy of such a composition that the molar ratio of the more noble component is x and that of the less noble metal is $(1 - x)$ is desired, then assuming that Faraday's law is obeyed,

$$x = \frac{n_A}{n_A + n_B} = \frac{j_A/p}{j_A/p + j_B/q} \tag{4}$$

and

$$(1 - x) = \frac{n_B}{n_A + n_B} = \frac{j_B/q}{j_A/p + j_B/q} \tag{5}$$

so that

$$\frac{j_A}{j_B} = \frac{p}{q} \frac{x}{(1 - x)} \tag{6}$$

The desired current density ratio can be achieved by properly adjusting three essential variables: the concentration of the depositing ions at the electrode/solution interface, where the discharge takes place, the electrode potential, and the temperature. This will be discussed in detail in subsequent sections.

III. EQUILIBRIUM (REVERSIBLE) POTENTIAL OF AN ALLOY IN CONTACT WITH A SOLUTION

When an alloy composed of two elements, A and B, is immersed in a solution containing ions corresponding to its metal constituents, it takes up a potential, termed the "static potential" in the electroplating literature. It is of interest to establish the conditions under which no net process would take place, so that this potential could be considered as the equilibrium (reversible) potential of the alloy. One should also bear in mind that an alloy may (and very often does) consist of several different phases. As each phase may have different thermodynamic properties, it should be expected to have a different reversible potential as well.

This is essentially a nonequilibrium situation. Formation of nonequilibrium phases was interpreted by Fischer[10] and later by Banerjee[11] as resulting from the state of metal adatoms at the surface, resembling a high-temperature melt. Different phases may arise from different local compositions of that "melt." Nevertheless, the thermodynamic properties of each phase, and hence the problem of its reversible potential, can be treated as though it were the only phase present in a given situation. (The problem of stability of a phase in the presence of other phases will be discussed later.) On the positive side of its reversible potential, the phase should dissolve, while on the negative side, under proper conditions, it should continue to grow.

The problem of the reversible potential of alloys has been treated in the literature[12-14] in terms of the Nernst equation applied to the "less noble" metal (that having a more negative standard potential), the "more noble" component of the alloy being considered as an inert matrix. This, however, is not an appropriate approach, as is reflected in the fact that it yields unrealistic results. Thus, if an alloy containing a divalent less noble metal exhibits a reversible potential more positive than its standard potential by, for example, 0.6 V, the activity of the metal in the alloy, as calculated from the Nernst equation, should be of the order of 10^{-20}, which has no physical meaning.

In a proper thermodynamic approach, one could not assign a thermodynamic property to an individual component of the alloy. Instead, a phase should be treated as composed of a chemical entity of stoichiometric composition corresponding to the composition of the alloy. Thus, a phase can be described as $A_xB_{(1-x)}$, the formation of one mole of the substance characterizing the phase being represented by the following chemical reaction:

$$xA + (1-x)B \Leftrightarrow A_xB_{(1-x)} \tag{7}$$

1. Effect of Gibbs Energy of Phase Formation

The Gibbs energy change in the formation of the phase can be described in terms of the standard partial molar Gibbs energies (or standard chemical potentials) as

$$\Delta_f G^{\ominus}(A_xB_{(1-x)}) = \mu^{\ominus}(A_xB_{(1-x)}) - x\mu^{\ominus}(A) - (1-x)\mu^{\ominus}(B) \tag{8}$$

The entity $A_xB_{(1-x)}$ can also be formed in an electrochemical cell from the ions of both metals, A^{p+} and B^{q+}, present in solution, the cell being composed of an electrode made of the alloy phase as cathode and a standard hydrogen electrode as anode. The cell reaction is then

$$xA^{p+} + (1-x)B^{q+} + \left[\frac{xp + (1-x)q}{2}\right]H_2 = A_xB_{(1-x)} + [xp + (1-x)q]H^+ \tag{9}$$

The standard Gibbs energy change in this reaction should be

$$\Delta G^{\ominus}_{\text{cell}} = \mu^{\ominus}(A_xB_{(1-x)}) + [xp + (1-x)q]\mu^{\ominus}(H^+)$$
$$- x\mu^{\ominus}(A^{p+}) - (1-x)\mu^{\ominus}(B^{q+}) - \left[\frac{xp + (1-x)q}{2}\right]\mu^{\ominus}(H_2) \quad (10)$$

By convention,

$$[xp + (1-x)q]\mu^{\ominus}(H^+) - \left[\frac{xp + (1-x)q}{2}\right]\mu^{\ominus}(H_2) = 0 \quad (11)$$

so that the standard Gibbs energy change in the cell reaction is

$$\Delta G^{\ominus}_{\text{cell}} = \mu^{\ominus}(A_xB_{(1-x)}) - x\mu^{\ominus}(A^{p+}) - (1-x)\mu^{\ominus}(B^{q+}) \quad (12)$$

The electromotive force of this cell, which in this case is identical with the electrode potential of the alloy phase on the standard hydrogen scale, is given by

$$E(A_xB_{(1-x)}) = E^{\ominus}(A_xB_{(1-x)}) + \frac{RT}{[xp + (1-x)q]F} \ln a(A^{p+})^x a(B^{q+})^{(1-x)} \quad (13)$$

where $a(A^{p+})$ and $a(B^{q+})$ are the activities of the corresponding ions in solution. The standard electrode potential of the alloy phase, $E^{\ominus}(A_xB_{(1-x)})$, is related to the standard Gibbs energy change in the cell according to

$$E^{\ominus}(A_xB_{(1-x)}) = \frac{-\Delta G^{\ominus}_{\text{cell}}}{[xp + (1-x)q]F} = \frac{-\mu^{\ominus}(A_xB_{(1-x)}) + x\mu^{\ominus}(A^{p+}) + (1-x)\mu^{\ominus}(B^{q+})}{[xp + (1-x)q]F} \quad (14)$$

as $[xp + (1-x)q]$ is the total number of electrons exchanged in one act of the cell reaction (Eq. 9).

The standard Gibbs energies of formation of the ions relative to that of the hydrogen ion (taken as zero), which are equal to the standard chemical potentials of the ions, are related to the standard potentials of the corresponding metals on the standard hydrogen scale as

$$\Delta_f G^{\ominus}(A^{p+}) = \mu^{\ominus}(A^{p+}) = pFE^{\ominus}(A^{p+}/A) \quad (15)$$

and

$$\Delta_f G^{\ominus}(B^{q+}) = \mu^{\ominus}(B^{q+}) = qFE^{\ominus}(B^{q+}/B) \quad (16)$$

Substituting Eqs. (15) and (16) into Eq. (14) and then substituting the resulting $E^{\ominus}(A_xB_{(1-x)})$ into Eq. (13), one obtains the reversible potential of the alloy phase as

$$E(A_xB_{(1-x)}) = \frac{xp}{[xp+(1-x)q]}\left[E^{\ominus}(A^{p+}/A) + \frac{RT}{pF}\ln a(A^{p+})\right]$$

$$+ \frac{(1-x)q}{[xp+(1-x)q]}\left[E^{\ominus}(B^{q+}/B) + \frac{RT}{qF}\ln a(B^{q+})\right] - \frac{\mu^{\ominus}(A_xB_{(1-x)})}{[xp+(1-x)q]F} \quad (17)$$

where $\mu^{\ominus}(A_xB_{(1-x)}) = \Delta_f G^{\ominus}(A_xB_{(1-x)}) - x\mu^{\ominus}(A) - (1-x)\mu^{\ominus}(B)$ is the standard chemical potential of the alloy phase relative to those of the metal constituents. The terms in brackets correspond to the potentials of the pure metal constituents in solutions of their ions.

It is seen that (a) the reversible potential of an alloy depends on the activities of ions of both metals in solution and (b) the dominant role in determining the position of the standard potential of the alloy relative to those of the pure metals is played by the standard molar Gibbs energy of formation of the alloy.

An example can be taken of an alloy phase in which the constituents are in a 1:1 molar ratio and are obtained by reduction of divalent metal ions. Thus, $x = (1-x) = 0.5$ and $p = q = 2$, so that

$$E^{\ominus}(AB) = \frac{E^{\ominus}(A^{p+}/A) + E^{\ominus}(B^{q+}/B)}{2} - \frac{\mu^{\ominus}(A_{0.5}B_{0.5})}{2F} \quad (18)$$

The first term is seen to represent a mean between the standard potentials of the two metals. Suppose that the standard potentials of the two metals are $E^{\ominus}(A^{p+}/A) = 0.3$ V and $E^{\ominus}(B^{q+}/B) = -0.7$ V. The mean potential is then -0.2 V. The solid line in Fig. 2 depicts the position of the standard potential of the alloy phase as a function of the energy of formation the phase, represented by $\mu^{\ominus}(AB)$.

It is seen that for any value of the Gibbs energy of alloying smaller than 100 kJ/mol, the standard potential of the alloy phase is more negative than the standard potential of the more noble component. Hence the statement found in the literature that the potential of an alloy is more noble than that of both constituents is correct only under special circumstances. The crossing point at which this becomes true depends on (a) the valency of the ions, which determines both the slope and the intercept of the $E^{\ominus}$ versus $\mu^{\ominus}$ dependence, and (b) the difference between the standard potentials of the constituents. The smaller the difference, the smaller the Gibbs energy of phase formation that is required for the crossing point to be reached. Thus, for $p = 2$ and $q = 1$ the intercept is shifted to the positive side by 0.17 V and the slope is increased to $(3F)^{-1}$ so that the crossing

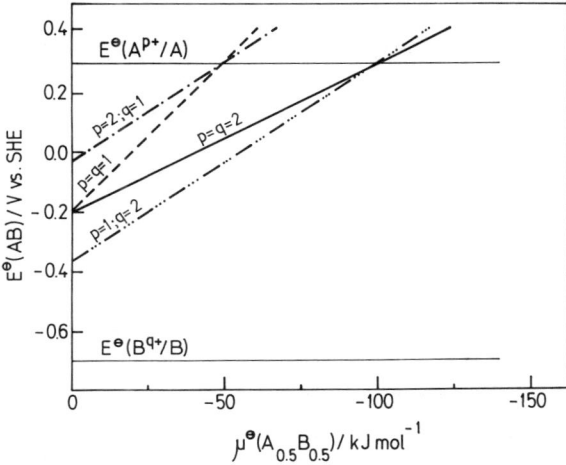

Figure 2. Standard potential of an alloy phase consisting of an equimolar ratio of metals A and B as a function of the energy of formation of the phase for different combinations of metal valencies.

point is shifted to –50 kJ/mol. This and other possible cases are also shown in Fig. 2.

2. Stability of an Alloy Phase in Solutions of the Corresponding Ions

When ions corresponding to metal constituents of an alloy phase are present in solution, it is necessary to take into consideration their tendency to make other possible alloy phases or undergo reduction to pure metals. Thus, if another phase or a pure metal would yield a more noble reversible potential, thermodynamic conditions are created for anodic dissolution of the existing phase ("replacement reaction"). Hence, the phase must be considered unstable under those conditions, tending to undergo corrosive degradation. It can be shown that the instability depends on the activity of the metal ions in solution.

Equation (13) defines the reversible potential of any alloy phase, or, for that matter, also of the pure more noble metal ($x = 1$). Assuming a constant activity of the ions of the less noble constituent [e.g., $a(B^{q+}) = 1$], it is seen (Fig. 3) that the slope of a plot of E_r versus $\ln a(A^{p+})$ is always larger for the pure more noble metal than for the alloy phase, as $\{x/[xp + (1-x)q]\} < 1/p$ for any $x < 1$.

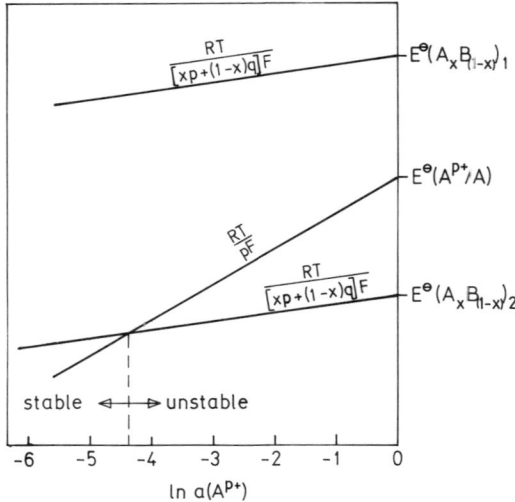

Figure 3. Dependencies of reversible potentials of alloys 1 and 2 as well as that of the pure more noble constituent on the activity of the corresponding ions in solution.

If $E^{\ominus}(A_xB_{(1-x)})_1$ of alloy 1 turns out to be more positive than that of any other alloy phase or of the pure metal, then, as seen in Fig. 3, the alloy will remain stable over the entire range of activities as it can never be turned into an anode by some other depositing phase. Yet, as seen in the same figure, if, for example, for alloy 2, $E^{\ominus}(A_xB_{(1-x)})_2 < E^{\ominus}(A^{p+}/A)$, there is a crossing point between the two functions. At larger ionic activities of A^{p+} than that at the crossing point, the metal would tend to precipitate on account of dissolution of the alloy phase. Conversely, if the activity is smaller than that at the crossing point, the phase would be stable (unless there is some pure metal phase to dissolve, resulting in precipitation of some additional amounts of the alloy phase).

In electrochemical investigations of alloys in electrolytic solutions, it is advisable not to introduce into the solution any ions of the more noble constituent but only those of the less noble one. In the first measurements of potentials of alloys in solutions done by N. A. Pushin as early as 1907,[15] this was understood, so that the electrolyte used contained ions of the electronegative component only and the electronegative metal served as a reference electrode. In such a case, the potential of the more noble constituent (or any of the more noble alloy phases) is theoretically $E(A^{p+}/A) = -\infty$. Hence, the alloy phase can never be more negative than the pure metal and act as an anode, and thus it will be stable.

This ideal situation cannot of course be maintained because an infinitely negative potential is tantamount to an infinite affinity for creating the more noble metal ions out of the metal present in the phase. Hence, some of the alloy must dissolve, on account of precipitation of the less noble metal. As the electrolyte does not contain those ions, they will keep diffusing away into the bulk of the solution. The situation is typical of the "floating potential" established whenever a metal is immersed into solutions not containing corresponding ions.[16] Nevertheless, unless the volume of solution is infinitely large, minute (negligible) dissolved amounts of the more noble metal will bring the system to equilibrium.

3. Effect of Gibbs Energies of Formation in View of the Type of Alloy

All alloys can be classified as one of four different types:

(a) Eutectic alloys
(b) Solid solutions
(c) Intermediate phases
(d) Intermetallic compounds

Each type has distinct thermodynamic characteristics and, hence warrants a separate treatment.

(i) Eutectic Alloys

Many elements do not mix in the solid phase. Hence, they crystallize, either from a melt or upon cathodic reduction, as pure metal crystals. Such behavior is exhibited by, for example, binary mixtures of Sn and Pb, Cd and Zn, Sn and Zn, and Ag and Cu. In such a binary system, there is always a single composition (eutectic point) that provides for simultaneous crystallization of both components, although in separate phases, at a defined temperature or current density of deposition.

Characteristic of a eutectic are extremely fine crystal grains, so that the alloy appears microscopically homogeneous. Even X-ray techniques cannot discern the phase structure as the grain size is beyond Bragg's limits so that no clear reflection pattern can be obtained. Nevertheless, the electrochemical technique of phase identification using anodic linear sweep voltammetry [cf. Section VII.2(i)] leaves no doubt that such alloys are composed of two pure metal phases.

Each of the phases establishes its own equilibrium with the corresponding ions in solution. As there is no chemical energy of mixing, which could change the standard potentials of each phase from the values of the standard potentials of the corresponding bulk metals, one could expect the latter to be relevant for the two phases in a eutectic alloy as well.

One should note, however, that there is a physical energy of dispersion of the metals into such small grains as exist in the eutectic. This is expressed as the surface energy that must be introduced into the system to create such a dispersion. This surface energy raises the Gibbs energy of each phase compared to that of the bulk metals and, hence, must shift the corresponding standard potential into the negative. Hence, a standard potential of a bulk metal can be taken only as an approximation to the standard potential of a eutectic alloy constituent.

The phenomenon is similar to that of small liquid droplets having a higher vapor pressure than large ones, which makes them less stable and causes their disappearance on account of the growth of larger droplets. This is known as the Kelvin effect, and the Gibbs energy change due to this effect is given by

$$\Delta G_{\text{cell}}^{\ominus} = \frac{2\sigma V}{RTr} \tag{19}$$

where σ is the surface tension, V is the molar volume, and r is the radius of the metal grains.

Hence, the standard potential of, for example, the metal A in the eutectic should be

$$E^{\ominus}(A^{p+}/A)_{\text{eut.}} = E^{\ominus}(A^{p+}/A) - \frac{2\sigma_A V_A}{pFRTr_A} \tag{20}$$

the same equation being applicable to the metal B. For a typical range of σ values characteristic of solid metals, one can calculate that deviations of the standard potential from that of the bulk phase are in the microvolt range and hence can be neglected for all practical purposes.

The eutectic alloy is the only type of two-phase alloy that can attain a true thermodynamic equilibrium with the solution. In general, if an alloy consisted of two or more phases, each containing, for example, the metal

A, the latter would have a different chemical potential in each phase. Hence, if the solution comes to equilibrium with one phase, it cannot be at equilibrium with the other one(s). However, in the case of a eutectic alloy, equilibrium can be attained as each of the metals is present in one phase only.

Still, this can only be achieved at a single value of the activity of the more noble metal ions in solution relative to that of the less noble metal. This can be found from the condition that the reversible potentials of the two phases be equal as a prerequisite for reaching equilibrium. Thus,

$$E^{\ominus}(A^{p+}/A)_{\text{eut.}} + \frac{RT}{pF} \ln a(A^{p+}) = E^{\ominus}(B^{q+}/B)_{\text{eut.}} + \frac{RT}{qF} \ln a(B^{q+}/B) \quad (21)$$

so that

$$\ln a(A^{p+})_{\text{eq.}} = \frac{[E^{\ominus}(B^{q+}/B)_{\text{eut.}} - E^{\ominus}(A^{p+}/A)_{\text{eut.}}]pF}{RT} + \frac{p}{q} \ln a(B^{q+}) \quad (22)$$

At all conditions of metal ion activities different from those specified by Eq. (22), the alloy is unstable and one phase must dissolve on account of growth of the other phase.

It is noteworthy that in the processes of deposition of eutectic-type alloys carried out in practice, some relatively coarse grains of one of the metals (most often the less noble one) appear. It is stated in the literature that this occurs in the high-current-density region. This phenomenon, however, should not be related to current density. One must take into account that in any eutectic binary system, as pointed out above, the eutectic point is a strictly defined point in terms of composition. It is also known that eutectic alloys obtained by a metallurgical process have embedded in them coarse crystals of the metal constituent that is in excess over the content defined by the composition of the eutectic. Hence, a pure eutectic can be obtained by electrodeposition only at that potential at which the partial currents of deposition of the two components obey Eq. (6), x pertaining to the eutectic composition.

In the above considerations, it is assumed that the two phases indeed consist of pure metal constituents of the alloy. However, especially in electrodeposition processes and despite the thermodynamic nature of the system, it often happens that each phase incorporates a certain amount of the other metal, to form a "supersaturated solid solution" [cf. Section III.3(ii)]. This is found, for example, in the deposition of such eutectic alloys as Cu–Pb, Cu–Tl, Cu–Sn, and Cu–Cd.[17] In the eutectic alloy Ag–Pb, up to 8% of Pb is found in Ag, leading to significant expansion of the crystal lattice.[18–20] Skirstymonskya[21] found a linear dependence of the lattice parameters of Ag on Pb content up to 7% Pb. It is obvious that in such an expanded lattice Ag cannot have the same standard chemical potential as in pure Ag. This must be reflected in the equilibrium potential of that phase immersed into solution.

(ii) Solid Solutions

At the other extreme from eutectic alloys are systems in which the components are miscible in the solid at the atomic level over the entire range of compositions. Many systems of practical interest, such as Ni–Co, Ni–Cu, Fe–Co, and Fe–Ni, are of this type, termed "solid solutions."

One could define an ideal solid solution as one in which the interatomic forces between unlike atoms are the same as those between the like atoms. In such a case no heat should evolve upon mixing. Thus, the standard enthalpy change, $\Delta H^\ominus$, is zero, and the Gibbs energy change upon the formation of such a solid solution, that is, the standard chemical potential of the alloy phase, is determined entirely by the entropy of mixing. The latter is known to be dependent on the content of the components in the alloy and temperature only. Thus,

$$\mu^\ominus(A_x B_{(1-x)}) = -RT[x \ln x + (1-x) \ln (1-x)] \qquad (23)$$

The largest entropy effect is obtained at $x = 0.5$. For $T = 298$ K, it amounts to 1717 J/mol. For an alloy phase composed of monovalent metals, this amounts to an effect on the standard potential of about 18 mV, while for one composed of divalent metals, the effect is half this value.

One could thus conclude that the entropy effect is a small one indeed and that for an ideal solid solution there is a virtual additivity of the standard potentials of the metal constituents, which is indicated by Eq. (17) with $\mu^\ominus(A_x B_{(1-x)}) \to 0$.

A phenomenon very often encountered in electrochemically deposited alloys is that of supersaturated solid solutions. This is, of course, not expected in systems that give a continuous series of solid solutions over the entire composition range (complete miscibility), although in some electrodeposited Cu–Au and Cu–Ni alloys, two phases are found. It occurs in systems characterized by limited miscibility or immiscibility of the components in the solid phase, such as the Ag–Pb alloy mentioned in the previous section, but also in many other systems. Thus, in Sn–Cu deposits, Lainer[22] found a solid-solution phase supersaturated in Cu. Raub and Engel found supersaturated solid solutions in electrodeposited Cu–Pb and Ag–Bi[23] whereas Raub and Sautter[24] found such phases in Ag–Tl alloys.

Limited miscibility is found in systems in which the differences in the atomic radii of the components make it difficult to build a stable crystal lattice. Still, in the cathodic reduction of ions, the atoms are thrown upon the entire surface at a rate imposed by the current density, which may be too high for the atoms to have time to diffuse across the surface to a site where they could be incorporated into a corresponding stable (equilibrium) lattice. A portion of them then get trapped on the spot by a growing phase, creating distortions and other type of faults in its crystal lattice, which otherwise has no affinity to accommodate them. This phase thus becomes a supersaturated solid solution or, else, an intermediate phase [cf. Section III.3(*iii*)] that is not predicted by the phase diagram for that particular composition. One should note that, in eutectic systems, partial dissolution of one component into the phase of the other component to form a supersaturated solution in electrodeposited alloys is the rule rather than the exception. The distorted lattice of the supersaturated solid solution possesses a higher chemical potential than that of the corresponding equilibrium solid solution. The value of the distortion energy can be evaluated by the method developed by Pines.[25] Hence, such a phase is less noble, that is, has a more negative standard potential, than the equilibrium phase.

(iii) Intermediate Phases

A large number of systems form a solid-solution type of lattice within a limited range of compositions. When the composition is altered to fall outside that range, a new type of lattice can accommodate atoms of both components to form a new equilibrium phase. Thus, $\alpha, \beta, \gamma, \delta, \eta$, etc., phases are characteristic of equilibria in different systems over the entire range of compositions.

It is noteworthy that, according to Hume-Rotery and Raynor,[26] phase transitions are connected with electron concentration. This was supported by Jones[27] and Mott,[28] who related change of phase energy to gradual filling of Brillouin zones. The zone theory predicts limiting concentrations of components in any phase. Thus, for an α phase to exist in Cu or Ag alloys, the content of divalent metals (Zn, Cd) should not exceed 38% that of trivalent metals should not exceed 19%, and that of tetravalent metals

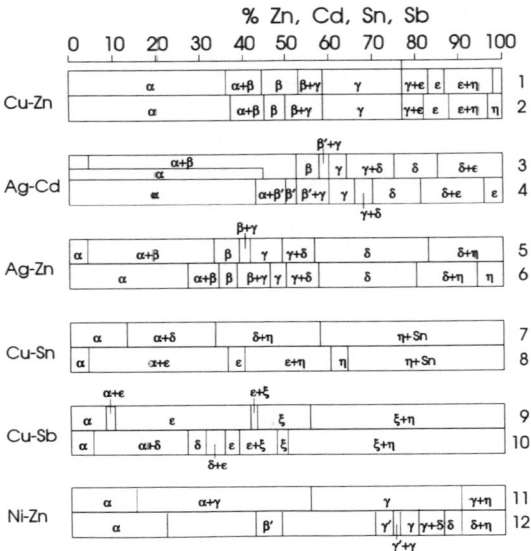

Figure 4. Distribution of phases in electrodeposited and recrystallized alloys. 1, 3, 5, 7, 9, and 11, Electrodeposited alloys; 2, 10, and 12, after recrystallization at 400°C; 4, after recrystallization at 200°C; 6, after recrystallization at 280°C; 8, distribution of phases according to the phase diagram of the alloy at 200°C.

should not exceed 13%. Volume effects (lattice distortions) can make these solubilities even smaller, but never larger, than predicted by the zone theory. Figure 4 shows phase fields at ambient temperature of different alloys forming intermediate phases. The bottom rows represent phase distributions in thermally equilibrated (recrystallized) alloys, while the top rows give phase distributions in alloys obtained by electrodeposition. It is seen that some electrodeposited alloys, such as Cu–Zn, follow rather closely the equilibrium diagram. Others, however, in some composition ranges crystallize into structures that are unexpected on the basis of phase diagrams and hence must be considered as nonequilibrium phases. From the point of view of the standard potentials of these alloys, one could maintain that the situation is similar to those of saturated and supersaturated solid solutions. In electrodeposited alloys that have phase fields comparable to the equilibrium ones, one does not expect an excess Gibbs energy in the lattice. Hence, one could expect additivity of standard potentials of the alloy constituents in terms of Eq. (17) with negligible deviations caused by $\mu^{\ominus}(A_xB_{(1-x)})$. Conversely, shifts of standard potentials in the negative direction should be expected in systems in which the intermediate phases found in experiments do not correspond to those required by the phase diagram.

(iv) Intermetallic Compounds

A number of systems tend to exhibit one or several intermetallic compounds of well-defined stoichiometry. Thus, for example, Cu–Cd forms a whole series of compounds in a sequence of increasing Cd content: Cu_2Cd, Cu_4Cd_3, Cu_5Cd_8, and $CuCd_3$.

The standard equilibrium potential of each of these phases is affected by the chemical energy of formation of the compound, and hence significant deviations from additivity of the standard potentials can be expected. A good summary of thermochemical data (heats of formation) exists in the literature.[29]

Characteristic of electrodeposition is that, depending on the current density, several intermetallic compounds can deposit simultaneously. Obviously, they cannot all be at equilibrium with the solution. Hence, the more noble phases would be expected to grow on account of dissolution of the less noble ones.

IV. CATHODIC PROCESSES IN ELECTRODEPOSITION OF ALLOYS

In the process of electrodeposition of any two (or more) metals forming an alloy, a route consisting of several steps is followed. These are:

(a) Transport of the depositing material from the bulk of the solution to the surface of the cathode (or, more precisely, to the outer Helmholtz plane of the electrical double layer)
(b) Possible transformation of the dominant species carrying the metal ion into an electroactive one undergoing discharge (electron transfer)
(c) Possible adsorption of the dominant or the electroactive (or both) species at the surface (or more precisely, at the inner Helmholtz plane)

(d) Discharge with the transfer of the metal ion/atom across the total or the inner part of the double layer
(e) Surface diffusion of adatoms to the point of incorporation into the lattice of a certain alloy phase
(f) Incorporation into the lattice
(g) Participation in nucleation leading to formation of a new grain
(h) Entrapment into a noncorresponding (nonequilibrium) grain

Step (d), i.e., discharge of polyvalent ions is usually a complex reaction consisting of several electron transfer steps with possible chemical transformation in between.

The rate constants in the rate equations for each of these steps may vary over a wide range of values. If one of them is smaller by at least one order of magnitude than those of all other steps (which most often is the case), that step determines the rate of the overall reaction [the "rate-determining step" (rds)]. All the other steps can then be considered to be at equilibrium, implying that in each step the rates in the two possible directions are virtually equal. In most cases of steady-state metal deposition, it is the discharge step or the transport of the dominant species that is rate-determining.

All types of cathodic processes leading to codeposition of two metals to form an alloy were placed by Brenner[3] into one of five categories:

1. Regular codeposition
2. Irregular codeposition
3. Equilibrium codeposition
4. Anomalous codeposition
5. Induced codeposition

The basis for such a division is the relation between the composition of the deposit and the "metal ratio", that is, the ratio of the total concentrations of the two metals in solution in whatever ionic form they are ("stoichiometric concentrations"). Thus, regular codeposition is that in which the composition of the deposit corresponds to the metal ratio. Irregular codeposition represents the situation in which the more noble metal is obtained in a higher percentage and the less noble one in a lower percentage than would be indicated by the metal ratio, whereas anomalous codeposition corresponds to the reverse situation. Equilibrium codeposition and induced codeposition represent special cases to be discussed below. As it will be seen in the following sections, the application of fundamental electrochemical principles required a somewhat different order in this division.

1. Equilibrium Codeposition

Equilibrium codeposition implies a common reversible potential for both metal constituents so that a deviation of the potential in the negative direction would result in cathodic reduction of both metal ions. It was seen in Section III.1 that electrode potentials of the pure metals as well as that of the alloy can be manipulated by changing the activity of the discharging species in solution. However, to close the gap between the standard potentials of the depositing metals, it is necessary in most cases to make the concentration of simple salts (undergoing complete dissociation) of the more noble metal impractically low and of the less noble one impractically (or impossibly) high.

From the very first experiments in electroplating of alloys,[1] the basic way of overcoming this problem was found in complexation of the metal ions, which usually changes the activity of the resulting species in solution by many orders of magnitude, while keeping the total amount of one or the other metal in solution sufficiently high for a good supply of plating material to the cathode.

The activity of a simple metal ion in solution in the presence of a complexing ligand is determined by the equilibrium in the reaction of complex formation. Thus, for example, in the case of cyanide complexes of Ag, a series of reactions takes place:

$$Ag^+ + CN^- \Leftrightarrow AgCN \qquad (24)$$

$$AgCN + CN^- \Leftrightarrow Ag(CN)_2^- \qquad (25)$$

$$Ag(CN)_2^- + CN^- \Leftrightarrow Ag(CN)_3^{2-} \qquad (26)$$

and

$$Ag(CN)_3^{2-} + CN^- \Leftrightarrow Ag(CN)_4^{3-} \qquad (27)$$

The resulting equilibria can be described by the following set of equations:

$$\frac{a[AgCN]}{a[Ag^+]a[CN^-]} = K_1^{\ominus} \qquad (28)$$

$$\frac{a[\text{Ag(CN)}_2^-]}{a[\text{AgCN}]a[\text{CN}^-]} = K_2^{\ominus} \tag{29}$$

$$\frac{a[\text{Ag(CN)}_3^{2-}]}{a[\text{AgCN}_2^-]a[\text{CN}^-]} = K_3^{\ominus} \tag{30}$$

and

$$\frac{a[\text{Ag(CN)}_4^{3-}]}{a[\text{Ag(CN)}_3^{2-}]a[\text{CN}^-]} = K_4^{\ominus} \tag{31}$$

One can relate the activities of the species to their concentrations by introducing the corresponding activity coefficients. For a constant ionic strength of the solution, the activity coefficients can be considered constant, that is, independent of the concentrations of the species. Hence, a parallel set of equations can be written in terms of concentrations of the species involved, whereby the so-called formal stability constants, K_{1-4}, will have somewhat different values from the standard thermodynamic ones employed in Eqs. (28)–(31), as they contain the activity coefficients of all the species involved.

Material balance of total Ag^+ and total CN^- requires that the following equations are obeyed:

$$C^{\circ}(\text{Ag}^+) = C(\text{Ag}^+) + C[\text{AgCN}] + C[\text{Ag(CN)}_2^-]$$
$$+ C[\text{Ag(CN)}_3^{2-}] + C[\text{Ag(CN)}_4^{3-}] \tag{32}$$

$$C^{\circ}(\text{CN}^-) = C(\text{CN}^-) + C[\text{AgCN}] + 2C[\text{Ag(CN)}_2^-]$$
$$+ 3C[\text{Ag(CN)}_3^{2-}] + 4C[\text{Ag(CN)}_4^{3-}] \tag{33}$$

With known stability constants, which can be found in the literature,[30,31] and for given C° values, the six equations (28)–(33) can provide values for the concentrations of all the six species in solution. Figure 5 shows the calculated distribution of the species as a function of the total ligand concentration. It is seen that even at the lowest practical concentration of CN^-, free Ag^+ ions are virtually nonexistent. In the cyanide concentration range used in plating practice (2–3M, marked in the figure), the prevailing species should be the complex $Ag(CN)_3^{2-}$.

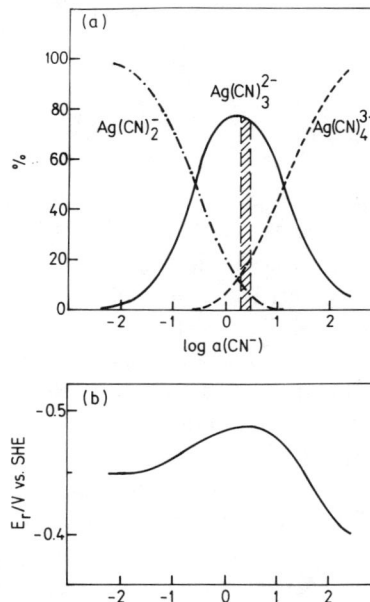

Figure 5. Calculated distribution of different cyanide complexes (a) and the reversible potential (b) of silver as functions of the concentration of cyanide in solution for $C°(Ag^+) = 0.001M$.

This situation can be generalized for any number of metal ions and ligands in solution. In such a case a system of equations must be set:

$$\frac{C(ML_{i+1})}{C(ML_i)C(L)} = K_i \tag{34}$$

where i can have any value between 0 and I and

$$C°(M) = \sum_{i=0}^{I} C(ML_i) \tag{35}$$

$$C°(L) = \sum_{i=1}^{I} v_i C(ML_i) + C(L) \tag{36}$$

where v_i is the number of ligands in the ith complex.

Such a system of 19 equations with 19 unknowns was solved, for example, for the pyrophosphate–oxalate bath used in brass plating in order to assess the dominant species when the concentrations of both metal ions and ligands are those used in practice.[32] With the ligands $P_2O_7^{4-}$, $C_2O_4^{2-}$, and OH^-, the likely species existing in the bath should be $CuP_2O_7^{2-}$, $Cu(P_2O_7)_2^{6-}$, CuC_2O_4, $Cu(C_2O_4)_2^{2-}$, $CuOH^+$, $Cu(OH)_2$, $Cu(OH)_3^-$, $Cu(OH)_4^{2-}$, $ZnP_2O_7^{2-}$, $Zn(P_2O_7)_2^{6-}$, ZnC_2O_4, $Zn(C_2O_4)_2^{2-}$, $Zn(C_2O_4)_3^{4-}$, $ZnOH^+$, $Zn(OH)_2$, $Zn(OH)_3^-$, and $Zn(OH)_4^{2-}$. It was found that in $0.123M$ pyrophosphate, $0.1M$ in oxalate at pH 10.3, representative of the bath used in practice, the dominant species should be $Cu(C_2O_4)_2^{2-}$ and $Zn(P_2O_7)_2^{6-}$.

The results of the above type of calculations should be taken as an approximate indication of the structure of a plating bath because of the limited applicability of stability constants found in the literature. They usually refer to ionic strengths (often equal to zero) different from those of the bath. Hence, shifts in abundance maxima in one or another direction should be expected in a real situation. Nevertheless, it is useful to have a rough picture of the solution composition before speculating about possible mechanisms of cathodic reduction.

Adjusting the equilibrium potentials of two metals to a common value for the equilibrium codeposition requires that a complexing agent be found whose complexes with the less noble metal ions are less stable than those with the more noble ones, to the extent that a well-defined relationship between activities of the two free metal cations of the type of Eq. (22) can be satisfied.

The extremely small activities of free metal cations that are most often obtained in this way could not possibly have any kinetic significance, as they could not sustain any practical current density. Hence, the reduction is quite likely to involve directly the dominant species or some other complex into which the prevailing one is transformed at a high rate. It is actually the total concentration $C^\circ(M)$ of each metal that is relevant for the supply of the depositing material to the electrode surface. In the usual case, one of the complexes prevails to the extent that all the other concentrations in Eq. (35) can be neglected. If the dominant species for the complexes of the metal ions A^{p+} and B^{q+} contain m and n ligands, respectively, the activities of the free metal ions in Eq. (22) can then be replaced by

Electrodeposition of Alloys and Metal Composites

$$a(A^{p+}) = \frac{a(AL_m)}{\left(\prod_{i=1}^{m} K_i\right) a(L)^m} \quad (37)$$

$$a(B^{q+}) = \frac{a(BL_n)}{\left(\prod_{j=1}^{n} K_j\right) a(L)^n} \quad (38)$$

The resulting relationship is

$$\ln a(AL_m) \approx \ln C^o(A) = \left[\frac{pF}{RT} E^\ominus(B^{q+}/B) - \frac{p}{q} \ln\left(\prod_{j=1}^{n} K_j\right)\right] - \left[\frac{pF}{RT} E^\ominus(A^{p+}/A) - \ln\left(\prod_{i=1}^{m} K_i\right)\right] - \frac{p}{q} \ln a(BL_n) - \left(\frac{m}{p} + \frac{n}{q}\right) \ln a(L) \quad (39)$$

The first two terms of Eq. (39) represent redefined standard potentials of the two metals, relating the standard potentials to the activities of the dominant complexes of the two metals in solution. In the case of a prevailing single complex of B, $a(BL_n) \cong C^o(B)$, and if a solution contains a large excess of L with respect to A^{p+} and B^{q+}, Eq. (39) yields a condition for the equality of equilibrium potentials of the two metals in terms of total amounts of metal ions and ligands in solution, as a prerequisite for equilibrium codeposition.

It is noteworthy that there is an inherent contradiction in the term "equilibrium codeposition." "Equilibrium" implies that no net deposition process takes place, whereas codeposition can occur only if, in the processes defining the equilibrium, the one occurring in the cathodic direction occurs somewhat faster than the opposite one resulting in anodic oxidation. Hence, the term should be understood as meaning that the codeposition occurs at very low polarization, that is, at a potential close to the common reversible potential.

Yet, because any shift of potential from the equilibrium one can have a different effect on the rates (partial currents) of deposition of the two metals, nothing can be said *a priori* about the relation between alloy composition and the metal ratio in solution. In fact, the situation is similar to that of "irregular deposition," which is described in the next section.

2. Activation-Controlled (Irregular) Codeposition

The previous case can ideally be achieved only in systems with infinite rate constants of all the processes involved in the mechanism of discharge and electrocrystallization. In reality, in any electrochemical system the relationship between the rate of deposition, represented by cathodic current density, and the independent variables by which the system can be manipulated (i.e., concentration of depositing species in solution, electrode potential, and temperature) would most often be described by the following rate equation, common to the kinetics of all electrode processes:

$$\frac{j}{j^\circ} = \frac{j_0}{j^\circ}\left[\exp\left(\frac{\alpha_a F}{RT}\eta\right) - \left(\frac{C}{C^\circ}\right)\exp\left(\frac{-\alpha_c F}{RT}\eta\right)\right] \quad (40)$$

where the overpotential η represents the deviation of the actual electrode potential from the equilibrium one:

$$\eta = E - E_r \quad (41)$$

C and C° represent the concentration of the discharging species at the surface of the electrode and in the bulk of the solution, respectively. The parameters characteristic of the depositing metal are j_0, the "exchange current density," related to the rate constant of the process, and the anodic and cathodic transfer coefficients α_a and α_c, related to the mechanism of discharge.[†]

Equation (40) is applicable to (a) systems in which the processes following discharge (surface diffusion of adatoms, incorporation into the existing lattice, or nucleation) are so fast compared to the discharge that they require no significant additional energy gradient, and (b) systems in which, in the possible sequence of discharge steps, one step has by far the lowest rate constant (the "rate-determining step") so that all other steps can be considered to be at equilibrium.

[†]Note that for a cathodic process both η and j acquire negative signs. Also note that the current densities are divided by (arbitrarily chosen) unit current density j° in order to obtain dimensionless values for further use.

The transfer coefficients are related to the mechanism of discharge[33] so that, for example, for a simple sequence of electron transfer or chemical steps,

$$\alpha_a = (1 - \beta)z_d + \frac{z''}{\mu_d} \qquad (42)$$

$$\alpha_c = \beta z_d + \frac{z'}{\mu_d} \qquad (43)$$

where β is the "symmetry factor," related to the energy barrier (in most cases having a value close to 0.5), z', z_d, and z'' are the number of electrons exchanged before, in, and after the rds, respectively, and μ_d is the stoichiometric number, denoting the number of times the rds must occur for one occurrence of the overall discharge process. Typical values for the rate-determining direct discharge of univalent ions from solutions are $\alpha_a = \alpha_c = 0.5$. For divalent ions, $\alpha_a = 1.5$ and $\alpha_c = 0.5$, or vice versa, depending on whether the first or the second step in a two-step discharge is rate-determining. If the discharge occurs in a single step, or if there is a rate-determining chemical reaction between two discharge steps (e.g., transformation of one complex species into another one more suitable for discharge in the second step), $\alpha_a = \alpha_c = 1$.

Numerous examples can be found in the literature on deposition of single metals. Thus, the deposition of Cu from simple salts occurs in two steps, the first one being rate-determining[34,35] so that $\alpha_a = 1.5$ and $\alpha_c = 0.5$. The discharge of Zn from zincate solutions[32] occurs in four steps, the third one being a rate-determining chemical transformation of a hydroxo complex Zn(I)X into Zn(I)Y, releasing some OH⁻ ions, so that $\alpha_a = \alpha_c = 1$.

In the case in which the discharge is sufficiently slow that the supply of depositing species to the electrode surface occurs without difficulty, the concentration C virtually does not deviate from C°, so that $C/C^\circ = 1$. Such a case is termed "activation-controlled" deposition, as the rate-determining factor is the activation energy of the discharge process.

At any cathodic overpotential larger than –40 mV, the first term in Eq. (40) becomes negligible, so that the equation can be transformed into a simpler one known as the Tafel equation; i.e.,

$$\eta = a' - b' \ln\left(-\frac{j}{j^\circ}\right) = a - b \log\left(-\frac{j}{j^\circ}\right) \qquad (44)$$

where the Tafel constant a' is

$$a' = \frac{RT}{\alpha_c F} \ln\left(\frac{j_0}{j^o}\right) \quad \text{or} \quad a = \frac{2.3RT}{\alpha_c F} \log\left(\frac{j_0}{j^o}\right) \qquad (45)$$

and the slope of the linear dependence obtained from a plot of η versus $\ln(-j)$ or $\log(-j)$ (Tafel slope) is

$$b' = \frac{RT}{\alpha_c F} \quad \text{or} \quad b = \frac{2.3RT}{\alpha_c F} \qquad (46)$$

The above reasoning applies equally and independently to both metals codepositing into an alloy, the current density j becoming partial current densities j_A and j_B and the total current density being $j = j_A + j_B$.

One should note here that the concept of overpotential, being strictly defined by Eq. (41), is related to the reversible potential of a pure metal in a given solution. In the case of codeposition of two metals into a phase of the type $A_x B_{(1-x)}$, this potential has no physical meaning and just represents an arbitrary point to which j_0 is related. For codeposition processes, it is practical to select a point that would be common to both depositing metals. Introducing Eq. (41) into Eq. (44) and rearranging, one obtains, for deposition of, for example, the metal A, a Tafel equation of the form

$$E = a_A - b_A \log\left(-\frac{j_A}{j^o}\right) \qquad (47)$$

where the new value of the Tafel constant is

$$\begin{aligned} a_A &= E_r(A^{p+}/A) + \frac{RT}{\alpha_c F} \ln\left[\frac{(j_0)_A}{j^o}\right] \\ &= E^\ominus(A^{p+}/A) + \frac{2.3RT}{pF} \log a(A^{p+}) + \frac{2.3RT}{(\alpha_c)_A F} \log\left[\frac{(j_0)_A}{j^o}\right] \end{aligned} \qquad (48)$$

Note that the Tafel constant is the main parameter of the deposition process, as it contains both the thermodynamic and the kinetic characteristics of the system. Its physical meaning is that of a potential at which the cathodic deposition runs at unit current density; i.e.,

$$a = E_{(-j/j^o)=1} \qquad (49)$$

As the choice of the unit current density $j°$ is arbitrary, for a proper definition of a it must always be stated. In what follows, $j° = 1$ mA/cm^2 will always be used as that unit. Also, the potential scale on which E is measured must be stated. One should also note here that at ambient temperature and with the α_c values cited above, the highest value of the Tafel slope b is 0.12 V/dec.

Although the above relationships are derived for deposition from simple salt solutions, they apply equally to cases in which complexation takes place, as described in Section IV.1. In such a case, the standard potential in Eq. (48) should be replaced by that in Eq. (39), and all the other parameters in Eq. (48) pertain to discharge of the dominant species in solution.

One should note here an interesting phenomenon in the potential region in which the discharge of the more noble metal prevails to the extent that virtually pure metal is deposited. In such a case, one would expect the rate of discharge to be independent of the presence of the ions of the less noble metal in solution, that is, to be the same as that in their absence. However, in some systems, for example, deposition of brass from the pyrophosphate–oxalate solution, addition of other metal species to the solution, at one and the same electrode potential, results in a significant increase of the rate of discharge of the first metal. Formally, the phenomenon could be viewed as catalysis. However, the underlying cause is the effect of the added species on the structure (distribution of complexes) of the solution. Thus, in the cited example the added species shifts the complexing equilibrium from prevailing CuL$_2$ to prevailing CuL. The latter exhibits a significantly higher electrochemical activity (primarily by lowering the Tafel slope, as discussed in Section IV.3). Hence, the phenomenon has been termed "pseudo-electrocatalysis."

Typical cases of activation-controlled codeposition of the metals A and B are shown schematically in Fig. 6, in which the polarization curves, that is, the current density versus potential relationships, are shown for the total current as well as for the two partial currents. The Tafel functions (Fig. 7) indicate linear relationships between the log of the partial current densities and the electrode potential. When this is the case, the total current density j cannot be a linear function in the region in which the two partial currents are comparable, as $\log (j_A + j_B) \neq \log j_A + \log j_B$. Conversely, it is in the region in which the total current does not give rise to a linear Tafel function that the alloy is deposited. When the Tafel function of the total current merges with one or the other partial current line, then one or the

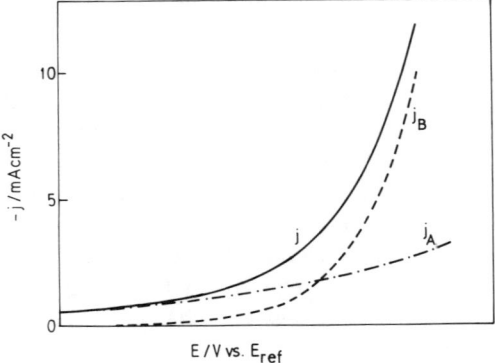

Figure 6. Polarization curve for an activation-controlled deposition of an alloy as a sum of the two partial current densities j_A and j_B.

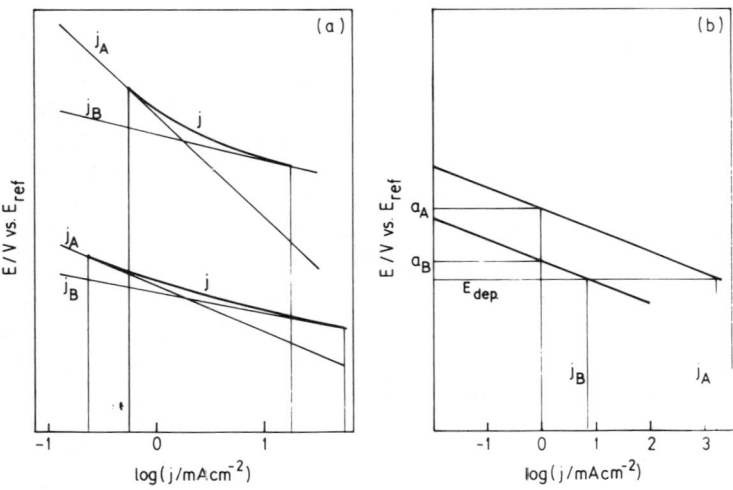

Figure 7. Tafel functions of the two partial current densities j_A and j_B and the total current density j for the case in which the slope b_A is larger than b_B (a) and for the case of equal Tafel slopes (b).

other metal is obtained virtually pure. One should note (Fig. 7a) that the span of the current density region in which metals are deposited in comparable amounts, varying from $x \to 1$ to $x \to 0$, depends on the difference between the Tafel slopes b_A and b_B. The larger the difference is, the narrower the current density range, and vice versa. As is seen in Fig. 7b, in the extreme case in which the Tafel slopes for the two deposition processes are equal, the difference between $\log j_A$ and $\log j_B$ remains constant, that is, independent of either the potential or the overall current density of deposition of the alloy, implying a constant ratio of the partial current densities of deposition of the two metals. In view of Eq. (6), this is tantamount to a constant overall composition of the alloy. (This, of course, does not mean a constant phase structure, as the latter depends on the actual values of the current density.) The actual composition in such a case depends on the difference between the a values for the two metals only.

The considerations given above explain cases of irregular codeposition in Brenner's terms, as the composition is seen to depend not only on the metal ratio [reflected in the activities $a(A^{p+})$ and $a(B^{q+})$] but also on other characteristics of the system given in Eq. (47).

Finally, there could be a case in which the Tafel slope of the more noble metal is smaller than that of the less noble metal. In such a case, the more noble component is likely to be deposited alone except for the special situation of anomalous deposition (cf. Section IV.5).

On the basis of the above derivations, one can analyze the factors determining the deviation of the alloy composition from the metal ratio in solution. Using Eq. (47), at any constant potential of codeposition,

$$a_A - b_A \log\left(-\frac{j_A}{j_0}\right) = a_B - b_B \log\left(-\frac{j_B}{j_0}\right) \qquad (50)$$

and, in view of Eq. (6), one can derive

$$\log \frac{x_A}{x_B} = \log\left(\frac{pj_A}{qj_B}\right) = \left[\left(\frac{a_A}{b_A} - \frac{a_B}{b_B}\right) - \left(\frac{b_B - b_A}{b_A b_B}\right) E\right] \frac{p}{q} \qquad (51)$$

Introducing the Tafel constants and Tafel slopes from Eqs. (48) and (46), rearranging, and returning to the linear coordinates, one obtains

$$\frac{x}{1-x} = \frac{x_A}{x_B} = \frac{(j_0)_A p a(A^{p+})^{(\alpha_c)_A/p}}{(j_0)_B q a(B^{q+})^{(\alpha_c)_B/q}}$$

$$\times \exp \frac{F}{RT}\left\{[(\alpha_c)_A E^\ominus(A^{p+}/A) - (\alpha_c)_B E^\ominus(B^{q+}/B)] - \frac{(\alpha_c)_A - (\alpha_c)_B}{(\alpha_c)_A^2 (\alpha_c)_B^2}\right\} \qquad (52)$$

It is seen that the composition of the alloy follows a complex dependence on the metal ratio, involving all the thermodynamic and kinetic parameters determining activation-controlled codeposition.

The Tafel lines shown in Fig. 7 are meant to pertain to the deposition of pure metals. Hence, when they are applied to the deposition of the alloy, there is an implicit assumption that alloying does not change the Tafel constants. In many cases, this assumption should be justified as, in such activation- (or transport-) controlled processes, the metal side of the double layer serves just as a source of electrons. The Fermi energy of metals is not very sensitive to their nature when they are as close as an alloy and its constituents. In such a case, the partial currents at a given potential can be considered similar to those of pure metals, and the additivity principle should be obeyed.

However, in practice, deviations from additivity have been recorded in certain systems, for example, some eutectic alloys. In explaining this phenomenon, Gorbunova and Polukarov[7] assumed that each metal ion deposits on grains of the corresponding metal only. Thus, the partial current density of each of the metals calculated on the basis of the geometric surface area must be equal to the true current density for this metal multiplied, by the fraction of the surface occupied by its grains. In this way, these authors arrived at the following expression for the partial current density, of, for example, the metal A, j'_A:

$$j'_A = \frac{j_A^2}{j_A + \left(\dfrac{\varepsilon_B \rho_A}{\varepsilon_A \rho_B}\right) j_B} \qquad (53)$$

where the ε's and the ρ's are the electrochemical equivalents and the densities of the metals, respectively.

Such an expression, however, can apply only to an extreme case in which surface diffusion of A across the grains of B and nucleation of new grains of A are strongly inhibited. Otherwise, and if the deposition is under complete activation control (i.e., if surface diffusion or nucleation is sufficiently fast), the entire metal surface acts as a homogeneous source of electrons, and adatoms can deposit at any place. In such a case, $j'_A = j_A$. If ions happen to be discharged at a grain of the other metal, the adatoms would then either have to diffuse across the surface to the nearest grain of the like metal or, if a sufficient degree of supersaturation by adatoms is reached, create a nucleus as a site of growth for a new grain.

Additivity should be achieved. This, of course, is again an ideal situation, and in any real situation the partial currents should be in between those for these two extreme cases.

One should also note that in a relatively dilute solution (e.g., $10^{-2}M$) one would not expect the ions of the metal A to affect directly the discharge of those of the metal B, and vice versa, or the corresponding adatoms to affect each other in surface diffusion or nucleation. One must bear in mind that, in such a case, discharge onto a flat surface is a relatively rare event, occurring simultaneously at distances of the order of 10^2 ionic diameters. Exceptions to this rule can be expected only if the surface is not homogeneous, that is, if discharge occurs only or predominantly at some limited number of sites. In concentrated solutions, however, such effects could be real.

Another reason for the absence of additivity could be differences in the adsorption of either additives or species participating in the deposition at the surface of alloy phases compared to adsorption on pure metals (cf. Section IV.3).

3. Effects of Adsorption on Irregular Deposition

A major change in both Tafel constants can arise if some neutral substance added to the solution (additive), or a species participating in the discharge, is adsorbed at the surface of the deposit.

(i) Adsorption of Additives

A number of agents are purposely added to plating baths to improve the quality of the deposit (leveling agents, brighteners, etc.) or affect some of its properties (e.g., hardness). They all act by being adsorbed at the growing deposit. Investigations have shown, for example, in the case of the addition of gelatin, that the presence of such agents inhibits the deposition process by decreasing the deposition current at a given potential or requiring an increase of cathodic polarization in order to maintain the current density of deposition.

A primary reason for such an effect of an additive can be found in its covering the entire surface or occupying surface sites that are the most active in the process of discharge. In covering the surface, the molecules of the additive drive away the water molecules and, hence, change the environment through which the electron exchange takes place. Thus, organic substances usually have a much lower dielectric permittivity than water molecules. This has a direct effect on the properties of the electrical double layer at the surface. Also, hydrated simple metal ions or complex species cannot approach the surface for electron transfer, as closely as in the absence of the adsorbed layer, or else, they must push their way through the adsorbed molecules to properly approach the electrode.

Examples of the presence of surfactant entirely changing the picture are found in the case of electrodeposition of Cu–Pb alloys.[37] In the absence of surfactants, virtually all the Pb crystallizes as a separate phase, as should be the case in that eutectic system. However, in the presence of disulfide formamidine (which is the product of the oxidation of thiourea added to the bath), all the Pb enters a supersaturated solid solution. Raub and Engel[23] found that Cu can thus accommodate up to 12% Pb, resulting in a sharp increase in density. A similar effect was found by Raub[39] upon addition of "Turkish red" to the bath for Ag–Cd codeposition. One could visualize that adsorption of these agents results in inhibition of surface diffusion, which prevents adatoms landing at the surface from wandering around searching for their own crystal lattice, so that they are incorporated more or less at the landing site.

These are the likely reasons for inhibited discharge in the presence of additives. The theory, however, did not advance to the level of treating these effects in quantitative terms.

(ii) Adsorption of Discharging Species

Complex species containing metal ions tend to adsorb at cathode surfaces when their ligands are surface-active substances. Contrary to the inhibitory effect of additives described above, this adsorption is likely to make the discharge easier, as the discharging species comes closer to the electrode surface, at the distance of the so-called inner Helmholtz plane (IHP). This occurs on account of the removal of the firmly attached first layer of water molecules at the surface. Thus, an increase in the value of the Tafel constant a is expected in the case of an activation-controlled process.

The complex species most often carries some negative charge as the ligands are mostly in anionic form and their number often overrides the number of positive charges carried by the metal cations. It is noteworthy that in such a case the Tafel slope, b, can be significantly affected. Indeed, the values of b larger than 0.12 V/dec that are often found are difficult to explain in any other way than by the effect of adsorption of the anionic form of the discharging species. This is based on the following reasoning. Negatively charged species must experience electrostatic repulsion from a negatively charged cathode, assuming, as is usually the case, that the electrode at the given cathodic potential is on the negative side of the potential of zero charge (pzc). The repulsion must be stronger, the larger the number of negative charges carried by the discharging species and the more negative the electrode potential. Thus, extra energy must be supplied to the system over the activation energy for a similar neutral species.

This can be expressed in mathematical form. Within the Helmholtz layer, it is generally assumed that there is a constant electric field, that is, a linear change of potential from the metal surface to the outer Helmholtz plane (OHP). As potentials are expressed on some scale relative to a reference electrode potential, the potential difference across the layer should be, to a first approximation,

$$\Delta E = E - E_{pzc} \tag{54}$$

where E_{pzc} is the potential of zero charge on the given potential scale. (The approximation consists in neglecting the unknown surface potential that makes for the difference between E_{pzc} and the true absolute zero of the electrode potential.) With m negative charges on each molecule of the species, the electrical energy needed to bring 1 mol of the species from the OHP to the IHP, at which the particles are adsorbed, against the field should be

$$\Delta G_{el} = -\gamma m F \Delta E \tag{55}$$

where γ represents the portion of the Helmholtz layer between the OHP and IHP.

The relevant concentration of the discharging species to be used in the rate equation (Eq. 40) is now its surface concentration, C_{ads}, resulting from adsorption. This can be related to the bulk concentration using an adsorption isotherm in the linear region (at low coverage) as

$$\frac{C_{ads}}{C^o} = \exp\frac{-(\Delta G_{ads}^\ominus - \Delta G_{el})}{RT} = \exp\frac{-(\Delta G_{ads}^\ominus + \gamma m F E_{pzc}) + \gamma m F E}{RT} \tag{56}$$

As the chemical free energy of adsorption is negative and so is E, the repulsive electrostatic energy works against adsorption, the more so the larger the E.

Introducing Eq. (56) into Eq. (40) and rearranging, one obtains

$$\frac{j}{j^\circ} = \frac{j_0}{j^\circ} \left\{ \exp\left(\frac{\alpha_a F}{RT}\eta\right) - \exp\left[\frac{\Delta G^\ominus_{ads} + \gamma mF(E_{pzc} + E_r)}{RT}\right] \exp\left[\frac{-(\alpha_c + \gamma m)F}{RT}\eta\right] \right\} \quad (57)$$

Following the course taken in transforming Eq. (40) into the Tafel equation (Eq. 47), one arrives at new values for the Tafel constant and the Tafel slope; i.e.,

$$a_A = \left(1 - \frac{\gamma mF}{RT}\right) E_r(A^{p+}/A) - \frac{\Delta G^\ominus_{ads} + \gamma mFE_{pzc}}{RT} + \frac{RT}{\alpha_c F} \ln\left[\frac{(j_0)_A}{j^\circ}\right] \quad (58)$$

and

$$b_A = \frac{RT}{2.3(\alpha_c - \gamma m)F} \quad (59)$$

It is seen that, because the γm value is subtracted from α_c, b_A is larger than with α_c alone and hence can take any value above 0.12 V/dec, depending on how deeply the charged particle penetrates into the Helmholtz layer.

An example of such an effect of adsorption can be found in Cu deposition from pyrophosphate solutions.[32] Cu forms a number of complexes with the pyrophosphate anion, the one prevailing in large excess of pyrophosphate being $Cu(P_2O_4)_2^{6-}$, carrying six negative charges ($m = 6$). In solutions containing pyrophosphate at a concentration comparable to that of Cu^{2+}, the mono ligand complex $CuP_2O_7^{2-}$ is dominant with $m = 2$. Indeed, a significantly larger slope of the Tafel line for Cu deposition is found in the first case than in the second case. The calculated γ values (assuming $\alpha_c = 0.5$) of 0.065 and 0.144, respectively, indicate that only a small portion of the Helmholtz layer is penetrated by either species, the one with the lower charge penetrating deeper, as expected on the basis of the above reasoning.

4. Transport-Controlled (Regular) Codeposition

Ionic material for deposition is transported from the bulk of solution to the electrode surface by molecular diffusion and electromigration. In the concentrated electrolytes normally used in practice, with large amounts of ions not participating in deposition, both the electric field in the electrolyte and the transport number of the depositing species are so small that the contribution of electromigration to the motion of the particles can be neglected.

Molecular diffusion requires a concentration gradient, implying that the concentration of the depositing species must be smaller at the electrode surface than in the bulk of the solution. However, as long as the rate of deposition is small, this difference can be neglected and the concentration factor in the second term of Eq. (40) can be considered as unity. At increasing current densities, however, an ever larger deviation takes place.

Under steady-state conditions of deposition, the diffusion is governed by Fick's first law, and it can be derived[36] that

$$\left(\frac{C}{C^\circ}\right) = \frac{j_D + j}{j_D} \tag{60}$$

where j_D is the diffusion limiting current, related to the concentration of the depositing species, C°, as

$$j_D = \frac{zFD}{\delta} C^\circ \tag{61}$$

where D is the diffusion coefficient and δ is the Nernst diffusion layer thickness.

Introducing Eq. (60) into Eq. (40) and rearranging, one obtains

$$j = \frac{j_0 \left[\exp\left(\frac{\alpha_a F}{RT}\eta\right) - \exp\left(-\frac{\alpha_c F}{RT}\eta\right)\right]}{1 + \frac{j_0}{j_D}\exp\left(-\frac{\alpha_c F}{RT}\eta\right)} \tag{62}$$

At increasing values of overpotential, the second term in the denominator becomes overwhelming, and the current density tends to a potential-independent diffusion limiting current; i.e.,

$$j \to -j_D \tag{63}$$

When the diffusion control is becoming overwhelming, a plot of E versus $\log[(-j_D/j) - 1]$ should be made instead of the regular Tafel plot. In such a case, the relationship should be linear with a slope of $(2.3RT/\alpha_c F)$.

In the codeposition of two metals, the ions of the more noble metal are usually at lower concentration in the bath than those of the less noble one. Hence, the diffusion limiting current of the former is smaller, and hence the condition of rate-limiting transport is more easily reached by

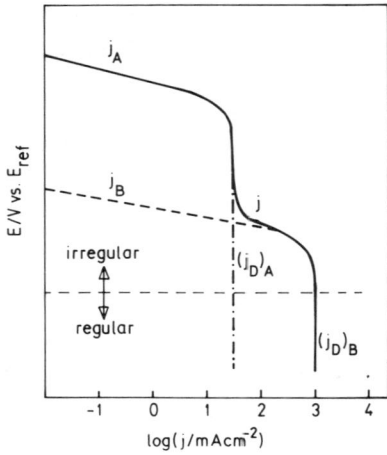

Figure 8. Tafel plots of partial current densities in the case of increasingly inhibited transport of electroactive species from solution. $(j_D)_A$ and $(j_D)_B$ are the diffusion limiting currents of species carrying metals A and B, respectively.

the partial current of deposition of the former than that of the latter. The situation is shown schematically in Fig. 8.

As the potential is driven negative, the partial current density j_A at one point becomes constant. As j_B continues to rise, the content of the metal B in the alloy continues to increase. It is at a potential where the limiting current of deposition of B is reached that the composition of the alloy becomes determined by the metal ratio only. Codeposition comes under total transport control and does not depend anymore on the mechanism and kinetics of discharge or any other step beyond it. From there on, "regular codeposition" in Brenner's terms takes place, and further increase of the cathodic potential does not change the composition of the alloy.

A typical example of the situation described above is found in the deposition of brass from the alkaline pyrophosphate–oxalate bath. The polarization curve obtained by linear sweep voltammetry, shown in Fig. 9, indicates that in the potential region more positive than –0.6 V, virtually pure Cu is deposited under transport control, with $j_D = 10$ mA/cm^2. At some more negative potential, Zn starts depositing, and past –1.2 V regular deposition of brass occurs, as $(j_D)_{Zn}$ is reached.

The diffusion limiting currents are well-defined entities when hydrodynamic conditions in the electrolyte are well defined and controlled. The latter depend on the manner and rate of stirring. The best defined limiting currents are obtained at a rotating disk electrode, as the diffusion layer

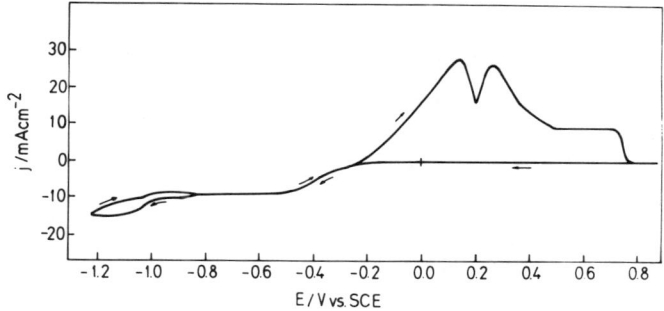

Figure 9. Linear sweep voltammogram for a pyrophosphate–oxalate bath of the composition 0.05M ZnSO$_4$ + 0.02M CuSO$_4$ + 0.12M Na$_4$P$_2$O$_7$ + 0.1M H$_2$C$_2$O$_4$ + 0.12M Na$_2$CO$_3$ + 0.09M H$_3$BO$_3$ + 0.13M NaOH.

thickness δ in Eq. (61) is very precisely determined by the Levich equation,

$$\delta = 1.612 D^{1/3} v^{1/6} \omega^{-1/2} \tag{64}$$

where v is the kinematic viscosity, and ω is the rotation speed.

In systems in which there are several species involving one or the other metal ion which can be transformed into each other by a chemical reaction (dissociation or association) and in which the dominant species is not the electroactive one undergoing discharge, the current–potential relationship is somewhat more complex (combined diffusion and reaction polarization). Such a case has been treated in detail in the literature.[40]

5. Anomalous Codeposition

Cases are met in practice[3] in which the nominally less noble metal deposits before the more noble metal as the potential is driven cathodic. Two situations can be envisaged to explain this phenomenon.

The deposition of the more noble metal can be so strongly inhibited by some substance that, as the potential is driven negative, no current is recorded at any practically measurable level. Hence, as shown in Fig. 10, at some potential the less noble metal starts depositing and causes the current to rise. At some even more negative potential, there may come a breakthrough in the deposition of the more noble metal. This may happen as a result of reduction or removal of the additive or by opening some

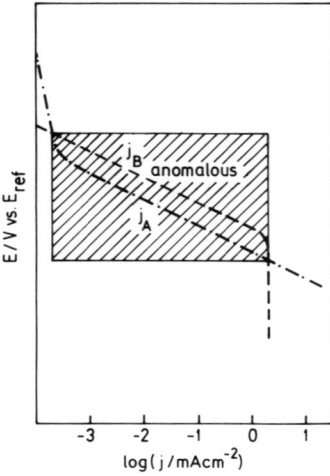

Figure 10. Tafel plots of partial currents j_A and j_B of deposition of metals A and B, respectively, yielding anomalous codeposition.

alternative mechanism of discharge not operative at less negative potentials.

A good example of such a case is found in the deposition of Ni–Zn alloys.[38] A schematic representation of the Zn content as a function of current density is given in Fig. 11. It is seen that normal codeposition occurs up to a current density j_t. In the lower current density range, the deposition is of the irregular type, while at j_t it falls into the category of regular codeposition. The current density j_t causes a sharp transition into abnormal deposition, in which the Zn content is much higher than indicated by the metal ratio in solution. This corresponds to entering the potential–current density region at the first crossing point of the Tafel diagram of Fig. 10. Beyond that, at increasing current density the composition remains virtually constant (Fig. 11, region II), implying equality of the Tafel slopes for j_A and j_B. The second crossing point arises when the limiting current for B is attained, whereupon the content of Zn starts falling (Fig. 11, region III).

Different explanations have been forwarded for such behavior, but the most likely one appears to be that suggested by Dahms and Croll[41] and supported by Higashi et al.[42] as well as by Horkans,[43] known as "the hydroxide suppression mechanism." According to this concept, coevolution of hydrogen, occurring at the negative potentials of deposition, causes an increase of pH at the electrode/solution interface. The latter produces

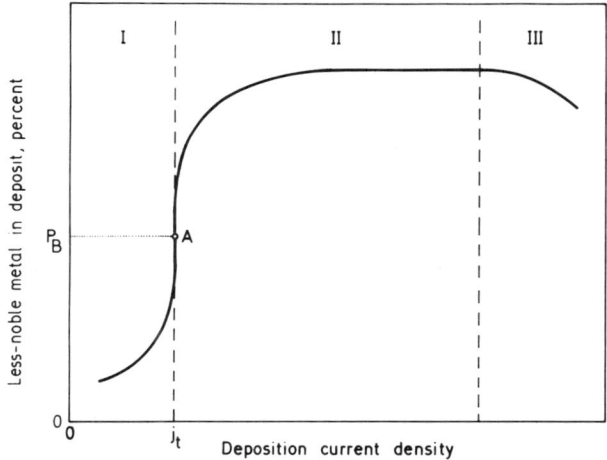

Figure 11. Schematic representation of the relationship between deposit composition and deposition current density for anomalous codeposition. p_B is the percentage of the less noble metal in the bath.[41]

hydrolysis of Zn species and precipitation of a layer of solid Zn hydroxide. The hydroxide provides a good supply of ionic Zn for discharge and deposition but suppresses the transport of the Ni species from the bulk of solution to the electrode surface.

One should have in mind also another possible cause. The complexing of the more noble metal can be so much stronger than that of the less noble one that the difference in the standard potentials is overcome and the reversible potential of the "more noble" metal (which it is not anymore) becomes more negative than that of the "less noble" one. Although this case may be considered trivial, deposition could again be taken as anomalous in Brenner's terms.

If this is the case and if, for some reason, the Tafel slope for the deposition of the "more noble" metal proves smaller than that for deposition of the less noble one, as shown in Fig. 12, with the potential going cathodic, j_B is larger than j_A only in a current density–potential region up to the crossing point of the two Tafel lines. At that point, again a breakthrough to normal deposition takes place.

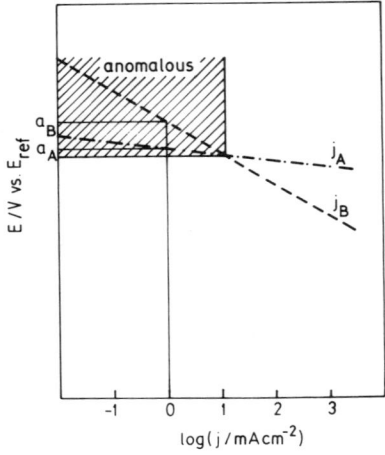

Figure 12. Tafel plots for deposition of metals A and B leading to anomalous codeposition in a certain potential and current density range (shaded area).

If either of the two cases is found to be operative in a system exhibiting anomalous codeposition, the equations derived in the previous sections can be used for a semiquantitative treatment of the problem.

6. Induced Codeposition

The ions of some elements, such as W, Ti, Mo, and Ge, cannot be electrochemically reduced to the metal in aqueous solution as they are usually involved in complex molecules that cannot approach the electrode sufficiently closely for a complete charge transfer. They are often seen to deposit, but the deposit consists usually of some lower valency species (mostly oxides or hydroxides) that often possess electronic conductivity. Hence, in such a case the electrons from a metal cathode pass through the deposit without further reduction that would reduce species in solution. Hence, inhibition of further reduction is of a kinetic rather than a thermodynamic nature.

In such systems, reduction to the metal state is, however, possible if an appropriate partner is found in solution, so that both codeposit to form a binary alloy. This type of reaction was termed induced codeposition by Brenner.[3] Good examples are found in codeposition of Mo, W, or Ge with Fe, Co, or Ni.

Each induced codeposition represents a special case. However, a feature common to all these cases is that the second partner acts as an intermediary, capable of a redox-type charge transfer well known in homogeneous catalysis.

V. LAMINAR ELECTRODEPOSITION OF METALS

1. Spontaneous Formation of a Layered Deposit

It has been observed in quite a few instances of electrodeposition of alloys that deposits acquire a laminated structure. Lamination was found to occur parallel to the cathode surface. Thus, Meyers and Phillips[44] found such a phenomenon in codeposition of Cu with other metals, Aotani[45] found it in codeposition of Fe and Ni, and Mikhalev[46] observed layered growth in Zn–Cd codeposition. The authors recorded fluctuations of potential occurring simultaneously with layer formation, ranging from only 0.1 mV up to 0.1 V.

Any fluctuations of periodic nature (oscillations) must be due to superposition of two counteracting processes. As an example, the following model can be offered, for example, for the deposition of a Cu–Pb layered deposit.

Suppose that, in the formation of a eutectic alloy, each metal must crystallize in a separate phase. There is a need of a continuous nucleation of new grains for the growth to be possible. In order for nucleation to become fast, a certain critical supersaturation of the surface by adatoms of the given metal is required. The probability of formation of a nucleus is known to increase suddenly by orders of magnitude at a certain degree of supersaturation. Hence, at the very first moments of deposition, the rate of deposition of Cu adatoms is sufficient for Cu nuclei to be formed and Cu grains to grow. However, as the diffusion layer becomes depleted of Cu^{2+} species below a certain critical concentration, the rate of formation of new nuclei drops dramatically, and deposition of Cu virtually comes to a halt. Under galvanostatic conditions, that is, at a constant deposition current, the discharge of Pb^{2+} species must compensate for the decrease of the Cu partial current. This requires a certain increase in cathodic overpotential. As nucleation of new Pb grains starts leading to the growth of a Pb layer, and as the latter covers most of the surface, the concentration of Cu^{2+} species at the surface recovers to the point where Cu nucleation can start again at the same time as Pb growth subsides because of depletion of Pb^{2+} species in the solution. Hence, the process starts all over again.

This, of course, is a possible mechanism giving a qualitative picture of the events. Proving this mechanism requires more detailed investigation, which so far has not been done.

2. Formation of Laminar Deposits by Pulsating Current

Instead of letting a laminar structure appear spontaneously during alloy deposition, there was significant motivation to obtain well-defined and controlled microlayered deposits. Experiments in another field—that of physical vapor deposition—indicated that a layered structure obtained by alternating deposition of two different metals can exhibit unique and valuable properties different from those of either the parent metals or their alloys. Thus, such structures expressed unusual flow stress. At a wavelength (sum of two individual layer thicknesses) of 2–3 nm, very high biaxial modulus could be obtained. Magnetic properties showed a marked deviation from bulk alloy behavior of the same metals.

As early as 1921, Blum[47] obtained microlayered deposit of Cu and Ni by electroplating, in a procedure consisting of alternating the conventional baths. Such a procedure, however, was impractical for obtaining a large number of thin layers in a relatively thick deposit. The possibility of obtaining microlayered structure from a single bath was first demonstrated by Brenner.[3] Among the early work, one should cite that of Cohen *et al.*,[48] who alternated the current density in order to obtain an Ag–Pd alloy with a periodically changing composition, so as to improve the wear performance of electrical contacts. Their work led to layers as thin as 30 nm. Pulsating current regimes were used first by Tench and White[49] and Ogden[50] to obtain microlayered Cu–Ni deposits and to show enhancement in tensile strength and microhardness, respectively, as the layer thickness decreased below 300 nm. A theoretical treatment of the variation of the composition of a deposit obtained by pulsating current from a bath containing two different metal cations was worked out by Verbrugge and Tobias.[51] Using a significantly simpler mathematical approach, Despić and co-workers, in 1987, published the first of a series of papers developing a quantitative theory of laminar deposition based on fundamental concepts of electrode kinetics.[52–54]

Thus, Despić and Jović showed[52] that a train of current pulses can produce, under special circumstances, a well-defined laminar deposit. The conditions to achieve this are

(a) that the deposition potentials of the two metals are sufficiently different at a given current density so that metal B virtually does not deposit during deposition of metal A until complete concentration polarization with respect to the ions of metal A takes place, and

(b) that, within the duration of the current pulse, Sand's equation for diffusional polarization is obeyed with respect to concentration change, resulting in transition to deposition of metal B after a well-defined transition time.

The first requirement is opposite to the condition for a codeposition of two metals to form a homogeneous alloy. The second essential condition is that the current density in a pulse is sufficiently high to cause, after a certain time of deposition of the more noble metal of the two, concentration polarization involving its ions to a potential level at which the less noble metal is deposited. This can be done by tailoring the current pulse to match the concentration of metal A in solution in such a way as to cause total concentration polarization (drop of its concentration at the surface to zero) at the moment, τ, when the desired thickness of the layer of metal A is reached. After that, metal B starts plating out, together with some metal A, into a second layer of complex composition. Based on Faraday's law, the thickness of the first layer of pure metal A that can be obtained is given by

$$d(\text{I}) = \left(\frac{M_A}{z_A \rho_A F}\right) j \tau_A \tag{65}$$

where M_A, ρ_A, and z_A are, respectively, the atomic weight, density, and charge on the ions of metal A, j is the current density in the pulse, and τ_A is the transition time with respect to the ions of metal A.

Provided the current density is constant and convection or migration of ions in the electrolyte is negligible, the transition time is given by Sand's equation[36]:

$$\tau_A = \frac{(z_A F)^2 \pi D_A}{4} \left(\frac{C_A^o}{j}\right)^2 \tag{66}$$

where C_A^o and D_A are, respectively, the bulk concentration and diffusion coefficient of the ions of metal A. Introducing Eq. (66) into Eq. (65) and rearranging, one obtains

$$\frac{(C_A^o)^2}{j} = \frac{4\rho_A}{z_A F \pi D_A M_A} d(\text{I}) \tag{67}$$

It is seen that to achieve a desired thickness of the layer of metal A, it is necessary to match precisely the ionic concentration and the current density in the pulse, depending on some inherent properties of metal A and its ions, so as to satisfy the derived relationship (Eq. 67).

There are, however, some factors imposed in practice that limit the maximum thickness achievable in this manner. Sand's equation is valid only as long as the changes in concentration occur within a stagnant layer of solution undisturbed by convection, that is, as long as the diffusion layer boundary δ stays within the hydrodynamic layer boundary Δh. The Nernst diffusion layer is known to expand with time according to

$$\delta = (\pi D_A t)^{1/2} \tag{68}$$

Hence, the largest transition time, τ_A^{max}, obtainable in the system is

$$\tau_A^{max} = \frac{(\Delta h)^2}{\pi D_A} \tag{69}$$

Thus, in view of Eq. (66), for a given concentration C_A^o, there is a minimum current density that must be employed for the transition to take place and the second layer to be deposited. This is given by

$$j^{min} = \frac{zFD_A\pi}{2\Delta h} C_A^o \tag{70}$$

and, from this, the maximum achievable thickness of the first layer is

$$d(I)^{max} = \frac{M_A C_A^o \Delta h}{\rho_A} \tag{71}$$

Taking the likely values of the parameters in Eqs. (69)–(71), it was calculated that τ_A^{max} could not be larger than 89 s so that a value for j^{min} of 76 mA/cm^2 must be used in order to achieve the maximum attainable value of $d(I)$ of about 3 μm.

Plating of the second layer occurs after τ_A is reached and the reversible potential of the metal B is overcome by polarization, as shown schematically in Fig. 13.

It is seen that metal A continues to deposit but at a decreasing partial current density j_A while the partial current of the metal B, j_B increases to make a constant overall current density. If the concentration of the ions of metal B in the solution is made much larger than that of the metal A (e.g.,

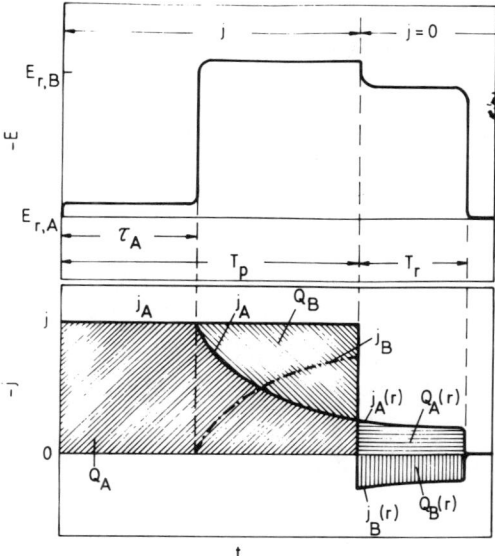

Figure 13. Schematic representation of the potential response and corresponding partial current changes during the deposition of the two layers of metals A and B by constant current density, j, and during the replacement reaction, after cutting off the current. j_A, Diffusion limiting current density of deposition of metal A after transition time τ_A; Q_A, charge for deposition of metal A during the pulse; j_B, current density of deposition of metal B; Q_B, charge for deposition of metal B during the pulse; $j_A(r)$, current density for deposition of metal A during the replacement reaction; $Q_A(r)$, corresponding charge of metal A; $j_B(r)$, current density for dissolution of metal B during the replacement reaction; $Q_B(r)$, corresponding charge of metal B.

2–3 orders of magnitude), the growth of the second layer could be virtually unlimited, as, to a first approximation (neglecting j_A after some time), it should follow the equation

$$d(\text{II}) = \frac{M_B j}{z_B \rho_B F}(t - \tau_A) \tag{72}$$

The composition of the second layer is not constant as the ratio between the two partial currents changes continuously. The content of metal A in any increment of the second layer is

$$x_A^{\partial d(\mathrm{II})} = \frac{z_B j_A}{z_A j + (z_B - z_A) j_A} \quad (73)$$

so that the integral content must be

$$x_A^{d(\mathrm{II})} = \frac{z_B \int_{\tau_A}^{t} j_A \, dt}{z_A j t + (z_B - z_A) \int_{\tau_A}^{t} j_A \, dt} \quad (74)$$

Using a known pattern of change of j_A with time, it was shown that, for a selected duration of the pulse T, the content of metal A in the second layer should be

$$x_A^{d(\mathrm{II})} = \frac{2 z_A z_B F C_A^o D_A^{-1/2} \pi^{-1/2} [(T - 0.595 \tau_A)^{1/2} - 0.636 \tau_A^{1/2}]}{z_A j (T - \tau_A) + 2(z_B - z_A) z_A F C_A^o D_A^{-1/2} \pi^{-1/2} [(T - 0.595 \tau_A)^{1/2} - 0.636 \tau_A^{1/2}]} \quad (75)$$

The equations are significantly simplified for $z_A = z_B$. Figure 14 shows the overall content of metal A in the second layer as a function of the thickness of the deposit. It is seen to be dependent on the concentration-to-current density ratio; for a sufficiently low value of that ratio, the second layer can consist of virtually pure metal B.

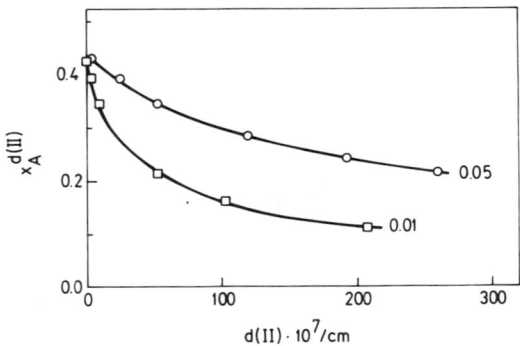

Figure 14. Dependence of calculated content of metal A in the second layer, $x_A^{d(\mathrm{II})}$, on the thickness of the second layer, $d(\mathrm{II})$, for different C_A^o/j ratios (0.01 and 0.05), at $z_A = z_B$.

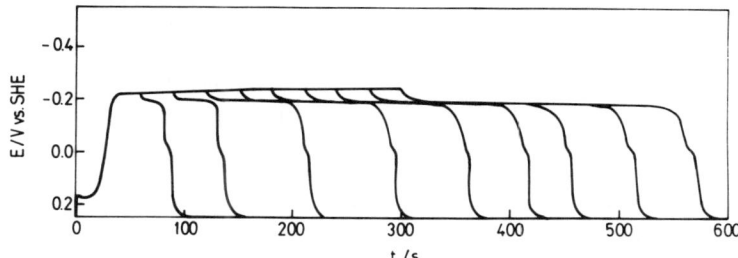

Figure 15. Potential responses to a series of pulses of constant current density, $j = 1$ mA/cm^2, but of different duration, obtained in a solution containing $0.01M$ Cu(CH$_3$COO)$_2$ + $0.01M$ Pb(CH$_3$COO)$_2$ + $1M$ HBF$_4$.

The problem of obtaining a multilayered laminar deposit arises when the current is interrupted between the pulses. At that moment, the concentration of the ions of metal A at the surface starts reverting to the bulk value. As virtually pure metal B is facing the solution, the potential would tend to attain its reversible potential value. That value is necessarily significantly more negative than the deposition potential of metal A (cf. Fig. 14). Hence, unless the cathodic or the anodic process is strongly inhibited, a *replacement reaction* must take place. If the interval between the pulses is sufficiently long, the second layer could completely dissolve.

Replacement (cementation) reactions have been extensively studied in the literature.[55] A good example of such a case was found in the laminar deposition of Cu and Pb. As seen in Fig. 15, after the interruption of the current the potential stays for some time at the level of the reversible potential of Pb, after which it falls to the potential of Cu. That time is seen to be proportional (virtually equal) to the pulse duration, signifying that the dissolution process takes place all the time. An analysis of the result showed that virtually all the lead is dissolved in the process. Inhibition of the anodic dissolution can occur if, during replacement, metal A forms a compact nonporous layer, which is not a likely situation, or else if passivation of metal B takes place.

This is what was found in the laminar deposition of Cu and Ni,[49,50,56,57] which could be effected by the dual- or triple-pulse method (cf. Section V.3) without difficulty. Although some of the authors claimed that no sign of Ni passivation could be found,[49] it is difficult to see how else the absence of replacement could be explained.

3. Dual-Current Pulse Laminar Deposition

As shown above, laminar deposition can be achieved with a series of constant-current pulses, with some interruption time in between, sufficient for a full recovery of the concentration of the more noble metal at the surface. However, this occurs in a relatively narrow range of well-defined current densities in the pulses and pulse durations. This imposes limitations on both the thickness of the layers of the more noble metal A and on the content of that metal in the layer consisting predominantly of the less noble metal B.

A method more suitable for manipulation of both $d(I)$ and $x_A^{\partial d(II)}$ than that of the pulsating current is the dual-current pulse methods.[53, 56,57] Each pulse consists of two parts with current densities $j(I)$ and $j(II)$ and durations $T(I)$ and $T(II)$, respectively. The method thus offers four independent variables for adjusting the conditions of deposition to obtain the desired lamination.

The only condition that both $j(I)$ and $j(II)$ must satisfy is that they be smaller than the values given by Eq. (70) for the bulk concentrations of the two kinds of metal ions, C_A^o and C_B^o, respectively. In such a case, transition times of total concentration polarizations with respect to ions of either metal A or metal B are not reached within the corresponding parts of the pulse, so that both layers can be deposited to any desired thickness.

Assuming the current efficiency of deposition to be 100%, the thickness of the first layer should be

$$d(I) = \frac{M_A}{\rho_A z_A F}[j(I)T(I) + j(II)\tau_A] \tag{76}$$

where the second term inside the brackets corresponds to deposition of the pure more noble metal in the second part of the pulse before its transition time is reached. The latter is virtually the same as in the single-pulse method, if $j(I)$ is sufficiently smaller than j^{min}, and is also defined by Eq. (66).

In the dual-current pulse method it is also desirable to have

$$C_B^o \gg C_A^o \tag{77}$$

the best choice being that which satisfies the equality

$$\frac{C_B^o}{j(II)} = \frac{C_A^o}{j(I)} \tag{78}$$

The thickness of the second layer is given by

$$d(\text{II}) = \left\{ \frac{M_A}{z_A} j_{D,A} + \frac{M_B}{z_B} [j(\text{II}) - j_A] \right\} \frac{T(\text{II}) - \tau_A}{\rho F} \tag{79}$$

where $j_{D,A}$ is the diffusion limiting current density of deposition of metal A, and ρ is the density of the alloy forming the second layer. The former is defined by Eq. (70), as $j^{\min}$ is in fact equal to $j_{D,A}$. If, for example, a rotating disk was used as a cathode for laminar deposition, the hydrodynamics should obey the Levich dependence on the rotation speed[36]:

$$j_{D,A} = 0.62 z_A F D_A^{2/3} v^{-1/6} C_A^o \omega^{1/2} \tag{80}$$

The content of metal A in a thin section $\partial d(\text{II})$ of the second layer will be

$$x_A^{\partial d(\text{II})} = \frac{z_B F D_A C_A^o}{\Delta h j(\text{II}) + (z_B - z_A) F D_A C_A^o} \tag{81}$$

which simplifies for $z_B = z_A = z$ to

$$x_A^{\partial d(\text{II})} = \frac{z_B F D_A C_A^o}{\Delta h j(\text{II})} \tag{82}$$

The integral composition $x_A^{d(\text{II})}$ remains the same as the differential one since the ratio between j_A and $j(\text{II})$ remains virtually constant after a short time, and so $x_A^{\partial d(\text{II})}$ is independent of time.

It is of interest for $x_A^{d(\text{II})}$ to be as small as possible, so that the second layer can be considered as almost pure metal B. When the limitation imposed on $j(\text{II})$ [cf. Eq. (70)] that

$$j(\text{II}) < \frac{(z_A D_A C_A^o + z_B D_B C_B^o) F}{\Delta h} \tag{83}$$

is introduced into Eq. (78), the lowest attainable content of metal A in the layer is obtained (for $z_A = z_B$) as

$$\left(x_A^{d(\text{II})} \right)_{\min} = \frac{D_A C_A^o}{D_A C_A^o + D_B C_B^o} \tag{84}$$

That is, it depends only on the ratio between the concentrations of the two kinds of metal ions in solution.

It is seen that the hydrodynamic layer thickness plays an important role in determining the conditions of deposition leading to a desired laminar deposit. As this depends on the rate of stirring, it was shown[49,53,56] to be possible to manipulate the deposition additionally by varying that rate. Thus, if the rate of stirring is increased during the first part of the pulse, so is $j_{D,A}$, and hence a larger $j(I)$ can be employed, leading to a shorter deposition time as seen in Fig. 16. For the second part of the pulse, the rate of stirring is reduced back to the original one, and thus the low content of metal A in the second layer is provided.

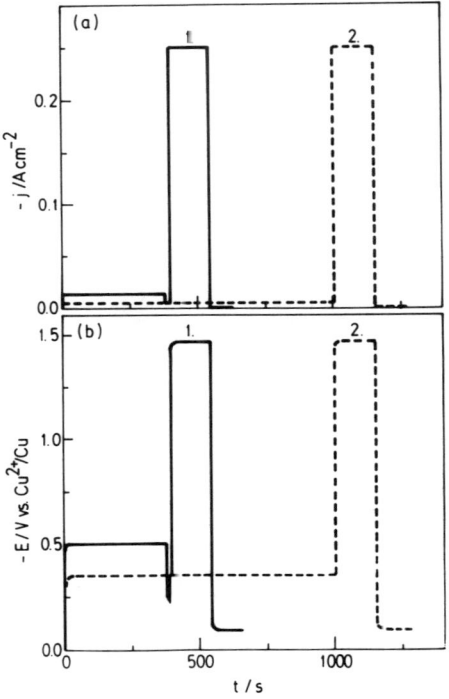

Figure 16. Current pulse train (a) and corresponding potential response (b) for two-layer deposition, the first layer being 2 μm of copper and the second layer being 12 μm of Cu–Ni alloy in the same solution containing 0.05M CuSO$_4$ + 2M NiSO$_4$ + 0.5M NaC$_6$H$_5$O$_7$ with a rotation speed in the second pulse of 500 rpm and $j(II) = 250$ mA/cm^2. 1, $j(I) = 13.8$ mA/cm^2, $T(I) = 396$ s, 5000 rpm; 2, $j(I) = 5.4$ mA/cm^2, $T(I) = 1007$ s, 500 rpm.

In the deposition of microlaminar deposits with layers in the nanometer range, the transition between deposition of the two layers becomes important. A transition layer is likely to interpolate between the two with a thickness comparable to those of the two layers and characterized by a changing composition between those of the first and the second layer. The transition layer may play an important role in determining the properties of the entire deposit. In some cases, it may be desirable to achieve a sharp and coherent boundary between two layers and, hence, to have the transition layer as thin as possible. In other cases, the transition layer may be a key factor determining properties different from those of the parent metals, and, hence, it may be desirable to make it thick compared to the two layers of constant composition. Thus, it is of interest to analyze the factors that determine the sharpness of the transition layer.

Such an analysis was carried out[54] assuming that the main cause of a continuous transition of composition is a slow change in the partial currents of deposition, j_A and j_B, the first one decaying and the second one rising as the total current is stepped up from $j(I)$ to $j(II)$. The slow change is due to two factors:

(a) The need to change the potential imposes charging of the double layer at the metal/solution interface, which possesses a certain capacitance, C_{dl}, and this takes a part of the constant total current j.
(b) The partial current j_A, reaching the value of the diffusion limiting current $j_{D,A}$ at the beginning (until $\delta < \Delta h$), is not constant but decays continuously with time, because of the expansion of the Nernst diffusion layer δ according to Eq. (68) until the latter does not reach the hydrodynamic boundary layer, so that

$$j_{D,A} = z_A F \pi^{1/2} D_A^{1/2} C_{o,A} / 2 t^{1/2} \tag{85}$$

The following equation was derived relating $j(II)$ and the electrode potential, E, with time, applicable equally to a galvanostatic mode of operation (constant-current step with changing potential) and the potentiostatic mode (constant potential and changing current):

$$j(II) = -C_{dl}[E + R_\Omega j(II)]/dt + $$

$$\frac{j_{o,A} \exp[(\alpha_{c,A}F/RT)E_{r,A}]\exp\{(-\alpha_{c,A}F/RT)[E + R_\Omega j(II)]\}}{1 + (2t^{1/2}/z_A F \pi^{1/2} D_A^{1/2} C_A^o) j_{o,A} \exp[(\alpha_{c,A}F/RT)E_{r,A}]\exp\{(-\alpha_{c,A}F/RT)[E + R_\Omega j(II)]\}}$$

$$+ j_{o,B} \exp[(\alpha_{c,B}F/RT)E_{r,B}]\exp\{(-\alpha_{c,B}F/RT)[E + R_\Omega j(II)]\} \tag{86}$$

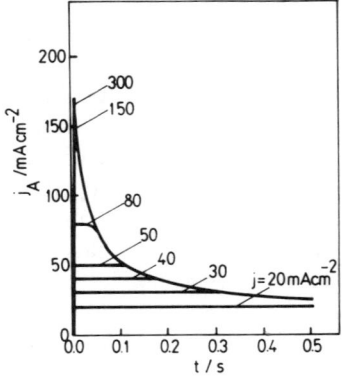

Figure 17. Partial current of deposition of metal A, j_A as a function of time, after imposition of different constant-cathodic-current steps, j.

where R_Ω is the resistance of the solution between the electrode surface and the tip of the Luggin capillary, and $j_{o,A}$, $\alpha_{c,A}$, and $E_{r,A}$ are, respectively, the exchange current density, the transfer coefficient, and the reversible potential of deposition of metal A, the same quantities with index B pertaining to metal B.

To obtain the partial current, j_A, as a function of time for a constant $j(II)$, Eq. (86) could be integrated from $t = 0$ at the beginning of the second part of the pulse and from the corresponding potential. For a set of reasonable values of the parameters, such as those expected, for example, in the deposition of Cu and Ni, the time dependence was obtained as shown in Fig. 17. It is seen that $j(II) = 0.020$ A/cm^2 was obviously smaller than j^{min} so that the transition time is never reached, while at 0.3 A/cm^2 the transition time is less than 0.01 s.

The content of metal A in the deposit at any moment is given by Eq. (73) whereas the thickness of the deposit is obtained from Eq. (79) by replacing T with running time t. By combining the two, the dependence of $x_A^{\partial d(II)}$ on $d(II)$ is obtained, as shown in Fig. 18. Dots on the lines denote the thickness of the transition layer d_τ within which there is a changing composition of the second layer.

On the basis of the above equations, it was derived that

$$d_\tau = \frac{\pi M_A C_A^o}{4\rho} \frac{\Delta h}{x_{A,max}^{d(II)}} \quad (87)$$

where $x_{A,max}^{d(II)}$ is the maximum allowable content of metal A in the second layer. The result indicates that the transition layer thickness does not depend on the current density of deposition. As d_τ strongly depends on

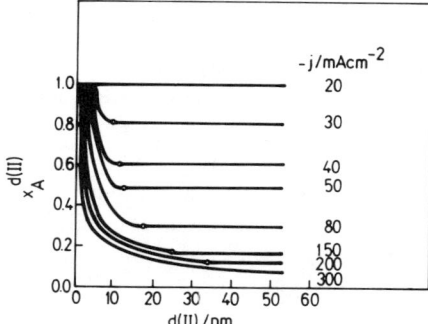

Figure 18. The content of metal A as a function of thickness of layers deposited at different constant current densities, j.

Δh, it is sensitive to the rate of rotation of a rotating disk cathode. Figure 19 shows this dependence to be determined by the maximum allowable content of metal A. If $x_{A,\max}^{d(\mathrm{II})}$ is to be about 1%, there is no practically acceptable rate of rotation (<3000 rpm) that could make the transition layer thinner than 100 nm.

It is noteworthy that the major effect is that of slow extension of the diffusion layer into the solution, while the contribution of double-layer charging is negligible for all the systems met in practice.

In order to decrease d_τ, Lashmore and Dariel[57] introduced pulses consisting of three parts: a high-cathodic-current-density part, a brief interruption of the current, and a low-cathodic-current-density part. It is claimed that more coherent thin layers with sharp boundaries are obtained in this way.

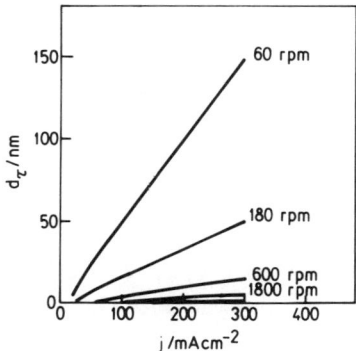

Figure 19. Transition layer thickness as a function of deposition current density at different constant rates of rotation of a rotating disk cathode.

A similar system was applied also by Tench and White,[58] but with significantly longer current interruptions between the two parts of the pulse (2 s) as well as between pulses (3 s), combined with high rotation speed in the low-current-density part and absence of rotation in the high-current-density part of a pulse. They found it advantageous to maintain a constant potential rather than a constant current density in the low-current-density part and constant current in the high-current-density part. Very reproducible results of tensile strength measurements were obtained in this way down to very low thicknesses of individual layers (10 nm). It is questionable, however, to what extent coherence of individual layers was maintained under such conditions. Also, it is questionable whether the deposit could be considered as laminar or rather as a composition modulated one.

VI. DEPOSITION WITH INCLUSION OF NONMETALLIC MATERIALS

Another type of composite materials, besides laminar metal plates, that can also be produced by electrochemical deposition are metal deposits with inclusion of nonmetallic particles.

The phenomenon of electrolytic codeposition of inert, nonmetallic particles (cermet deposits) has been known since conventional electrolysis was first developed (formation of rough deposits was attributed to the presence of "impurities" in the plating bath). During the past three decades, however, interest in commercial application of such deposits has increased since it was found that such coatings could be used to solve some persistent engineering problems. Despite the potential interest in such deposits and many years of work in this field, only a few large-scale commercial baths have been developed so far for industrial production of such composites. Examples of industrial applications include the use of composite Ni–SiC and Co–SiC coatings for some engine parts[59] and the production of decorative (mainly Ni) coatings.[60,61] Composite coatings have also been used in solving some antifriction problems[61] and for improving corrosion protection.[62] In general, isolated, highly specialized, and quite small baths for production of such composites are in use at present.[60]

A comprehensive overview of the characteristics and properties of inorganic composite materials, elaborating on metal deposits containing uniformly dispersed nonmetallic particles, was recently given by Saifullin.[63] According to this review, numerous metals (Ni, Co, Fe, Cu, Ag, Au, Pd, Zn, Cd, In, Sn, Pb, Cr, and Al) and metal alloys (Cd–Bi, Sn–Zn, Ni–Mn, Pb–Sn, Co–Ni, Cr–Mo, Cu–Sn, and Fe–Zn) can be codeposited with inert particles such as SiC, TiO_2, Al_2O_3, Cr_2O_3, SiO_2, and ZrO_2.

In the early days of research on such composite coatings, attention was directed primarily toward the determination of optimum conditions for their production as well as toward obtaining as much data on their properties as possible. Few efforts were made to elucidate the mechanism of codeposition of metals and inert particles. The first mechanism of codeposition, based on the process of electrophoresis, was proposed[64] in 1962. Somewhat later, mechanisms involving mechanical entrapment[65] and adsorption of particles at the cathode[66] were forwarded (in 1964 and 1967). In 1963 a hypothesis was offered[67] that codeposition of particles into the metal matrix is a result of adsorption of metal cations at the surface of inert particles, imposing a positive surface charge to the latter, which makes them attracted to the cathode. Some years later, in 1972, the importance of hydrodynamics for the process of electrochemical codeposition of metals and inert particles was first mentioned.[68]

1. Description of the Processes Involved in the Codeposition of Metals and Inert Particles

Possible mechanisms of codeposition of inert particles and their inclusion into the metal matrix, which were generally accepted in the above-mentioned references, are[69]:

1. Particles can be transferred to the cathode by an electrophoretic action to be keyed into the matrix during electrodeposition of a metal.

2. Particles can be thrown against the cathode by bath agitation and are embedded there into the electrolytically deposited metal matrix (mechanical entrapment).

3. Particles can be attached to the cathode by van der Waals type attraction forces.

Actually, all three mechanisms play an important role in the process of codeposition. In view of this, the process depends on several factors, such as pH, temperature, current density, type of particles used, their size and concentration, the metal being deposited, and possible presence of additives.

It seems that two types of adsorption processes are involved in the mechanism of codeposition: (a) adsorption of metal cations onto the

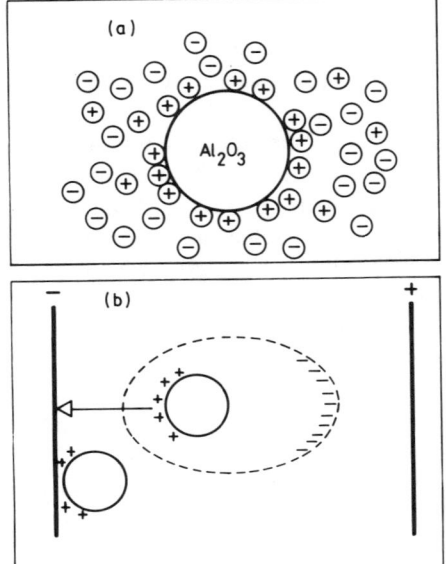

Figure 20. Schematic representation of important adsorption phenomena in codeposition[69] on an inert Al_2O_3 particle (a) and on a cathode surface (b).

surface of inert particles, imposing a positive surface charge on them and causing formation of an ionic cloud around them, and (b) adsorption of such complex structures onto the cathode surface.[69] A schematic representation of these two types of adsorption is shown in Fig. 20.

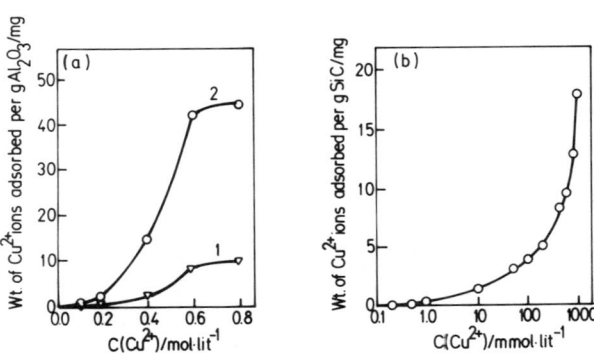

Figure 21. Adsorption isotherms[71] for adsorption of Cu ions onto Al_2O_3 (a) and SiC (b). (a) 1, $CuSO_4$; 2, $CuSO_4$ and Tl_2SO_4 ($Cu^{2+}:Tl^+$ = 635:1).

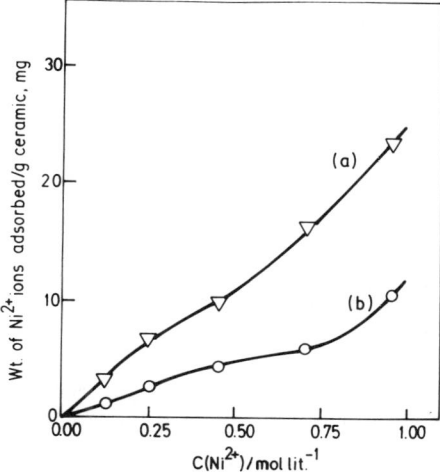

Figure 22. Adsorption isotherms for adsorption of Ni^{2+} ions onto TiO_2 (a) and SiC (b) from a nickel sulfamate bath.[71]

Adsorption of metal cations onto the inert particles was investigated by determining their adsorption isotherms.[70] It was shown that the amount of adsorbed cations increases with their increasing concentration in the solution and that addition of heavy monovalent cations (such as Tl^+) promotes adsorption of other metal cations. Figure 21 gives examples of adsorption isotherms for Cu^{2+} cations on Al_2O_3 and SiC, while Figure 22 shows adsorption of Ni^{2+} onto TiO_2 and SiC particles.[71]

With the use of a rotating disk electrode assembly, which provides for constant and reproducible conditions of hydrodynamic mass transport (cf. Section IV), the influence of bath agitation was investigated.[68,69] It was shown (Fig. 23) that in the region of laminar flow a constant amount of inert particles is embedded. At the start of the transition zone, a marked decrease in the amount of embedded particles is obtained. Further increase of the rotation speed leads to a significant increase in the inclusion of particles. It is interesting to note that the maximum amount of embedded particles appears to be reached at the end of the transition zone and the beginning of the turbulent zone. It was concluded that the increased amount of codeposited alumina in the transition zone is a result of formation of some Al_2O_3 agglomerates and subsequent codeposition of

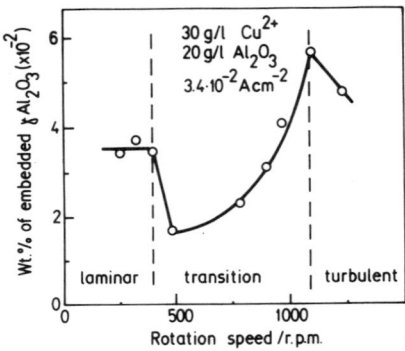

Figure 23. The influence of hydrodynamics on the incorporation of γ-Al$_2$O$_3$ in a copper deposit on the rotating Riddiford type electrode.[69]

such agglomerates.[72] It was also shown by calculation[73] that the particles can not reach the surface of a rotating disk electrode by inertial forces alone as long as the following condition is fulfilled:

$$3.52\rho_p r_p^2 \nu^{-1}\omega < 3/4 \qquad (88)$$

where ρ_p and r_p are, respectively, the density and the radius of the inert particle.

One of the functions characterizing a codeposition process is the dependence of the weight percent of embedded particles on applied current density. In almost all cases this dependence shows a sharp maximum.[69,74,75,80–87] An exception is found in the case of codeposition of gold

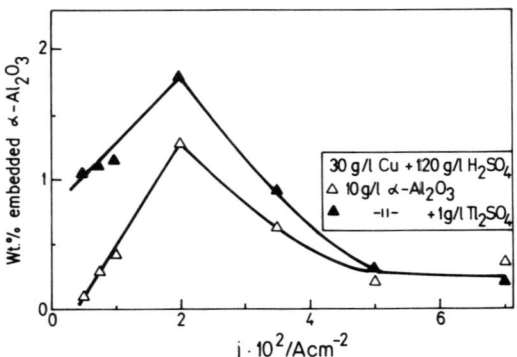

Figure 24. Plot of wt. % of embedded alumina against current density for deposits obtained from an acidified copper sulfate solution on a flat vertical cathode.[69,74]

with alumina from a cyanide bath, where two maxima were found.[69] The shape of this dependence is shown in Fig.24 for the Cu–Al$_2$O$_3$ system. It was shown that the increase of the amount of embedded particles at increasing current density appears in the region where the reduction of metal cations is under charge transfer control and where the reduction of adsorbed metal cations onto the inert particles is the rate-determining step. Once the reduction of cations is under diffusion control ($j > 0.02$ A/cm^2 in Fig. 24), the amount of codeposited inert particles decreases gradually at increasing current density.

2. The Mechanism of Codeposition of Inert Particles and Metals

The first mathematical model describing the process of codeposition of inert particles and metals was derived by Guglielmi in 1972.[76] This model is based on the similarity between the curves relating the volume percent of codeposited particles to the volume percent of particles suspended in the solution and a well-known adsorption isotherm. It consists of two successive adsorption steps. The first step is a physical process whereby a layer of loosely adsorbed particles is formed, providing a rather high degree of coverage of the cathode surface by particles. A subsequent strong adsorption, which is thought to be field-assisted and therefore electrochemical in nature, permits permanent bonding and entrapment of the inert particles.

The process of codeposition was described by the equation

$$\frac{C_p}{(v/o)_p} = \left(\frac{1}{k^*} + C_p\right)\frac{M_m j_0 \exp(B/A)}{zF\rho_m v_0} j \exp\left[1 - \frac{B}{A}\right] \qquad (89)$$

where C_p is the concentration of suspended particles in solution (vol. %); $(v/o)_p$ is the volume fraction of particles in the deposit; M_m is the atomic weight of the deposited metal; ρ_m is the density of the deposited metal; k^* is the constant of the Langmuir adsorption isotherm, which depends essentially on the intensity of interaction between the particles and the cathode; j_0 and A are parameters related to the metal deposition (constants in the Tafel equation); and v_0 and B are constants related to the deposition of the inert particles. The validity of this model was verified by several authors for different codeposition systems.[75-78] However, the other important parameters controlling the process of codeposition of inert particles with metals, such as hydrodynamics, effect of size and type of particles, effect of bath constituents, conditions of electrolysis (pH), and bath temperature, are not taken into account by this model.

Two years later, in 1974, a new mathematical expression that could describe the effect of hydrodynamics was developed by Kariapper and Foster.[71] The rate of codeposition (dV/dt) was expressed by the equation

$$\frac{dV_p}{dt} = \frac{N^* h C_p}{1 + h C_p} \qquad (90)$$

with N^* being the number of collisions of particles suitable for codeposition per second and V_p being the volume fraction of particles embedded in the deposit. The parameter h depends on several factors and is given by

$$h = h^*(q\Delta E + Lj^2 - MN) \qquad (91)$$

where h^* is a constant. The first term inside the parentheses describes the electrostatic attraction, which is a function of the adsorbed charge density on the particles (q) and the potential field at the cathode surface (ΔE). The second term represents a physical bond and is a function of the rate of metal deposition (j) and the bond strength of the metal–particle interaction (L). The third term accounts for mechanical factors such as the size, shape, and density of the particles (M) and the rate of agitation (N). The validity of this model has not as yet been proven by experiments, because of the complex interrelationship between some of these parameters.

In 1987, another model, based on a statistical approach and containing measurable parameters of the process of incorporation of inert particles in the metal matrix, was developed by Celis *et al.*[79] This model is based on two fundamental postulates:

1. An adsorbed layer of ionic species is created around inert particles at the time the particles are added to the plating solution, or during the pretreatment of these particles in ionic solution.

2. The reduction of some of these adsorbed ionic species is required for incorporation of the inert particles into the metallic matrix.

To be incorporated into a metallic matrix, the inert particles must proceed through five active stages (schematically represented in Fig. 25) on their way from the bulk of the solution to the cathode surface: (i) adsorption of ionic species on the particle (formation of ionic cloud); (ii) convection of particles from the bulk of the solution to the hydrodynamic boundary layer; (iii) diffusion through the diffusion layer; (iv) adsorption of particles, together with their ionic cloud, at the cathode surface; and (v)

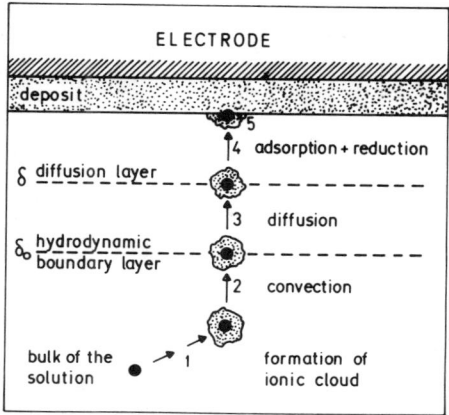

Figure 25. Schematic representation of five stages in the process of codeposition of inert particles and metals.[79]

reduction of some adsorbed ionic species and their incorporation into the metallic matrix.

Assuming that inert particles are spherical and that steady-state conditions for electrodeposition are fulfilled, a hypothesis was postulated: A particle will only be incorporated when a certain number, k, of K adsorbed ionic species is reduced, depending on the combination of plating solution characteristics and particle characteristics. As a consequence of this hypothesis, still another hypothesis should be added: No distinction is made between free and adsorbed ionic species, and both may be considered equal with respect to transport and reduction processes.

By the application of a statistical approach using a binomial distribution formula to calculate the probability for the incorporation of one particle based on the reduction of k ions out of K ions $[P_{(k/K,i)}]$ at a constant current density j, a final expression for the weight percent of embedded spherical particles (w/o) is derived:

$$(\text{w/o}) = \frac{4\pi r_p^3 \rho_p N_{\text{ion}} \dfrac{C_p^*}{C_{\text{ion}}^*} \left(\dfrac{j_{\text{tr}}}{j}\right)^\alpha HP_{(k/K,i)}}{\dfrac{3Mj}{zF} + 4\pi r_p^3 \rho_p N_{\text{ion}} \dfrac{C_p^*}{C_{\text{ion}}^*} \left(\dfrac{j_{\text{tr}}}{j}\right)^\alpha HP_{(k/K,i)}} \qquad (92)$$

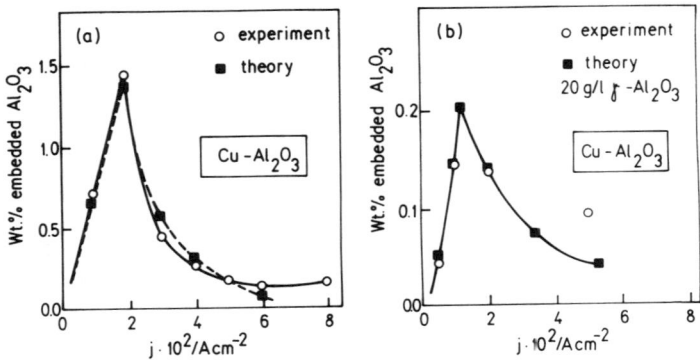

Figure 26. Comparison between experimentally obtained and theoretically predicted results[79] for Cu–Al$_2$O$_3$ codeposition on a rotating disk electrode (a) and on a vertical cathode (b).

where N_{ion} is the number of ions crossing the diffusion layer; C_p^* is the number of particles in the bulk of the solution; C_{ion}^* is the number of ions in the bulk of the solution; j_{tr} is the transition current density from charge transfer to diffusion control; α is a measure of the interaction between free and adsorbed ions due to current density effects; and H is a parameter describing the hydrodynamics; $H = 0$ under high turbulent flow when the particles collide frequently and can hardly enter the very thin diffusion layer, and $0 < H < 1$ in the transition and early turbulent regimes.

The main parameter of the model is j_{tr}. It determines indirectly the value of the maximum codeposition level, and extremely accurate measurement or calculation of its value is necessary for the prediction of any codeposition curve.

In the application of this model to the experimentally obtained results, two parameters, k and α, have still to be determined. The parameter α is determined by the type of overvoltage control ($\alpha = 0$ for charge transfer control), whereas the parameter k may be evaluated from a comparison between experimental and theoretical results.

Such an analysis was performed for the process of codeposition of Al$_2$O$_3$ with Cu from acid sulfate plating baths, which has been extensively investigated.[69,75,80–87] In Fig. 26, results from this application of the model are presented. It can be is seen that a very good agreement is obtained, indicating that this model could be successfully applied to the process of codeposition of inert particles with metals.

VII. CHARACTERIZATION OF PHASE STRUCTURE OF ALLOYS BY ELECTROCHEMICAL TECHNIQUES

As early as 1928, an electrochemical technique, based on the dissolution characteristics of plated samples, was developed for the determination of the thickness of metallic coatings.[88] Using this method of analysis,[89] it was possible to determine the thickness of the coating and that of the intermediate layer between the coating and the base metal (usually composed of various phases) as well as to obtain data about the phase structure of the coating. That was the first attempt to determine alloy phase structure by an electrochemical technique.

Since then, electrochemical techniques have been extensively used for investigating processes of alloy dissolution. In all of these cases, bulk alloys have been used.[90–100]

Relatively recently, three laboratories reported attempts to determine the phase structure by complete dissolution of a thin layer (up to 10 μm) of an alloy.[101–115] Different electrochemical techniques, such as galvanostatic techniques,[101,103] potentiostatic techniques,[108,109,112] and, in most cases, anodic linear sweep voltammetry (ALSV),[101–115] were applied. In all cases, the thin layers of binary alloys were obtained by electrodeposition.

For such an analysis to be successfully performed, several conditions must be fulfilled, as discussed below.

1. Conditions for Identification of Phases in Thin Layers of Alloys

(a) For a quantitative determination of the phase composition, it is important to dissolve the entire alloy sample, as only in such a case the diffusion of alloy constituents in the solid phase cannot influence the process of alloy dissolution (masking of some phases by other ones). If there are some diffusional limitations, they can be overcome by using long time polarization (galvanostatic or potentiostatic techniques) or a very low sweep rate in the ALSV technique. These techniques are particularly convenient for electrodeposited samples, as such samples often cannot be characterized by X-ray techniques [cf. Section VII.2(iii)(c)].

(b) Processes of passivation or replacement must be prevented during the alloy dissolution. This can be achieved by a proper choice of the solution for alloy dissolution. Solutions of simple salts (Na_2SO_4, $NaClO_4$, NaCl) with pH values in the range $4 \leq pH \leq 6$ are preferred. For a better determination of the dissolution potential of different phases, addition of cations of the less noble component to the solution is recommended.[102]

(c) Techniques in which the potential is controlled rather than the current, particularly ALSV, are preferred[102,104–115] because the galvanostatic responses are not always well defined.[101,103]

2. Dissolution Characteristics of Different Alloys

From a consideration of the phase diagrams of alloys and the Gibbs energies of phases appearing therein, it is possible to predict the shape of the ALSV or of responses to the galvanostatic dissolution of the alloys. The characteristics of these responses will differ depending on the type of alloy concerned.

(i) Eutectic Type Alloys

In the case of eutectic type alloys, where the interaction (miscibility) between two components in the solid phase is absent or negligible, the Gibbs energy of each component should not be notably different from that of the corresponding pure metal [cf. Section III.3(i)]. Hence, the reversible potential of each component in the alloy should be virtually the same as that of the pure component of the corresponding grain size. The ALSV should in such a case be characterized by two separate dissolution peaks, each of them reflecting the dissolution of the corresponding pure component.[101,102]

Examples are shown in Figs. 27 and 28 for Cu–Pb[102] and Zn–Cd[115] eutectic alloys, respectively. In both cases, alloys of different compositions were electrodeposited on a rotating disk electrode made of silver at 1000 rpm. As can be seen in Figs. 27 and 28, when sweeping the potential at a low sweep rate from the reversible potential of the pure less noble component in the positive direction, the components dissolve independently of each other, giving well-defined and well-separated ALSV peaks (curves 3). However, in both cases the dissolution of the less noble component from the alloy takes place at significantly more positive potentials than that for dissolution of the corresponding pure component (curves 2) and is characterized by two peaks, one of them being less pronounced.[102,115] The dissolution of the more noble component proceeds through a single sharp peak at a potential somewhat more negative [cf. Section VII.2(ii)] than that for dissolution of the corresponding pure component (curves 1).[102,115]

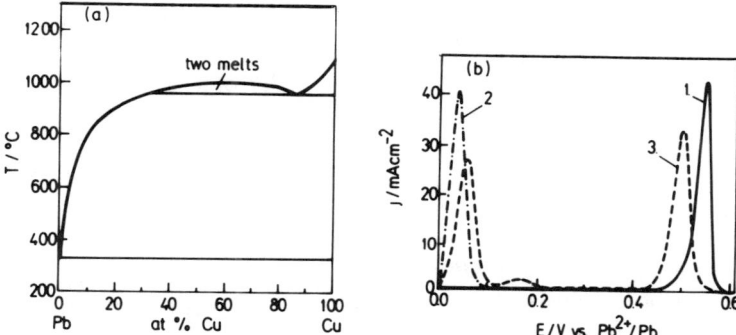

Figure 27. (a) Phase diagram of Cu–Pb binary system. (b) Voltammograms obtained by ALSV for dissolution ($v = 2$ mV/s) of pure copper (curve 1), pure lead (curve 2), and an alloy containing 55 at.% Cu and 45 at.% Pb (curve 3). Solution: $0.01M\,Pb^{2+} + 1M\,HBF_4$; 1000 rpm.

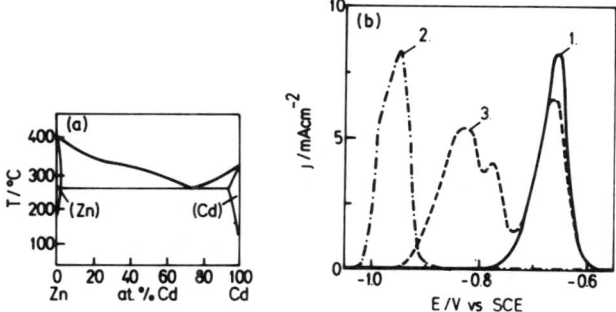

Figure 28. (a) Phase diagram of Zn–Cd binary system. (b) Voltammograms obtained by ALSV for dissolution ($v = 1$ mV/s) of pure cadmium (curve 1), pure zinc (curve 2), and an alloy containing 58 at.% Zn and 42 at.% Cd (curve 3). Solution: $0.2M\,Na_2SO_4$, pH 5; 1000 rpm.

This is not unexpected, since it is known that both alloys, Cu–Pb[23] and Zn–Cd,[116] if obtained by electrochemical deposition, form some supersaturated solid solution; that is, the miscibility of the components in the solid phase extends to 10 wt. % for Pb in Cu[23] and 4 wt. % for Zn in Cd.[116] Hence, the appearance of two peaks for the dissolution of the less noble component could be a result of dissolution of pure metal and some solid solution formed in the electrodeposited alloys.[102,115] Inhibition of dissolution of the less noble component could be the result of transport difficulties, since it must diffuse through the alloy to enter the solution (cf. Section VII.4.).

(ii) Solid-Solution Type Alloys

When considering alloys that make solid solutions, one should be very careful in trying to predict the shape of ALSV or other electrochemical responses reflecting their dissolution. It is known from the literature[117] that metal atoms in ideal solid solutions should be quite randomly arranged but that in reality such alloys can also contain some short- and long-range order. Some alloys, for example, those in otherwise totally miscible Cu and Au, possess "superlattice" structures (Cu_3Au, CuAu, and $CuAu_3$ superlattice). Each of these types of solid solution could have different dissolution characteristics.

As is usual in the dissolution of binary alloys,[101,102] if the Gibbs energy of mixing is not very high and if the potential sweep is sufficiently slow, the less noble component should dissolve completely at more negative potentials than the more noble component. In such a case one should expect at least two ALSV peaks, the last peak in the voltammogram representing dissolution of the pure more noble component. All the preceding peaks should represent dissolution of the less noble component only, assuming that, even if the two metals dissolved simultaneously, as is likely to happen with a solid solution, the more noble metal should immediately reprecipitate, regaining the electrons released on oxidation (cf. Section VII.4.). It is noteworthy that a theory predicting the appearance of two peaks, one corresponding to preferential dissolution of the less noble component, has been proposed.[101]

(a) The Cu–Ni system—ideal solid solution

It is well known that in the Cu–Ni system, atoms of Cu and Ni are randomly arranged, making an ideal solid solution[118] (Fig. 29a). The voltammograms obtained by ALSV for Cu–Ni alloy dissolution by differ-

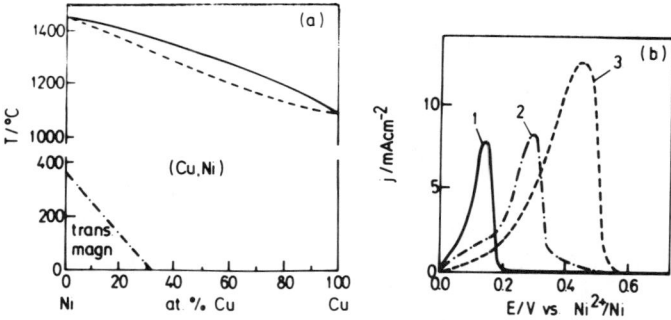

Figure 29. (a) Phase diagram of Cu + Ni binary system. (b) Voltammograms obtained by ALSV for dissolution ($v = 2$ mV/s) of pure copper (curve 1), pure nickel (curve 2), and an alloy containing 20 at.% Cu and 80 at.% Ni (curve 3). Solution: $0.01M$ Ni^{2+} + $0.5M$ NH$_4$Cl, pH 5; 1000 rpm.

ent authors[102,107] are almost identical. An example is shown in Fig. 29b. As can be seen, the dissolution of the solid solution gives a single ALSV peak, indicating the existence of a single phase. This peak occurs at more positive potentials than those for dissolution of pure Cu or pure Ni. Hence, it represents an exception to the expected behavior.

It is interesting to note that dissolution of a layer (0.2 μm) of pure Ni (curve 2 in Fig. 29b), electrodeposited from an appropriate bath, takes place at somewhat more positive potentials than that of a layer of pure Cu (of the same thickness), although their standard potentials differ by several hundred milivolts, $E^{\ominus}(\text{Cu}^{2+}/\text{Cu})$ being the more positive one.

Obviously, such behavior is a result of kinetic limitations, that is, of high irreversibility in the process of Ni deposition and dissolution (cf. Section VII.4.). Accordingly, if all the Ni is dissolved together with Cu (simultaneous dissolution) at more positive potentials than that for deposition of the latter, there is no possibility for either of the pure metals to be reprecipitated. In such a case only one ALSV peak should be expected.

(b) The Ag–Pd system

Most of the data on the phase diagram of the Ag–Pd binary system,[118] obtained by X-ray analysis, indicate that it consists of a homogeneous solid solution over the entire composition range. However, some workers[119,120] have concluded, after measuring the microhardness and specific electric resistivity as a function of composition and temperature, that some ordered structures may be formed in the solid phase. The investigation of this binary system by the ALSV technique has recently been performed.[114] The results of this investigation are presented in Figs. 30–32.

Figure 30 shows that even low contents of Pd cause a significant shift of the potential of the first peak (denoted as A) compared to that of pure Ag. This indicates that Ag in this alloy is not only intimately mixed with Pd, forming a homogeneous solid solution, as predicted by the phase diagram (Fig. 31a), but also is rather strongly bound.

The dissolution of the more noble component (Pd) exhibits a well-defined sharp peak (denoted as C). The shift of the peak potential in the negative direction with decreasing Pd content, seen in Fig. 30, is larger than expected on the basis of the Pd content. The increasing ease of dissolution of Pd that this implies could be due either to formation of finer Pd crystallites upon reprecipitation or to loosening of the crystal structure after extraction of Ag.[114]

Finally, the appearance of a third peak B (clearly seen after the deconvolution in Fig. 31b), in addition to the two peaks (A and C) expected

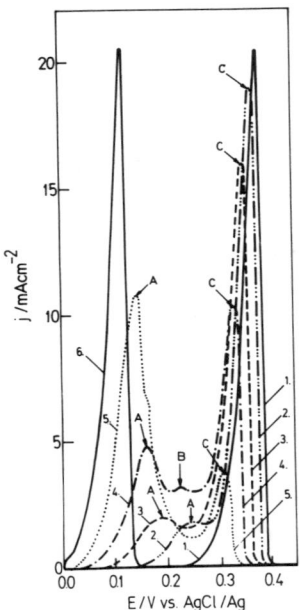

Figure 30. Voltammograms obtained by ALSV for dissolution of Ag–Pd alloys, deposited to $Q_d = 1.1$ C/cm^2 onto a gold electrode, in a solution containing 0.001M AgCl + 12M LiCl + 0.1M HCl at 1000 rpm and $T = 60°$C with sweep rate $v = 1$ mV/s. Composition of deposited alloys in at.% Ag–at.% Pd: 1, 0–100; 2, 20–80; 3, 40–60; 4, 60–40; 5, 80–20; 6, 100–0.

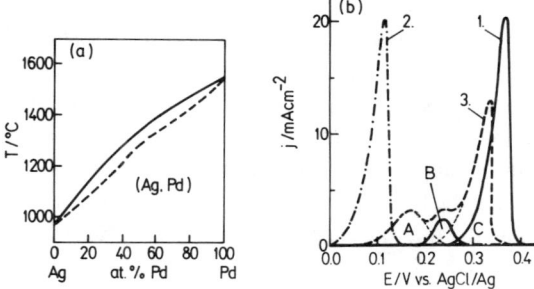

Figure 31. (a) Phase diagram of Ag–Pd binary system. (b) Voltammograms obtained by ALSV for dissolution of pure palladium (curve 1), pure silver (curve 2), and an alloy containing 50 at.% Ag and 50 at.% Pd (curve 3), all deposited to $Q_d = 1.1$ C/cm^2.

on the basis of the phase diagram, implies the existence of another phase. As seen in Fig. 32, its formation, by deposition of the alloy on a gold substrate, appears to depend strongly on the thickness of the deposit. In a deposit about 4 μm thick, this phase overwhelms the homogeneous solid solution (peak A).

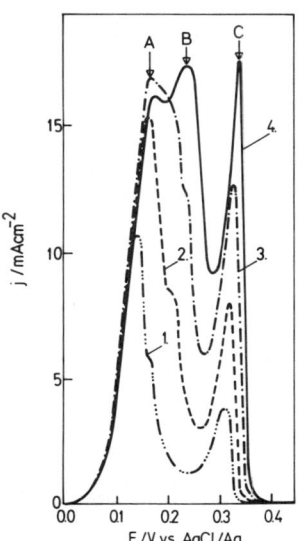

Figure 32. Voltammograms obtained by ALSV for dissolution of an alloy containing 80 at.% Ag–20 at.% Pd deposited onto a gold electrode. Conditions of dissolution as in Fig. 30. 1, $Q_d = 1.1$ C/cm^2; 2, $Q_d = 2.08$ C/cm^2; 3, $Q_d = 3.0$ C/cm^2; 4, $Q_d = 4.15$ C/cm^2.

This finding[114] indicates that at room temperature the alloy is not a simple homogeneous solid solution.[119,120] Confirmation of this finding is provided by the fact that microhardness of electrodeposited Ag–Pd alloys increases with increasing content of Ag.[121,122] Such behavior could only be explained by the presence of some phase (most probably an intermetallic compound) that possesses a higher value of microhardness than that of pure Pd. Since the microhardness of pure Pd is three times higher than that of pure Ag, the microhardness value of their solid solution should be somewhere in between.[122]

(iii) Alloys Containing Intermediate Phases and/or Intermetallic Compounds

It is known that the configuration of atoms that has minimum Gibbs energy after the mixing of two metals often does not have the same crystal structure as either of the pure components. This new structure may be an intermediate phase or an intermetallic compound. The difference between an intermediate phase and an intermetallic compound is not in their structures, but in the change of their Gibbs energies with composition,[117] as is schematically represented in Fig. 33. Since the change of the Gibbs energy of an intermetallic compound as a function of composition is very sharp, it seems reasonable to assume that its reversible potential is a singular point and that, accordingly, the dissolution peak of an intermet-

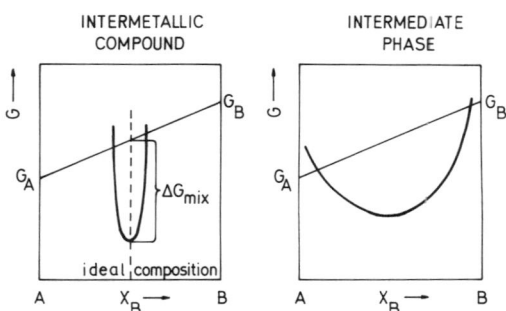

Figure 33. Schematic representation of change of the Gibbs energy as a function of composition for intermediate phases and intermetallic compounds.

allic compound should be better defined than that of an intermediate phase.[102]

One of the best examples of the power of the ALSV technique in studying the phase composition of alloys is found in the work described below on the Cu–Cd binary alloy (containing four intermetallic compounds), which could be successfully electrodeposited to any desired composition from a solution of simple $CuSO_4$ and $CdSO_4$ salts.[108,109,112,113]

(a) Electrodeposition of Cu–Cd alloys

Electrochemical deposition of copper and Cu–Cd alloys on a rotating disk electrode made of silver at a constant potential is characterized by flat j–t transients, that is, by a constant deposition current, as shown in Fig. 34a [provided that crystallization phenomena are overcome by applying sufficiently high (negative) potential]. The polarization j–E curve, shown in Fig. 34b, was obtained using the values from the flat portions on the j–t transients.[109]

In the potential region between -0.2 V and -0.6 V versus Cu^{2+}/Cu, the diffusion limiting current of copper deposition $j_L(Cu)$ is reached. Hence, at potentials more negative than -0.6 V versus Cu^{2+}/Cu, codeposition of cadmium commences, and the deposition current starts to increase. Since the diffusion limiting current is the maximum current of copper deposition in the given electrolyte, the excess current over $j_L(Cu)$ represents the partial current of cadmium deposition. Hence, the alloy constituents should deposit in relative amounts corresponding to ratios of diffusion limiting current to excess current (the "current ratio"). The phase diagram of the Cu–Cd system (Fig. 35a) indicates the existence of four different intermetallic compounds with increasing Cd content: Cu_2Cd, Cu_4Cd_3, Cu_5Cd_8, and $CuCd_3$. This is the reason why corresponding current ratios were chosen in preparing the samples. Ratios of 1:6 and 1:15, for depositing excess Cd, were also tested.[109]

(b) ALSV response to phase composition

Up to six ALSV peaks, two reflecting dissolution of pure Cd and pure Cu and four corresponding to different Gibbs energies of formation of the four intermetallic compounds, should be expected. Also, the series of peaks of increasingly positive potential (with respect to the reversible potential of pure Cd in a solution of the corresponding ions, Cd^{2+}/Cd)

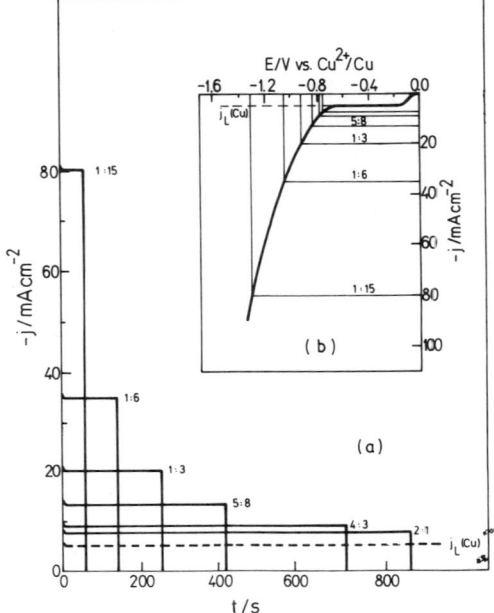

Figure 34. (a) Potentiostatic $j - t$ transients of Cu–Cd alloy deposition to a thickness of 3.2 μm at current ratios marked on the figure. (b) Polarization diagram obtained from flat portions of j–t transients. Solution: $0.01M$ $CuSO_4 + 0.5M\ CdSO_4 + 0.2M\ Na_2SO_4 + 0.01M\ H_2SO_4$; 1000 rpm.

should follow the order of decreasing Cd content in the compounds. Indeed, in one case (Fig. 35b, curve 3) all the expected peaks are recorded.[108,109] As can be seen, the peaks are well separated, except for peaks D and E, the latter covering the former to a large extent. In order to separate peaks D and E, a simple deconvolution, schematically represented in Fig. 35b can be used.

The voltammograms obtained by ALSV for the dissolution of layers of various Cu–Cd alloys of the same thickness, 3.2 μm, are shown in Fig. 36. Multiple peaks were obtained for each alloy composition, and the number of ALSV peaks was found to depend on the content of Cd in the alloy. For the alloy deposited at a current ratio of 2:1, only two peaks

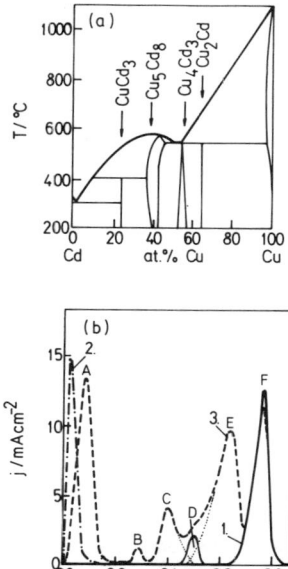

Figure 35. (a) Phase diagram of Cu–Cd binary system. (b) Voltammograms obtained by ALSV for dissolution ($v = 0.5$ mV/s) of pure copper (curve 1), pure cadmium (curve 2), and an alloy containing 75 at.% Cd and 25 at.% Cu (curve 3). Solution: $0.01M$ CdSO$_4$ + $0.2M$ Na$_2$SO$_4$ + $0.01M$ H$_2$SO$_4$; 1000 rpm.

appear on voltammogram, whereas for alloys deposited at current ratios of 5:8 and 1:3, there are five (six) peaks.

The peaks in the voltammograms shown in Figs. 35 and 36 can be attributed to either the presence of different intermetallic compounds or the existence of different crystal structures of cadmium in the electrode-posited alloys.[109] Taking into account that the number of peaks depends on the current ratio of deposition and that the most positive peak (E) of alloy dissolution is positive by some 0.6 V with respect to the reversible potential of cadmium, it is not likely that the four peaks B, C, D, and E represent dissolution of different crystal structures of cadmium. Hence, it seems reasonable to attribute the peaks B, C, D, and E to the intermetallic compounds CuCd$_3$, Cu$_5$Cd$_8$, Cu$_4$Cd$_3$, and Cu$_2$Cd, respectively.

It seems that using the ALSV technique, Cu$_4$Cd$_3$ and Cu$_2$Cd cannot be detected separately[108,109] in the alloys containing less than 75 at.% Cd, although they both exist in the deposit (cf. Table 1). However, at higher content of Cd in the alloy, this can be done by deconvolution (cf. Fig. 35b).

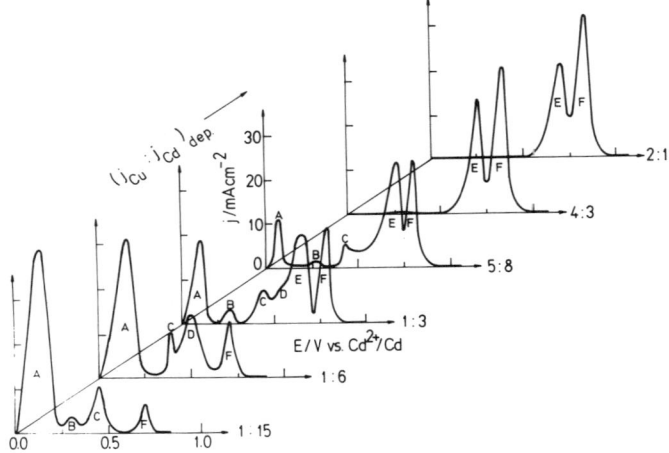

Figure 36. Voltammograms obtained by ALSV for dissolution of 3.2 μm-thick Cu–Cd alloys of different compositions obtained by electrodeposition at current ratios marked on the figure. The alloys were kept for 60 min in an inert atmosphere before dissolution. Conditions of dissolution as in Fig. 35.

The general conclusion can be drawn that, in spite of adjustment of the current to the values required for the formation of particular compounds (phases), as revealed by ALSV, in most cases more than one compound is obtained; that is, single-phase equilibrium alloys are rarely obtained, a fact known in electrochemical practice.[7,8]

(c) Use of X-ray analysis for determining phase composition

In order to evaluate the effectiveness of the ALSV method for phase structure analysis, it was necessary to compare the results with those obtained by using conventional techniques. Thus, an attempt was made to detect intermetallic compounds in electrodeposited thin layers (up to 20 μm) of Cu–Cd alloys obtained at various current ratios, using X-ray (Guinier) analysis without previous treatment of the electrodeposited layers. Taking into account that copper was deposited at its diffusion limiting current and that the grain size of the alloy constituents was very small, it was not surprising that the results of X-ray analysis were inadequate. Apparently, the grain size of the deposits was less than that required for discernible X-ray peaks.[109]

Table 1
Comparison of the Results of the Guinier Method of Phase Structure Analysis for Cu–Cd Alloys Electrodeposited at Different Current Ratios and Submitted to Subsequent Thermal Treatment and the Results of ALSV Analysis

Current ratio of deposition (j_{Cu}:j_{Cd})	Interplanar spacing, d (nm)	Relative intensity, I	Intermetallic compound (ASTM tables)	Peak	Intermetallic compound
4:3	2.5062	50	Cu_2Cd	E	Cu_2Cd
	2.1700	20	Cu_4Cd_3		
	1.5060	15	Cu		
5:8	2.5062	10	Cu_2Cd	E	Cu_2Cd
	2.2801	10	Cu_4Cd_3	C	Cu_5Cd_8
	2.2651	100	Cu_5Cd_8	B	$CuCd_3$
				A	Cd
1:3	3.0540	60	Cu_5Cd_8	E	Cu_2Cd
	2.6859	80	$CuCd_3$	C	Cu_5Cd_8
	1.8924	30	Cd	B	$CuCd_3$
				A	Cd
1:6	3.0434	20	Cu_5Cd_8	D	Cu_4Cd_3
	2.5714	40	Cu_5Cd_8	C	Cu_5Cd_8
	2.0488	40	Cu_5Cd_8	A	Cd
1:15	2.2629	80	Cu_5Cd_8	C	Cu_5Cd_8
	2.7310	40	$CuCd_3$	B	$CuCd_3$
	2.5679	60	Cd	A	Cd

In order to increase the grain size of the deposits, samples were submitted to thermal treatment at a temperature of 200°C for 50 h in vacuum. The results obtained using Guinier analysis after the thermal treatment[109] are given in Table 1.

(d) Comparison of X-ray (Guinier) and ALSV analysis

In spite of the fact that the X-ray analysis was performed on samples thermally treated after deposition, comparison of the results of ALSV with those of the Guinier phase structure analysis (Table 1) shows relatively good agreement between them. As a result of the thermal treatment, there is a tendency for a reduction in the number of compounds (phases) found in corresponding samples. As some solid-state reactions are expected to take place at the elevated temperature of the treatment, this could be explained in terms of the tendency toward equilibrium.[109]

As can be inferred from the above results, the alloys deposited electrochemically are not in an equilibrium state. When a voltammogram of the type shown in Fig. 35b (curve 3) is recorded by ALSV, the question arises as to whether the dissolution peaks represent phases (intermetallic compounds) that are all present in the alloy from the start or reflect the transformation of the phases initially present into compounds ever richer in Cu during the process of dissolution,[101] as cadmium is dissolved, for example, from a cadmium-rich alloy of the composition $CuCd_3$. If at the start only $CuCd_3$ existed and subsequently transformed successively into compounds richer in copper, the charges under the peaks should follow a rational sequence defined by stoichiometry. Thus, by calculation, the charges under peaks C, D, and E should be equal to 0.607, 0.178, and 0.357, respectively, of that under peak B. This is obviously not so. Indeed, the more likely event would be the formation of different compounds in accordance with the standard Gibbs energy per atom sequence, but this is not the case either. Hence, it can be concluded that all the compounds form simultaneously.[109] This conclusion is in accordance with the results of the Guinier phase structure analysis.

(e) Microstructural analysis

From transmission electron microscopy (TEM) analysis of a sample electrodeposited at a current ratio of 1:3 (without thermal treatment after deposition) to a thickness of 5 μm and subsequently exposed to argon-ion erosion to make a thin foil, it would appear (Fig. 37b) that the intermetallic compound $CuCd_3$ predominates, its crystals being about 20 nm in size (Fig. 37a). This result is not in agreement with the results of ALSV analysis. A possible explanation could be that a nonhomogeneous distribution of various intermetallic compounds is originally present in the electrodeposited alloy, so that the foil investigated, obtained by erosion, contained predominantly $CuCd_3$. It is obvious that more detailed microstructural (TEM) investigations should be carried out on very thin (nanometers) films of alloys of different compositions, obtained by electrodeposition, without subsequent argon-ion erosion.[109]

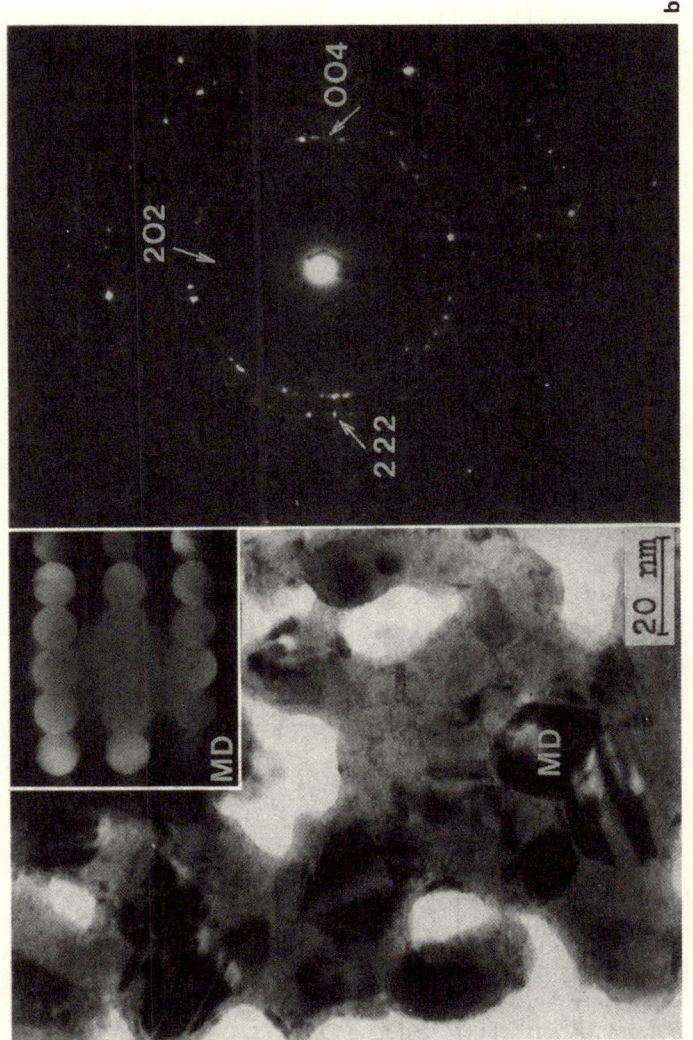

Figure 37. Microstructural analysis (TEM) of electrodeposited Cu–Cd alloy containing 75 at.% Cd and 25 at.% Cu. (a) Bright-field image of very thin foil (~50 nm); inset is micro-micro diffraction (MD) of region of several nanometers of a selected area; (b) diffraction pattern of very thin foil of alloy, showing polycrystalline diffraction image corresponding to $CuCd_3$ intermetallic compound.

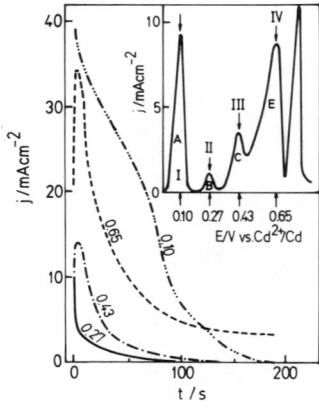

Figure 38. Potentiostatic current density–time $(j-t)$ transients of dissolution of various compounds from Cu–Cd alloy electrodeposited to 4.4 μm thickness on silver RDE at current ratio of 1:3 (corresponding ALSV is shown as inset). Solution for dissolution as in Figure 35.

(f) Potentiostatic step dissolution of individual intermetallic compounds

Instead of using deconvolution for accurate determination of the amount of charge under the ALSV peaks (cf. Fig. 35b), $j-t$ transients recorded upon the application of potentiostatic pulses that match the potentials of ALSV peaks should give a more precise result. Such pulses should be applied in sequence beginning at the most negative peak on the voltammogram. When the current of the corresponding $j-t$ transient decays to zero, one can assume that the phase is completely dissolved. Another pulse, at the potential of the next ALSV peak, can then be applied.

The results of the application of this stepwise potentiostatic pulse technique to the Cu–Cd system are shown in Fig. 38. It can be seen that the $j-t$ transients obtained exhibit different characteristics. The dissolution of pure Cd (peak A, $E_p = 0.10$ V vs. Cd^{2+}/Cd) is seen to take place smoothly with a shoulder; this shape is characteristic of the dissolution of pure metals. A small amount of $CuCd_3$ (peak B, $E_p = 0.27$ V vs. Cd^{2+}/Cd) is seen to dissolve with a fast decay of the current, while $j-t$ transients of the components richer in Cu (peaks C and E) exhibit maxima. It seems that diffusion of Cd through matrices containing excess Cu becomes a limiting factor after Cd atoms in the surface layer have been dissolved, causing a decrease of dissolution current and the appearance of maxima on the $j-t$ transients.[109,112]

Comparison of the results of deconvolution of the voltammograms with those of integration of $j-t$ transients recorded in the potentiostatic pulse sequence showed that a somewhat better resolution between the charges under peaks C, D, and E was obtained by the latter technique.

3. Quantitative Aspects

The ALSV records can be used for quantitative assessment of alloy coatings and their chemical and phase composition. Thus, if the current scale is reduced to 1 cm^2 of the electrode surface and the potential scale converted into the corresponding time scale, the integral of a record yields the total quantity of metal in the alloy layer per square centimeter. In an alloy composed of a more noble p-valent metal A and a less noble q-valent metal B, and if the peak with the most positive peak potential pertains to dissolution of the more noble component, the total quantity of metal [n(A + B) in moles per square centimeter] must be

$$n(A + B) = (Q_A/pF) + (Q_B/qF) \tag{93}$$

where Q_A is the integral charge under the most positive peak, and Q_B is the integral charge under all the preceding peaks. The thickness of the deposit can be calculated approximately as

$$d = \frac{n(A + B)[x_A M_A + (1 - x_A)M_B]}{x_A \rho_A + (1 - x_A)\rho_B} \tag{94}$$

assuming additivity of the densities of the two components, which in most cases is a good approximation.

The overall composition of the alloy is obtained by comparing the integral charge Q_A or Q_B with the total charge under the voltammogram; i.e.,

$$x_A = \frac{Q_A}{pFn(A + B)} \quad \text{or} \quad x_B = \frac{Q_B}{qFn(A + B)} \tag{95}$$

The amount of the ith phase can be obtained by integration of the corresponding peak; i.e.,

$$n_i = Q_i/qFm \tag{96}$$

where m is the number of atoms of the less noble metal in a "molecule" of the phase that is oxidized under the peak.

Having in mind that the peak reflects dissolution of the less noble component only (cf. Section VII.4.), the content of the ith phase in the alloy can be obtained as

$$x_i = n_i / \sum_{i=1}^{k} n_i \qquad (97)$$

if there are k phases in the alloy.

Hence, the ALSV method should provide for a complete description of an alloy coating.

4. Mechanism of Dissolution of Binary Alloys

When a single-phase binary alloy is anodically dissolved in an aqueous solution, assuming that all the necessary conditions (cf. Section VII.1) are fulfilled, the mechanism of dissolution may be[91]:

1. Simultaneous dissolution of both components of the alloy, with the more noble one being reprecipitated on the spot.

2. Partial dissolution of the less noble component, causing restructuring of the remaining matrix into a phase richer in the more noble component.

3. Complete dissolution (ionization) of the less noble component, with the atoms of the more noble component aggregating by surface diffusion (formation of patches and monolayers of the more noble component with the possibility of formation of small three-dimensional crystals, which may prevent further dissolution of the less noble component).

4. Complete dissolution (ionization) of the less noble component, leaving atoms of the more noble one in an unstable matrix. Atoms of both metals then move in the solid phase by volume diffusion, one kind dissolving, and the other kind restructuring the matrix so that it corresponds to the crystal lattice of the pure metal.

On the basis of diffusivities extrapolated from high-temperature measurements, the participation of the fourth mechanism should be negligible. However, this extrapolation assumes that the equilibrium concentration of vacancies is established and that the contribution of diffusion along small grain boundaries and dislocations is negligible. This is not the case in reality, since the diffusion coefficients at room temperature may be considerably larger than the extrapolated ones (in the case of Cu, monovacancies cannot account for this mechanism, but divacancies, with $D = 1.3 \times 10^{-12} \text{cm}^2/\text{s}$, may make the mechanism operative[91]).

There is a significant inconsistency in the results reported earlier from investigations of the mechanism of alloy dissolution, particularly for results obtained with bulk alloy samples.[90–100] For example, for one and the same system (Cu–Ni) it was found both that components dissolve simultaneously[95] and that Ni dissolves preferentially.[96] In the case of Cu–Au and Cu–Zn systems, two different mechanisms were observed, preferential dissolution of the less noble component and simultaneous dissolution of both components.[91–94] Actually, it was concluded that in the Cu–Zn system, enrichment of the surface with copper is a necessary condition for simultaneous dissolution of both components to start. Hence, there exists an induction period of disproportionate dissolution of the less noble component, preceding simultaneous dissolution.[94]

It is likely that the mechanism depends on the composition of the alloy. Thus, at low contents of the more noble metal, the dissolution of the less noble one leaves the atoms of the former loose and ready to oxidize. Conversely, at high contents of the more noble metal, the atoms of the less noble one are likely to tend to squeeze out of the lattice without producing much change in the latter.

It is obvious that in bulk alloy samples diffusional limitations (diffusion of metal atoms from the bulk of the alloy to the surface) influence the process of anodic dissolution. In the case of dissolution of thin layers of alloys, where the entire amount of the alloy is dissolved, such limitations (if they exist) can be overcome (cf. Section VII.1). However, two phenomena should be noted:

(a) Simultaneous dissolution creates locally an increased concentration of the positive constituent, corresponding to a more positive reversible potential of the pure metal phase than that attained at a particular moment in the course of the anodic sweep. Hence, that potential turns out to be cathodic with respect to the reversible potential, so that this metal constituent must deposit on the spot either in the form of another alloy phase with a more positive reversible potential or in the form of pure metal crystallites. Hence, the net recorded current reflects the dissolution of the negative constituent only.

(b) Dissolution of one alloy phase from a mixture with other phases, or through redeposited positive constituent, must run into transport difficulties. Hence, the dissolution current comes under diffusion control, which makes the current–potential (time) pattern similar to that obtained in such a situation for a cathodic metal deposition process; that is, it exhibits a maximum and subsequent decay.

As it was mentioned earlier[102,108,109] [cf. Sections VII.2(ii)(a) and VII.2(ii)(b)], the mechanism of simultaneous dissolution of both components, with immediate reprecipitation of the more noble one, is the most probable one. Such a conclusion appears to be confirmed by the results for Zn–Ni alloy dissolution presented in Fig. 39b. It is known that the potentials of pure Ni deposition and dissolution differ by several hundred millivolts (irreversible deposition/dissolution) and that a layer of pure Ni thicker than 1 μm cannot be quantitatively dissolved in any solution except concentrated HCl[107,115] (dissolution of Ni starts at about –0.15 V vs. SCE[107]). As can be seen in Fig. 39b, pure Ni (curve 1) does not dissolve in the solution investigated whereas pure Zn dissolves quantitatively[28] (curve 2). An alloy containing approximately 54 at. % Ni (curve 3) also dissolves quantitatively, indicating the presence of different phases/intermetallic compounds (cf. phase diagram in Fig. 39a). Taking into account that the Ni dissolved together with Zn from the alloy cannot reprecipitate (since the potential of alloy dissolution is more positive than the potential of pure Ni deposition) and that the entire amount of the alloy is quantita-

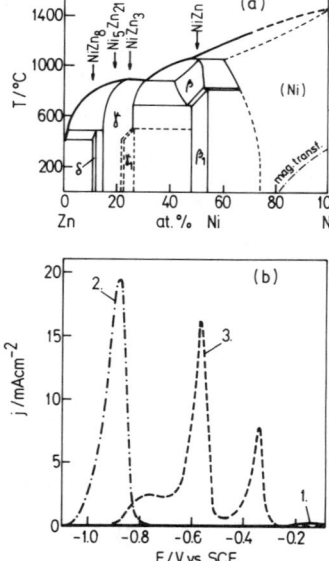

Figure 39. (a) Phase diagram of Zn–Ni binary system. (b) Voltammograms obtained by ALSV for dissolution ($v = 1$ mV/s) of pure nickel (curve 1), pure zinc (curve 2), and an alloy containing 54 at.% Ni and 46 at.% Zn (curve 3). Solution: $0.5M$ Na_2SO_4 + $0.1M$ EDTA, pH 5; 1000 rpm.

tively dissolved, the only explanation for such behavior is that both metals dissolve simultaneously.[115]

5. Kinetics of Dissolution of Binary Alloys

The peak potentials are complex quantities depending on the amount of the component in the alloy, the sweep rate, and diffusion limitations. Hence, the use of another characteristic of the ALSV curves is preferred. Thus, if it is assumed that at low current densities ($j \ll j_p$) the alloy component giving rise to the peak dissolves under activation control and that the cathodic partial current is negligible, the current density–potential relation should follow a Tafel type equation.

The problem is treated in detail in Section IV.2. Hence, Eqs. (40)–(48) are applicable. It was shown that the Tafel constant a is an important characteristic of the deposition/dissolution process as it reflects both a thermodynamic property, $E^\ominus$, and a kinetic property, j_0, of the alloy. Although the two cannot be separated and, hence, neither the Gibbs energy of formation of the alloy nor the rate constant of its electrochemical dissolution can be evaluated without some additional assumption, a can be considered as the electrochemical characteristic of the alloy, defining the potential range in which the alloy undergoes active dissolution. Having in mind (Eq. 49) that $a = E_{(j/j_0 = 1)}$, its value can easily be determined experimentally.

Such an analysis was performed for the binary system Ag–Pd[114] (Fig. 30). Table 2 lists the a values for Ag in a series of alloys covering a broad range of compositions. In the case of pure Ag, $E^\ominus$ can be calculated from the known values $E^\ominus(Ag^+/Ag) = 0.80$ V and $E^\ominus(Pd^{2+}/Pd) = 0.83$ V versus SHE and the stability constants of the $AgCl_4^{3-}$ and $PdCl_4^{2-}$ complexes present in solution. From the value of a and using Eq. (48), one can calculate $j_0 = 0.9$ A/cm^2. This value is consistent with the known fast electron exchange of Ag in aqueous solutions.[123]

The change in a if the alloy is an ideal solid solution can be predicted. In this case Eq. (17) should be used to calculate the alloy potentials

$$j_0 \text{ (alloy)} = -j_0(Ag)\, x(Ag) \qquad (98)$$

where $x(Ag)$ is the mole ratio of Ag in the alloy. The calculated values of a (ideal), neglecting $\mu^\ominus$, are also shown in Table 2.

It can be seen that the difference between experimental and calculated a values increases with increasing Pd content. This may be due to either

Table 2
Characteristic Parameter a for the Dissolution of Ag from Ag–Pd Alloys[a]

Composition (at.% Ag)	a	a(ideal)
20	0.207	0.148
40	1.140	0.100
60	0.093	0.063
80	0.057	0.041
100	0.025	0.025

[a] $j^o = 1$ mA/cm^2 (cf. eq. 49); $E_{ref} = E$(Ag/AgCl).

deviations from ideality in the Gibbs energy of mixing or kinetic inhibitions in the release of Ag from the crystal lattice (diffusion). However, some discrepancy may occur because the condition $j \ll j_p$ is not fulfilled, particularly at low Ag contents (curve 2 of Fig. 30).

6. ALSV Investigation of Phase-Transformation Kinetics

It is well known that equilibrium alloys can hardly be obtained by electrochemical deposition.[7,8] In the case of Cu–Cd binary alloy, it was found that in most cases more than one intermetallic compound is present in the electrodeposited alloys and that the shape of the voltammogram obtained by ALSV changes with time when a freshly deposited alloy is held in an inert atmosphere of purified nitrogen at room temperature. It has been assumed that such a change is a result of some solid-state reaction caused by the tendency to establish a minimum Gibbs energy of formation of intermetallic compounds.[109,117]

Recently, an attempt was made to investigate the phase-transformation kinetics in electrodeposited Cu–Cd alloys using the ALSV technique.[113] The results of this investigation are shown in Figs. 40–42.

The change of the shape of an ALSV a freshly deposited Cu–Cd alloy held in an inert atmosphere at room temperature with time of holding is shown in Fig. 40.

A well-defined shoulder, which vanishes within 5 h, is seen at the position of peak B (cf. Fig. 35b). On the other hand, the peak C, which initially does not exist, develops into a well-defined one after the same period. Peak D grows at the same time, with some shift of the peak

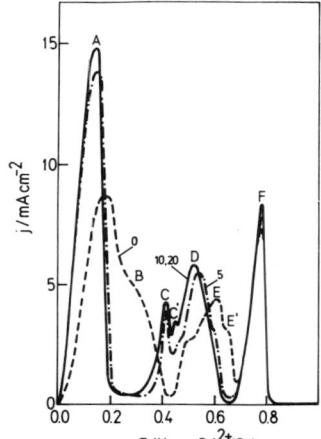

Figure 40. Voltammograms obtained by ALSV at different times (indicated in hours on the curves) after deposition of Cu–Cd alloy containing 86 at.% Cd. Conditions of dissolution as in Fig. 35.

potential and a narrowing of the shape, while peak E vanishes. Moreover, some additional small peaks are seen to vanish (E′) or appear (C′). The time dependence of these changes was followed in a quantitative manner using shorter time intervals for recording the ALSV. As the sweeps were rather slow (1 mV/s), the time that a particular peak was recorded had to

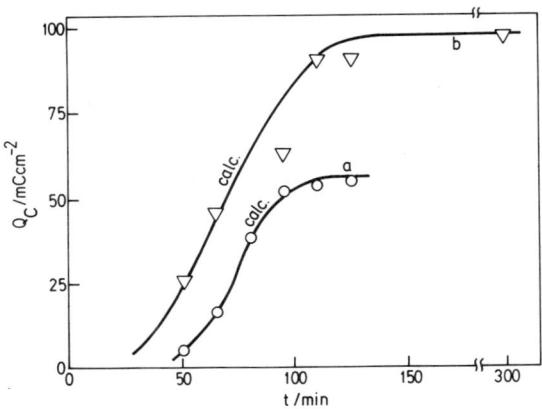

Figure 41. Time dependence of the charge under peak B of Cu–Cd alloys containing 67 at.% Cd–33 at.% Cu (a) and 86 at.% Cd–14 at.% Cu (b) shortly after deposition.

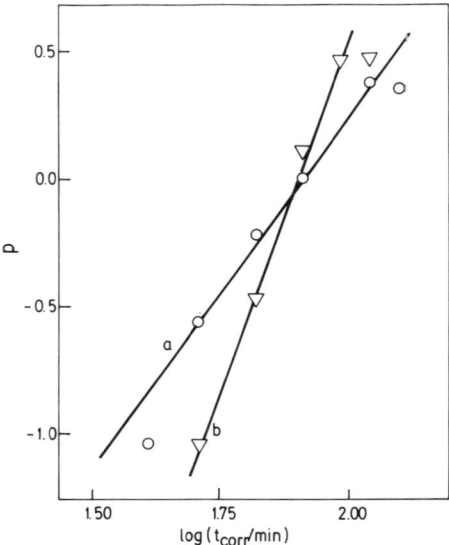

Figure 42. Time dependence of the complex variable p (Eq. 102) in alloys containing 67 at.% Cd–33 at.% Cu (a) and 86 at.% Cd–14 at.% Cu (b).

be corrected for the time that had elapsed between starting the ALSV and reaching the peak.

The most dramatic change was the appearance and development of peak C, as well as the vanishing of peak B. The change of charge under peak C with time is shown in Fig. 41. As can be seen, most of these changes take place between 30 min and 2h.

Quantitative treatment of the kinetics can be performed by using the empirical Johnson–Mehl equation[124]

$$y = 1 - \exp(-kt^n) \qquad (99)$$

where y is a reaction variable which can be defined as

$$y = x_t / x_{t \to \infty} \qquad (100)$$

where x_t and $x_{t \to \infty}$ are the contents of the compound in the alloy at time t and after stabilization, respectively.

In order to determine k and n from the experimental data of Fig. 41, Eq. (99) is transformed into

$$p = \ln\left[\ln\left(\frac{1}{1-y}\right)\right] = n \ln t + \ln k \qquad (101)$$

Plots of p versus $\log t$ are shown in Fig. 42. It can be seen that in the case of an alloy with a lower amount of Cd (Fig. 42a) the value of n is very high (4.4), reflecting the delay in the appearance of the peak in the voltammogram and then the relatively sudden rise to complete formation of the phase. Such a rapid change is characteristic of slow nucleation producing a burst of nuclei after an induction period.[125] The slower change with $n = 2.7$ in the case of a larger excess of Cd over Cu is likely to indicate that a slow release of Cd is rate-determining. The time dependences calculated using the values obtained for n and k are seen in Fig. 41 to provide a fairly good fit of the experimental data.

ACKNOWLEDGMENTS

The authors are indebted to their co-worker Rade Stevanović (M.Sc.) for technical assistance in the preparation of the manuscript.

REFERENCES

[1] M. de Ruolz, *C.R. Acad. Sci.* **15** (1842) 1140.
[2] R. D. Srivastava and R. C. Mukerjee, *J. Appl. Electrochem.* **6** (1976) 321.
[3] A. Brenner, *Electrodeposition of Alloys: Principles and Practice*, Academic Press, New York, 1963.
[4] V. V. Bondar, V. V. Grimina, and V. N. Pavlov, *Itogi nauki i tehniki,* Elektrokhimiya, Vol. 16, Izd. Viniti, Moscow, 1980.
[5] F. Spitzer, *Z. Elektrochem.* **11** (1905) 345.
[6] M. Schlötter, in *Die Elektrolytische Darstellung von Legierungen aus wässerigen Lösungen* (Sammlung Vieweg, Tagesfragen aus den Gebieten Naturwiss.u.der Technik, No. 19), Ed. by R. Kremann, Vieweg, Braunschweig, 1914.
[7] K. M. Gorbunova and Yu. M. Polukarov, in *Advances in Electrochemistry and Electrochemical Engineering*, Vol. 5, Ed. by C. W. Tobias, John Wiley & Sons, New York, 1976.
[8] N. P. Fedoteev, N. N. Bibikor, P. M. Vyacheslavov, and C. Ya. Grilihes, *Elektroliticheskie Splavy*, MASHGIZ, Moscow, 1962.
[9] C. L. Faust, in *Modern Electroplating*, Ed. by F. A. Lowenheim, John Wiley and Sons Inc., New York, 1963, Chapter 18.
[10] H. Fischer, *Z. Elektrochem.* **59** (1955) 612.
[11] T. Banerjee, *Symposium on Electroplating and Metal Finishing, India*, No. **29**, 1952.
[12] W. Reinders, *Z. Physik. Chem.* **42** (1902) 225.
[13] R. Kremann and R. Müller, in *Ostwald–Drucker Handbuch der algemeine Chemie*, Vol. 1, Ed. by P. Walden and C. Drucker, Akademische Verlagsgesellschaft, Leipzig, 1930, pp. 626–628, 644–648.
[14] F. A. Krüger, *J. Electrochem. Soc.* **125** (1978) 2028.
[15] N. Pushin, *Z. Anorg. Chem.* **56** (1907) 1.

[16] G. E. Kimbal, *J. Chem. Phys.* **8** (1940) 199.
[17] Yu. M. Polukarov, K. M. Gorbunova, and V. V. Bondar, *Zh. Fiz. Khim.* **36** (1962) 1870.
[18] A. L. Rotinyan and E. N. Molotkova, *Zh. Prikl. Khim.* **11** (1959) 2502.
[19] F. Spitzer, *Z. Elektrochem. Angew. Physik. Chem.* **23** (1905) 345.
[20] N. D. Tomashov, *Teoriya korrozii i zashchity metallov*, M., AN SSSR, 1960, p. 591.
[21] B.I. Skirstymonskya, Dissertation, LTI im. Lensoveta, 1958.
[22] V. I. Lainer, Sbornik Nauchnikh Rabot, Metallurgizdat, Moscow, 1950, p. 70.
[23] E. Raub and A. Engel, *Z. Metallk.* **41** (1950) 485.
[24] E. Raub and F. Sautter, *Metalloberfläche* **9B** (1955) 145.
[25] B. Ya Pines, *Izv. Sekt. Fiz. Khim. Analy. Inst. Obshch. Neorg. Khim. Khim. Tekhnol. Akad. Nauk SSSR* **16** (1943) 63.
[26] W. Hume-Rotery and G. V. Raynor, *The Structure of Metals and Alloys*, Institute of Metals, London, 1956.
[27] H. Jones, *Phil. Mag.* **41** (1950) 663.
[28] N. F. Mott, *Prog. Metal Phys.* **3** (1952) 76.
[29] C. Wagner, *Thermodynamics of Alloys*, Addison-Wesley, Cambridge, Massachusetts, 1952.
[30] J. A. Dean, *Lange's Handbook of Chemistry*, 13th ed., McGraw-Hill, New York, 1985.
[31] K. V. Yuatsimirsky, V. P. Vasilev, *Konstanty Nestojkosti Kompleksnyh Soedinenij*, AN ASSR, Moscow, 1959.
[32] A.R. Despić, V. Marinović, and V. D. Jović, *J. Electroanal. Chem.* **339** (1992) 473.
[33] A. R. Despić, D. Jovanović, and T. Rakić, *Electrochim. Acta* **21** (1976) 63.
[34] F. Mattson and J. O'M. Bockris, *Trans. Faraday Soc.* **55** (1959) 1586.
[35] J. O'M. Bockris and H. Kita, *J. Electrochem. Soc.* **109** (1962) 928.
[36] P. Delahay, *New Instrumental Methods in Electrochemistry*, Interscience, New York, 1954.
[37] M. D. Allen, *Organic Electrode Processes*, Reinhold, New York, 1958.
[38] D. E. Hall, *Plating Surf. Finish.* (November 1983) 59.
[39] E. Raub and B. Wullhorst, *Z. Metallforsch.* **2** (1947) 33.
[40] K. Vetter, *Elektrochemische Kinetik*, Springer-Verlag, Berlin, 1961.
[41] H. Dahms and J. Croll, *J. Electrochem. Soc.* **112** (1965) 771.
[42] K. Higashi, H. Fukushima, T. Urakawa, T. Adaniya, and K. Matsuko, *J. Electrochem. Soc.* **128** (1981) 2081.
[43] T. Horkans, *J. Electrochem. Soc.* **126** (1979) 1861; **128** (1981) 45.
[44] W. R. Meyer and A. Phillips, *Trans. Electrochem. Soc.* **73** (1938) 377.
[45] K. Aotani, *J. Electrochem. Soc. Jpn.* **21** (1953) 180.
[46] P. F. Mikhalev, *C. R. Acad. Sci. URSS* **24** (1939) 899.
[47] W. Blum, *Trans. Electrochem. Soc.* **40** (1921) 307.
[48] U. Cohen, F. B. Koch and R. Sard, *J. Electrochem. Soc.* **130** (1983) 1987.
[49] D. Tench and J. White, *Metall. Trans.* **15A** (1984) 2039.
[50] C. Ogden, *Plating Surf. Finish.* **73** (1986) 130.
[51] M. W. Verbrugge and C. W. Tobias, *J. Electrochem. Soc.* **132** (1985) 1298.
[52] A. R. Despić and V. D. Jović, *J. Electrochem. Soc.* **134** (1987) 3004.
[53] A. R. Despić, V. D. Jović, and S. Spaić, *J. Electrochem. Soc.* **136** (1989) 1651.
[54] A. R. Despić and T. Trišović, *J. Appl. Electrochem.* **23** (1993) in press.
[55] G. P. Power and I. M. Ritchie, in *Modern Aspects of Electrochemistry*, No. 11, Ed. by B. E. Conway and J. O. M. Bockris, Plenum Press, New York, 1975, Chapter 5.
[56] J. Yahalom and O. Zadok, *J. Mater. Sci.* **22** (1987) 499.
[57] D. S. Lashmore and M. P. Dariel, *J. Electrochem. Soc.* **135** (1988) 1218.
[58] D. Tench and J. White, *J. Electrochem. Soc.* **137** (1990) 3061.
[59] J. R. Roos, J. P. Celis, and J. A. Helsen, *Trans. Inst. Met. Finish.* **55** (1977) 113.
[60] T. W. Tomaszewski, *Trans. Inst. Met. Finish.* **54** (1976) 45.

[61] M. A. Belen'kii and F. A. Ivanov, *Elektroosazhdenie metallicheskih pokrytiya*, Izd. Metallurgiya, Moscow, 1985.
[62] I. G. Habibullin and R. S. Saifullin, *Anticorrosion Coatings*, Meeting on Thermostable Coatings, Proceedings, Nauka, Leningrad, 1983, p. 84.
[63] R. S. Saifullin, *Neorganicheskie kompozitsionnie materiali*, Izd. Himiya, Moscow, 1983.
[64] J. C. Whithers, *Prod. Finish* **8** (1962).
[65] P. W. Martin and R. V. Williams, *Proceedings Interfinish '64*, British Iron and Steel Research Associates, London, 1964, pp. 182–188.
[66] E. A. Brandes and D. Golthorpe, *Metallurgiya* **1967** 195.
[67] T. W. Tomaszewski, R. J. Claus, and H. Brown, *Proc. Am. Electroplaters Soc.* **50** (1963) 169.
[68] D. W. Snaith and P. D. Groves, *Trans. Inst. Met. Finish.* **50** (1972) 95.
[69] C. Buelens, J. P. Celis, and J. R. Roos, *J. Appl. Electrochem.* **13** (1983) 541.
[70] J. Foster and A. M. J. Kariapper, *Trans. Inst. Met. Finish.* **51** (1973) 27.
[71] A. M. J. Kariapper and J. Foster, *Trans. Inst. Met. Finish.* **52** (1974) 87.
[72] J. R. Roos, J. P. Celis, H. Kelchtermans, M. Van Camp, and C. Buelens, in *Proceedings of the 10th World Congress of Metal Finishing*, Ed. by S. Hawyama, The Metal Finishing Society of Japan, Kyoto, 1980, p. 203.
[73] M. Van Camp, Engineering Thesis, K.U. Leuven, 1979.
[74] J. P. Celis, Ph.D. Thesis, K. U. Leuven, 1976.
[75] J. P. Celis and J. R. Roos, *J. Electrochem. Soc.* **124** (1977) 1502.
[76] N. Guglielmi, *J. Electrochem. Soc.* **119** (1972) 1009.
[77] N. Masuko and K. Mushiake, *J. Met. Finish. Soc. Jpn.* **28** (1977) 534.
[78] M. K. Totlani and S. N. Athavale, *J. Electrochem. Soc. India* **31** (1982) 119.
[79] J. P. Celis, J. R. Roos, and C. Buelens, *J. Electrochem. Soc.* **134** (1987) 1402.
[80] T. W. Tomaszewski, L. L. Tomaszewski, and H. Brown, *Plating* **56** (1969) 1234.
[81] E. S. Chen, G. R. Lakshminarayanan, and F. K. Sautter, *Metall. Trans.* **2** (1971) 937.
[82] C. White and J. Foster, in Proceedings of the Annual Technical Conference IMF, Torquay, 1978.
[83] J. E. Hoffmann and R. C. Ernst, INCRA—Project No. 31, 1964.
[84] A. A. Wragg, Claire George and R. M. Hooper, 41st ISE Meeting, Book of Abstracts, Vol. II, p.Th-Ms10/1, Prague, 1990.
[85] V. D. Stanković, B. Stanojević, and D. Marković, Book of Abstracts, 38th ISE Meeting, Maastricht, the Netherlands, 1987, Vol. II, p. 893.
[86] V. D. Stanković and M. Gojo, Book of Abstracts, 42nd ISE Meeting, Montreux, Switzerland, 1991, pp. 3–24.
[87] V. D. Stanković and M. Gojo, *J. Mater. Sci.*, in press.
[88] U. R. Evans, *J. Inst. Metals* **40** (1928) 99.
[89] A. Glazunov, *Metallic Protective Coatings* **1953** 1262.
[90] R. F. Steigerwald and N. D. Greene, *J. Electrochem. Soc.* **109** (1962) 1026.
[91] H. W. Pickering and C. Wagner, *J. Electrochem. Soc.* **114** (1967) 698.
[92] H. W. Pickering, *J. Electrochem. Soc.* **115** (1968) 690.
[93] H. W. Pickering and P. J. Byrne, *J. Electrochem. Soc.* **118** (1971) 209.
[94] J. E. Holliday and H. W. Pickering, *J. Electrochem. Soc.* **120** (1973) 470.
[95] J. O'M. Bockris, B. T. Rubin, A. Despić, and B. Lovreček, *Electrochim. Acta* **17** (1972) 973.
[96] A. Mance and A. Mihajlović, *J. Appl. Electrochem.* **10** (1980) 967.
[97] H. P. Lee and K. Nobe, *J. Electrochem. Soc.* **131** (1984) 1236; **11** (1981) 205; **11** (1981) 299.
[98] I. Petro, T. Mallat, A. Szabo, and F. Hange, *J. Electroanal. Chem.* **160** (1984) 289.
[99] E. T. Shapovalov, L. I. Baranova, and G. O. Zektser, *Elektrokhimicheskie metody v metallovodenii i fazovom analize*, Izd. Metallurgiya, Moscow, 1988.

[100] J. Stevanović, L. Skibina, M. Stefanović, A. Despić, and V. D. Jović, *J. Appl. Electrochem.* **22** (1992) 172.
[101] S. Swathirajan, *J. Electrochem. Soc.* **133** (1986) 671.
[102] V. D. Jović, R. M. Zejnilović, A. R. Despić, and J. S. Stevanović, *J. Appl. Electrochem.* **18** (1988) 511.
[103] S. Swathirajan, *J. Electroanal. Chem.* **221** (1987) 211.
[104] P. C. Andricacos, J. Tabib, and L. T. Romankiw, *J. Electrochem. Soc.* **135** (1988) 1172.
[105] P. C. Andricacos, C. Avana, J. Tabib, J. Duković, and L. T. Romankiw, *J. Electrochem. Soc.* **136** (1989) 1336.
[106] K. H. Wong and P. C. Andricacos, *J. Electrochem. Soc.* **137** (1990) 1087.
[107] J. Horkans, I-Chia Hsu Chang, P. C. Andricacos, and E. J. Podlaha, *J. Electrochem. Soc.* **138** (1991) 411.
[108] V. D. Jović, A. R. Despić, J. S. Stevanović, and S. Spaić, *Electrochim. Acta* **34** (1989) 1093.
[109] V. D. Jović, S. Spaić, A. R. Despić, J. S. Stevanović, and M. Pristavec, *Mater. Sci. Technol.* **7** (1991) 1021.
[110] L. Skibina, J. Stevanović, and A. R. Despić, *J. Electroanal. Chem.* **310** (1991) 391.
[111] R. Stevanović, I. Kovrigina, and A. R. Despić, *J. Serb. Chem. Soc.* **56** (1991) 217.
[112] A. R. Despić, *Electrochemistry in Transition from the 20th to the 21st Century*, Ed. by O. J. Murphy, S. Srinivasan, and B. E. Conway, Plenum Press, New York, 1992, p. 453
[113] J. S. Stevanović, V. D. Jović, and A. R. Despić, *J. Electroanal. Chem.* **349** (1993) 365.
[114] V. D. Jović, B. M. Jović, and A. R. Despić, *J. Electroanal. Chem.* **357** (1993) 357.
[115] V. D. Jović, A. R. Despić, J. S. Stevanovći, and V. Jevtić, Book of Abstracts, 44th ISE Meeting, Berlin, Germany, 1993, p. 240.
[116] N. P.Fedotev and P. M. Vyacheslavov, *Plating* **1970** 700.
[117] D. Porter and K. A. Easterling, *Phase Transformations in Metals and Alloys*, Van Nostrand Reinhold, Wokingham, United Kingdom, 1980.
[118] M. Hansen and K. Andrenko, *Constitution of Binary Alloys*, McGraw-Hill, New York, 1958.
[119] Ch. Thierer, Dissertation, Stuttgart Technische Hochschule, 1935; *Gmelin Handbuch der anorganischen Chemie*, System 65A(5), Weinheim/Bergstrasse, Verlag Chemie, 1951, p. 653.
[120] E. M. Savitskii and N. L. Pravoverov, *Zh. Neorg. Khim.* **76** (1961) 499.
[121] B. Sturzenegger and J. Cl. Puippe, *Platinum Met. Rev.* **28** (1984) 117.
[122] V. D. Jović, B. N. Grgur, M. V. Stojanović, and B. M. Jović, Extended Abstracts, 25th Meeting of Miners and Metallurgists, Bor, Yugoslavia, 1993, Vol. II, p. 828.
[123] A. R. Despić and J. O'M. Bockris, *J. Chem. Phys.* **32** (1960) 389.
[124] W. A. Johnson and R. F. Mehl, *Trans. AIME* **135** (1939) 416.
[125] A. R. Despić and M. V. Djorović, *Electrochim. Acta* **29** (1984) 131.

Electrochemical Aspects of Stress Corrosion Cracking

José R. Galvele

*Gerencia de Desarrollo, Comisión Nacional de Energía Atómica,
1429-Buenos Aires, Argentina*

I. INTRODUCTION

Stress corrosion cracking, or environmentally induced cracking, is a challenging subject for very many reasons. It stands as an equal challenge to physicists, chemists, electrochemists, metallurgists, mechanical engineers, etc. It has been suffered by the human race since as long ago as the Bronze Age, but no convincing explanation has been found so far for this phenomenon. Many theories have been suggested, but, in spite of the optimistic claims of their authors, among which the present writer is included, Nature has inevitably condemned them to a poor, short life.

The study of this subject involves such a wide spectrum of scientific disciplines that no scientist at present can cover it all with equal expertise.

Since many of the cases of stress corrosion cracking take place in the presence of electrolytic environments, electrochemical reactions are involved in the process. The present chapter will concentrate on the electrochemical aspects of stress corrosion cracking. The important contributions made by electrochemistry to a better understanding of the problem will be reviewed. The limitations of electrochemistry in terms of providing a complete description of the problem will also be pointed out.

1. Definitions

As the subject of this chapter is a fully interdisciplinary one, we will present here a series of definitions, which, while completely obvious to some readers, are unknown to or sometimes misused by others.

Modern Aspects of Electrochemistry, Number 27, edited by Ralph E. White *et al.* Plenum Press, New York, 1995.

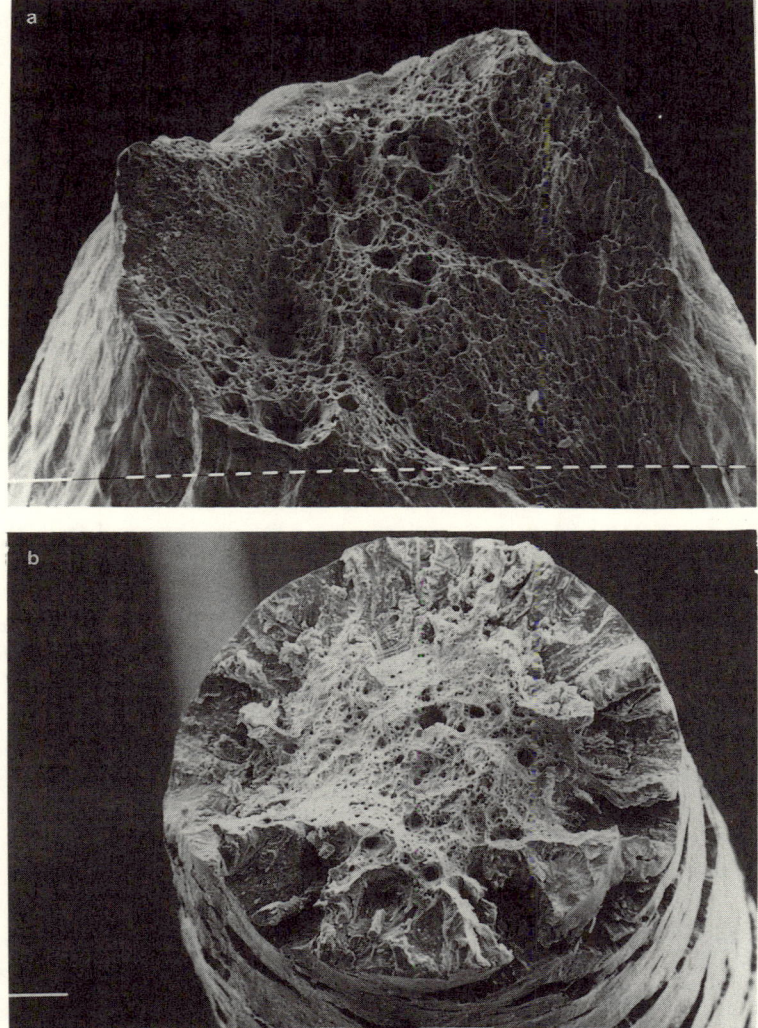

Figure 1. (a) Brass (64% Cu–26% Zn) alloy bar broken by overloading in air. Typical ductile failure. (Dashes = 10 μm.) (b) and (c) Samples of the same alloy as in (a), strained in a NaNO$_2$ solution at 0.2 V vs. NHE. Typical stress corrosion cracking fractures. (Dashes = 100 μm.)

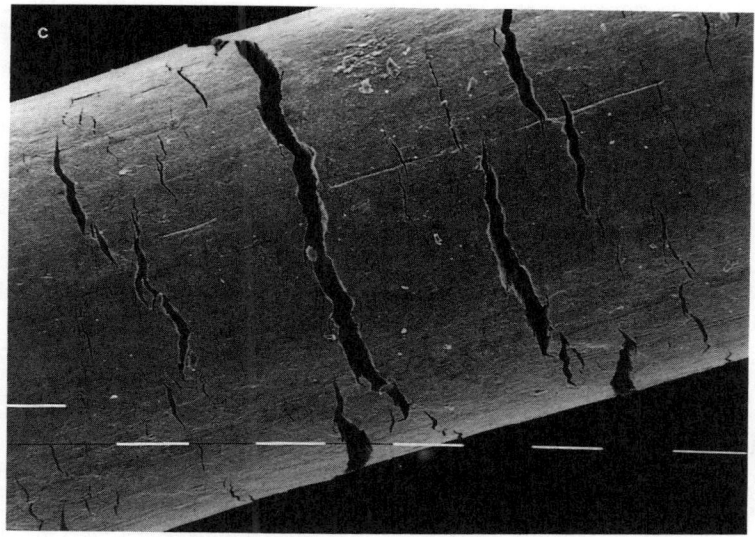

Figure 1. (continued)

(i) Environmentally Induced Cracking

Our everyday experience with materials leads to the ideas of plasticity, ductility, fragility, and brittleness. The idea of plasticity is easily conveyed by the example of materials such as lead or plaster. Ductility is measured by the amount of plastic deformation a material shows before fracture. Ductile materials are exemplified by mild steel, aluminum, and lead. On the other hand, we get easily acquainted with the idea of fragility when we work with glass or ceramics. We say that a material is brittle when it breaks with almost no plastic deformation.

For many practical applications, brittle materials are avoided. In the construction of the load-sustaining parts of innumerable devices, only ductile materials are used. This choice is made either for safety reasons or for the sake of durability of the structure.

Over a century of experience in the use of many of these materials has shown that very often, as a result of exposure to the environment, ductile materials break unexpectedly in a brittle-like fashion (Fig. 1).

Stress corrosion cracking (SCC), or environmentally induced cracking, is a phenomenon by which ductile materials, when exposed to certain environments, behave in a brittle-like fashion, cracking at stress values well below the fracture stress.

The environmental degradation of materials is not only restricted to ductile metals. It has also been observed that, when exposed to certain environments, some materials such as glass, plastics, and ceramics show a severe degradation of their mechanical properties.

While the causes of degradation are as many as the nature of the affected materials is diverse, many points of coincidence can be found in the various processes of environmentally induced cracking.

(ii) Crack Aspect Ratio

Before we consider the various mechanisms of SCC, it is worthwhile to point out one important property of the SCC process, which is the *crack aspect ratio*. It is the ratio between crack length and crack opening, which in environmentally induced cracking is always high. Its value is well above 100 and usually even above 1000. Such a high aspect ratio means that while the crack is propagating, the sides of the crack show neither corrosion nor a significant plastic deformation. Any environmentally induced cracking mechanism proposed has to take this fact into account.

(iii) Stress Distribution Heterogeneity

Another important property of a cracked material subjected to stress is shown in Fig. 2. When the crack is observed from the inside of the

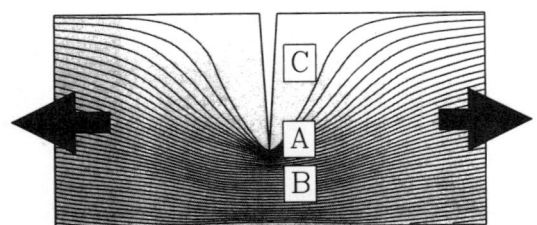

Figure 2. Cracks in a solid act as stress concentrators, having a strong influence on the fracture strength of the material. From the material point of view, the stress concentration in **A** is higher than that in **B**. On the other hand, from the environment point of view, the metal shows a very high stress concentration at the tip of the crack, **A**, whereas there is no stress at the sides of the crack, **C**.

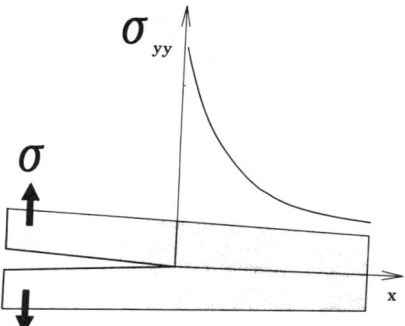

Figure 3. Stress distribution in a pre-cracked brittle material. The stress value shows a sharp increase at the tip of the crack, reaching eventually that of the fracture strength of the material.

material, there is a stress concentration induced by the presence of the crack. However, when the crack is observed from within the environment, a very important point is noticeable: the walls of the crack show practically no stress, while a very high stress concentration is found at the tip of the crack. From the point of view of stress distribution, the environment is in contact with a strongly heterogeneous material. The metal is practically stress-free, with a very high stress concentration located at the tip of the crack. Computer simulation of cracks[1,2] shows that this heterogeneity covers a very short distance, amounting only to several interatomic distances.

(iv) Stress Value at the Tip of the Crack

Another important parameter is the value of the tensile stress at the tip of the crack. In brittle materials (Fig. 3), the stress will show a sharp increase at the tip of the crack, although it will remain below the ideal strength value, which can range, for metals, between 10 and 100 GPa.[3] In the case of ductile materials (Fig. 4), on the other hand, the stress at the tip of the crack will increase up to the yield strength value, leading to a plastic zone in front of the crack. The maximum stress value would be a function of the hardness of the metal and could range from around 5 up to 2000 MPa.[4] The size of this plastic zone, as shown by fracture mechanics, is a function of the applied stress and the geometry of the sample.

(v) Induction Time

In most cases of environmentally induced cracking, an *induction time* is clearly distinguished. From a practical point of view, this parameter

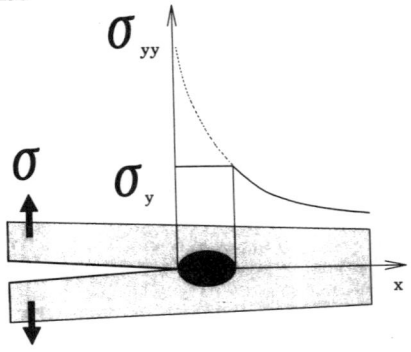

Figure 4. Stress distribution in a pre-cracked ductile material. The stress value increases as the tip of the crack is approached, reaching eventually the yield strength of the material and producing a plastic zone in front of the crack.

could be so relevant as to allow the use of materials under conditions in which they are susceptible to environmentally induced cracking. From a mechanistic point of view, the induction time is related to the time it takes to produce, at the metal interface, the environmental conditions leading to cracking. In some systems, cracking will only occur at certain acidic pH values, found only in pits or crevices. In such cases the induction time will be consumed in the formation of a pit or in the initiation of crevice corrosion. This explains why in those systems the cracking susceptibility measured on smooth samples is smaller than that measured on precracked samples.[5,6]

(vi) Failure Time

This parameter was very frequently measured in the past. It involves the induction time plus the time for crack propagation. Unless these two components can be evaluated independently, the measurement of the failure time has in itself very little significance in mechanistic studies.

(vii) Rate-Controlling Step

The reactions involved in SCC are heterogeneous and consequently involve a series of steps, the rate-controlling step being the slowest one. Its relevance in environmentally induced cracking has been pointed out by various authors.[7-11]

The steps involved could be numerous and will be different for different mechanisms. A hypothetical, arbitrary, and by no means exhaustive selection could be:

1. Propagation of an elastic wave in the crystal
2. Rate of movement of dislocations

3. Diffusion of the environment-active species
4. Molecular-level crack propagation step [such as (i) dissolution of a metal atom, (ii) decohesion of two metallic atoms, (iii) arrival of a vacancy to the tip of the crack, or (iv) breaking of a chemical bond, etc.]
5. Diffusion of the reaction products
6. Relaxation of the crack

Several other steps could be suggested, and their arrangement could be either in series and/or parallel. From a mechanistic point of view, step 4 is the one that attracts most of the researchers' attention. If the experimental conditions are such that step 4 is the slowest one, the nature of this step could be determined by analyzing how its rate is changed by a series of variables, such as temperature, stress, material composition, environment composition, etc.

In environmentally induced cracking, changes in the rate-controlling step have very often been detected and have to be taken into account in the analysis of the mechanisms involved. In SCC in aqueous environments (Fig. 5),[12] for example, the maximum crack velocities reported, for a wide variety of metals and environments, are on the order of 10^{-5} m/s. No higher crack velocities can be obtained, under conventional testing techniques, because the slowest step is that of diffusion of the reactive species to the tip of the crack. A good example is the case of Ag–15Pd in various environments.[12] It was found that SCC starts only above the formation potential of the respective silver compounds. The presence of such compounds was a necessary condition for SCC. For such a compound to be

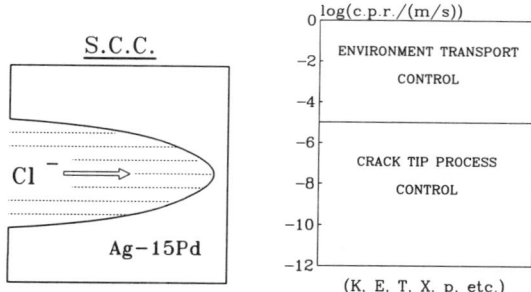

Figure 5. Example of stress corrosion cracking where high crack velocities are controlled by the rate of transport of chloride ions from the bulk solution to the tip of the crack. The example is an Ag–15Pd alloy in NaCl solutions.[12]

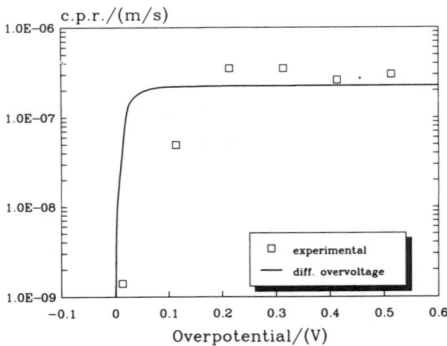

Figure 6. Maximum crack velocity allowed by diffusion overvoltage (—) compared with the crack propagation rate values measured for an Ag–15Pd alloy in a $1M$ KBr solution (□). The maximum c.p.r. measured is shown to be limited by the diffusion overvoltage.[12]

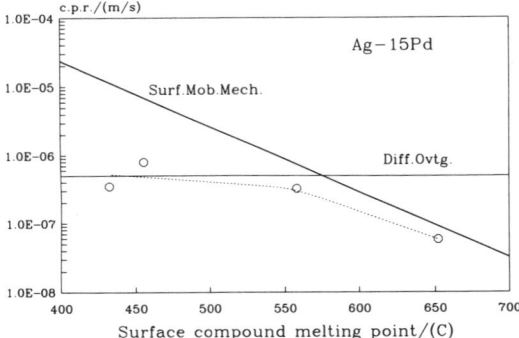

Figure 7. Crack propagation rate as a function of the surface compound melting point for silver alloys, $\sigma = 62.4$ MPa. Temperature = 25°C. Comparison between the values predicted by the surface-mobility SCC mechanism (—) and the c.p.r. values measured for an Ag–15Pd alloy in $1M$ KCl, $1M$ KBr, $1M$ KI, and $1M$ Na$_2$SO$_4$ solutions (○).[12] The measured values follow the predicted values up to the point where diffusion overvoltage becomes the rate-controlling step.

present at the tip of the crack, the respective anion had to diffuse through the solution inside the crack and reach the crack tip. If the crack velocity was increased, for example, by increasing the electrode potential, a point was reached when the rate of diffusion of the active species became the rate-controlling step. This is shown in Fig. 6. It is observed that the crack velocity increased up to a limiting value where the process became controlled by diffusion overvoltage. Figure 7 shows the transition from a diffusion-overvoltage-controlled process to a process controlled by the reaction at the tip of the crack.

The diffusion-controlled crack velocity can be changed by changing either the temperature or the viscosity of the solution,[7,13] but the results are of little significance if the aim is to investigate the process at the tip of the crack. These measurements only give the rate of migration of the reactive species along the crack, but give no information about the mechanism by which the damaging action of such species takes place at the tip of the crack.

In some other cases, as in liquid metal embrittlement (LME) (Fig. 8),[13,14] very high crack velocities can be reached before the arrival of the liquid metal at the tip of the crack becomes the rate-controlling step. In solid-metal-induced cracking (SMIC; Fig. 9),[15] on the other hand, the embrittling species—a solid metal—has to reach the tip of the crack by surface diffusion, and the range of crack velocities available for the study of the reaction at the tip of the crack is very narrow.

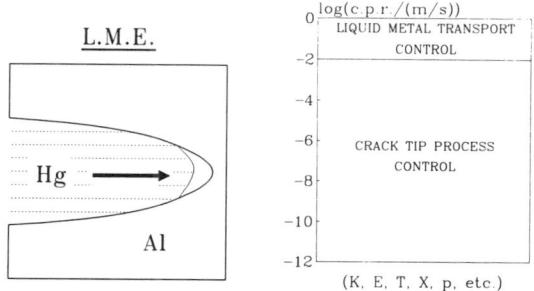

Figure 8. Example of liquid metal embrittlement, where high crack velocities are controlled by the rate of arrival of the liquid metal at the tip of the crack, as is observed for aluminum alloys and titanium alloys in mercury.[13,14]

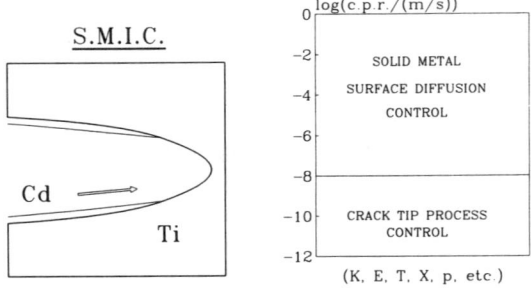

Figure 9. Case of solid-metal-induced cracking.[15] The embrittling species—a solid metal—has to reach the tip of the crack by surface diffusion, and the range of crack velocities available for studying the critical step at the tip of the crack is very narrow.

Finally, in hydrogen embrittlement (HE), the range of crack velocities useful for mechanistic studies varies with the metal and the mechanisms involved. Figure 10 depicts a case where the arrival of hydrogen atoms at the tip of the crack could be the rate-controlling step.

(viii) Stress Intensity Factor

The ideal strength of solids is reported to be from 10 to 100 GPa.[3] The experimental values found in brittle materials, such as glass, are less

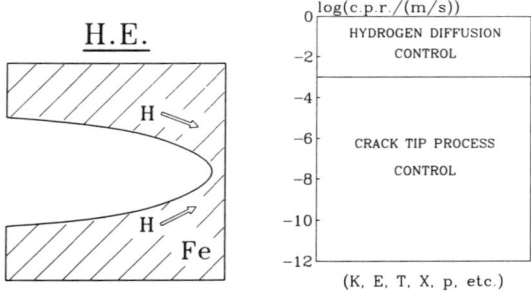

Figure 10. Case of hydrogen embrittlement (HE). Here the range of crack velocities useful for mechanistic studies varies with the metal and the mechanisms involved. The figure depicts a case where the arrival of hydrogen atoms at the tip of the crack is the rate-controlling step.

than one-hundredth of the ideal value. To account for this difference, Griffith[16] assumed the existence of small cracks that, acting as stress concentrators (Fig. 2), lead to early fractures of the material. According to Griffith, the critical stress for cracking, σ_c, in the presence of those cracks, would be

$$\sigma_c = \left(\frac{2E \cdot \gamma}{a \cdot \pi}\right)^{1/2} \tag{1}$$

E being the elastic modulus, γ the surface free energy, and a the size of the crack. The above equation was developed on the assumption that in order for the crack to propagate, the energy supplied to it should be equal to that necessary to create the new crack surfaces. This equation only applies to ideal brittle materials, and no plastic deformation is taken into account. In many solids, and particularly in metals, on the other hand, besides the energy consumed to create a new surface, there is an important amount of energy consumed by plastic deformation at the crack tip. To account for this, Orowan[17] modified the above Griffith equation as follows:

$$\sigma_c = \left(\frac{2E\,(\gamma + \gamma_p)}{a \cdot \pi}\right)^{1/2} \tag{2}$$

γ_p being the energy involved in plastic deformation. Both these equations assume that sharp cracks are involved. The condition for a crack to propagate is that the $\sigma_c \cdot (a \cdot \pi)^{1/2}$ value, characteristic of each material, is surpassed. The expression $\sigma \cdot (a \cdot \pi)^{1/2}$, very important in fracture mechanics, is called the *stress intensity factor*, and it is identified by the letter K.

As pointed out by Thomson,[18] cracks play a similar role in the understanding of fracture to that played by dislocations in plastic deformation. The dislocations were introduced as lattice defects to explain the plastic deformation of crystals at stresses lower than those expected for a perfect lattice. In the same way, cracks should also be treated as lattice defects.

Equations (1) and (2) refer to the propagation of sharp cracks, in the absence of a corrosive environment. In environmentally induced cracking we are also dealing with sharp cracks (high aspect ratio), but the applied stresses are well below σ_c, and there is no crack propagation in the absence of the environment. The stress distribution to be expected, in a precracked sample, is shown in Figs. 3 and 4.

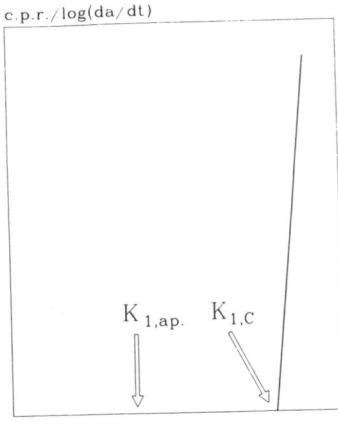

Figure 11. Crack-propagation-rate vs. crack-tip-stress-intensity diagram. In the absence of aggressive environments, cracks will propagate spontaneously only when the stress intensity factor reaches the critical value $K_{1,C}$. If lower stresses are applied, as indicated by $K_{1,ap.}$, no crack propagation will be detectable.

As indicated by Eq. (2), cracks will propagate spontaneously only when the applied stress is above σ_c. When stress intensity factors are considered, this critical value will be related to an also critical value for K, $K_{1,C}$ in Fig. 11. If a lower stress intensity is applied, as indicated by $K_{1,ap.}$ in the same figure, no crack propagation will be detectable. In the presence of an aggressive environment, on the other hand, cracks propagate at much lower stress intensity values, as shown in Fig. 12. There is

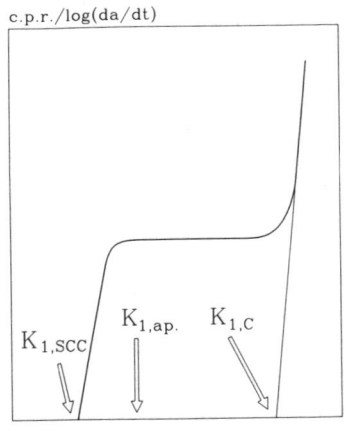

Figure 12. Crack-propagation-rate vs. crack-tip-stress-intensity diagram. In the presence of an aggressive environment, cracks propagate at much lower stress intensity values than those shown in Fig. 11. There is still a K value, indicated by $K_{1,SCC}$, below which no cracking is found, but it is considerably lower than $K_{1,C}$.

still a K value, indicated by $K_{1,SCC}$, below which no cracking is found, but it is considerably lower than $K_{1,C}$. The aim of the studies of environmentally induced cracking mechanisms is to understand why such degradation in the mechanical properties of the materials takes place and how to keep it under control.

2. Range of the Phenomenon

Practically all the alloys, many pure metals, and a great number of crystalline and amorphous nonmetallic solids can be subject to environmentally induced cracking. When the problem was first detected, cracking was believed to occur only in specific environments. Early literature on the subject mentioned that mild steel showed stress corrosion cracking only in hot alkaline solutions and in hot nitrate solutions; stainless steels suffered stress corrosion only in hot chloride solutions; copper alloys apparently were susceptible to it only in ammoniacal solutions, etc. Later work[19,20] showed that a great number of other environments could produce cracking, and, in many cases, even pure water could cause failure. Consequently, the idea of specificity lost ground. Nevertheless, even though many different environments can cause failure in a given alloy, each one will lead to different crack velocities: crack velocities ranging from less than 10^{-11} m/s up to over 10^{-2} m/s have been reported in various environments.[21,22] On the other hand, there are still numerous environments where no cracking is found. From a mechanistic point of view, although specificity is no longer relevant as regards susceptibility, it becomes relevant when the kinetics of the process is considered. Different crack velocities are measured in the various environments, and any mechanism proposed to explain the SCC process should account for this crack velocity variety.

3. Experimental Techniques

From a mechanistic point of view, the range of experimental techniques available for the study of SCC kinetics is scarce and disappointingly poor.

In the early literature constant-load experiments were used.[23] In these experiments metal samples, of different shapes, were exposed to the environments to be studied while subjected to a constant load. The load was applied either in the form of a tensile charge or by bending the samples. Usually, besides an idea of the degree of susceptibility to SCC, only a value of the failure time could be obtained. Although the technique could have some practical applications, it was abandoned as a method for studying SCC mechanisms.

In the 1960s Parkins and co-workers introduced the constant-potential slow-strain-rate technique.[24] By this technique the samples are strained, at a slow strain rate (typically between 10^{-7} s^{-1} and 10^{-4} s^{-1}) and in the environment to be studied. It was observed that in these experiments the induction time is usually absent. So cracking is believed to start as soon as the mechanical straining begins. At the end of the experiment, the size of the cracks can be measured, either by looking at the fracture surface with a scanning electron microscope or by a metallographic sectioning of the corroded metal samples. From the length of the cracks and the exposure time, a mean crack velocity can be calculated.

With the advent of fracture mechanics, the use of precracked specimens was introduced in SCC studies.[20,21,25,26] Thus, a correlation between crack velocities and stress intensity values could be made. Unfortunately, from a mechanistic point of view, there is no clear correlation between the stress intensity value, K, and the stress at the tip of the crack. The calculation of K is based on the elastic strain in the material, around the crack. The plastic zone at the tip of the crack is ignored. Correlations between K values and stresses at the tip of the crack, based on computer simulation, were published.[27] Unfortunately, when those correlations are applied to macroscopic cracks, the stress values calculated have no physical meaning, since they are well above the ideal strength for solids.

As described below, there is a series of electrochemical techniques that are used in SCC research, such as measurements of repassivation rate and of bare metal current density during straining experiments. Nevertheless, all these techniques must be referred to the crack velocity measurements, which are based on the techniques described above.

There is great confidence at present[28] in techniques such as scanning tunneling microscopy (STM), which could allow the observation of the propagation of a crack at an atomic level. However, at the time of writing of this chapter, there is no available information of any successful observation of this kind.

4. Historical Background

The environmental failure of metals must have been known, though not understood, very early by man. For example, reports have been published[29] of probable stress corrosion cracks in primitive bronze screws manufactured during the Bronze Age.

Yet it was only toward the middle of last century, as a result of the industrial revolution, that the problem was systematically observed. One of the first important difficulties appeared with the extensive use of steam engines. A problem, later described as "caustic cracking," caused frequent explosions in steam boilers. According to Keating,[30] between 1865 and 1870, 288 boiler explosions were reported in England.

The following description was published, in the *Scientific American* issue of March 1881[31]:

> The records kept by the Hartford Steam Boilers Inspection and Insurance Company show that 170 steam boilers exploded in the United States last year, killing 259 persons and injuring 555. The classified list shows the largest number of explosions in any class to have been 47, in sawing, planing, and woodworking mills. The other principal classes were in order: paper, flouring, pulp and grist mills, and elevators, 19; yachts, steam barges, dredges and dry docks, 15; portable engines, hoisters, thrashers, pile drivers and cotton gins, 13; ironworks, rolling mills, furnaces, foundries, machine boiler shops, 13; distilleries, breweries, malt and sugar houses, soap and chemical works, 10 . . .

The first evidence that a metal with internal stresses would crack in the presence of certain aggressive environments was reported by Roberts-Austen[32] in 1886. This author used ferric chloride solutions to detect internal stresses in 13-carat gold. The stress corrosion cracking of brass, or "season cracking," was seriously taken into account for the first time during the Boer War.[33] Logan[23] reported that the first detailed description of brass "season cracking" was published by Sperry in 1906. Numerous examples of this type of cracking in small arm cartridge cases could be found in the General Discussion on Failure of Metals of 1921[34] and in the Symposium on Stress Corrosion Cracking of 1944.[35]

Stress corrosion cracking of aluminum alloys was first reported by Rosenhain and Archbutt[36] in 1919. Since then, stress corrosion cracking of aluminum alloys has been a problem in modern technology. Speidel[13] reported that failures due to stress corrosion cracking of aluminum were observed during the construction of the zeppelin in Germany 70 years ago, and more recently they were found in the Saturn rocket and in the lunar module during the NASA Apollo program.

The first case of transgranular stress corrosion cracking of austenitic stainless steel was observed in 1937, as reported by Ellis.[37]

A classic in the stress corrosion cracking literature was described by Johnson.[38] It refers to the unexpected stress corrosion cracking failures of titanium alloy pressure vessels containing nitrogen tetroxide. They were detected in January 1965 during the development of the Apollo Program and threatened to jeopardize it. Intensive research showed that minor amounts of nitric oxide in the nitrogen tetroxide were necessary to inhibit stress corrosion cracking of the titanium alloy.

Many other important examples of stress corrosion cracking failure are found in the literature. In a monograph on the subject, Brown[25] shows the impressive picture of "Silver Bridge" over the Ohio River, which collapsed on December 15, 1967, as the result of stress corrosion cracking. This accident caused the death of 46 people. An accident of similar proportions was the collapse of the Berlin Congress Hall, in May 1980, which was a case of stress corrosion cracking in a prestressed concrete structure.[39]

Leak[40] surveyed failure cases in military aircraft, most of them due to stress corrosion cracking and leading to accidents where people and/or serious economic loss were involved. Abundant examples of stress corrosion cracking could also be found in the nuclear industry, such as cracking in nickel alloy steam generators, in carbon steel pressure vessels,[41] and in zirconium alloy fuel elements.[42]

In recent times, such an avalanche of new cases of stress corrosion cracking have been reported that the exact dates have become irrelevant. The feeling is now that every metal or alloy can, under certain circumstances, suffer stress corrosion cracking. Apparently innocuous environments can lead to the cracking of otherwise highly corrosion resistant alloys, as was the case with thiosulfate solutions on stainless steels.[5]

As for the mechanistic point of view, from review of the literature,[8,43] it is found that since about 1919 numerous mechanisms have been suggested to explain the environment-induced embrittlement process. The number of mechanisms published in this lapse of about three-quarters of a century is overwhelming. There were authors who suggested that stress corrosion cracking was due to the plastic flow of an amorphous cement present along the grain boundaries[36] (Fig. 13). Some other authors[30,44] suggested that stress corrosion cracking propagation was a combination of steps which included plastic deformation, brittle fracture, and localized corrosion. Others[45] suggested that new solid faces were formed at the bottom of the crack, leading to fast localized dissolution. There were theories[46–50] that suggested that hydrogen was responsible for the crack

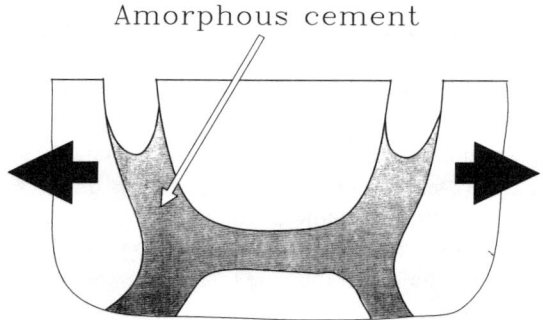

Figure 13. Amorphous cement theory of stress corrosion cracking proposed by Rosenhain and Archbutt in 1919.[36]

propagation process, either by hydride formation or by hydrogen-induced metal decohesion. Some authors[51,52] supported a mechanism based on surface energy reduction induced by ion adsorption on the metal surface. Finally, the existence of a yield-assisted anodic dissolution was also suggested.[53–55] This dissolution was proposed to be due either to the formation of new active sites at the bottom of the crack or to a film breakdown process.

Many of these theories have been forgotten. Only a few have survived to the present day and have strong supporters. However, if the validity of a mechanism is evaluated by its ability to predict new environment-induced embrittlement cases, we have to admit that none of them has succeeded in doing so.

Most cases of environment-induced embrittlement known so far are the result of failures observed in practice. It is difficult to find cases of susceptibility predicted by scientists working in their laboratories, and the few examples we find were the result of chance rather than of scientifically based knowledge.

It is enlightening to analyze the reasons why scientific research has failed so badly in this area. It cannot be attributed to lack of interest in the subject. Already in 1921[34] such an important scientific society as The Faraday Society, in London, published, together with The Institution of Mechanical Engineers, The Iron and Steel Institute, The Institute of Metals, and other scientific societies, the proceedings of a meeting on the subject, in which undoubtedly the best available scientists took part. The number of meetings and international conferences on the subject since then has been well above a hundred. As for the scientific papers published on the matter, they easily amount to several thousand.

The amorphous cement mechanism theory is most convenient to point out the difficulties in developing working theories on the subject. It was proposed by Rosenhain and Archbutt[36] (see Fig. 13) in 1919. According to this theory, the "season cracking" of metals was due to a viscous or viscoelastic movement of the amorphous metal present in the intercrystalline boundaries. It was accepted that corrosion could accelerate the process, but it was believed at the time that, in the absence of corrosion, cracking could still happen.

In the 1921 meeting mentioned above,[34] Rosenhain was in charge of the introductory lecture. He strongly defended the amorphous cement theory, and concluded that "at this time the theory appears to be overwhelmingly strong as compared to any other explanation." The mechanical engineers present at the meeting wanted to know how to stop "season cracking." However, to their disappointment, most of the discussion was about the thickness of the amorphous layer. Was it two atoms thick or was it several atoms thick?

If the amorphous cement theory is analyzed in the context of the other related disciplines, such as metallurgy and electrochemistry, it is found that it was *impossible* at the time to have any good stress corrosion cracking theory. To center most of the argument on the property of the grain boundaries was correct. All the stress corrosion cracking cases known at the time were intergranular. It was only in 1937 that the first case of transgranular cracking was reported. On the other hand, the idea of an amorphous cement is not surprising, because in 1919 Beilby's theories of amorphous or vitreous conditions produced on metals by straining or by polishing were very popular.

The reason for the failure of the theories must be found in the nature of the problem.[8,43] Environmentally induced cracking is the result of the combined action of the environment and the mechanical stresses, acting simultaneously on the material. To study the problem, we have to resort to three different disciplines: metallurgy, mechanics, and electrochemistry. In metallurgy, for example, the crystal defect theories were developed between 1929 and 1934.[56,57] The Frank–Read source for multiplication of dislocations was postulated only in 1950,[57] and the first analysis of a correlation between dislocations and transgranular stress corrosion cracking was reported in 1962.[58–60] Incidentally, the first papers on this subject seemed to prove that stress corrosion cracking was due to stacking fault energies and dislocation pileups. It was found later[61] that, while the existence of these was a necessary condition, it was not a sufficient one; some other important parameters had to be considered.

Mechanics is a very old branch of physics. Nevertheless, the study of fractured solids is rather recent.[62] Between 1953 and 1956, when catastrophic failures of aircrafts and gas pipelines were reported, great confusion about this subject was found in the literature. The study of the problem, presently known as fracture mechanics, dates from the sixties, and the first publication where fracture mechanics was applied to stress corrosion cracking is from 1965.[63]

Classical electrochemistry was developed during the last century. However, the understanding of electrochemical reactions at an atomic level is something that has been achieved, at least partially, during the last 30 years. Many doubts are still found in this area, and computer simulation of the metal/solution interphase, even in the absence of reactions, presents many unsolved problems.

It is well known that, in general, the rate of a chemical reaction is doubled when the temperature is increased some 20 to 25°C. Hence, no chemical kinetics study is conceivable without an appropriate temperature control. In electrochemical reactions, on the other hand, the rate of a reaction can be increased 10^9 times just by changing the electrode potential by less than one volt. Consequently, no electrochemical process can be reasonably studied without a precise control of the potential. Galvanostatic techniques were known at the beginning of the century.[64] About 1920, galvanostatic curves were used in the study of corrosion processes. In 1932 Evans and Hoar[65] explained for the first time the wet corrosion of iron by means of polarization curves. In 1938 Wagner and Traud[66] introduced the principle of superposition of polarization curves. Nevertheless, the first equipment for potential control, the potentiostat, was described in 1942,[67] and the first paper where such equipment was used for corrosion studies was published 10 years later.[68] It was thanks to the use of a potentiostat that Humphries and Parkins,[69] in 1967, were able to reproduce in the laboratory the "caustic cracking" of mild steel. It took almost 100 years to make it possible to reproduce in the laboratory a process that was the cause of serious concern in the boiler industry.[8,43]

The first experimental information about solution chemistry inside the stress corrosion cracks was published by Brown and co-workers in 1969.[70] Until then, the composition of the solution inside a crack was guessed to be different from that in the bulk environment. Nevertheless, no exact information about the grade and the significance of such differences was available.

Under the above circumstances, it is clear that the knowledge of the subject was ready for the development of a reasonable theory of environmentally induced cracking only a few years ago. In recent years, to the scientific advances reported above, we have to add the introduction of some sophisticated surface analysis techniques, such as ESCA or Auger electron spectroscopy, the introduction of scanning tunneling microscopy, the analysis of the transport processes inside cracks, and the invaluable help of computer simulation. Now scientists are beginning to visualize the processes taking place at the tip of the crack at an atomic level. It is to be hoped that the conditions are almost set for a reasonable understanding of the mechanisms responsible for environmentally induced cracking.

II. CRACK TIP CHEMISTRY

1. Crack Chemistry Measurements

Making abstraction of the mechanisms involved, when a stress corrosion crack propagates, new metal surface is being exposed to the environment. In the particular case of aqueous solutions, the metal is usually unstable in contact with water, and a reaction of the type

$$Me + nH_2O = Me(OH)_n + nH^+ + ne^- \tag{3}$$

takes place. If the aqueous solution contains anions, X^-, with more affinity for the metal than the OH^- ions, instead of the hydrolysis of the metal ions (Eq. 3), the formation of soluble or insoluble complexes will take place:

$$Me + nX^- = MeX_n + ne^- \tag{4}$$

As the crack has an occluded geometry, the exchange of ions between the crack and the bulk solution will be restricted and slow. As a consequence, reactions like those represented by Eq. (3) or Eq. (4) will lead to local composition changes inside the crack. In the case of reaction (3) a localized acidification should be expected. In the case of reaction (4), on the other hand, either a depletion of X^- should take place, if MeX_n is insoluble, or an accumulation of MeX_n in the crack solution will appear, if this compound is soluble. Eventually, a supersaturation of the soluble MeX_n complex could be reached, and a salt film will precipitate at the tip of the crack.

The interest in the composition changes inside stress corrosion cracks appeared in the corrosion literature some 30 years ago.[71] Numerous authors developed various techniques to measure the composition of the solution inside the cracks. Most of these measurements were described in a review by Turnbull.[72]

The first experimental observations on solution chemistry within stress corrosion cracks were due to Brown *et al.*[70] They studied the composition of the solution inside cracks in aluminum and titanium alloys and in carbon steels. They used rectangular specimens about 3 mm thick. A saw cut was made in the middle of the long dimension, and a metal wedge was pressed into the saw cut. In this way tensile stresses were produced at the tip of the cut. The samples were exposed to a corrosive environment, such as an aqueous NaCl solution, and they were allowed to corrode, until the crack propagated for about one centimeter. Afterward, the specimens were removed from the solution and immersed in liquid nitrogen, to freeze the solution retained in the crack. The crack was opened by mechanically breaking the sample, and the microvolume of solution retained on the sides of the crack was analyzed through microanalysis techniques. The pH and the presence of ions, and eventually the concentration of the ions present, were determined.

Other tests, as described by Turnbull,[72] included direct, *in situ* measurements; solution extraction and analysis; and use of artificial cracks. The pH values reported for structural steels in chloride solutions were about 4; for stainless steels the pH values measured ranged between 0 and 3. For aluminum alloys in chloride solutions, the pH values found were between 3 and 4; and for titanium alloys values between 1 and 2 were reported. As for potential drops, no generalized conclusions could be drawn from Turnbull's review. In most cases, at an open-circuit potential, the potential drop was suggested to be small, almost negligible. On the other hand, if the samples were polarized, high potential drops, above 1 V, could be expected.

2. Model Calculations for Pits, Crevices, and Cracks

As soon as fresh metal is exposed to the solution, a current will flow through the metal/solution interface. Assuming that soluble species are produced, this current will generate a time-dependent local concentration buildup of metal cations, until steady-state conditions are reached.

The mathematical treatment for the time-dependent concentration changes mentioned above was solved a long time ago by Sand[73] and by

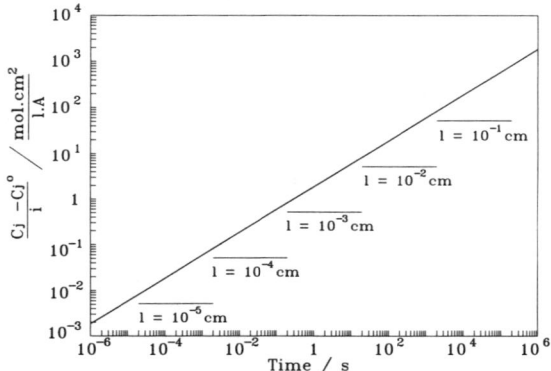

Figure 14. Changes of concentration at the tip of the crack, brought about by diffusion, following Rosebrugh and Lash Miller calculations.[74] C_j^0 is the initial concentration and C_j is the concentration at the time t for the species j in moles per liter. Reaction considered: Me → Me^{2+} + $2e^-$. Diffusion coefficient $D_j = 10^{-5}$ cm²/s; current density in A/cm². Horizontal lines indicate stationary values for diffusion paths 1 cm long.[80]

Rosebrugh and Lash Miller.[74] Figure 14 shows the rate of the metal ion buildup at the metal/solution interface for a semi-infinite unidirectional pit or crack. For this calculation, the following equation, due to Sand, was used:

$$C_j = C_j^0 + \frac{2}{\sqrt{\pi}} \cdot \frac{i \cdot v_j}{n \cdot F \cdot \sqrt{D_j}} \cdot \sqrt{t} \tag{5}$$

In this equation, C_j^0 is the bulk concentration of the species j, C_j is the concentration at the electrode surface, in the present case the tip of the crack, F is the Faraday constant, v_j is the stoichiometric factor of j in the overall electrode reaction, t is the time in seconds, and D_j is the diffusion coefficient of the species j.

For the calculations in Fig. 14, D_j was assumed to be 10^{-5} cm²/s, $n = 2$, and $v_j = 1$. The current density i is in amperes per square centimeter. The values of the concentrations in Eq. (5) are given in moles per cubic centimeter, whereas in Fig. 14, for the sake of simplicity, they were plotted in moles per liter.

The horizontal lines in Fig. 14 are the stationary-state values ($t = \infty$) for various crack lengths (l) and were calculated with the following equation[74]:

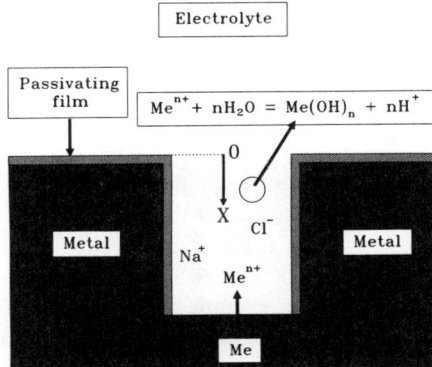

Figure 15. Unidirectional pit model,[80] used as a simplified description of a crack.

$$\frac{C_j - C_j^0}{i} = \frac{l}{n \cdot F \cdot D_j} \qquad (6)$$

From Fig. 14 several interesting conclusions can be drawn. From the interception of the concentration versus time curve in Fig. 14 with the horizontal lines, pairs of length and time values are obtained. From these values a hypothetical crack velocity can be calculated, and this could be applied to real SCC conditions. Cracks propagating at velocities lower than this hypothetical crack velocity, would be under steady-state conditions, while the others would not. The first conclusion is that for cracks 0.01 cm deep a crack velocity of up to 1.11×10^{-5} m/s would propagate under steady-state conditions. Smaller cracks could allow higher crack velocities and still remain in the steady-state condition. For example, a crack 0.001 cm deep would allow a crack velocity of 1.11×10^{-4} m/s; on the other hand, for a 1-cm-deep crack, only crack velocities up to 1.11×10^{-7} m/s would propagate under steady-state conditions.

Another important conclusion is related to the maximum ion concentration expected at the tip of the crack. A crack like the one shown in Fig. 15 assumes a unidirectional system. If there was a crack opening of a certain angle, the concentration gradients would be slightly smaller, since a transition to a radial distribution of fluxes begins. Taking a very conservative case, and assuming that the diffusion path is only 0.1 cm, according to Fig. 14, the steady-state value for $(C_j - C_j^0)/i$ would be 51.8 mol l^{-1} cm^2 A^{-1}. Let us see how this value compares with well-known SCC cases. It is customary, in the SCC literature, to correlate the anodic current density at the tip of a crack with a certain crack velocity (see Section VI.2). For this purpose, the following equation is used:

$$c.p.r. = \frac{i \cdot E}{F \cdot d} \tag{7}$$

where *c.p.r.* is the crack propagation rate, i is the anodic current density at the tip of the crack, E is the equivalent weight of the dissolving metal (for alloys a mean value of E is calculated taking into account the atomic composition of the alloy and the valence at which each element is dissolving; for example, for type AISI 304 stainless steel a value of $E = 26.0$ g/equiv is found, assuming that chromium dissolves as Cr^{3+}, nickel as Ni^{2+}, and iron as Fe^{2+}), F is the Faraday constant, and d is the alloy density.

In the case of SCC of mild steel in $2M$ $Ca(NO_3)_2$, it was reported that, if cracks propagate by anodic dissolution, a current density of 2 A/cm² should be expected at the tip of the crack.[75] Consequently, the molar concentration of metal ions at the tip of the crack should be $103.6M$. In the case of austenitic stainless steel in $11.8N$ LiCl, the current density at the tip of the crack was reported to be 0.6 A/cm².[76] Such a current density would lead to an ionic concentration at the tip of the crack of $31.08M$. If reaction (4) takes place, conditions for oversaturation of the species MeX_n should be expected. This explains why some authors, such as Beck[7,77] in his model for SCC of titanium or Speidel[78] for SCC of aluminum, assume that a salt film is present at the tip of the crack.

Let us return now to reactions like that shown in Eq. (3). As reviewed by Turnbull,[79] several mathematical models have been published to describe the mass transport and electrochemical processes inside cracks or similar occluded cells, such as pits or crevices.

One of the relationships used in the mathematical modeling of mass transport and electrochemistry in cracks is that giving the flux (J) of each dissolved species[79]:

$$J_j = C_j \cdot v - D_j \cdot \nabla (C_j) - z_j \cdot u_j \cdot F \cdot C_j \cdot \nabla (\Phi) \tag{8}$$

where v is the fluid velocity, C_j is the concentration and D_j is the diffusion coefficient of species j, $\nabla (C_j)$ is the concentration gradient, $\nabla (\Phi)$ is the electric field, z_j is the charge u_j is the mobility of the ion, and F is the Faraday constant.

The mass conservation of species j is given by

$$\frac{\delta C_j}{\delta t} = -\nabla J_j + R_j$$

which gives the rate of change of concentration with time and where R_j is the rate of production, or depletion, of the species j by chemical reaction in solution. In addition, the electrical neutrality condition is included:

$$\sum z_j \cdot C_j = 0$$

Another relationship gives the current in the electrolyte solution:

$$i_d = F \cdot \Sigma z_j \cdot J_j \tag{9}$$

where i_d is the current density.

A first estimate of the chemical changes taking place inside a crack could be obtained by direct application of the calculations made for unidirectional pits,[80] as shown in Fig. 15. For the present purposes the crack will be assimilated to a unidirectional pit. In such a treatment, a simplification in Eq. (8) is made, by assuming that no convection is taking place, $v = 0$, and that a supporting electrolyte is present, that is, the last term in Eq. (8) is also zero.

The sides of the crack are assumed to be passive, and the following assumptions are made:

1. The metal dissolves only at the bottom of the pit or crack, by the following general reaction:

$$Me = Me^{n+} + ne^- \tag{10}$$

2. Reaction (10) is assumed to occur in the presence of a supporting electrolyte. The pH of the solution could have any value and is given as a boundary condition.

3. It is assumed that reaction (10) is followed by hydrolysis equilibrium of the type:

$$Me^{n+} + H_2O \rightleftharpoons Me(OH)^{(n-1)+} + H^+$$

$$Me(OH)^{(n-1)+} + H_2O \rightleftharpoons Me(OH)_2^{(n-2)+} + H^+$$

$$\cdots\cdots\cdots\cdots \tag{11}$$

$$Me(OH)_{n-2}^{2+} + H_2O \rightleftharpoons Me(OH)_{n-1}^+ + H^+$$

$$Me(OH)_{n-1}^+ + H_2O \rightleftharpoons Me(OH)_n + H^+$$

Each step in Eqs. (11) occurs simply by the loss of successive protons, and reactions of this type are invariably rapid.[81] The equilibrium conditions would be reached after only a few microseconds.[82] Polynuclear complexes have been reported in numerous cases. Nevertheless, in a first approach, they can be ignored, mainly because while the formation of mononuclear species is very fast, the formation of polynuclear species is a much slower process.[81,83]

As the hydrolysis reactions are very fast, the ions considered will always be in equilibrium. As a first approximation, the system will be assumed to be under steady-state conditions. As an example, the case of a divalent metal, leading by hydrolysis to soluble and insoluble species, will be considered.[80] In order to be able to calculate the ion concentration profiles inside the crack, the following species have to be accounted for:

$$3Me^{2+} + 3H_2O + 2OH^- \rightleftharpoons Me(OH)^+ + 3H^+ + Me(OH)_{2(aq)} + Me(OH)_{2(s)} \quad (12)$$

$S_1 \quad S_2 \quad S_3 \quad S_4 \quad S_5 \quad S_6 \quad S_7$

As a first approximation, it is assumed that the species S_7 is at the very initial steps of precipitation and that its flow properties will be similar to those of the other species present. From Eq. (12) we find that inside the crack there are seven species (S_1–S_7), the concentrations of which should be calculated. The detailed mathematical treatment has been reported in the literature.[84] The seven unknown concentrations are calculated by resolution of the seven following equations. The flow of the species containing Me atoms will be given by

$$D_1 \frac{dC_1}{dx} + D_4 \frac{dC_4}{dx} + D_7 \frac{dC_7}{dx} = \frac{i}{2 \cdot F} \quad (13)$$

If the conditions for precipitation of S_7 are reached, the concentration of the species S_6 can increase only up to a certain limit, C_6^*, above which the precipitation of solid $Me(OH)_2$ will start. The limiting concentration value is given by the equilibrium[85]

$$Me(OH)_{2(s)} \rightleftharpoons Me(OH)_{2(aq)}$$

with the equilibrium constant[†]:

$$K_{s2} = C_6^* \quad (14)$$

[†]In the present chapter the notation system used both in the "Stability Constant" table by Sillen and Martell[86,87] and in Butler's textbook[85] is followed.

The value of this equilibrium constant can be calculated with the following relation:

$$K_{s2} = \beta_2 \cdot K_{s0}$$

The flow of the species containing O atoms will be given by

$$D_2 \frac{dC_2}{dx} + D_3 \frac{dC_3}{dx} + D_4 \frac{dC_4}{dx} + 2D_7 \frac{dC_7}{dx} = 0 \tag{15}$$

The flow of the species containing H atoms will be given by

$$2D_2 \frac{dC_2}{dx} + D_3 \frac{dC_3}{dx} + D_4 \frac{dC_4}{dx} + D_5 \frac{dC_5}{dx} + 2D_7 \frac{dC_7}{dx} = 0 \tag{16}$$

Then the hydrolysis equilibrium constants leading to S_4 and S_7 species are considered. The concentration of S_4 in equilibrium with solid $Me(OH)_2$, S_7, will be given by

$$Me(OH)_{2(s)} \rightleftharpoons Me(OH)^+ + OH^-$$

$$K_{s1} = K_{s0} \cdot \beta_1 = C_3 \cdot C_4 \tag{17}$$

S_1 will also be in equilibrium with solid $Me(OH)_2$:

$$Me(OH)_{2(s)} \rightleftharpoons Me^{2+} + 2\,OH^-$$

$$K_{s0} = C_1 \cdot C_3^2 \tag{18}$$

Another equilibrium constant to be taken into account is K_w:

$$K_w = C_3 \cdot C_5 \tag{19}$$

We thus have that the concentrations of the seven species in Eq. (12) are correlated with the seven equations (13)–(19). By solving this system of equations,[80] the concentrations of the various species inside the crack are found. The above equations were solved as functions of the $x \cdot i$ parameter and are valid for any combinations of crack depths, x, and current densities, i. Figure 16 shows the concentrations expected in a crack in iron. Figure 17 shows the percentage of each of the iron species present along the crack. There is a critical $x \cdot i$ value above which the solid species are replaced by soluble ones. In the particular case of pitting,[80] this is the point at which pitting starts. In the case of cracks, no explicit analysis of this

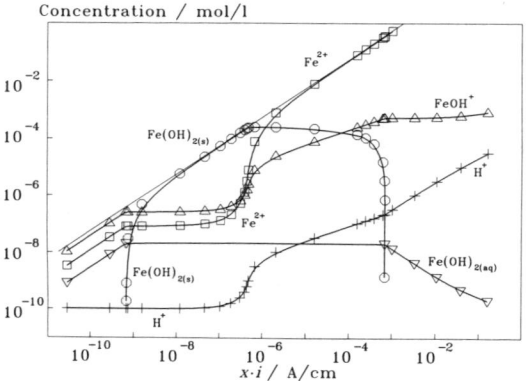

Figure 16. Concentrations of Fe^{2+}, $FeOH^+$, $Fe(OH)_{2(aq)}$, $Fe(OH)_{2(s)}$, and H^+ as a function of the depth x and the current density i in a unidirectional pit or crack, for iron in a pH 10 solution.[80]

point has been made, but it could be the value at which the passive film at the sides of the cracks is replaced by a salt film.

A more general calculation model was described by Gravano and Galvele,[88] assuming the absence of supporting electrolytes and solving

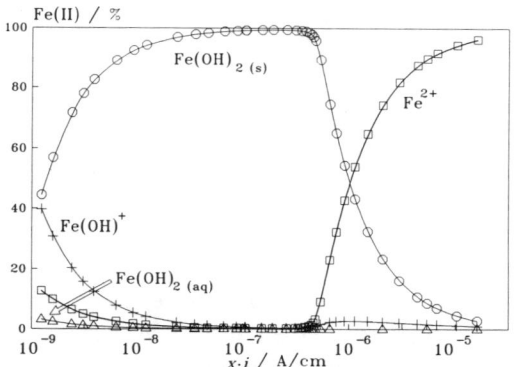

Figure 17. Distribution of iron corrosion species as a function of the product of the depth x and the current density i in a unidirectional pit or crack in a pH 10 solution.[80]

Eq. (8) also for Φ. In addition, some other models were developed taking into consideration different cathodic reactions.[79] Shuck and Swedlow[89] developed a transport model in cracklike regions for various crack geometries. Searching for an explanation for the very high crack velocities found for a titanium alloy in molten LiCl–KCl eutectic, Smyrl and Newman[90] developed a model that includes convective liquid flow and a variable crack angle. In this model mass transfer of minor components, such as traces of H_2O, was analyzed. The crack velocity found in the plateau region of the c.p.r. versus K diagram (see Fig. 12) was 10^{-2} m/s. The limiting crack velocity could not be explained by diffusion of traces of impurities in the molten salt, but the model was generalized and could be used for aqueous solutions. Other approaches to the calculation of crack compositions can be found in the literature.[79] One important limitation in most of these models is the assumption that a dilute solution treatment is valid.

As discussed below, the composition of the solution inside the cracks has various degrees of relevance, depending on the SCC mechanisms involved. As shown in Fig. 16, when pitting takes place, localized acidification is developed at the metal/solution interphase. In the case of hydrogen embrittlement, this drop in pH leads to the thermodynamic conditions for hydrogen evolution, and, in susceptible metals, to hydrogen embrittlement. There are other cases, such as brass in nitrite solutions, where the onset of pitting is also related to SCC initiation.[91] In the case of surface mobility mechanisms, as shown below, the formation of salt films at the tip of the crack is critical for crack propagation. On the other hand, for mechanically based SCC mechanisms, such as the cleavage mechanism, the composition of the solution inside the crack appears to be of little relevance.

III. PREEXISTING LOCALIZED PATHS

Stress corrosion cracks in metals could either propagate along grain boundaries (intergranular cracking) or cross the grains (transgranular cracking). Eventually, a mixture of both morphologies could be found on the same specimen.[92] In certain alloys, such as austenitic stainless steels in chloride solutions, mixtures of both types of cracks have been observed.[92,93] In numerous cases, particularly with homogeneous alloys, there is no evidence that cracks would follow a preexisting path. On the other hand, there is an important number of cases where chemical heterogeneities are present in the alloy and cracks follow these preexisting localized paths. A quick review of those cases is given in the following sections.

1. Aluminum Alloys

Practically without any exception, SCC of aluminum alloys follows intergranular paths.[92] Most of these aluminum alloys are subject to some kind of heat treatment, usually aging. This treatment improves their mechanical resistance but frequently leads to conditions of high susceptibility to SCC. Let us see how this localized susceptibility originates.

It is well known that solubilities in metals are a function of temperature. For metallic alloying elements, the most common observation is that solubilities are higher, the higher the temperature, as shown in the equilibrium diagram in Fig. 18. A reduction in solubility leads to the formation of new phases in the alloy matrix. In the case of Fig. 18, the solid solution α could retain more than 5% copper at temperatures of about 550°C. Copper solubility in the α phase drops drastically as the temperature is decreased, and the excess of copper precipitates as an intermetallic θ, with an approximate composition of Al_2Cu, which is close to Al–33% Cu.

Such changes can only take place by a diffusion process, by which the excess of solute moves along the solid matrix of the alloy to the nucleus where particles of the new phases start to grow. Since this diffusion takes place in a solid phase, the diffusion rate is relatively slow and strongly temperature dependent. For most metals, particularly when diffusion of substitutional atoms is involved, the rate of diffusion, while important near

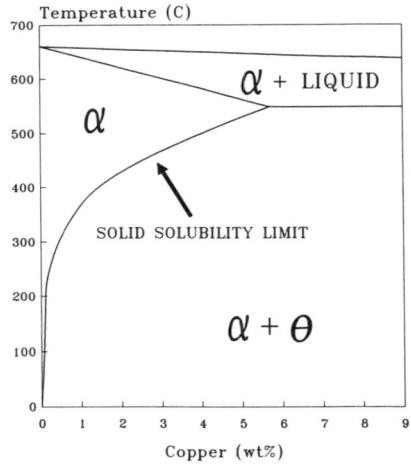

Figure 18. Schematic view of the aluminum-rich zone in the aluminum–copper equilibrium diagram.

the melting point, becomes almost negligible at temperatures below 0.5Tm (Tm being the absolute temperature melting point of the metal).

By quenching an alloy from high temperature to room temperature, unstable supersaturated structures can be produced. By controlled precipitation of these structures, materials of high technological interest are obtained. The strength of the alloy increases with the aging time, until a maximum is reached. If aging is continued, the strength eventually starts to decrease, and macroscopic precipitates start to form. Nevertheless, undesirable structures are also produced in this way. Since diffusivity along grain boundaries is considerably higher than that in the bulk of the grains, during the age-hardening process the rate of aging near the grain boundaries is faster than that in the grains. Consequently, heterogeneities are produced in the alloy. While the grains, from an electrochemical point of view, are still a homogeneous solid solution, along the grain boundaries second phases are formed surrounded by a solute-depleted zone. Consequently a three-phase system is formed, as shown in Fig. 19. This heterogeneous structure, as a result of differences in the chemical stability of the various phases, is very frequently susceptible to localized corrosion.

In the case of single-phase Al–Cu alloys, the pitting potential of the alloy is higher, the higher the content of copper.[94] During the precipitation process, incipient copper-rich precipitates are formed, while the surround-

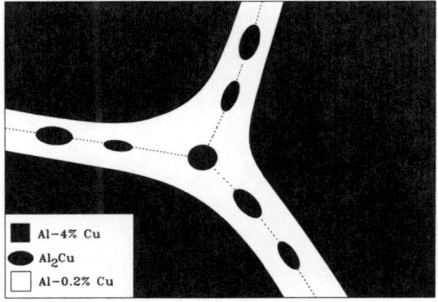

Figure 19. Schematic representation of the microstructure of a partially aged Al–4% Cu alloy. At this stage the grains show small changes, and their electrochemical behavior is similar to that of solubilized Al–4% Cu. Precipitates of the intermetallic Al_2Cu are formed along the grain boundaries, surrounded by a copper-depleted alloy.

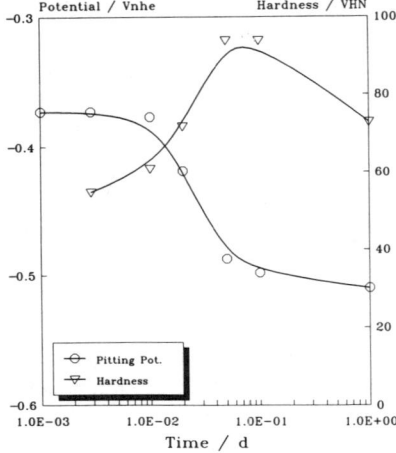

Figure 20. Effect of aging time on hardness and on the pitting potential of Al–3.3% Cu in a deaerated 1M NaCl solution at 25°C. Aging temperature, 240°C.[94]

ing matrix is copper depleted. Consequently, the presence of these copper-depleted regions induces a drop in the pitting potential of the alloy, as shown in Fig. 20.[94] The pitting potential starts to decrease as soon as precipitation begins and reaches its minimum value immediately after the maximum hardness of the Al–Cu alloy is obtained. A similar process takes place along the grain boundaries, except that it is a much faster one. As a result of this difference in precipitation rates between the

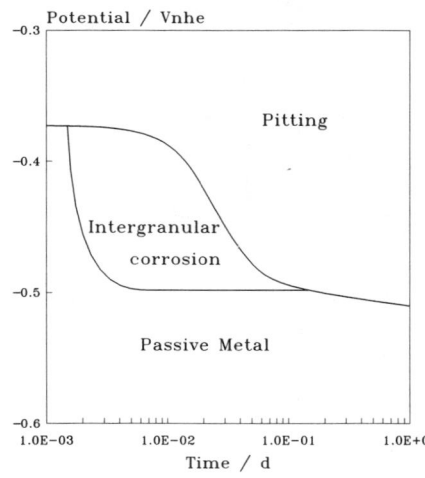

Figure 21. Effect of aging time on the corrosion behavior of Al–3.3% Cu in a deaerated 1M NaCl solution at 25°C. Aging temperature, 240°C.[94]

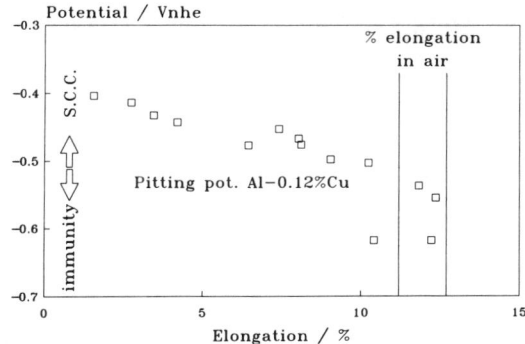

Figure 22. Effect of the potential on ductility of aged Al–4% Cu, during potentiostatic straining in a $1M$ NaCl solution. Aging time and temperature, 3 h at 210°C. Initial strain rate, 2.6×10^{-4} s^{-1}.[96]

grain bodies and the grain boundaries, the pitting potential of the grain boundaries shows a considerable drop whereas that of the grain bodies has practically not changed. This difference in pitting potential values explains the susceptibility to intergranular corrosion of aged Al–Cu alloys.[95] Since this difference in pitting potentials is a function of the aging time, there is only a restricted region of aging time where this susceptibility is present,[94] as shown in Fig. 21. During constant-potential slow-strain-rate experiments, SCC of aged Al–Cu alloys has been found to appear only above the pitting potential of the copper-depleted zone, at the grain boundaries,[96] as shown in Fig. 22. This observation indicates a close relation between SCC and intergranular corrosion in these Al–Cu alloys. Other aluminum alloys could show a slightly different behavior, since the relative susceptibility to corrosion of the grain bodies, the intermetallic phases, and the solute-depleted zones changes in the different alloys, as shown in Table 1.

Following the description in Fig. 19, the grain bodies could be represented by the solubilized alloy, while the grain boundaries would be represented by two phases: the intermetallic one and that of solute-depleted aluminum surrounding it. The former would be Al$_2$Cu, for Al–Cu alloys, Al$_3$Mg for Al–Mg alloys, and Zn for Al–Zn alloys. The nature of the expected corrosion, as shown in the literature,[97,98] would be the result of the relative pitting potential values and of the distribution of each of these phases.

Table 1. Pitting Potentials in Aluminum Alloy Components

Phase	Pitting potential in a 1M NaCl solution (V vs. NHE)	Reference(s)
Al	−0.53	95
Al–2% Cu	−0.41	94,96
Al–4% Cu	−0.35	94,96
Al–1% Zn	−0.62	94
Al–3% Zn	−0.75	94
Al–3% Mg	−0.57	94
Al–5% Mg	−0.58	94
Zn	−0.86	99
Al_3Mg_2	−0.75^a	100
Al_2Cu	−0.41	95

^{a}Pitting potential in a 0.5M NaCl solution.

2. Sensitized Stainless Steels

The corrosion resistance of stainless steels is due to their chromium content, which has to be over 12% to be effective. The carbon content of these alloys is usually lower than 0.1%. Nevertheless, at room temperature even this relatively low carbon content is substantially above the solubility limit of carbon in the alloy. In addition, when carbon precipitates in the presence of chromium, carbides of the type $Cr_{23}C_6$ are formed. When the weight percent composition is analyzed, these carbides contain 94.3% chromium, which has to be captured from the surrounding alloy.

Usually, stainless steels are used with the carbon retained in a soluble state. For this purpose, the steels are heated up to temperatures of about 1100°C and then water quenched. Unfortunately, stainless steels, in technological applications, have to be subjected to heat treatments that lead to localized precipitation along the grain boundaries. It is said that these stainless steels are sensitized.

The diffusivity necessary for precipitation is reached when the steel is heated at high temperatures, for example, during welding. As was the case with the aluminum alloys described above, the precipitation takes place faster along the grain boundaries than in the grain bodies. The carbide particles precipitate on the grain boundaries and become surrounded by a chromium-depleted zone. The degree of depletion is a function of the carbon content of the alloy and of the temperature reached.[101] For an 18% Cr type 304 stainless steel, with a carbon content of 0.02%, heated at 500 to 600°C, the equilibrium content of chromium at the carbide interface was found to be only 6%. Since this low content of chromium is not enough to retain steel passivity, a continuous path for localized corrosion along the grain boundaries is produced. The SCC susceptibility of sensitized stainless steels has become a subject of great concern since cracks were found, around 1974, in AISI type 304 stainless steel piping in boiling water reactors.[101] It is also a source of problems in the petrochemical industry, when polythionic acids are formed,[102] a problem known since the 1950s,[103] as well as in the nuclear industry, when the same material is exposed to the action of thiosulfate solutions, as was the case in the Three Mile Island unit.[5,103]

Was and Rajan[103] measured the chromium concentration gradient along grain boundaries of sensitized Ni–16Cr–9Fe alloys and compared it with the susceptibility of the alloys to intergranular cracking in sodium tetrathionate solutions. They found that the slope of the chromium concentration profile at the grain boundary was not relevant, but only the chromium concentration at the grain boundary was important. From slow-strain-rate experiments in a $0.017M$ $Na_2S_4O_6$ solution at room temperature, they found that, no matter what the extension of the chromium depletion—20 nm or 200 nm— intergranular cracking was found only in those samples with a grain boundary chromium concentration

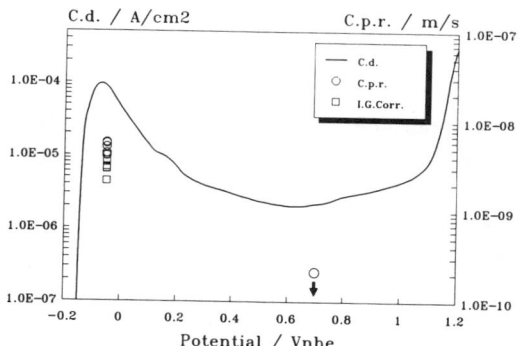

Figure 23. Corrosion behavior of sensitized type AISI 304 stainless steel in a 5% H_2SO_4 solution. Intergranular penetration on static samples (□) and samples strained at 2×10^{-6} s^{-1} (○), after an equivalent exposure time. SCC is found on the straining samples.[104]

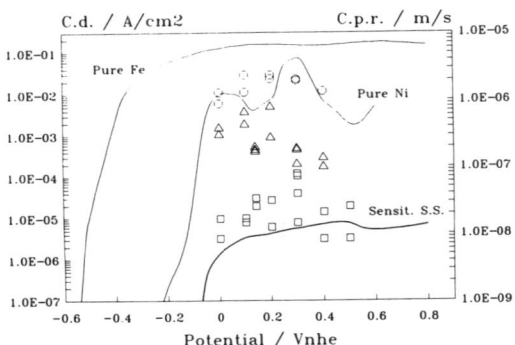

Figure 24. Corrosion behavior of sensitized type AISI 304 stainless steel in a $0.5M$ $Na_2S_2O_3$ solution. Intergranular penetration on static samples (□) and samples strained at 2×10^{-6} s^{-1} (○) and at 6.5×10^{-5} s^{-1} (△), after an equivalent exposure time. SCC is found on the straining samples. For the sake of comparison, the polarization curves of pure Ni, pure Fe, and stainless steel are included.[104]

below 5 wt.%, whereas no cracking was observed when the grain boundary chromium concentration was above 8.3 wt.%.

The presence of intergranular corrosion does not necessarily imply that SCC susceptibility will be present. As shown in Fig. 23, type AISI 304 sensitized stainless steel is susceptible to intergranular corrosion in sulfuric acid solutions, at −0.05 V versus NHE, but the use of straining wires does not increase the crack propagation rate. In other words, no evidence is found of an SCC process.[104] On the other hand, exposure of static or straining wires in the passive zone, 0.7 V versus NHE, show neither intergranular corrosion nor SCC. Nevertheless, if instead of sulfuric acid solutions, thiosulfate solutions are used (Fig. 24),[104] the straining samples show a crack propagation rate almost one order of magnitude higher than the rate for intergranular corrosion.

The intergranular SCC under sensitized conditions is not exclusive to austenitic stainless steels. High-nickel alloys, such as Alloy 600, show intergranular corrosion and SCC in 30% NaOH + 10% Na_2SO_4 at 350°C.[105] The susceptibility to localized corrosion has proved to be related to the presence of $M_{23}C_6$ carbides along the grain boundaries.

3. Grain Boundaries

Grain boundaries appear also as localized paths even in homogeneous alloys. One of the reasons for this behavior is that impurities have a tendency to segregate to grain boundaries.

Hondros and Seah[106] reviewed this process. They pointed out that the redistribution of solutes by equilibrium segregation is reversible and is analogous to the chemisorption or adsorption phenomenon on solid surfaces.

The field of structural perturbation of a grain boundary should not extend beyond a few atom distances, and it is now experimentally confirmed that the space occupied by the segregation is localized within one or two atom distances of the interface plane (Fig. 25).

While pure iron is resistant to SCC in hot aqueous nitrate solutions, commercial carbon steels fail intergranularly in such environments. The importance of the impurity contents of these commercial steels was shown in a study by Lea and Hondros.[107] These authors prepared samples from a high-purity mild steel stock, each sample containing a single impurity at a dilute level comparable to that found in commercial material. The free residual sulfur was rendered harmless by the addition of manganese. Each impurity element was added separately at a bulk level predicted to give, following an adequate heat treatment, a grain boundary segregation content of about 20% of a monatomic layer. The samples were subjected to a slow-strain-rate test in a $5N$ ammonium nitrate solution at 80°C. In each

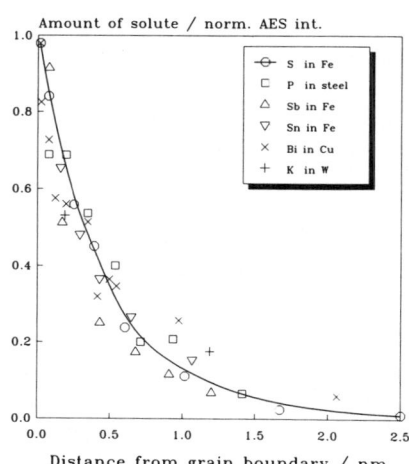

Figure 25. Localization of segregant atoms at the grain boundaries in various systems by Auger electron spectroscopy (AES) with argon etching.[106]

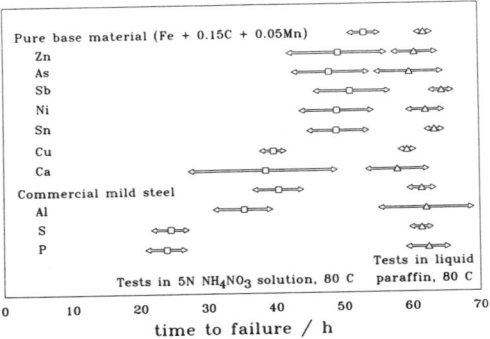

Figure 26. Failure time measurements for carbon steels, with controlled amounts of contaminants in the grain boundaries. Comparison between tests in an ammonium nitrate solution and in an inert environment. Strain rate, 2×10^{-6} s^{-1}.[107]

case, a control sample was tested under exactly the same conditions but in liquid paraffin at 80°C. Every specimen exhibited SCC in the nitrate solution, failing in a shorter time than in the control environment. The observations made by Lea and Hondros are shown in Fig. 26. The data on the right-hand side of the figure refer to the tests in paraffin. In the aqueous ammonium nitrate solution the fracture was intergranular, while in the control environment ductile fracture was observed. It was found that, under SCC conditions, all other variables being identical, life reduction depends on the impurity present.

These same authors noted the effect of impurities in the grain boundaries on hydrogen embrittlement. In this case impurities again provided a preferential path for cracking.

IV. DELETERIOUS ANODIC FILMS

Parkins and co-workers[108] studied SCC susceptibility of ferritic steels in various environments at various combinations of potential and pH. They plotted their observations on the Pourbaix diagram for Fe–H$_2$O and found that the upper limit for SCC is met when the phase γ-Fe$_2$O$_3$ becomes stable. At those potentials where this phase is stable, only ductile failure could be found. This applied to a wide range of environments and to temperatures ranging from 20 up to 288°C. In all the systems studied,

Fe_3O_4 was formed when SCC was found, although associated with other phases, such as $FeCO_3$ in the case of cracking in carbonate–bicarbonate solutions or $Fe_3(PO_4)_2$ for cracking in phosphate solutions. In all those cases where the potential was high enough for Fe_2O_3 to be stable, only ductile failures were found. The only exception seemed to be the case of SCC in nitrate solutions, where cracking was found at very high potential values and Fe_2O_3 was stable. However, in these cases, cracking initiated in the pits, and it was observed that inside the pits, and also inside the cracks, the phase present was Fe_3O_4. Although the external pH and potential conditions were favorable for the formation of Fe_2O_3, inside the pits and cracks the electrochemical conditions favored the formation of Fe_3O_4. From these observations it could be concluded that, from the SCC point of view, and for ferritic steels, Fe_3O_4 was a deleterious film whereas Fe_2O_3 was a protective film. According to Parkins,[108] the reasons for such a correlation were not established.

A similar attempt was made to correlate SCC of copper alloys with the copper E–pH diagram, but the results were not as conclusive as with ferritic steels.

According to Staehle,[109] no SCC is found in the passive region of austenitic stainless steels. In this case the protective film would be Cr_2O_3.

Although these observations were known for some time, no detailed elaboration was made on their significance. This was most probably because no SCC mechanism, among those discussed in Sections V–VII, gave any clue on this matter.

Duffó and Galvele[12,110] reported similar observations for silver–palladium alloys. They found that stress corrosion cracking of an Ag–15Pd alloy in KCl, KI, KBr, and Na_2SO_4 solutions was found only at potentials above the formation potential of the respective silver compounds: AgCl, AgI, AgBr, and Ag_2SO_4 (Fig. 27). It was concluded that for the silver alloy studied, the above-mentioned compounds acted as deleterious species for SCC susceptibility.

From Carranza and Galvele's work,[111,112] reported in Section VI, it could be concluded that deleterious anodic films could be those that grow by ionic diffusion whereas protective films would be those formed by a high-field ion migration process. As described in Section VIII, the former films would be deleterious because, during their growth, they provide a source of vacancies to the stressed metal surface, while the second type of films is protective because it hinders such process.

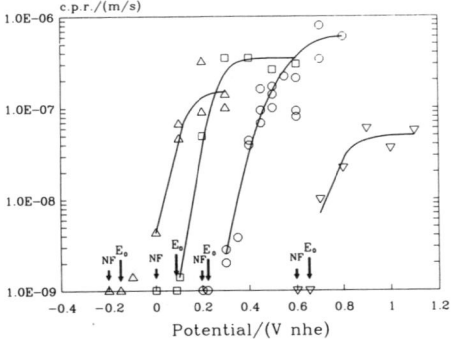

Figure 27. Crack propagation rate as a function of the electrode potential for Ag–15Pd wires strained in various solutions: △, $1M$ KI; □, $1M$ KBr; ○, $1M$ KCl; ▽, $1M$ Na$_2$SO$_4$. Strain rate, 2.6×10^{-6} s^{-1}. E_0 is the formation potential of the respective silver compound. NF: No cracks detectable after the straining test.[12]

Some authors, particularly those supporting the discontinuous cleavage mechanism, would include dealloyed layers among the deleterious anodic films. Nevertheless, as discussed in Sections VII and VIII, there is no general agreement about this point.

V. MECHANISMS BASED UPON ACTIVE DISSOLUTION

1. Introduction

As pointed out by Oltra,[113] by Staehle,[114] and by Parkins[115] in recent reviews on stress corrosion cracking, there are three mechanisms in the present-day literature which try to explain this phenomenon: (1) the anodic dissolution mechanism, discussed in the present section and in Section VI, (2) the discontinuous cleavage mechanism, which will be treated in Section VII, and (3) the surface-mobility mechanism described in Section VIII.

In the literature, strong disputes are found between the supporters of the various alternative explanations. This is particularly true in the case of the oldest approaches to the problem, namely, anodic dissolution and discontinuous cleavage. The history of science has shown that usually the most acrimonious disputes ended with no winners, for the simple reason that both views were, at least partly, right. Since the supporters of the alternative views are still actively working to try to prove their points of view, the present author will attempt to make, most probably without much success, an impartial description of the subject.

As shown in Section I (Fig. 2), when a precracked metal is subjected to the action of tensile stresses, there is a considerable lack of homogeneity in the stress distribution along the crack. A very large stress concentration is produced at the tip of the crack. Since the processes taking place at that point are responsible for the environmentally induced cracking, one of the first approaches to the study of the problem was to see to what extent stresses affect the electrochemical behavior of a metal.

From a mechanical point of view, a stress applied to a material induces a wide variety of changes. In the present review only two extreme cases will be considered: (a) low stresses, which induce only reversible changes in the shape of the metal, described as elastic strains, and (b) high stresses, which, on the other hand, lead to permanent changes in the shape of the material and are described as plastic strains. Since both types of mechanical action give rise to different types of electrochemical behavior, they will be treated separately.

The effects of both elastic and plastic strains on the electrochemical behavior of a film-free metal are discussed in the present section. Finally, as an example of a model of an SCC mechanism for active metal anodic dissolution, the one developed by Doig and Flewitt[116-118] is analyzed. The effect of stresses on filmed metal surfaces will be discussed in Section VI.

2. Elastic Strains

The effect of an elastic strain on the electrochemical behavior of a metal has been studied by various authors.[119-121] Tan and Nobe[121] studied the electrode potential of silver, brass, and steel wires under stress, in silver nitrate, copper sulfate, and ferrous sulfate solutions, respectively. The measurements on silver wires were made in $1N$, $0.1N$, $0.01N$, and $0.001N$ silver nitrate solutions. The authors reported that with the application of the load, the potential of the wires became more positive with relation to the unstressed reference electrode. Similar types of electromotive force (emf) versus time responses were reported by other authors in various systems, and Fig. 28 shows a typical example of this type of behavior.[119] After loading, the electrode potential increases monotonically with time and approaches a steady-state value. With the removal of the load, the electrode potential returns to the original no-load value in the same

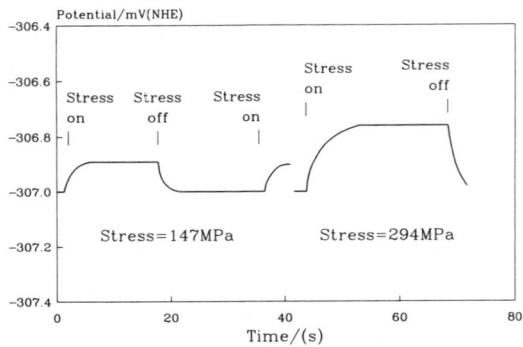

Figure 28. Typical changes of corrosion potential with time for iron exposed to deaerated $0.1N$ H_2SO_4 and subject to elastic strains. After Despic et al.[119]

manner. The emf values measured by Tan and Nobe[121] were approximately a linear function of the tensile stress, as shown in Fig. 29. The slope for $0.1N$ silver nitrate solution was 0.97 mV per 100 MPa. The measurements for mild steel (99.8% Fe) in $1N$ $FeSO_4$ gave potential–time curves similar to those in Fig. 28 and an emf–stress slope of 0.187 mV per 100 MPa.

The behavior of brass was different from that reported above for silver and iron. Upon loading of the samples, in copper sulfate solutions, the brass wire potential shifted to a more negative value with respect to the reference electrode. During the transient period, the emf attained a maxi-

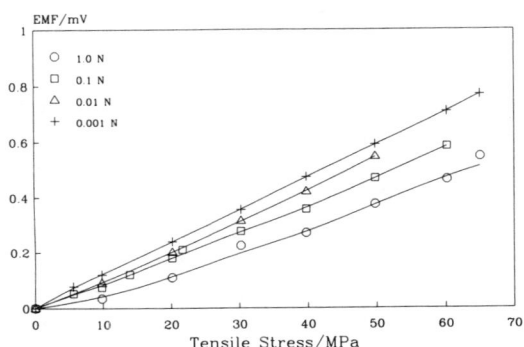

Figure 29. Electromotive force of silver electrodes stressed in $AgNO_3$ solutions.[121]

mum value almost instantaneously and decayed exponentially to a steady-state value. The removal of the load resulted in a mirror image of the emf–time curves for the loading process.

Nobe and Tan[120] also studied the electrode potential responses of silver, steel, and brass in sodium chloride solutions. In these cases the authors found that the potential shifted to more negative values, not only on loading the samples, but also on unloading them. This behavior was attributed to the corroding action of the chloride ions in the solution.

Despic et al.[119] studied the effect of elastic strains on the electrochemical behavior of iron in deaerated, $0.1N$ H_2SO_4 solutions. The study was made under conditions where the metal was in the active state, that is, free of surface films. Iron, in this acid solution, undergoes corrosion; hence two reactions are taking place simultaneously: (a) the anodic dissolution of iron and (b) the cathodic evolution of hydrogen. The authors studied the effect of the elastic stress on both the corrosion potential, under open-circuit conditions, and the electrochemical reactions, under a small, either anodic or cathodic, polarization of the samples.

Figure 28 shows a typical example of the shift in the corrosion potential upon application of the elastic stress. Despic et al. observed that the corrosion potential changed approximately 0.8 mV per 1000 MPa. The same authors studied the changes in both the anodic and the cathodic reactions when elastic strains were applied under a constant overpotential. The overpotential used was only 2 mV. From the measured current densities and the Tafel slopes for both the hydrogen evolution and the iron dissolution reactions, they calculated the effect of the elastic stress on the anodic and cathodic reactions. As shown in Fig. 30, they found that the elastic stress affected only the cathodic reaction of hydrogen evolution, but had no effect on the anodic reaction of iron dissolution.

To calculate the values in Fig. 30, the authors made the following deduction: (a) Cathodic and anodic current densities at a constant potential close to the corrosion potential, E_{corr}, represent differences in the iron-dissolution and hydrogen-evolution current densities, i_{Fe} and i_H, and (b) Tafel relations are assumed to be followed by both processes in that potential region:

$$E = \mathbf{a} + \mathbf{b} \cdot \ln i \qquad (20)$$

In the above equation, E is the electrode potential, i is the current density, and $\mathbf{a}$ and $\mathbf{b}$ are constants, $\mathbf{b}$ being the Tafel constant usually expressed as $\mathbf{b} = RT/\alpha F$.[122]

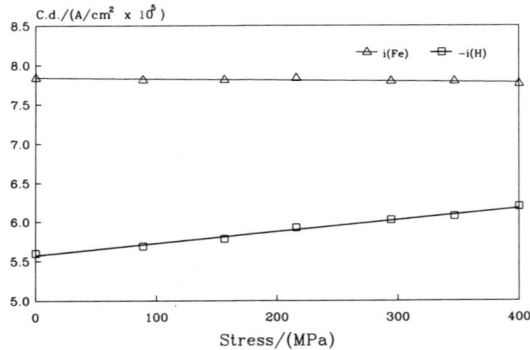

Figure 30. Effect of elastic strains on hydrogen-evolution and metal-dissolution current densities, for iron exposed to deaerated 0.1N H$_2$SO$_4$. Anodic and cathodic polarization = ±2 mV. After Despic et al.[119]

These assumptions lead to the following equations for the pure hydrogen-evolution current density and for the pure iron-dissolution current density at the respective cathodic and anodic potentials:

$$(i_H)_{\eta_c} = \frac{i_c + i_a \exp[(-\eta_c + \eta_a)\alpha_{Fe} F/RT]}{1 - \exp[(-\eta_c + \eta_a)(\alpha_H + \alpha_{Fe}) F/RT]} \quad (21)$$

$$(i_{Fe})_{\eta_a} = \frac{i_a + i_c \exp[(-\eta_c + \eta_a)\alpha_H F/RT]}{1 - \exp[(\eta_c + \eta_a)(\alpha_H + \alpha_{Fe}) F/RT]} \quad (22)$$

where i_c and i_a are the observed current densities, F is the Faraday constant, R is the gas constant, T is the temperature in degrees Kelvin, $\eta_c = E_{cathodic} - E_{corr}$, and $\eta_a = E_{anodic} - E_{corr}$. With both cathodic and anodic polarizations of about 2 mV, $\alpha_{Fe} = \alpha_H = 0.5$, and Tafel constants of 0.110 and 0.040 for hydrogen-evolution and iron-dissolution reactions, respectively, the values in Fig. 30 were calculated for unstrained wire and for the different stresses applied.

These observations could be extrapolated to the above-mentioned results reported by other authors. The equilibrium potential of a metal is the point where the anodic reaction of metal dissolution is balanced by an equal rate for the cathodic reaction of metal deposition,

$$Me \rightleftharpoons Me^+ + e^- \quad (23)$$

By extrapolation of the results obtained by Despic et al., we could infer that the potential changes reported by Tan and Nobe[121] were observed because the stresses affected only the cathodic branch of the equilibrium reaction. Nevertheless, these are mere speculations, since no experimental measurements are available on this matter.

A thermodynamic approach to the study of the effect of elastic stresses on the equilibrium electrode potential of a metal was made by Flood.[123] This author developed a general equation for the electrode potential changes in the metal electrodes exposed to both hydrostatic stresses (either positive or negative) and shear stresses. He concluded that the electrode becomes more negative (less noble) under a positive hydrostatic pressure, while, on the other hand, the electrode becomes positive (more noble) when the hydrostatic pressure is negative. When shear stresses are present, the shear strain energy always makes the electrode less noble. For elastic solids following Hooke's law, Flood reports that his equation can be written approximately:

$$E - E_o = -\frac{v}{nF}\left\{\left[\frac{p_x + p_y + p_z}{3}\right] + \left[\frac{(p_x - p_y)^2 + (p_y - p_z)^2 + (p_z - p_x)^2}{12 \cdot M}\right]\right\} \quad (24)$$

where E_o is the electrode potential in the absence of stress, E is the electrode potential with normal stresses p_x, p_y, and p_z, M is the modulus of rigidity, v is the volume per mole, n is the number of charges, and F is the Faraday constant. The first term in Eq. (24) is for hydrostatic stresses while the second term is the contribution due to shear stresses.

No information about calculations of the emf changes expected for a complex stress distribution, such as exists at the tip of a crack, has been found in the literature.

As shown above, elastic strains have a small effect on the electrochemical behavior of metals. In the particular case of equilibrium electrode potentials, the SCC literature[124] dismissed these effects as too small to be relevant. Nevertheless, recent publications[125] indicated that these values could be important when considered as the driving force for a diffusion path along a few atomic distances.

It is worthwhile to keep in mind that for an anodic dissolution SCC mechanism, tensile stresses should be expected to accelerate the dissolution of the metal.[13] Nevertheless, the equilibrium potential measurements on metals subjected to homogeneous elastic tensile stresses seem to show the opposite. Apparently, if the metal was initially at the equilibrium potential, the presence of a tensile stress would favor the cathodic reaction of metal deposition. It is also true that there are two points which were not considered. The first one is that in SCC, the tensile stresses at the tip of the crack are not homogeneously distributed whereas they were in the experiments reported above. The second is that, in all SCC cases, the metal is exposed at potentials well above the equilibrium potential. In any case, this is an area that would require further study.

3. Plastic Strains on Film-Free Metals

When Hoar and Hines[54] suggested an SCC mechanism by which the anodic dissolution of the metal at the tip of the crack was enhanced as a result of localized straining, the electrochemical behavior of straining metals became of interest. Hoar and West[55] studied the behavior of 18% Cr–8% Ni austenitic stainless steel strained in boiling, saturated $MgCl_2$ solutions. At the time it was believed that no films could be present on stainless steels in such an aggressive environment as a saturated, boiling $MgCl_2$ solution. Therefore, Hoar and West centered the interpretation of the data obtained with straining wires on a hypothetical increase in the active sites for dissolution. Later work by Staehle et al.[126] and by Wilde[127] showed the presence of films under those experimental conditions. Therefore, film breakdown had to be taken into account in the analysis of these experiments. These measurements will be discussed in the next section, when the effect of plastic strain on filmed metals is considered.

The effect of plastic straining on film-free metals was studied by various authors.[119,128,129] Despic et al.[119] published a full study of the anodic behavior of iron, nickel, copper, and molybdenum during straining, at a constant potential. The solutions chosen were as simple as possible. Hence, these authors used a $0.1N\ H_2SO_4$ solution for iron and nickel wires, a $0.1N\ H_2SO_4 + 0.1N\ CuSO_4$ solution for copper, and a $1N$ KOH solution for molybdenum. All measurements were made in a purified argon atmosphere, and all the wires were annealed before straining.

As reported above (Fig. 30), Despic et al. found that the anodic current density hardly changed with strain in the elastic region. However, when yielding began, there was a large sudden increase in the current density. A marked difference was observed in the rate of increase of the current density on straining between iron and molybdenum on one side and nickel and copper on the other (Fig. 31).

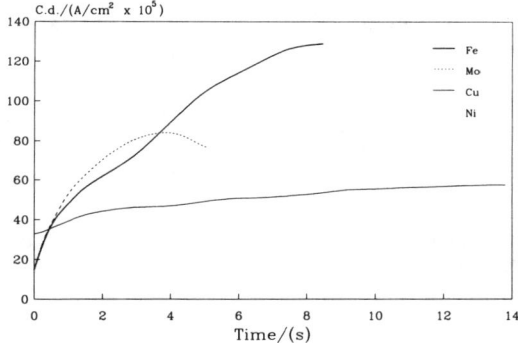

Figure 31. Anodic-current-density–time relationships for iron (0.1N H$_2$SO$_4$ solution, $E = -0.290$ V), molybdenum (1N KOH solution, $E = -0.340$ V), nickel (0.1N H$_2$SO$_4$ solution, $E = -0.060$ V), and copper (0.1N H$_2$SO$_4$ + 0.1N CuSO$_4$ solution, $E = +0.310$ V). Strain rate, 1.67×10^{-2} s^{-1}. Initial current density: for Fe, Mo, and Ni, 1.5×10^{-4} A/cm^2; for Cu, 3.3×10^{-4} A/cm^2. After Despic et al.[119]

The maximum current density was a function of the strain rate, but again the effect was more pronounced for iron and molybdenum than for nickel (Fig. 32).

In the analysis of their experimental results, Despic et al. pointed out the following facts. As plastic deformation starts, slipping in the metallic

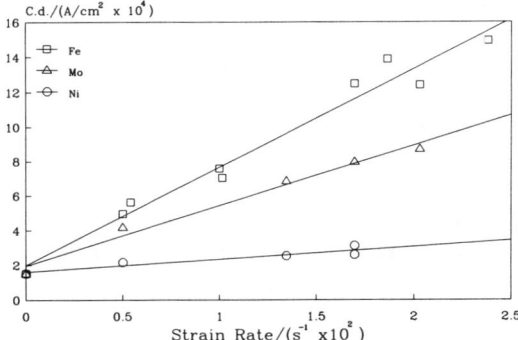

Figure 32. Maximum measured current density as a function of the strain rate for iron, molybdenum, and nickel. Experimental conditions as in the caption to Fig. 31. After Despic et al.[119]

crystals occurs. Small areas of slip planes emerge on the electrode surface, forming slip steps exposed to the solution. One of the first electrochemical processes taking place on this newly exposed surface is the charge of the double layer from the zero-charge potential to the potential maintained by the rest of the sample. The authors calculated the contribution of the double-layer charge (of the order of 10^{-7} A/cm^2) and found that it was negligible as compared to the other factors involved.

Another point taken into account was the increase of the surface roughness of the electrode. The roughness factors found were between 2.3 and 2.5 and led to a minor contribution to the straining metal current density.

A third contribution was the increased activity of dissolution at the edges of the slip steps. Despic et al. concluded that the current density at the edges could be taken as about 10 times larger than that on the flat surface, and the region of increased current density was assumed to be about 10^{-7} cm wide.

The three above-mentioned contributions gave a quantitative account for the anodic behavior of nickel and copper during straining. In the case of iron and molybdenum, there was a very important fourth parameter to be taken into account.

The slip planes for face-centered cubic (fcc) metals, such as copper and nickel, are known to be (111) planes. On the other hand, the (110), (112), and (123) planes are reported as slip planes for iron, whereas molybdenum slips along a (112) plane. Despic et al. suggested that high-index planes (112) and (123) were likely to be more active for the dissolution process owing to the lower binding energy of the surface atoms. The comparative atomic-surface densities are 0.892, 0.514, and 0.337 for (110), (112), and (123) planes, respectively, as compared to the surface density of the most closely packed structures. For copper, as quoted by the authors, the less close packed plane (110) has an exchange-current density for metal deposition and dissolution 10 times larger than that of the most closely packed (111) plane. Hence, in those systems where new high-index planes emerged on the surface of the wire during plastic deformation, they provide an increased dissolution activity. As the strain increased continuously, more of these new planes emerged on the electrode surface and there was a further increase of the anodic activity. However, there was a preferential dissolution of these new active planes. The surface in contact with the solution tended to become more stable, and it eventually showed dissolution rates similar to those of the un-

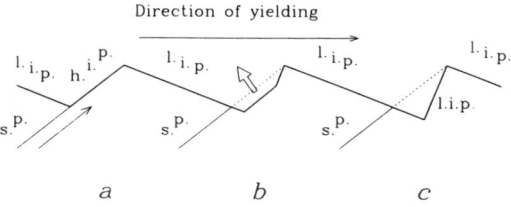

Figure 33. Schematic representation of development and preferential dissolution of the high-index planes. s.p., Slip planes; h.i.p., high-index planes; l.i.p., low-index planes. Sequence: (a) Emerging h.i.p.; (b) partly dissolved h.i.p. site; (c) completely dissolved h.i.p. site. After Despic et al.[119]

strained wire surface. There was, then, a sort of lifetime for the activity of the newly emerged surfaces, and Fig. 33 shows how Despic et al. envisaged the process taking place on iron and molybdenum during straining. From their measurements, they concluded that, on the whole, the current density at high-index planes for iron appeared to be 50–100 times larger than that for the low-index planes of an unstrained wire. In the case of molybdenum, the current density at the high-index planes was about 50 times larger than the current density at the low-index planes. The initial current density at an unstrained wire was 15×10^{-5} A/cm^2.

To sustain the above interpretation, Despic et al. compared their experimental measurements with theoretical calculations. Figure 34 shows the very good correlation that they found between theory and experiment.

Despic et al. did not attempt at the time an analysis of the relation between their measurements and the SCC process. In the light of what is known at present, an SCC model based on the anodic dissolution of a film-free strained metal can be disregarded. According to the measurements of Despic et al., no cracking should be expected in fcc alloys, since no high-index planes are produced during straining. Since the current increase found for fcc metals, such as nickel and copper in Fig. 31, is small, the metal/solution front, at the tip of the crack, if strained, will propagate at a rate only slightly higher than that at the sides of the crack. Therefore, no high aspect ratios should be expected, contrary to what is observed during SCC, as pointed out in Section I. Nevertheless, fcc alloys, such as α-brass or austenitic stainless steels, are known to suffer SCC, usually

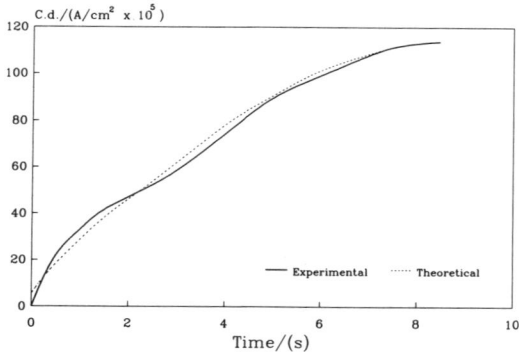

Figure 34. Increase in anodic current density upon straining past the yield point for iron in a $0.1N$ H_2SO_4 solution, $E = -0.290$ V. Strain rate, 1.67×10^{-2} s^{-1}. After Despic et al.[119] A very good correlation was found by the authors between theory and experiment.

with crack velocities higher than those for body-centered cubic (bcc) alloys, such as ferritic steels.

As for alloys with bcc structures, the fact that high-index planes dissolve from 50 to 100 times faster than low-index planes could account for the condition of a high aspect ratio, mentioned in Section I. Nevertheless, as shown in Fig. 35, the high aspect ratio would be the result of a macroscopic approach to the problem. When the process is analyzed at a microscopic level, a different conclusion is reached. Figure 35 shows the shape of the slip step surface both at the initial and at the final state. The

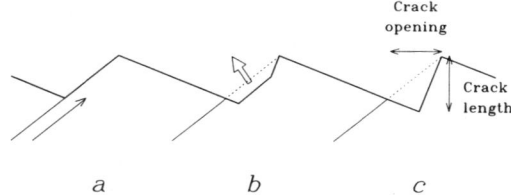

Figure 35. The shape of a dissolved high-index plane is characterized by very low aspect ratios. This is contrary to what is found in stress corrosion cracking. (a) Emerging h.i.p.; (b) partly dissolved h.i.p. site; (c) completely dissolved h.i.p. site.

comparison between the two states does not lead to a crack, but to a trench, with a low aspect ratio. Hence, the conditions for cracking are not fulfilled at a microscopic level, as should be expected for SCC. As further support of this point of view, no information is found in the literature on SCC occurring in the metal/environment systems studied by Despic *et al.* It must be concluded, then, that SCC is not due to the electrochemical anodic dissolution of nonfilmed, straining, metals.

4. Anodic Dissolution SCC Models

While Despic *et al.* obtained very useful information about the effect of stresses on the electrochemical behavior of metals, they did not elaborate a mechanism for SCC. At present, most of the anodic dissolution mechanisms for SCC involve filmed metals and a stress-induced film rupture process. These mechanisms will be discussed in Section VI. Nevertheless, one mechanism where active dissolution at the tip of the crack is assumed is that proposed by Doig and Flewitt.[118] This mechanism, which was developed from first principles of electrochemistry and mechanics, is a good example of what an SCC mechanism should be like, and it deserves a detailed analysis.

One unstated assumption is that a preexisting path, as described in the previous section, is present in the metal. Although not explicitly stated, Doig and Flewitt's model[118] assumes infinitely diluted solutions and the absence of convection.[130] The first of these two assumptions could seem an oversimplification of the problem, but it is frequently used when dealing with complex electrochemical processes.[131] As for the assumption of the absence of convection, it contradicts the assumptions made by West[132] for the propagation of cracks, but if the high aspect ratio of most stress corrosion cracks is taken into account, it seems reasonable.

As discussed in Section II, one of the relationships used in the mathematical modeling of mass transport and electrochemistry in cracks is that giving the flux (J) of each dissolved species,[79] that is, Eq. (8). Another relationship that gives the current in the electrolyte solution is Eq. (9).

When, as in the case of Doig and Flewitt's model, it is assumed that no concentration gradients are present in the crack, the above two relations are combined to give ohm's law[79]:

$$i_d = -k \nabla (\Phi) \qquad (25)$$

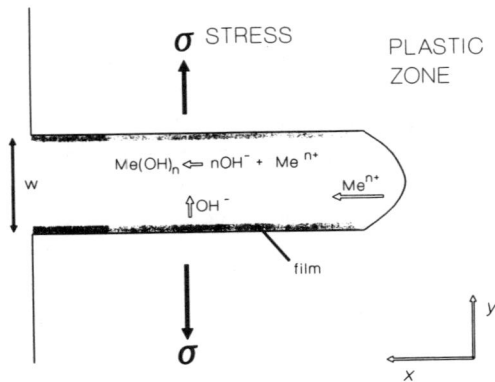

Figure 36. Schematic description of a crack and the reactions considered in the model by Doig and Flewitt.[118] The crack tip is anodically active, as a result of a continuous dynamic plastic strain, which exposes film-free metal surface to the local environment. The cathodic reaction takes place at the sides of the crack.

As Turnbull points out,[79] this is a commonly used approach, which is probably not an unreasonable approximation, provided the relevant values of the current density are used in the equation.

Figure 36 shows a schematic description of a crack and the reactions considered in the model. In this model[118] it is assumed that the crack tip is anodically active as a result of a continuous dynamic plastic strain, which exposes film-free metal surface to the local environment. Metal ions, Me^{n+}, are produced at a rate governed by the local electrochemical conditions, and they diffuse from the crack tip. The authors pointed out that in order to have an anodic ion current flow from the crack tip, an electrode potential gradient must exist along the crack. A cathodic current will flow from the sides of the crack to complement the anodic current at the crack tip (Fig. 37). The steady state is reached when the electrode potential at the crack tip produces an anodic current that is balanced by the cathodic current on the crack surfaces. Cathodic reactions outside the crack are not taken into account, a point which could be considered as a limitation of the mechanism.

Assuming Tafel relations for the anodic and cathodic reactions, Doig and Flewitt[133] found that the potential E_s, at the tip of the crack, at which

Figure 37. Doig and Flewitt pointed out that in order to have an anodic ion current flow from the crack tip, an electrode potential gradient must exist along the crack. A cathodic current will be flowing from the sides of the crack to complement the anodic current at the crack tip.

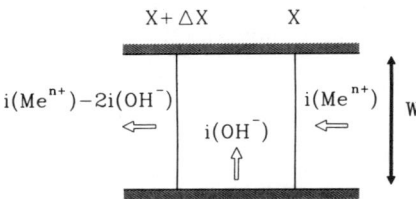

the balance between anodic and cathodic currents is reached is given by the following relation:

$$w i_F \exp[(E_s - E_F)/\alpha] = [\ 4\ i_c\ C\ w\ (E_s - E_c + \beta\ \{\exp[(E_c - E_s)/\beta] - 1\})]^{1/2} \quad (26)$$

where i_F is the anodic dissolution current density of the crack tip at potential E_F and α is the Tafel constant for the reaction; i_c and E_c are, respectively, the corrosion current density and the corrosion potential on the crack surfaces; β is the Tafel constant for the cathodic reaction on the crack surfaces; w is the width of the crack; and C is the ionic conductivity of the solution within the crack (assumed constant). The crack width was assumed to be equal to the crack opening displacement and hence given by

$$w = 0.5 \left(\frac{K^2}{\sigma_y E_y} \right) \quad (27)$$

where K is the applied stress intensity, σ_y is the material yield strength, and E_y is the Young's modulus of elasticity. Except for E_s, which is evaluated by Eq. (26), the other variables are obtained experimentally from conventional polarization measurements. Once the value of E_s is calculated, the crack velocity, v, can be predicted by the following equation:

$$v = i_F \exp\ [\ (E_s - E_F)/\alpha\ ]\ M/(\delta F n) \quad (28)$$

where M is the molar mass of the material, δ is the density, F is the Faraday constant, and n is the charge number of the anodic electrode reaction.

As pointed out by Doig and Flewitt,[116] this analysis predicts that the crack growth rate will be related to the electrochemical properties of the crack tip and surface, the solution conductivity, the applied stress intensity, and the bulk mechanical properties of the sample.

During crack growth, solid corrosion products are deposited on the crack surfaces by reaction of metal ions from the crack tip with the cathodic reaction products (Fig. 36). The presence of this film effectively reduces the width of the ion-conducting path in the crack, and Eq. (26) becomes

$$w\, i_F \exp[(E_s - E_F)/\alpha] = [\, 4i_c\, C\, (w - 2t)\, (E_s - E_c + \beta\, \{\exp[(E_c - E_s)/\beta] - 1\,\})]^{1/2} \quad (29)$$

where t is the limiting corrosion product film thickness. This reduces E_s and the associated growth rate by an amount which becomes more significant as the value of w decreases, that is, as the stress intensity is decreased. Doig and Flewitt[116] pointed out that $2t$ must be less than w for any realistic solution of this equation. The condition $w = 2t$ provides a criterion for a threshold stress intensity for continued crack propagation. This would be the explanation, according to Doig and Flewitt's model, for the existence of a K_{1SCC} value.

Doig and Flewitt applied the above analysis to the case of SCC of ferritic steels in 8M NaOH at 373 K and of aluminum base alloys in 1M NaCl, pH 1.1, at 298 K. As mentioned above, in relation to the evaluation of Eq. (26), they assumed that the electrode reaction kinetics could be approximated by Tafel type relationships. Table 2 shows the parameters used by Doig and Flewitt in their analysis.

Table 2. Electrochemical and Mechanical Parameters Used by Doig and Flewitt[a]

Parameter	Steel	Aluminum
σ_y (MPa)	500	200
E_y (MPa)	2×10^5	7×10^4
C ($\Omega^{-1}\,m^{-1}$)	50	50
E_F (V, SCE)	−1.150	−1.530
E_c (V, SCE)	−1.020	−1.050
α (V)	0.032	0.050
β (V)	0.064	0.090
i_F (A/m^2)	100	3000
i_c (A/m^2)	1	0.02
n	2	1
δ (kgm^{-3})	7.9×10^3	2.7×10^3
M (kg/mol)	5.6×10^{-2}	2.7×10^{-2}
F (C/equiv)	96500	96500

[a]Ref. 118.

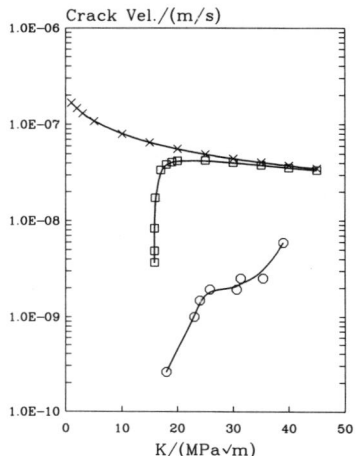

Figure 38. Crack velocities (○) measured by Singbeil and Tromans[134] for mild steel in 33% NaOH at an open-circuit potential (E_{corr} = –1.00 V vs. SHE) and 92°C, compared with theoretical values calculated with Eqs. (28) and (29). The calculations were made with the data in Table 2, assuming σ_y = 631 MPa, the same value as that for the materials used by Singbeil and Tromans. ×, $t = 0$ (no films); □, $t = 0.5\,\mu$m.

We will now analyze how the predictions of Doig and Flewitt's mechanism fit with the available experimental data. Figure 38 shows the values of crack propagation rate as a function of stress intensity, for iron in NaOH solutions, calculated with Eqs. (28) and (29). The calculations were made for two different cases. The first one assumes the absence of films on the crack walls, $t = 0$, and the second is meant for a film thickness of $t = 0.5\,\mu$m. The yield strength used was σ_y = 631 MPa, to allow a comparison with the experimental data published by Singbeil and Tromans.[134]

As shown in Fig. 38, there is reasonable agreement between theory and experiment. The calculations for a filmed crack adequately predict the existence of a threshold value for K. The difference in crack velocities, in region II, can be ignored, taking into account that the electrochemical values in Table 2 are only a rough approximation. The main objections to be made are the following: (a) whereas the theory does not predict any stress intensity dependence in region I, the experimental results show a strong dependence of crack velocity on K; and (b) whereas Doig and Flewitt suggest that in region II the crack velocity should show a slight decrease when K increases, the experiments show just the opposite. The theoretical prediction of a decrease in the crack velocity when K is increased originates from Eq. (27). The higher the value of K, the higher w will be. Hence, while the total cathodic area will remain constant, the anodic area will increase, and the crack propagation will be slower.

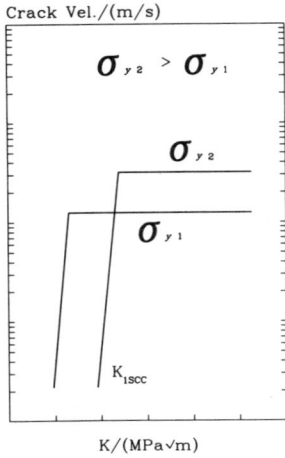

Figure 39. Schematic diagram showing the influence of material yield strength on the overall form of the crack velocity vs. applied stress intensity curves, according to Doig and Flewitt's theory.[118]

Doig and Flewitt considered the effect of yield stress on the crack velocity. The role of yield stress enters in the model through its effect on the width of the stress corrosion crack, w, given by Eq. (27). By increasing the yield stress, the crack opening displacement decreases, at any given applied stress intensity. This action increases the predicted crack growth rate and raises the threshold stress intensity, as schematically shown in Fig. 39. The increase in the crack velocity with the yield stress, according

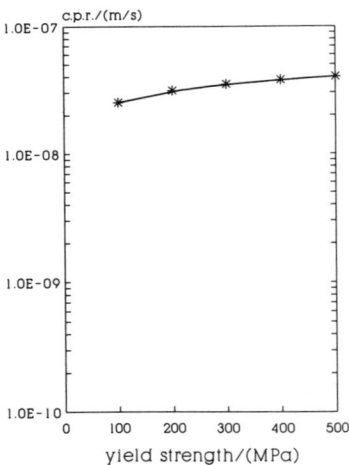

Figure 40. Effect of yield strength on the crack propagation rate, as predicted by Doig and Flewitt's theory, for iron in caustic solutions. Data as in Table 2; $K = 30$ MPa$\sqrt{m}$.

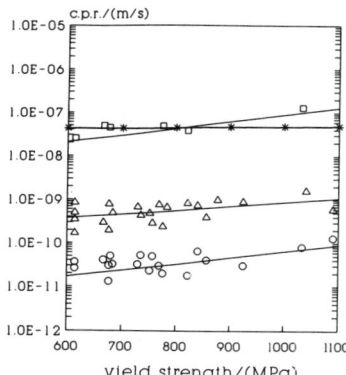

Figure 41. Comparison of the theoretical values for yield strength vs. crack propagation rate, as in Fig. 40, with the measurements reported by Magdowski and Speidel.[135] ○) 100°C; △) 160°C; □) 288°C; *) values calculated with the Doig and Flewitt Theory.

to this model, is due to the fact that while the area for the cathodic reaction is constant, that for the anodic reaction gets smaller. The authors report that there is a small increase in the crack growth rate of 1.5 times with increasing yield stress in the range of 100–500 MPa (Fig. 40).

Although the authors said that the verification of the effect of yield stress on crack velocity is not feasible, recently Magdowski and Speidel[135] published abundant information on this subject, applied to the case of steels for steam turbine rotors. Figure 41 shows the experimental results published by Magdowski and Speidel, as compared with the predictions of Doig and Flewitt. As is observed, the model predicts an effect of yield stress that is considerably lower than that found in practice. Models predicting the effect of yield strength on crack velocity are scarce, but in Section VIII a model with a better correlation between theory and experiment will be described.

The model predictions can also be checked on the behavior of aluminum alloys. There is abundant information both on the effect of heat treatment over yield stress and on crack velocities as a function of stress intensity. Figure 42 shows measurements reported by Speidel[13] on the effect of overaging on the SCC velocity of aluminum alloy 7079-T651. The overaging was made at 160°C, and the crack velocity was measured, for various stress intensity values, in a saturated aqueous NaCl solution, at 23°C and an open-circuit potential.

As reported by Speidel,[13] the yield strength for aluminum alloy 7079-T651 was 520 MPa and dropped to 275 MPa after 500 h overaging at 160°C. Figure 43 shows the values calculated with Doig and Flewitt's

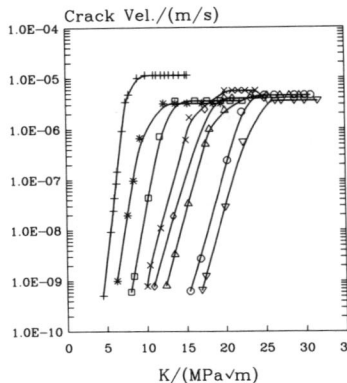

Figure 42. Effect of overaging of aluminum alloy 7079-T651 at 160°C on the velocity of stress corrosion cracks in a saturated aqueous NaCl solution, at an open-circuit potential and 23°C, after Speidel.[13] Aging time: +, 0 h; *, 6 h; □, 20 h; ×, 50 h; ◇, 84 h; △, 204 h; ○, 355 h; ▽, 500 h.

model, assuming a film thickness of $t = 0.25$ μm. The theory fails to predict the important change in the stress intensity threshold value produced by overaging. It does not predict, either, the stress intensity dependence of crack velocity in region I.

Figure 44 shows the case of another aluminum alloy. In this case it is the 7178-T651 alloy. The experimental values were reported by Speidel[13] and were measured in a saturated aqueous NaCl solution, at 23°C. The overaging was done at 160°C. According to Speidel,[13] the yield stress for alloy 7178-T651 was 607 MPa and dropped to 586 MPa after the alloy had been aged for 1 h at 160°C. Figure 45 shows the predictions for these values. Again, the discrepancy between theory and experiment is considerable.

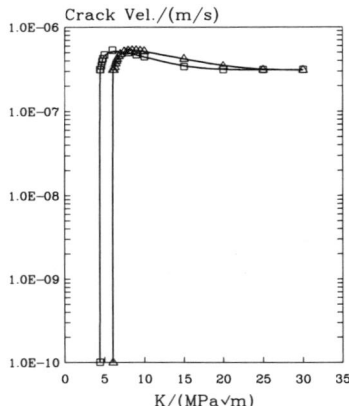

Figure 43. Theoretically predicted crack velocities for aluminum with two different yield strengths and a film thickness of $t = 0.25$ μm. △, Alloy 7079-T651, $\sigma_y = 520$ MPa; □, the same alloy after 500-h overaging at 160°C, $\sigma_y = 275$ MPa. Compare with Fig. 42.

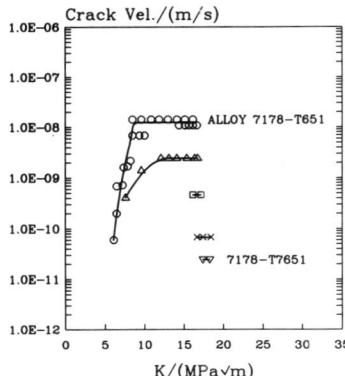

Figure 44. Effect of artificial aging at 160°C on the stress corrosion crack velocity vs. stress intensity for aluminum alloy 7178-T651, after Speidel.[13] Tests at open-circuit potential, in a saturated aqueous NaCl solution, at 23°C. Aging time: ○, 0 h; △, 1 h; □, 8 h; ×, 12 h; ▽, 15 h.

Most probably, the discrepancies between theory and practice, mentioned above, are due to the choice Doig and Flewitt made for the size of the anodically dissolving crack tip. According to these authors, w is given by Eq. (27) and is only a function of the stress intensity and the yield strength, as shown in Fig. 46. In systems with a preexisting path, as is the case with aluminum alloys, the size and the distribution of these paths are considerably affected by the heat treatment, a point that was ignored by the theory.

Presumably, the theory could be used to predict the effect of an external potential on crack velocity. Here again, as shown in Fig. 47, there is great discrepancy between theory and experiment.

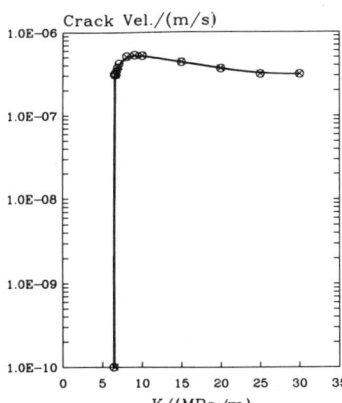

Figure 45. Theoretically predicted crack velocities for aluminum with two different yield strengths and a film thickness of $t = 0.25$ μm. ×, Alloy 7178-T651, $\sigma_y = 607$ MPa; ○, the same alloy after 1-h overaging at 160°C, $\sigma_y = 586$ MPa. Compare with Fig. 44.

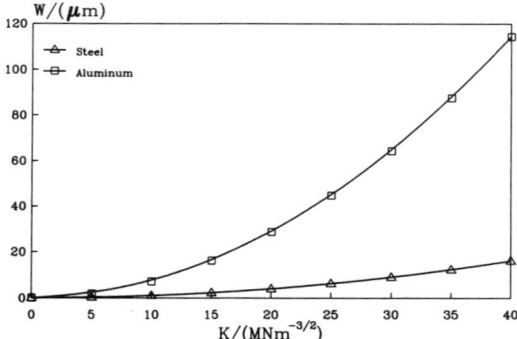

Figure 46. Effect of stress intensity on the size of the anodic front, at the tip of the crack, according to Doig and Flewitt, for aluminum and for steel.

Nevertheless, from a mechanistic point of view, the weakest point of this model, and probably of all the anodic dissolution models, is shown in Fig. 48. In this figure, crack velocities for aluminum alloy 7178-T651 in a 5M aqueous HI solution, at an open-circuit potential, reported by Speidel and Hyatt,[136] are shown. The alloy, in this acidic environment, shows intergranular corrosion. Then the active paths are already being corroded, and there is no film to occlude the cracks. Hence, the theoretical treatment should be that of a film-free crack, $t = 0$. The calculated values are shown

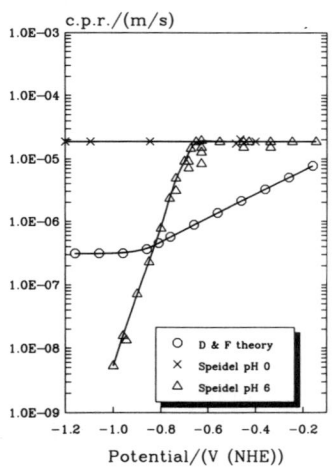

Figure 47. Effect of external potential on stress corrosion crack velocity of aluminum alloys. △ and ×, Experimental values for aluminum alloy 7079-T651 in a 5M aqueous iodide solution at 23°C and $K = 14$–16 MPa√m, after Speidel[13]; ○, theoretical values for aluminum in HCl solutions, for $\sigma_y = 520$ MPa (as for alloy 7079-T651), $K = 15$ MPa√m, and $t = 0$.

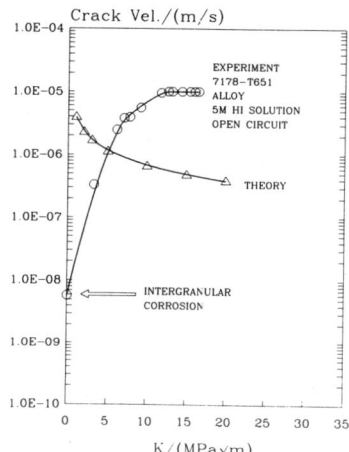

Figure 48. Effect of stress intensity on stress corrosion crack velocity. ○, High-strength aluminum alloy 7178-T651 in a $5M$ HI aqueous solution at open-circuit potential, after Speidel and Hyatt[136]; △, theoretical values for aluminum in HCl solutions, for $\sigma_y = 607$ MPa (as for alloy 7178-T651) and $t = 0$.

in the same figure. The theory predicts that the crack velocity will decrease when the stress intensity is increased. The experiments, on the other hand, show that the crack velocity increases by three orders of magnitude when the stress intensity is increased from zero up to 15 MPa√m. This indicates that, whatever the process at the crack tip, it is considerably accelerated by stress. This point is not taken into account in Doig and Flewitt's model. The active dissolution takes place even at $K = 0$; then we have to treat the system as an anodically dissolving active metal. As seen above, in Section V.3, in the discussion of plastic strains on film-free metals, there is no electrochemical process, presently known, that could explain such a stress-accelerated anodic dissolution. This is true in particular for fcc metals, such as aluminum alloys.

One of the most positive qualities of this mechanism is that it integrates a series of SCC properties starting from first principles. It sets the guidelines for future mechanisms to be developed. It could either be improved, to account for the discrepancies described above, or it could be assimilated to other mechanisms based on a different process at the tip of the crack.

5. Limitations of the Mechanisms

Some limitations of the mechanisms based on active metal anodic dissolution have been mentioned above. So far, there is no evidence showing that the dissolution rate of an active metal is increased by the presence of stresses. Elastic stresses induce only very slight changes in the electrode potential, and they have been dismissed as the cause of SCC. As for plastic straining, the work by Despic et al.[119] shows that it produces a moderate increase in the current density for bcc metals, but the effect is negligible for fcc metals. Anyway, the changes in dissolution rate observed could not account for the high aspect ratios observed in SCC.

The mechanism developed by Doig and Flewitt[116–118] is attractive (a) because, based on first principles, it gives a correlation between crack propagation rate, stress intensity, and electrode potential and (b) because it provides an explanation for the existence of a critical K_{1SCC}. Nevertheless, as shown above, the discrepancies between the predictions from this theory and experimental data are great.

There is another limitation to be taken into account in all the anodic-dissolution-based mechanisms, as recently pointed out.[137] The objection is centered on the description, at an atomic level, of the processes taking place at the tip of a sharp crack.

A simple description of the anodic dissolution of a metal[138,139] assumes that the metal atoms abandon the crystal lattice at the most exposed places, such as kink sites. Dissolution proceeds either by surface diffusion of the atom, as an adatom, followed by ionization plus dissolution, or by a direct ionization and dissolution from the kink site. The processes are schematically described in Fig. 49. It is generally accepted that the process

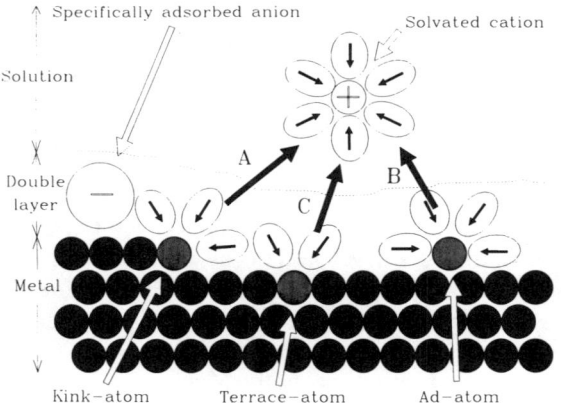

Figure 49. Schematic simplified description of the dissolution process of a pure metal. Steps **A** and **B** would be the most favored ones, because they would require a lower driving force. Step **C** would take place at a higher overpotential than steps **A** and **B**.

Electrochemical Aspects of Stress Corrosion Cracking

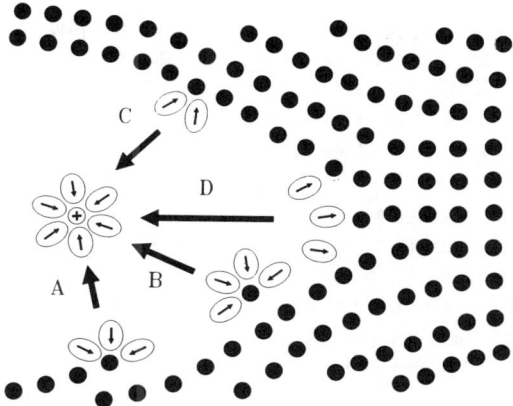

Figure 50. Electrochemical processes of metal dissolution, described in Fig. 49, applied to the tip of an atomically sharp crack undergoing anodic dissolution. Step **D**, which would be responsible for stress corrosion crack propagation, would be the least favored one.[137]

requiring the lowest overpotential is the dissolution of an adatom (step **B**) or that of an atom at a kink site (step **A**). The dissolution of a terrace atom (step **C**) would be the least favored and requires a higher overpotential.

If the tip of a crack, undergoing active dissolution, is considered (Fig. 50), the same steps will be present. However, a new one (step **D**) is to be included. It involves the dissolution of an atom, from the lattice, at the tip of the crack. Step **D** is the only one leading to crack propagation, and consequently it is the only one important for an anodic dissolution SCC mechanism. Nevertheless, as far as is known at present, it is the least favored process. The presence of stresses at the tip of the crack may be supposed to favor step **D** over the other three. Nevertheless, there is no experimental evidence to support this point. The work by Despic et al.[119] is conclusive about it. According to these authors, there are no evidences of a stress-assisted dissolution process. However, to be fair to those favoring active anodic dissolution mechanisms, it has to be pointed out that the measurements reported by Despic et al. apply only to homogeneously distributed stresses or strains. As shown in Fig. 2, such a distribution is far from being the real case with a stressed crack.

Speidel[13] points out that sharp cracks can be produced if dissolution is accelerated by stress. Nevertheless, there is no evidence so far that such a phenomenon occurs. As pointed out above, Hoar and West suggested such a mechanism for straining metals, but the work by Despic et al.[119] proved that this phenomenon does not occur.

Another point to be taken into consideration is the great similarity found in fractographs produced by SCC, hydrogen embrittlement, liquid metal embrittlement, etc., an aspect pointed out by various authors.[140–142] While anodic dissolution could be acceptable in some cases of SCC, it is clearly out of the question in the other cases of environmentally induced cracking. From a mechanistic point of view, this observation suggests that, besides the electrochemical reactions observed during SCC in electrolytic environments, there are some other processes, on the metal side of the crack tip, that are common to all these environmentally induced cracking processes, though they are ignored by the anodic dissolution mechanism.

A mechanism such as that developed by Doig and Flewitt, with some improvements, could be used for those systems where a preexisting path is present. This could be the case, for example, for an aged Al–Cu alloy in halide solutions, where stress accelerating effects could not be detected.[104] However, it is not directly applicable to those cases where stresses accelerate the crack propagation rate, as shown in Fig. 48.

In the presence of preexisting paths, as is the case with many intergranular SCC instances, anodic dissolution along the grain boundaries could be the operative mechanism. Some authors refer to these cases as stress-accelerated grain boundary corrosion. Probably this description is an oversimplification, since intergranular SCC is found even in cases where no intergranular corrosion can be detected. Other points remain unsolved in this description. It is not clear why there are critical potentials for cracking, there is no explanation of why certain species are so effective in accelerating cracking, and no description is available, yet, at an atomistic level, of what happens at the tip of the crack in an anodically controlled crack propagation process. The description by Doig and Flewitt assumes that the crack propagates by anodic dissolution over a front w, the size of which is a function of the stress intensity value K. On this assumption, these authors were able to explain the effect of stress intensity and yield strength on crack velocity. However, in the case of preexisting paths, the thickness of the susceptible material is a function of the composition and is not changed by the stress intensity. In the case of the aluminum alloys studied by Doig and Flewitt, for a moderate stress intensity value of 15 MPa$\sqrt{m}$, the dissolution front would be of nearly 20 μm and should be easily detectable on the fracture surface. The absence of evidences of anodic dissolution when fractographs are observed is a limitation to the mechanistic understanding of SCC, a limitation that was pointed out by Doig and Flewitt.[117]

As mentioned above, the anodic dissolution mechanism fails to explain the accelerating effect of stress on the supposed anodic dissolution process at the tip of the crack. This is the case with aluminum alloys, as shown in Fig. 48. A similar problem was reported by Was and Rajan,[103] who found that for intergranular cracking of Ni–Cr–Fe alloys in sodium tetrathionate solutions, the crack velocity was 11 times faster than that expected in anodic dissolution. These authors suggested a purely mechanical contribution of the stress.

It could be suggested that another important limitation of Doig and Flewitt's model is the fact that it ignores the current flow taking place from inside the crack to the external surface of the metal. Nevertheless, the most relevant limitations are those described above, since even if that extra current density were taken into account, the above limitations would not be solved.

VI. ANODIC DISSOLUTION AND FILMED METALS

1. Introduction

There are numerous reasons why several researchers believed that SCC was governed by an anodic dissolution process at the tip of the crack. One of the first observations made was that crack velocities were generally accelerated by anodic polarization and slowed down by cathodic polarization.[143] It was also found that in numerous instances certain critical potentials were necessary for SCC to take place.[144] As described below, Hoar and co-workers showed that, when straining metal samples in the corrosive environment at a constant potential, there was a good correlation between the current densities and the measured crack velocities. Following a similar approach, Parkins[145] found that anodic currents measured during potentiokinetic polarization curves were proportional to the crack velocities. All these measurements led numerous authors to believe that anodic dissolution was responsible for the SCC phenomenon and that the above observations were a clear proof of such a fact. As mentioned below, the present author shared this point of view.[146] Nevertheless, a more detailed analysis of the problem gives rise to numerous objections, and an alternative mechanism had to be sought.

2. Plastic Strains on Filmed Metals

In this subsection the constant-potential straining electrode measurements, which led to the idea that only anodic dissolution was responsible for numerous SCC cases,[108,147,148] will be described. For a full treatment of this subject, the reader is referred to a previous review.[146]

Only the straining electrode technique will be described, because Parkins and Holroyd,[149] when studying brass in acetate, tartrate, formate, and alkaline solutions, reported that the information on SCC susceptibility obtained by intermediate-strain-rate experiments was more reliable than that found by potentiokinetic measurements. One possible explanation of the difficulties found by Parkins and Holroyd with the potential scanning technique is that the voltage scanning should be started with a film-free metal. This type of limitation is not found with the intermediate-strain-rate technique.

Different straining electrode techniques, using various strain rates, have been reported in the literature. The constant-potential intermediate-strain-rate technique (ISRT) uses strain rates of 5×10^{-3} to 5×10^{-2} s^{-1}. It must be distinguished from the fast-strain-rate technique (FSRT)[150,151] used in transient electrode kinetics and in repassivation rate measurements, where strain rates from 5 to 60 s^{-1} are used, and from the slow-strain-rate technique (SSRT), developed by Parkins as an SCC test,[24] where strain rates between 10^{-4} and 10^{-8} s^{-1} are used.

The intermediate-strain-rate technique originated in Cambridge, England, in the sixties, with the work of Hoar and co-workers. The use of continuously straining metal for the study of SCC was introduced by Hoar and West in 1958.[152] According to an SCC mechanism due to Hoar and Hines,[54] the straining metal at the bottom of a crack was expected to dissolve much faster than the static metal, and this was supposed to lead to SCC. By using a straining wire, Hoar and West expected to reproduce on the whole sample the behavior of the metal at the bottom of the crack. Initially, the wires were strained manually; later, a modified Hounsfield Tensometer was used. Hoar and West[55] found strong depolarization when straining stainless steel wires in $MgCl_2$ solutions and suggested a mechanochemical dissolution mechanism where the number of active sites was increased by straining. Later, Hoar and Scully[153] reported current increases when straining stainless steel in $MgCl_2$ at a constant potential. In another work Scully and Hoar[154] found a qualitative correlation between the current increase of straining iron–nickel alloys and their susceptibility to SCC. Whenever an alloy was susceptible to SCC, high current increases were observed upon straining. On the other hand, alloys not susceptible to SCC would not show such current increases. This work was the first indication that the straining electrode technique could be used to detect SCC susceptibility.

In all the above experiments the straining samples, in $MgCl_2$, were assumed to behave as film-free metals. The current increases found were not high enough to support the mechanochemical anodic dissolution mechanism suggested by Hoar and Hines[54] and Hoar and West,[55] and this mechanism was later abandoned.

The analysis of a filmed straining metal, under intermediate-strain-rate conditions, was first done by Hoar and Galvele[75] in their study of SCC of mild steel in boiling nitrate solutions. These authors reported what seemed to be a quantitative correlation between the current densities found on the freshly exposed metal and those required for the propagation of stress corrosion cracks. The technique was further developed by Galvele and co-workers[8,43,155–158] as a test to predict SCC susceptibility, and it was successfully applied to the following cases of SCC: (i) mild steel in boiling nitrate solutions[75]; (ii) mild steel in boiling caustic solutions[159]; (iii) AISI 304 in a $1M$ HCl solution at room temperature[155,160]; (iv) AISI 304 in HCl+NaCl solutions at room temperature[160]; (v) AISI 304 in H_2SO_4 plus NaCl solutions at room temperature[157]; (vi) AISI 304 in boiling $20N$ NaOH solutions[161]; (vii) Incoloy 800 in a boiling $17.5N$ NaOH solution[161]; (viii) Inconel 600 in a boiling $17.5N$ NaOH solution[161]; (ix) AISI 304 in $MgCl_2$, $CaCl_2$, and LiCl solutions[93,158]; (x) yellow brass in $NaNO_2$ solutions[91,162]; (xi) high-purity copper in a $1M$ $NaNO_2$ solution[91]; (xii) yellow brass in a Na_2SO_4 + borate buffer solution[163]; and (xiii) yellow brass in Na_2SO_4 + Na_2S solutions.[163]

Let us see now a description of this experimental technique. The technique was used exclusively with systems where it was assumed that a brittle film rupture plus an anodic dissolution mechanism was involved. It was explicitly stated that it applied neither to hydrogen embrittlement nor to liquid metal embrittlement. When the technique was applied to filmed metals, no distinction as to the nature of the film was made. It could be a highly protective passive film, a porous anodic film, a film grown by dissolution and precipitation, a high-field anodically grown film, or a dealloyed metallic layer. The only requirements were the following: (a) the film had to be brittle, and (b) it should hinder the process of anodic dissolution.

According to the ISRT, if a filmed metal was strained while exposed to a corrosive environment, anodic currents flowed from the regions where the film was broken, and fresh metal was exposed to the corrosive environment. The current density being one of the easiest parameters to measure in electrochemistry, with the usual laboratory techniques a range of seven to eight orders of magnitude was easily covered. If the crack propagated by anodic dissolution of the freshly exposed metal, the crack propagation velocity could be estimated by Faraday's law, covering again a span of crack velocities of seven to eight orders of magnitude.

The ISRT assumed that when a filmed metal was exposed to a corrosive solution, at constant potential, a stationary current density circulated. If the metal was strained under tension, the film broke mainly at slip steps, exposing fresh metal to the solution. Since straining experiments last only a few seconds, the current density on the areas of nondisrupted film remained constant. Any change in current observed during straining was attributed to the contribution of the freshly exposed metal. The current density measured during straining, i_y, should be the result of the following contributions:

$$i_y = (i_s \times A_s) + (i_b \times A_b) \tag{30}$$

i_s being the current density on the film-covered metal, which was assumed to be equal to the stationary current density before straining; A_s is the fraction of area covered by a film; i_b is the current density on the freshly exposed metal; and A_b is the fraction of area of the freshly exposed metal. The crack velocity (Vp), in meters per second, is given by

$$\text{Vp} = \frac{i_b \cdot E}{100 \cdot F \cdot d} \tag{31}$$

i_b being the current density on the fresh metal in A/cm^2; E is the mean electrochemical equivalent weight of the alloy in $g \cdot (\text{mol}/z)^{-1}$, F is the Faraday constant in $C \cdot (\text{mol}/z)^{-1}$, and d is the alloy density in g/cm^3.

The mean electrochemical weight of the alloy, E, is given by the following summation:

$$E = \sum a_j \left(\frac{W_j}{z_j}\right) \tag{32}$$

where a_j is the atomic fraction of the element j in the alloy, W is its atomic weight, and z is the valence at which the element is dissolving. For example, for AISI 304 austenitic stainless steel, assuming that chromium is dissolving as Cr^{3+}, iron as Fe^{2+}, and nickel as Ni^{2+}, $E = 26.0$. For 70/30 copper–zinc brass, $E = 32.0$.

The samples used were wires, and the fractional area of the bare metal, A_b, was calculated by an equation developed by Bubar and Vermilyea[164]:

$$A_b = 1 - \left(\frac{L_0}{L}\right)^{1/2} \tag{33}$$

where L_0 is the initial specimen length, and L is the length at any time. This equation was derived from the assumption that the volume of both the metal and the film was constant, that the film cracked at $L = L_0$, and that the film remained in contact with the metal while the diameter of the wire was being reduced during its elongation.

If S_0 is the initial surface of the wire, and S is the value at any time, then

$$A_b = \frac{S - S_0}{S}$$

If r_0 is the initial radius of the wire, then the initial surface will be

$$S_0 = 2 \cdot \pi \cdot r_0 \cdot L_0$$

and the surface at any time will be

$$S = 2 \cdot \pi \cdot r \cdot L$$

and the volume of the wire, which is constant, will be given by

$$V = \pi \cdot r_0^2 \cdot L_0 = \pi \cdot r^2 \cdot L$$

By appropriate reordering and replacements, Eq. (33) is obtained.

The values for Eq. (30) were usually measured for elongations of 10 to 30%. The difference in values was small, and much lower than the scattering found in the measured crack velocities. For elongations lower than 10%, the system did not reach a steady state and poor reproducibility was found, whereas above 30% elongation the adherence of the film could be impaired by the deformation, and the validity of Eq. (33) becomes questionable. The value of A_s is given by

$$A_s = 1 - A_b \tag{34}$$

The following criterion was used to choose the appropriate strain rate for the ISRT. The strain rate should be one that gave, on freshly exposed metal, both electrochemical and mechanical conditions equivalent to those found at the bottom of the crack. Too fast strain rates would give bare metal current densities[150,151] and unrealistically high predicted crack velocities. On the other hand, too slow strain rates would lead to the deactivation of part of the exposed metal, and the application of Eq. (33) for the calculation of the fraction of bare metal would be invalidated. Another factor that conditioned the lowest strain rate to be used was the simultaneous growth of cracks. The use of Eq. (33) was possible only if no visible cracks were produced during straining. For high crack velocities, of the order of 5×10^{-7} m/s, cracks of 1 μm would be produced in a couple of seconds, and high strain rates should be used. In any case, the choice of the optimum strain rate has to be empirically made. For the first application of this technique,[75] in the study of mild steel in nitrate solutions, the fastest strain rates obtainable from a modified motor-driven Hounsfield Tensometer were used. The strain rates used were 5.1×10^{-2}, 2.6×10^{-2}, 5.2×10^{-3}, and 1.3×10^{-3} s^{-1}. All of them gave the same value of predicted crack velocity. Tests at a lower strain rate, 1.7×10^{-4} s^{-1}, showed evident signs of deactivation of part of the freshly exposed metal and had to be disregarded.[159,165] Later work showed that very good results were obtained with strain rates in the range between 5×10^{-3} and 5×10^{-2} s^{-1}.

For the ISRT to have a practical value, a criterion for a minimum crack velocity had to be established. It is evident that from an electrochemical point of view there should be a limit to the maximum Vp values, and it would be determined either by the ohmic drops or by concentration polarization effects, precipitation of solid corrosion products, salt films, etc. Apparently, according to this technique, there should not be a minimum Vp value below which no SCC would be expected.

Since the ISRT was intended to be used as a general test for SCC susceptibility, an empirical limit was necessary. When straining a passive metal, for example, current increases are observed due to the mechanical breakdown and anodic deformation of the passive film.[166] From Eqs. (30) and (31), a crack propagation rate value is obtained, in spite of the fact that no SCC is found in the passive region.[109] By straining AISI 304 in a Na_2SO_4 solution, in the region of passive potentials at room temperature, crack velocities of 1×10^{-10} up to 3×10^{-10} m/s were predicted. Consequently, from these measurements the range of 1×10^{-10} to 3×10^{-10} m/s was arbitrarily chosen as the minimum crack velocity. When the calculated

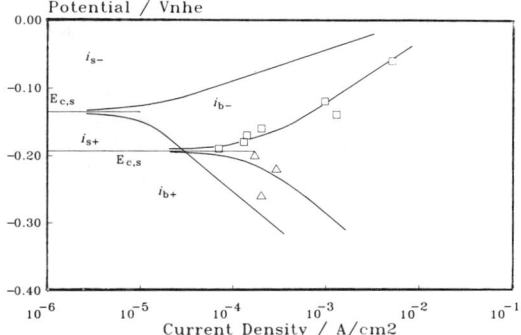

Figure 51. Calculated bare-metal polarization curve for straining AISI 304 stainless steel in a $1M$ HCl solution at 25°C. Notation: i_{b+}, Bare-metal net anodic current density; i_{b-}, bare-metal net cathodic current density; i_{s+}, stationary metal net anodic current density; i_{s-}, stationary metal net cathodic current density; $E_{c,s}$, stationary metal corrosion potential; $E_{c,b}$, bare-metal corrosion potential.[155]

Vp values were lower than these values, it was assumed that the metal was not susceptible to SCC.

The ISRT also allowed the prediction of the expected aspect ratio. As shown above, i_b will give the crack velocity, while i_s gives the corrosion rate of the film-covered metal. The idea of a film-covered metal applies not only to the external surface of the sample but also to the crack sides. As mentioned in the introduction, stress corrosion cracks always show a very high aspect ratio. If SCC is due exclusively to anodic dissolution, it should be concluded that the corrosion rate at the bottom of the crack should be several orders of magnitude higher than that at the sides. Consequently, from the intermediate-strain-rate technique it was possible to predict not only the crack velocity but also the aspect ratio, AR, which is given by the following relation:

$$AR = i_b / i_s \tag{35}$$

This relation was introduced by Galvele, Wexler, and Gardiazabal[155] in their study of AISI 304 in $1M$ HCl solutions, where the increase of i_b with the potential was slower than that of i_s, as shown in Fig. 51. Incidentally, it is interesting to notice that in this case the change induced by the strain on the corrosion potential is opposite to that found in Figs. 28 and 30.

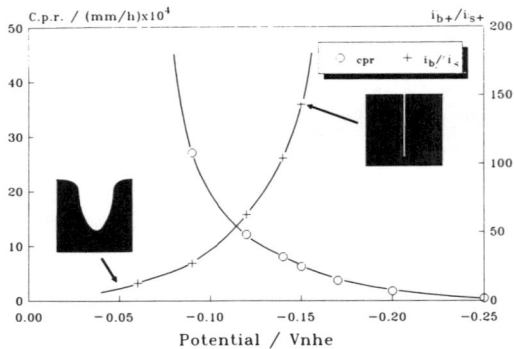

Figure 52. Calculated crack propagation rate and aspect ratio (i_{b+} / i_{s+}) as a function of the potential for stressed AISI 304 stainless steel in a $1M$ HCl solution at 25°C. Insets: Expected crack morphology.[155]

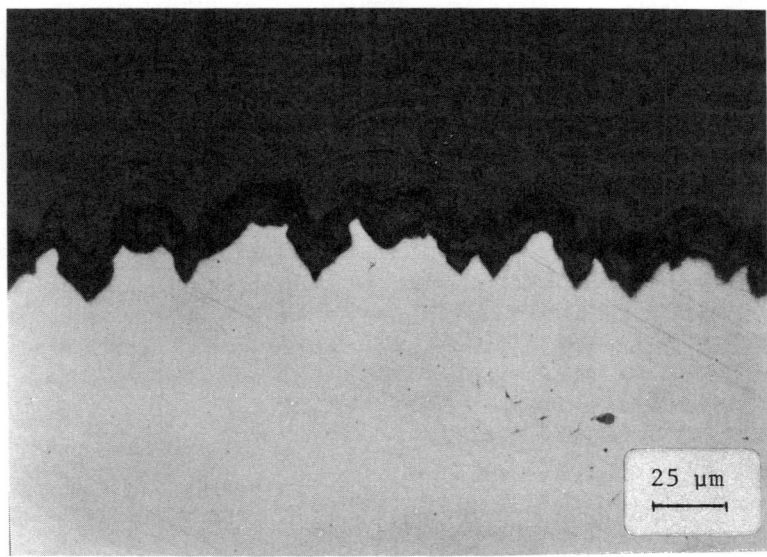

Figure 53. Cross section of stressed AISI 304 stainless steel in a $1M$ HCl solution at –0.050 V and 25°C after 30-h exposure, showing uneven general corrosion.[155] (See Fig. 52).

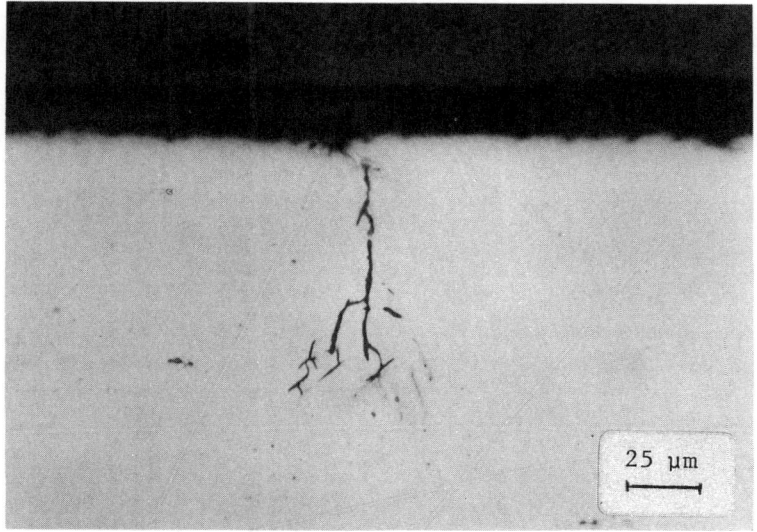

Figure 54. Cross section of AISI 304 stainless steel stressed in $1M$ HCl solution at –0.150 V and 25°C[155] for 206 hours, showing SCC. (see Fig. 52).

Equation (35) was used to explain the generalized corrosion observed near 0.0 V versus NHE.

The same approach was successfully applied to numerous cases of SCC.[8,43,156–158,160–162] The rule found was that SCC should be expected only when the aspect ratio is equal to or higher than 10. Aspect ratios lower than 5 are indicative of generalized corrosion, electropolishing, or pitting. In the transition between 5 and 10, both trenches and incipient cracking were observed.

Figure 52 shows the SCC diagram obtained by these authors[155] for AISI 304 stainless steel in $1M$ HCl solutions at room temperature. The inserts show the expected crack morphology, which, as shown in Figs. 53 and 54, was confirmed by experience.

Figure 55 shows another case of prediction of crack velocities by the ISRT. In this case, the system was yellow brass in a $1N$ NaNO$_2$ solution, at room temperature. Figure 56 shows the comparison between the predicted values and those measured with the slow-strain-rate technique. The correlation between both sets of measurements was very good. As men-

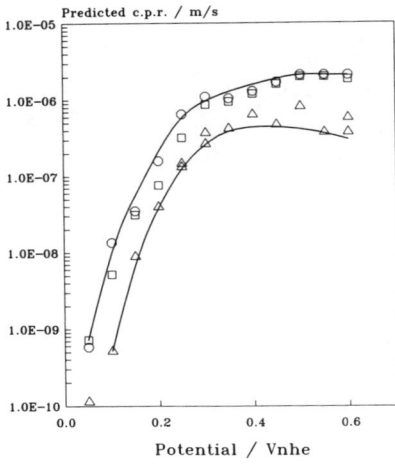

Figure 55. Crack velocities predicted from constant-potential intermediate-strain-rate experiments. The system is yellow brass in a deaerated $1M$ NaNO$_2$ solution. Strain rates used: △, 6.7×10^{-3} s^{-1}; ◻, 2.2×10^{-2} s^{-1}; ○, 4.3×10^{-2} s^{-1}.[162]

tioned above, such good correlations were found in an important number of SCC cases. It is not surprising, then, that in view of these results the anodic dissolution SCC mechanism had, and still has, strong supporters. We will return to this point later to show the limitations of the anodic dissolution mechanism. Meanwhile, a series of arbitrary conditions introduced in the above technique can be enumerated:

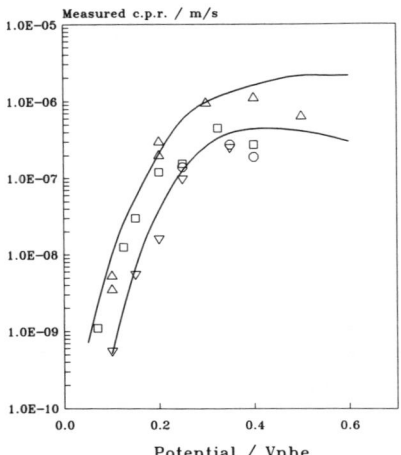

Figure 56. Crack velocity values measured for yellow brass in a deaerated $1M$ NaNO$_2$ solution. Strain rates used: △, 2.1×10^{-4} s^{-1}; ◻, 1.7×10^{-5} s^{-1}; ▽, 4×10^{-6} s^{-1}. ○, Crack velocities found after 10 min straining at 3×10^{-5} s^{-1}. The continuous lines show the crack velocity values predicted by the ISRT (see Fig. 55).[162]

(a) The elongation at which the measurements are made, 10 to 30%, is arbitrarily chosen and is not based on the assumed SCC mechanism.

(b) The recommended strain rates to be used, 5×10^{-3} s^{-1} and 5×10^{-2} s^{-1}, again are arbitrarily chosen, and this choice is not based on the mechanism.

(c) The technique predicts that SCC, with a crack propagation rate between 1×10^{-10} and 3×10^{-10} m/s, should be expected in the passive region. On the other hand, it has been reported that no SCC is found in that region.[109] Consequently, these crack velocities were taken as the minimum values, meaning that for these values no SCC should be expected. However, this is not correct, because Magdowski and Speidel[135] found crack velocities lower than 2×10^{-11} m/s for steel in water at high temperature, and crack velocities of 1×10^{-12} m/s were reported for steels in prestressed concrete.[167]

3. Repassivation and Slip Step Dissolution Models

The slip step anodic dissolution mechanism has been supported by numerous authors, as observed in the reviews by Staehle,[109] Parkins,[168] and Ford and Andresen.[147] In the following paragraphs, a description of this mechanism will be made. It should be pointed out that the microscopic level descriptions were developed by Scully[169,170] and by Smith and Staehle[171] and summarized by Staehle[109] over 20 years ago. Although the mechanism as a whole is still being supported, no updating of Staehle's[109] description could be found. Nevertheless, the general idea of film breakdown, mainly by slip steps emersion, followed by a characteristic repassivation process and the subsequent anodic dissolution is generally supported.[147,168]

As pointed out by Staehle,[109] Fig. 57 shows the essential unit process in the slip step dissolution mechanism. Staehle says that the nondissolving surface (as it is film covered or enriched in a noble component) is broken by emerging dislocations, and a local transient dissolution ensues. The propagation of the crack would correspond to a series of these events, as shown in Fig. 58. The metallographic observation of SCC fracture surfaces shows no evidence of anodic dissolution. According to Staehle,[109] in view of the metallographic evidence, the depth of the transients in SCC (Fig. 57) should be in the range of 5–100 nm. Galvele[172] pointed out that under such conditions there is a strong possibility that the conditions for a high aspect ratio of the cracks could not be fulfilled (Fig. 59). This is confirmed by the observations made by Silcock and Swann[173] for AISI type 316 stainless steel exposed to a 42% $MgCl_2$ solution at 150°C. These

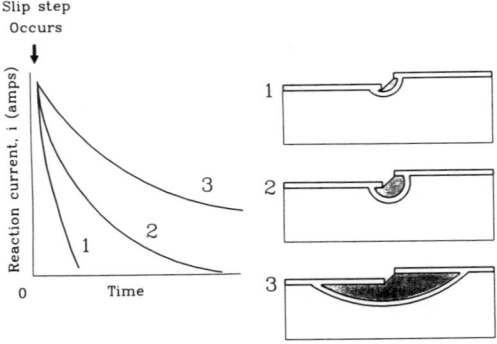

Figure 57. Schematic description of the relation between repassivation rates and the amount of material dissolved for a given slip step event. **1**, Immediate repassivation; **2**, intermediate repassivation rate favoring stress corrosion cracking; **3**, slow repassivation rate leading to nonrestricted lateral dissolution.[109]

authors observed cracks more than 10 μm long, with a cross section of only 10 nm. Such a low cross section allowed for very few events of slip emergence and dissolution, to keep such a high aspect ratio.

One of the conclusions to be drawn from Fig. 57 is that there is a typical, intermediate repassivation rate, characteristic of SCC. Too fast

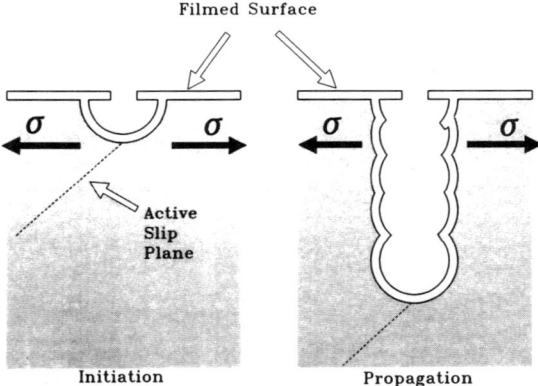

Figure 58. Schematic aspects of crack propagation by successive slip step emergence dissolution events.[109]

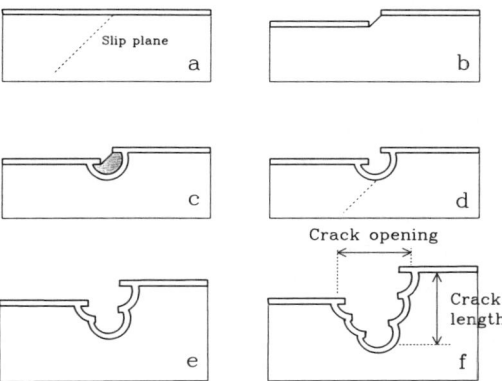

Figure 59. Aspect ratio expected for a slip step anodic dissolution process.[172]

repassivation rates would show no localized corrosion, and too slow repassivation rates would lead to generalized corrosion. As described below, Carranza and Galvele[111,112] showed that there is no characteristic repassivation rate for SCC. There is a fast repassivation rate for passive systems, where no SCC is found, and a slower repassivation rate, governed by diffusion, which appears under conditions for either SCC or pitting. The observations made by Carranza and Galvele on the repassivation rates could indicate that, rather than being related to the magnitude of corrosion involved in the slip step (Fig. 57), SCC susceptibility could be related to the formation of a deleterious anodic film, as described in Section IV.

By using an empirical approach, Ford[148] extended this mechanism to stainless steels and low alloy steels in water at 288°C. Although this model has received general acceptance, practically without any discussion, there is a weakness in it that should be pointed out. This mechanism is based on a generally accepted, but unproved, assumption. It is assumed that, in a cracked metal sample, the oxide film present at the tip of the crack should crack almost as soon as a stress is applied. No experimental evidence is available to support this basic principle. Intuitively, taking into account the brittleness of the films involved, the idea seems to be correct. Nevertheless, the fact that those films are subjected to a significant compressive stress at the tip of the crack is being ignored. The reason is that the oxides formed at the tip of the crack have a specific volume which could be up to two times higher than that of the metal consumed. Abundant information is available on this matter in the corrosion literature. This property gives rise to the phenomenon known as the wedging effect of the corrosion products, and it has even been considered as a possible cause for SCC. Pickering et al.[174] reported measurements of the stresses involved, and Nielsen[175] developed an SCC mechanism based on this phenomenon. The stresses involved could be, at least, of the order of the metal yield strength. Consequently, the external stresses, in order to crack the film, should be considerably high, and plastic deformation at the tip of the crack should be expected. If so, slip steps will be involved in the film fracture process, and the objections made above (Fig. 59) on the low aspect ratios expectable in Staehle's description would be applicable to Ford's model too.

4. Repassivation Rate Measurements

The possibility of a correlation between SCC susceptibility and the repassivation rate of metals in electrolytic media was mentioned above.

Repassivation kinetics were measured by numerous researchers[176] using a wide variety of experimental techniques. Beck[7,177] studied the repassivation of fresh metal surfaces produced by mechanical rupture of metallic samples. He reported measurements on zirconium, titanium, and sintered aluminum. Another technique used by various authors[7,178] was based on scratching a passive metallic surface and measuring the current transients. Finally, a third technique, extensively applied, was the use of very fast straining (above $10\ \mathrm{s}^{-1}$).[111,112,150,151,179] By this technique, wire samples were elongated, almost instantaneously, up to a certain percentage without breaking, and the current transient was subsequently measured.

The growth kinetics of anodic oxide films has been reviewed by Schultze et al.[180] As pointed out by these authors, usually the oxide formation starts with chemisorption of hydroxyl ions:

$$H_2O = OH^-_{ad} + H^+ \qquad (36)$$

After place exchange reactions with the metal ions in the surface, oxide molecules are formed. Further reactions yield monolayers, homogeneous multilayers, or thick islands of oxide, depending on the ratio of lateral to normal growth. In the case of passive films, usually homogeneous films with an almost constant thickness are formed, but due to local inhomogeneities, the thickness may change statistically. Sometimes, oxide film peel-off is found.

The growth rate is equivalent to the current density i_{ox} of the oxide formation, which will depend on the electrode potential E, the time t, and

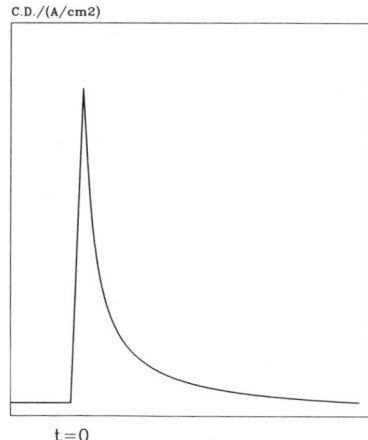

Figure 60. Schematic current–time transient observed after rapid straining of a passive metal at constant potential.

the concentration of OH⁻. As for the i/t transient, under potentiostatic measurements (Fig. 60) the current density i_{ox} will follow laws of the type:

$$i_{ox} = A \cdot t^b, \quad \text{with} \quad -1 \leq b \leq +3 \qquad (37)$$

From these laws, conclusions on the oxide growth mechanism can be drawn.

Schultze *et al.* pointed out that, in general, oxide formation takes place simultaneously with corrosion i_{cor}, oxygen evolution i_{O_2}, and oxidation of components of the electrolyte i_x:

$$i = i_{ox} + i_{cor} + i_{O_2} + i_x \qquad (38)$$

Then, the use of the total current density i instead of i_{ox} and also the assumption $i \approx i_{ox}$ can be erroneous, unless a careful analysis of the system is made.

The growth models for the anodic oxide films are the following:

(a) Unstable layer (flaking off):

$$i_{ox} = A \cdot t^0$$

(b) Homogeneous layer, slow diffusion:

$$i_{ox} = A \cdot t^{-1/2}$$

The diffusion, as pointed out by Evans,[181] could take place either in the solid oxide film or in liquid-filled pores in the film.

(c) Homogeneous layer, high field (Cabrera and Mott[182]): Here the ion transport within the oxide, which depends exponentially on the field strength, is rate determining. According to Beck,[177] the Eq. (37) would be

$$i_{ox} = A \cdot t^{-0.8}$$

(d) Homogeneous layer, high field (empirical):

$$i_{ox} = A \cdot t^{-1}$$

The discrepancy between the theoretical ($b = -0.8$) and the empirical ($b = -1$) values is attributed to secondary processes, such as aging of the films.[180]

Higher values of b, $+1 \leq b \leq +3$, are attributed by Schultze *et al.* to the growth of two- and three-dimensional nuclei of oxide films. Nevertheless, another source of such values, which is more frequently found in environmentally assisted cracking, is the nucleation and growth of pits.[183]

Carranza and Galvele[111,112,179] systematically studied the repassivation rate for austenitic stainless steel and α-brass in a series of environments and under different electrode potentials and pH values. The systems were chosen so as to include cases of SCC susceptibility, cases of pitting corrosion, and cases where stable passive films were formed. Some repassivation transients were simple, while others were the summation of various steps, as shown in Fig. 61. Two repassivation rates were found in the systems studied (Fig. 62): those with $b = -1$ and those with an initial

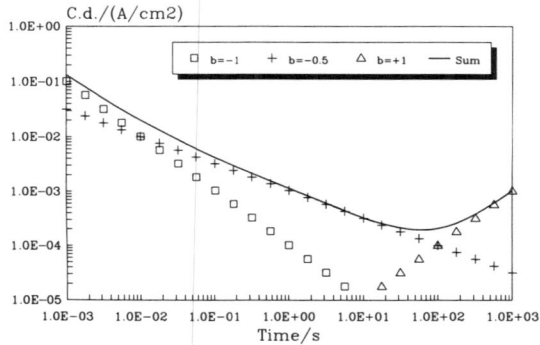

Figure 61. Typical current–time transient found after fast straining of an AISI 304 stainless steel wire, in an HCl solution, at constant potential.[111]

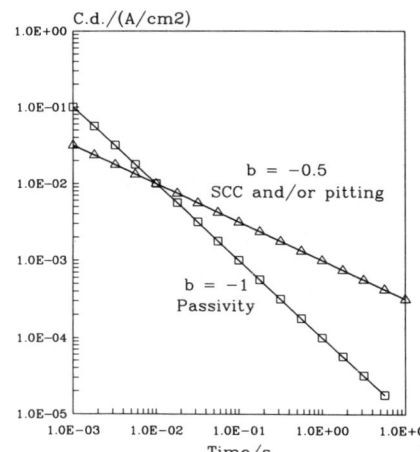

Figure 62. Typical repassivation rates found by Carranza and Galvele[111,112,179] for AISI 304 stainless steel and α-brass in various environments.

value of $b = -1$ followed by $b = -0.5$. In all those systems where passivity was present, and no evidence of localized corrosion could be detected, the slope was $b = -1$. In those cases where either SCC or pitting could be found, a slope of $b = -0.5$ was detectable. There was no typical b value that could be related to SCC susceptibility. The conclusion was that slow repassivation was a necessary but not a sufficient condition for SCC susceptibility. Two possible interpretations could be drawn from these experiments. One is that, as expected from the slip dissolution model, a slow repassivation process is a necessary condition. The other is that the difference between passive state and SCC susceptibility is that the metal is covered with films of different nature.

5. Limitations of the Mechanisms

The strongest argument given in support of the anodic dissolution slip step SCC mechanism is the good correlation between the crack velocities predicted by electrochemical methods and those found by conventional SCC tests.[147] We mentioned above the arbitrary choices involved in those measurements. They included the arbitrary choice of the experimental conditions and the arbitrary limits chosen for the non-SCC susceptibility. The electrochemical methods predict SCC velocities between 1×10^{-10} and 3×10^{-10} m/s for the passive metals. Since no SCC is found under those conditions, these velocities are chosen as the minimum values below which no SCC should be expected. Nevertheless, as mentioned above, crack velocities lower than 2×10^{-11} m/s for steel in water at high temperature[135] and crack velocities of 1×10^{-12} m/s for steels in prestressed concrete[167] have been reported. This criticism applies equally to the measurements made with the intermediate-strain-rate technique and to those made with the potentiokinetic polarization curves.

As mentioned above, the slip step anodic dissolution model has two weak points. One is that it has not been proved that the aspect ratio obtainable is as high as that found in practice. Apparently it is not so. The other is that there is no experimental proof that the oxide film at the tip of the crack is broken when applying a stress. Apparently no allowance was made for the compressive stresses present in such a film.

Numerous other objections, pointed out even by those who support this mechanism, are found in the literature. Doig and Flewitt[117] mentioned that the use of fractography in the development of the mechanistic understanding of growth processes, particularly under static loading, has introduced much controversy. This is a consequence of the observation that in ductile materials where slip-enhanced anodic-dissolution–repassivation models are proposed, there is rarely any fractographic evidence for either significant slip or anodic dissolution. Doig and Flewitt showed fracture surfaces of steels in aqueous environments that are indistinguishable from the surfaces resulting from a low-ductility hydrogen embrittlement fracture in a gaseous environment, where there is no possibility of anodic dissolution. They also mentioned the lack of evidence for slip or dissolution on the intergranular fracture surfaces.

Another objection pointed out by Doig and Flewitt[117] refers to the plastic deformation at the tip of the crack. They said that in addition to the lack of evidence for localized slip to produce enhanced anodic dissolution, the very "brittle" character of the crack path is inconsistent with the stress intensity values proposed to describe the stress state at the crack tip. The plastic zone associated with a crack results in local deformation immediately after the crack tip, producing an opening displacement of typically 1 to 100 μm. Where crack advance is associated with such stress intensity, considerable plastic flow (dislocation movement) must occur, and this would be expected to disrupt local microstructural features along the path. Such disruption is not observed even where the crack advance is considered to occur by interrupted cleavage. Here, the slow overall growth rate must allow time for the plastic strain to accumulate before the next cleavage event, and such strain should be evident on the fracture surface. This lack of macroscopic strain is confirmed by the observation of crack tips contained within thin foils, in the scanning transmission electron microscope.

Another important limitation of the mechanisms based on macroscopic removal of material from the tip of the cracks is the fact that they do not take into account the similarities found in various environments. Such similarities are seen, for example, in the case of aluminum alloys in aqueous environments, gaseous environments, liquid metals, etc., as pointed out by Speidel.[13] This author indicates that the shape of the crack velocity versus stress intensity curves, their changes with heat treatment, etc., are very similar in the various environments. Similar observations were reported by other authors.[140–142] While anodic dissolution could be an acceptable mechanism in some cases of SCC, it is clearly out of the question in the other cases of environmentally induced cracking. From a mechanistic point of view, this observation suggests that, besides the electrochemical reactions observed during SCC in electrolytic environments, there are some other processes, on the metal side of the crack tip, that are common to all these environmentally induced cracking processes but that the anodic dissolution mechanism ignores.

VII. MECHANISMS BASED UPON DISCONTINUOUS CLEAVAGE

1. Continuous versus Discontinuous Cleavage

Discontinuous SCC mechanisms appeared in the literature simultaneously with continuous anodic dissolution mechanisms. One of the reasons given in the literature[184] for looking for a mechanical step in the mechanisms of crack propagation was that stress corrosion cracks propagate at rates usually of the order of 1 cm/h. This rate was reported to be greater than the average rates of general corrosion by a factor of about 10^3. The continuous anodic dissolution mechanism remained popular because, as mentioned in the previous sections, very good correlations were found between the electrochemical measurements and the experimental crack velocities. On the other hand, the objection mentioned above has become unsustainable in view of our present knowledge of localized corrosion. The crack velocities indicated would require, for the crack to propagate by anodic dissolution, a current density lower than 10 A/cm^2. Studies on pitting corrosion initiation[131] have shown that current densities from 10 up to almost 100 A/cm^2 could be found. At first sight, this finding could

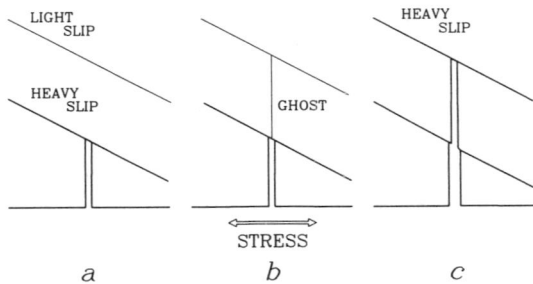

Figure 63. Schematic representation of crack propagation between slip bands. See text for details. After Edeleanu and Forty.[187]

have ended the dispute, but, as mentioned below, several authors observed apparent discontinuities in the process of crack propagation,[44,185,186] and the idea of a discontinuous crack propagation process remained alive up to the present day.

Edeleanu and Forty[187] studied the mechanism of cracking of α-brass single crystals in ammonia. The experimental procedure was as follows. The specimens were strained in bending with a residual stress applied through a compressed spring. Ammonia was added to the cell, and the surface of the specimen was continuously viewed through a glass window with a high-power water-immersion microscope objective. Their metallographic observations showed that cracking was discontinuous (Fig. 63). They observed that the crack appeared to run in short steps, each step starting and finishing on transverse slip lines. The crack halted temporarily at each new slip band, and there was a period of incubation here, which could be as short as one minute. When the next stage in the fracture did occur, it appeared suddenly as a very fine crack, barely visible under the microscope. The authors described it as a "ghost" crack. The "ghost," which terminated on a slip band, widened slowly into a visible crack by further slip.

According to these authors, this succession of events strongly suggested that fracture was occurring, not by the propagation of a single crack, but by a discontinuous process involving the spreading, stopping, and reinitiation of separate cracks.

In order to have an SCC mechanism, the action of the environment had to be accounted for, and the supporters of the discontinuous crack propagation processes gave several explanations. The mechanical aspects

Electrochemical Aspects of Stress Corrosion Cracking

of the discontinuous mechanisms will receive only a brief mention here, because the subject is outside the scope of the present chapter. The reader can find a full description of the mechanical aspects of every one of these mechanisms in the references given in each case. Most of the attention will be concentrated on the electrochemical aspects they involve.

2. The Two-Stage Model for Stress Corrosion Cracking

(i) Forty's Model

Based on the experimental results of Edeleanu and Forty,[187] the latter[188] proposed a two-stage mechanism for SCC.

Forty pointed out that the most important microscopic observation was that failure of the specimen was the result of a typical fracture process; that is, the cracking was largely mechanical in character. Yet this kind of failure occurred only in the presence of ammonia. Without the ammonia or with a sudden increase of stress, the crystal deformed plastically around the tip of the crack. The phenomenon of SCC then had to be considered as the process whereby a brittle-like fracture was triggered, in some way, by a chemical reaction between the brass and the ammonia.

Forty mentioned that the idea that brittle fracture might be possible in an fcc metal was somewhat surprising. Hence, he examined the conditions necessary for a propagation of cleavage through a ductile crystal to occur. As mentioned in Section I, following Orowan, it might be supposed that the criterion of fracture in a ductile material can be derived from the balance between the strain energy stored at the tip of a crack and the surface energy together with the plastic work required for growth of the fracture.

The plastic deformation associated with cleavage may be suppressed if the crack propagates faster than the material surrounding its tip can deform. In other words, if plastic deformation proceeds by the formation and movement of dislocations in the crystal, for a truly brittle fracture to occur, the velocity of the crack must exceed that at which dislocations move away from its tip.

Forty summarized his ideas on stress corrosion in systems such as α-brass in ammonia in the following way. The injection of vacancies by chemical dezincification of the alloy leads to surface embrittlement, either through the formation of pores or by interaction with dislocations. This sufficiently restricts the plastic deformation of the underlying crystal to allow the formation of a crack. Provided the crack velocity is high enough, it can propagate fairly readily through the alloy until it meets a particularly soft region such as a preexisting slip band. The crack can be reinitiated from here only after further embrittlement or selective corrosion at its tip, and it is this stage in the process that determines the overall failure rate.

(ii) Film-Induced Cleavage Model

According to Gerberich and Chen,[189] there are at least three types of film-induced cleavage models for SCC. The first one assumes that a passive film of ~5 nm suffers mechanical breakdown, leading to localized dissolution. This mechanism was suggested by Champion[190] and by Logan,[53] and, although Gerberich and Chen describe it as a cleavage-based mechanism, in the present review it was treated in Section VI on anodic dissolution and filmed metals.

The second one, suggested by McEvily and Bond[191] and by Pugh,[192] suggests that the formation of a ~100-nm tarnish film on the metal, because of the action of the environment, leads to SCC when mechanical stresses cause the fracture of the film. This mechanism was later abandoned when it was found that there was no tarnish film near the crack tip, for brass in ammoniacal solutions, as shown by Pinchback et al.[193]

The third mechanism, which has gained considerable support in recent years, is the one suggested by Sieradzki and Newman.[186,194] According to this mechanism, the brittle fracture of a ~30-nm surface film induces cleavage on ductile metal substrates. The surface film responsible for this action could be, according to the authors, either an oxide or a dealloyed coating.

3. Advantages and Limitations of the Mechanisms

Much of the support for the discontinuous mechanism stems from the observation of discontinuities during the crack propagation process.[44,185,186] Various authors[91,162,195] have given explanations for the above mentioned discontinuities. Figure 64 shows schematically the explanation given by Alvarez et al.[162] and by Rebak et al.[91] for the observation of discontinuities in the transgranular SCC of α-brass in $NaNO_2$ solutions. As shown by Sieradzki and Newman,[186] cracks nucleate on slip steps and propagate as isolated slots perpendicular to the applied stress (1 and 2 in Fig. 64). Those slots leave behind unbroken metal ligaments (a in Fig. 64). When cracks grow, the ligaments could be broken in two different ways: (1) in a continuous manner, leaving behind partially detached filaments on the fracture surface, like those reported by Silcock[196] during SCC of

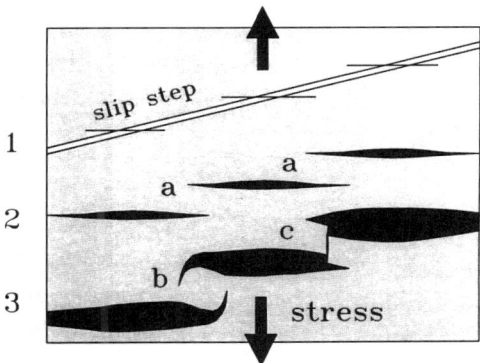

Figure 64. Schematic representation of ligament cracking during the crack propagation process, as described by Alvarez et al.[162] See text for details.

stainless steels (b in Fig. 64) and (2) by mechanical breaking (c in Fig. 64), leading to the crack arrest markings and acoustic emission reported by various authors. No general agreement on these explanations has been reached so far.

Gerberich and Chen[189] pointed out that there is some controversy regarding the experimental observations supporting Sieradzki and Newman's mechanism.[186,194] Although Gerberich and Chen did not reject the idea of film-induced substrate cleavage, they suggested that a 30-nm film generating a 2000-nm substrate crack would be overstating the effect for small cracks and that further theoretical and experimental work needs to be accomplished, because the most recent experiments imply that neither dissolution models nor the film-induced cleavage models alone account for the observed phenomenon.

Another limitation found is the lack of predictability of the models advanced so far. According to Forty,[188] α-brass would have the metallurgical requirements for SCC susceptibility. According to the same author, cracking will occur if the metal is exposed to any environment that produces surface dealloying. Corrosive environments with such properties are very numerous, but not all of them produce SCC, and even if only those agents producing SCC are considered, such as ammonia, methylamine, ethylamine, long-carbon-chain aliphatic amines, acetates, nitrites, and hydroxides, all of them lead to very different crack velocities. The two-step model, so far, does not give any clue as to how these different crack velocities could be predicted. There are qualitative explanations suggesting that the rate of the chemical step changes, but nothing has been said about how this change can be calculated. Consequently, even when everything mentioned so far by the supporters of this mechanism proves to be correct, there is still much work to be done to make it a working SCC model.

As mentioned above, Sieradzki and Newman[186,194] established the theoretical conditions under which a discontinuous cleavage process could take place in a ductile material. Nevertheless, most of the effort has been, so far, concentrated on showing the credibility of the mechanism. As mentioned above, no information is available on its predictive capability. Apparently, dealloying could lead to the surface films responsible for this type of process, but there is no clear criterion to be applied to cases where dealloying is absent. Besides, there is no indication as to the effect of temperature, yield strength of the material, presence of hydrogen, etc. As for the significance of the dealloying process, recent publications[197,198] shed some doubts on it.

Kelly et al.[199] published a work on SCC of Ag–20Au in a HClO$_4$ solution with results that were supposed to prove that SCC, in this system, was exclusively due to a discontinuous cleavage process. Recently, Duffó and Galvele[200] reproduced Kelly's work and extended the measurements to HClO$_4$, AgClO$_4$, and KCl solutions. Duffó and Galvele found a series of results that could not be explained by the cleavage mechanism; for instance, the crack sizes were a function of the charge circulated through

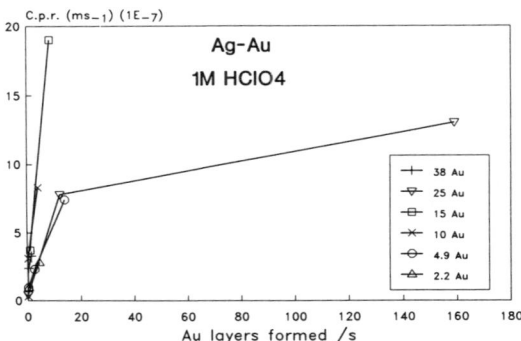

Figure 65. Crack propagation rate vs. the growth rate of atomic layers of gold on Ag–Au alloys in 1M HClO$_4$ during SCC.[172,201]

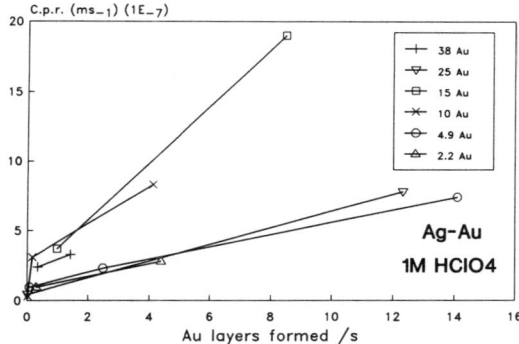

Figure 66. Crack propagation rate vs. the growth rate of atomic layers of gold on Ag–Au alloys in $1M$ HClO$_4$ during SCC (Fig. 65, detail).[172,201]

the system, and AgClO$_4$ specifically accelerated the SCC process. According to these authors, their results would have been better explained by the surface-mobility SCC mechanism (see Section VIII).

Maier et al.[172,201] have recently studied the SCC behavior of Ag–Au alloys, with gold contents ranging from 2.2 at. % up to 60 at. %. The environments studied included aqueous solutions of KCl, KI, KBr, Na$_2$SO$_4$, and HClO$_4$. The crack velocities for Ag–2.2Au were found to be almost as high as those for Ag–40Au. From the measurements in HClO$_4$ solutions, Maier et al. calculated the formation rate of atomic gold layers on the corroding alloys. According to the cleavage mechanism, there should be a linear relation between the crack velocity and the formation rate of the dealloyed layer. No such relation was found, as shown in Figs. 65 and 66.[172,201] Taking into account that the mechanisms discussed in the present section assume that the chemical reaction rate, in this case the gold layer formation rate, is the rate-controlling step, the results in Figs. 65 and 66 shed serious doubts on the validity of such mechanisms.

In conclusion, the cleavage-based mechanisms have not yet been developed as full workable mechanisms, and serious objections against their credibility can be raised.

VIII. MECHANISMS BASED UPON SURFACE MOBILITY

In view of the limitations of most of the available SCC mechanisms mentioned above, a new mechanism was suggested in 1986.[125] It is based on the assumption that the role of the environment is to enhance the surface mobility of the metal. The mechanism was fully developed in a publication in 1987[202] and has recently been improved.[22,172]

The idea of the mechanism[202] stemmed mainly from the publications of Rhead and co-workers,[203–208] and of other authors.[209] They reported that the surface self-diffusion of metals could be drastically changed by the presence of contaminants. These authors found that certain contaminants increased, by several orders of magnitude, the surface self-diffusion of the metals. These observations, combined with the fact that those contaminants were frequently responsible for both SCC and liquid metal embrittlement of the same metals, strongly supported the idea that the action of the environment in SCC is related to the change in the surface self-diffusivity of the metal.

1. Surface-Mobility SCC Mechanism

The mechanism is based on the following postulates:

First postulate: *The environment affects the metal by changing its surface self-diffusivity.* As mentioned above, this postulate is based on the observation that those environments reported to increase surface mobility are also responsible for environmentally induced cracking (Fig. 67).

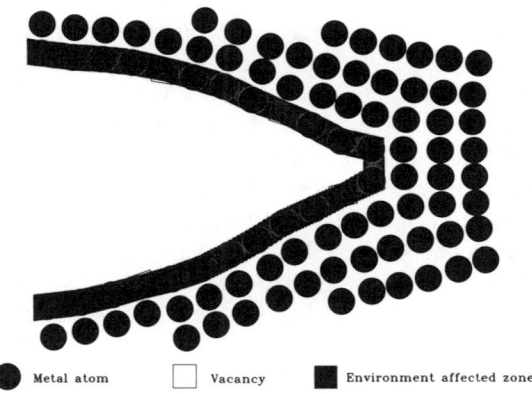

Figure 67. First postulate of the surface-mobility SCC mechanism: The environment affects the surface of the metal, by changing its self-diffusivity. An increase in the self-diffusivity of the metal surface will lead to SCC susceptibility.[202]

Second postulate: *The temperature at which SCC takes place is lower than 0.5Tm.* Tm is the absolute melting temperature for the metal or alloy considered. This postulate was supported by the available SCC data. The main consequence of this postulate is that volume diffusion in the metal can be ignored. That is, the movement of substitutional solute atoms, vacancies, etc., inside the metal does not practically exist.

Third postulate: *Only elastic stresses are relevant in the SCC process.* As mentioned above, very little plastic deformation was observed on the fracture surfaces. On the other hand, the aspect ratios of the cracks were very high, which again suggested that the plastic deformation was minimal. These observations led to the conclusion that plastic deformation, while probably important in SCC initiation, was not relevant, from a mechanistic point of view, in the crack propagation process.

Fourth postulate: *SCC takes place by capture of vacancies by the tip of the crack.* The capture of a vacancy at the tip of the crack (Fig. 68) leads to the propagation of the crack by an atomic step and to a partial relaxation of the stressed lattice. Consequently, as a result of this step, there is a reduction in the free energy of the system. The capture of a vacancy by the stressed lattice at the tip of the crack is the elementary step in the SCC process.

This fourth postulate has been questioned by an author[210] who suggested that vacancies should move in the opposite direction, that is, away

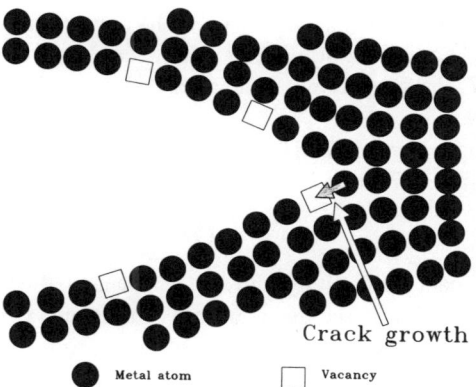

Figure 68. The capture of a vacancy by the stressed tip of the crack is the elementary step for crack propagation in the surface-mobility SCC mechanism.

from the tip of the crack. Accepting such a suggestion would mean ignoring the second postulate. The direction of vacancy movements suggested by the mentioned author[210] is typical of high-temperature creep fracture,[211] which does not occur in the present case. If his assumption was correct, then the tip of the crack would be the source of vacancies, and volume diffusion should be required. In other words, the temperature should be above 0.5Tm, contrary to what is stated in the second postulate.

From the above-mentioned postulates, an equation for the crack propagation rate can be developed in the following way. The tendency shown by the tip of the crack to capture vacancies should be a function of the crack tip tensile stress. As shown in Fig. 2, there is a high stress concentration at the tip of the crack. A tensile stress is known to reduce the free energy of formation of vacancies[212] by an amount equal to $\Delta F^o - \sigma \cdot a^3$, where σ is the tensile stress and a is the atomic size. This relation originates in a well-known thermodynamic equation:

$$(\delta F_f / \delta P) = V_f \qquad (39)$$

where F_f is the vacancy formation free energy, P is the pressure, and V_f is the vacancy formation volume (approximately equal to a^3). From these equations the equilibrium concentration of vacancies in the stressed region is found[212]:

$$C = C^0 \exp(\sigma \cdot a^3 / k \cdot T) \qquad (40)$$

As pointed out by Hirth and Nix,[212] C^0 is the concentration of vacancies in the standard state of the stress-free bulk metal. The other parameters in Eq. (40) are T, the temperature in degrees Kelvin, and k, the Boltzmann constant.

According to Hirth and Nix,[212] there is a relaxation time, τ, necessary to reestablish an equilibrated vacancy concentration:

$$\tau \approx \lambda^2 / D \qquad (41)$$

D being the vacancy diffusivity, and λ the distance to the nearest vacancy source or sink. Because the temperature is lower than 0.5Tm, the value for D is very low, and consequently the relaxation time is very long. As a result of this, a vacancy-deficient zone will be present at the tip of the crack (Fig. 69).

The environmentally affected surface of the crack (Fig. 67) will have a certain vacancy concentration, C_s^0, in the stress-free region. On the other

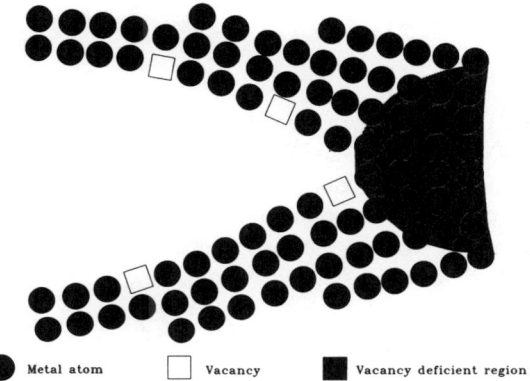

Figure 69. The presence of tensile stresses at the tip of the crack generates a vacancy-deficient zone. This zone cannot reach the new vacancy equilibrium concentration because of the low temperature of the metal (below 0.5Tm). Under this condition, volume diffusion is negligible, and the usual sources of vacancies, such as dislocations, grain boundaries, or metal surfaces, are inactive.

hand, at the tip of the crack, it will have a lower vacancy concentration, C_s, because of the vacancies captured by the vacancy-deficient zone. The maximum drop in the C_s value will be that required by the stressed metal to reach the equilibrium concentration given by Eq. (40). Consequently, a difference in vacancy concentration will appear on the surface of the crack. The maximum value of this difference in concentrations will be[202]

$$C_S^0 - C_S = C - C^0 = C^0 \left[\exp\left(\frac{\sigma a^3}{kT}\right) - 1 \right] \qquad (42)$$

To calculate the vacancy-concentration gradient, the distance L between the stressed and the stress free-regions must be known. As mentioned before, the stresses on the side surfaces of the cracks are found to drop substantially after only 5–7 atomic distances away from the tip of the crack. For most of the calculations, a conservative value of $L = 10^{-8}$ m has been used.[202] The vacancy concentration gradient will then be

$$\frac{dC_S}{dx} = \frac{C^0}{L} \left[\exp\left(\frac{\sigma a^3}{kT}\right) - 1 \right] \qquad (43)$$

and the vacancy flow, per unit of time, will be[202]

$$J_v = D_v \frac{dC_S}{dx} \tag{44}$$

For the sake of convenience, instead of the coefficient of surface diffusion of vacancies, D_v, the coefficient of surface self-diffusion, D_S / C^0, will be used.[213,214] In addition, the concentration of vacancies, given in Eq. (40) as the number of vacancies per unit of site, will be given in vacancies per unit of volume. The flow of vacancies per unit of area and unit of time will be given by[202]

$$J_v = \frac{D_S}{La^3}[\exp(\frac{\sigma a^3}{kT}) - 1] \tag{45}$$

If $v = a^{-2}$ is the number of atoms per unit of area, then every time that v vacancies move to the tip of the crack, the crack will advance a distance a. The crack velocity, V_P, will be given by[202]

$$V_P = \frac{D_S}{L}\left[\exp(\frac{\sigma a^3}{kT}) - 1\right] \tag{46}$$

V_P being the crack velocity, in m/s; D_S the surface self-diffusion coefficient, in m^2/s; L the diffusion distance of the vacancies, in m; σ the elastic surface stress at the tip of the crack, in N/m^2; a the atom size, in m; k the Boltzmann constant in J/K; and T the temperature in K.

2. Surface Self-Diffusion Coefficients

The surface self-diffusion of metals is a subject of high theoretical and practical interest, and numerous reviews have been published about it.[215–218] Several experimental techniques have been used in the measurement of D_S on metals.[215,216] Some of them, such as grain-boundary grooving or surface-scratch smoothing, are mass transfer techniques. They have been applied to a wide variety of metals, but they usually require high temperatures, above 400°C. Other techniques, such as field ion microscopy,[217,219] can be used at low temperatures, even below room temperature, but they are restricted to refractory metals, such as tungsten and platinum. With this last technique the movement of individual atoms can be followed.[217,219] Very recently,[220–222] room temperature measurements, under conditions that are directly applicable to the SCC processes, have been reported. Unfortunately, this last type of data is still very scarce.

Nevertheless, as shown below, these recent experiments allowed the confirmation of the fact that extrapolations of D_S data from high-temperature measurements are acceptable.

The initial D_S measurements were made either in vacuum or in the presence of a gas, such as hydrogen, argon, or helium. The results were systematized by Gjostein,[215,223] but he pointed out that strong deviations were observed in the presence of contaminants. The effect of contaminants was studied by Rhead and co-workers,[203–208] who were able to include the effect on contaminants in the systematization made by Gjostein.

Gjostein[215] showed that a plot of log D_S versus Tm/T (Tm: melting point; T: temperature; both in degrees Kelvin) correlates the surface diffusion data for most metals. The plot is nonlinear and can be represented by the sum of two Arrhenius factors. At high and low temperatures, respectively, the Arrhenius constants are given approximately by:

$$\left\{ \begin{array}{l} Q_S = 30\text{Tm cal/mol} \\ D_0 = 740 \text{ cm}^2/\text{s} \end{array} \right\} T/\text{Tm} \geq 0.75 \qquad (47)$$

$$\left\{ \begin{array}{l} Q_S = 13\text{Tm cal/mol} \\ D_0 = 0.014 \text{ cm}^2/\text{s} \end{array} \right\} T/\text{Tm} \leq 0.75 \qquad (48)$$

where Q_S is the activation energy, and D_0 is the Arrhenius preexponential factor.

Strong deviations from the predictions of Eqs. (47) and (48) were observed when impurities were present. While some impurities led to D_S values lower than those predicted, others gave much higher values.

Rhead[203,204] rationalized the effect of impurities by postulating that a constant value of D_S, ~ 3×10^{-8} m^2/s, was found at the melting point. He found that if, instead of the melting point of the bulk metal, the melting point of the surface-adsorbed product was used for Tm, the results obtained in the presence of impurities followed Eqs. (47) and (48). This conclusion was supported by measurements in the following systems: Ag–S, Cu–Pb, Cu–Tl, Cu–Bi, and Au–Pb. The above correlations were later extended by Delamare and Rhead[205] to Cu–Cl, Cu–Br, and Cu–I, by Pichaud and Drechsler[209] to W–Ni, and by Oda and Rhead[206] to Ni–Cl, Ni–Br, and Ni–I.

The amount of data available for D_S at the temperatures at which SCC takes place is, unfortunately, scarce. Consequently, an estimate of the data has to be made. In environmentally assisted cracking, the metal surfaces are always contaminated by impurities. There are no atomically clean surfaces. To estimate the most probable D_S values, the following assumption is made.[202] It is assumed that in all cases of environmentally assisted cracking, D_S will be given by Eqs. (47) and (48), with the modification introduced by Rhead:

$$D_S = 740 \times 10^{-4} \exp[-(30 T m/RT)] \\ + 0.014 \times 10^{-4} \exp[-(13 T m/RT)] \qquad (49)$$

where D_S is given in m^2/s, R is the gas constant (R = 1.987 cal/(mol · K), T is the absolute temperature in K, and Tm is the melting point of the surface-adsorbed impurity in K. In the work of Rhead *et al.*, the Tm values used were the melting point of Ag_2S for the Ag–S system, CuCl for Cu–Cl, CuBr for Cu–Br, CuI for Cu–I, NiI_2 for Ni–I, $NiBr_2$ for Ni–Br, etc.

Gjostein's equations [Eqs. (47) and (48)] were obtained from experimental data from fcc metals, while Rhead made no reference to the crystal structure of the adsorbed products. In the use of Eq. (49), no distinction is made between the crystalline structures of the bulk metal and the surface product.

The above approach to the calculation of D_S for SCC processes may seem very crude, but it is the best possible one at present and gives correct predictions when dealing with the phenomenon of specificity in SCC.[12,110,200,202,224] It will be seen that low-melting-point compounds are those that lead to fast SCC, whereas high-melting-point products are those that inhibit SCC. Empirical equations, such as Eqs. (47) and (48), using the melting temperature as a correlating parameter, have been frequently used for properties of materials in chemistry and physical metallurgy.[225] They were successfully used in the prediction of unknown parameters and in mastering large amounts of data. No conspicuous differences were observed between bcc structures and close-packed structures in any of the properties correlated by this kind of empirical equations.[225]

Equation (49) was established for high-temperature measurements, and, in order to use it in SCC applications, extrapolations to room temperature must be made. Recent surface diffusion coefficient values, measured for gold at room temperature, by Alonso *et al.*[220] using electrochemical techniques and by Oppenheim *et al.*[221] with STM techniques have shown that the D_s values calculated with Eq. (49) are very close to those measured experimentally. Figure 70 shows a comparison between the D_S values for gold, as a function of the temperature, calculated with

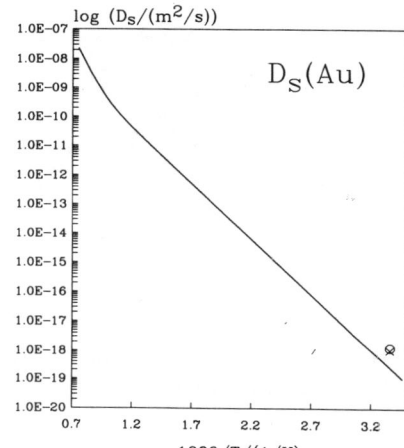

Figure 70. D_S values for gold, as a function of the temperature. The continuous line shows the values predicted by Eq. (49). The points show room temperature measurements recently reported.[220,221] Good correlation is observed between the predicted values and the experimental results.

Eq. (49), and recently reported[220–222] experimental results for room temperature. Good correlation is observed between the extrapolated values and the measured ones.

The theory of surface diffusion of pure metals, as well as that of foreign atoms contaminating a metal, has been studied in great detail in the last few years.[217,219] This is not the case with the self-diffusivity of surface-contaminated metals, such as those relevant for SCC studies. Except for the preliminary work by Rhead and others,[203–209] no work on this question is found in the recent literature.

The lack of theoretical information is particularly bad, because the idea of surface mobility remains rather vague, and the application of the SCC mechanism based on it[202] would require a better definition.

One possible explanation, given by Gjostein,[215] is depicted in Fig. 71. A metal adatom would move more easily on the contaminant layer than over the clean metal. This could be the case at high temperatures, close to the melting point, where, as suggested by Gjostein,[215] diffusion by surface adatoms would prevail, whereas a mechanism based on vacancies will predominate at lower temperatures. Rhead,[203,204] on the other hand, supports the idea of the formation of a two-dimensional liquid phase on the metal surface (Fig. 72). An alternative explanation was suggested by Galvele[226] (Fig. 73). The idea is that surface metal atoms are interchanged with the surface contaminant, giving an apparent step size of λ. The

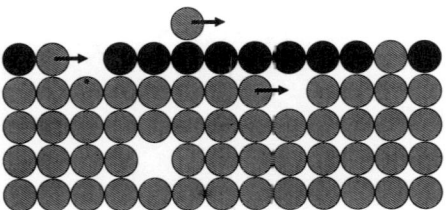

Figure 71. According to Gjostein,[215] the D_S values on a contaminated metal would be higher because a metal adatom would move more easily on the contaminant layer than over the clean metal surface.

restoration of vacancy concentration in the surface contaminant would take place spontaneously, owing to the contaminant's low melting point.

3. Effect of Hydrogen

Hydrogen is known to induce a degradation of the mechanical properties of metals. This action is considered to take place mainly through two mechanisms[227]: (a) by stress-induced hydride formation and cleavage, and (b) by hydrogen-enhanced localized plasticity. There is also the possibility for hydrogen to accelerate the surface-mobility SCC process. Although this could not be taken as a main cause for hydrogen embrittlement, it is quite possible that some cases of hydrogen embrittlement are due to this process.

Although hydrogen embrittlement would fall outside the scope of the present chapter, frequent references to similarities between LME and HE[142,228] and between SCC and HE[229,230] have been found in the literature. In the latter case, examples were found where the distinction between SCC

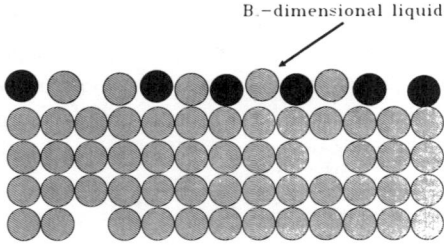

Figure 72. Rhead[203,204] supports the idea that the D_S values on a contaminated metal would be higher because of the formation of a two-dimensional liquid phase on the metal surface.

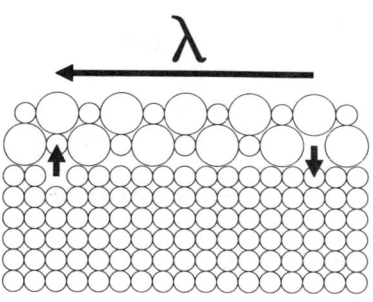

Figure 73. An alternative explanation for the high D_S values was suggested by Galvele.[226] The idea is that surface metal atoms are interchanged with the surface contaminant, giving an apparent step size of λ. Afterward, the restoration of vacancy concentration in the surface contaminant would take place owing to its low melting point.

and HE is subtle, and sometimes even unconvincing. We will analyze now what effect the hydrogen present in the lattice may have on the mechanism of crack propagation based on surface mobility.

It is well known that hydrogen, in metals, interacts with vacancies. Recent work[231,232] has shown that vacancies act as deep trap sites for hydrogen at room temperature. The binding energy reported[231] for H with monovacancies in nickel was $E_b = 0.43$ eV. For iron, the values reported were $E_b = 0.53$ eV for H with monovacancies and $E_b = 0.71$ eV for small vacancy clusters.

By comparison with carbon interstitials, which increase the self-diffusion in γ-iron, Gibala[233] suggested that hydrogen, interacting with vacancies, should also increase the self-diffusion of the metal. According to this author, the activation enthalpy for self-diffusion is reduced by an amount equal to the binding enthalpy of a vacancy interstitial. A similar effect should be expected for surface self-diffusion. In any case, the presence of low hydrogen concentrations in the metal lattice would induce only a moderate increase in the D_S value. Nevertheless, very large hydrogen concentrations are expected in elastically strained regions, such as the tip of a crack.[47,48,234–236] The high difference in concentrations between the tip and the sides of the cracks should lead to an effect similar to that described above for vacancies in stressed crystals. In other words, hydrogen would reduce free energy of the vacancy formation. To account for the influence of hydrogen in a crack, the following modification of Eq. (46) is suggested:

$$V_P = \frac{D_S}{L}\left[\exp\left(\frac{\sigma a^3 + \alpha E_b}{kT}\right) - 1\right] \quad (50)$$

where α is a dimensionless function that measures the difference in degree of saturation of the vacancies with hydrogen between the stressed and the stress-free regions. The value of α should range between 0 and 1 and will be a function of the stress. In the case of carbon in γ-iron (see Fig. 4 in Ref. 39), saturation of the vacancies is reached for a carbon content of 2 atom % in the metal. We have no equivalent information for hydrogen in metals.

According to Eq. (50), under maximum hydrogen availability ($\alpha = 1$), the H-vacancy binding energy for nickel could account for a crack velocity increase by a factor of 2.2×10^7, while in the case of iron the increment could be as much as 1.1×10^9 times. As previously shown,[202] Eq. (50) could account for the maximum crack velocities found both for iron and for nickel in the presence of hydrogen.

4. Experimental Observations

Since 1987 an abundant amount of experimental data has been collected in support of the surface-mobility SCC mechanism. In the following sections, that work is briefly reviewed.

(i) Gaseous Environments

Scarce information is available on the SCC of metals and alloys in hydrogen-free gaseous environments. Kerns and Staehle[237] studied the SCC of high-strength steels in chlorine. These authors were able to show that dry chlorine could induce SCC in high-strength steel, in the absence of hydrogen. Their observations were confirmed by Sieradzki and Ficalora.[238] According to Sieradzki,[239] there is no embrittlement of the high-strength steel for chlorine vapor pressures lower than 1.33×10^3 Pa. At higher pressures the embrittlement would be due to the formation of a thin brittle layer of $FeCl_2$. On the other hand, Galvele[202] explained the above observations by a surface-mobility mechanism, assuming the formation of $FeCl_3$ on the steel surface.

Zirconium alloys are susceptible to SCC in halogen vapors. Numerous papers are found on this subject,[42,240–242] including references to SCC of titanium alloys in halogen vapors.[42] Peehs et al.[243,244] concluded that a thermodynamic evaluation of their results reveals that SCC only occurs in the region of iodine concentration and temperature where stable zirconium iodides in the condensed phase are to be expected. Galvele[202] suggested that, due to its low melting point (499°C), ZrI_4 would induce SCC in zirconium and zirconium alloys by the surface-mobility mechanism. The surface-mobility SCC model predicts crack velocities as high as those found in practice.

In a recent work, Bianchi and Galvele[245] studied the susceptibility to SCC of pure copper in atmospheres containing CuCl and that of pure silver in the presence of both AgI vapor and iodine vapor. Delamare and Rhead[205] showed that pure copper suffers a pronounced increase in its surface self-diffusivity when it is exposed to an atmosphere containing CuCl vapor. On the other hand, Delamare[246] found that surface self-diffusion in the case of pure silver showed a considerable increase when a monolayer of AgI was present on the metal surface. A preliminary work, by Bianchi and Galvele,[247] confirmed the presence of SCC when pure copper was slowly strained under conditions of high surface mobility. In a more recent work,[245] those preliminary results were confirmed, and it has also been shown that pure silver is susceptible to SCC in the presence of both AgI vapor and iodine vapor. These findings confirm the expectations of the surface-mobility SCC mechanism.

(ii) Silver Alloys

In the development of the surface-mobility SCC mechanism,[202] it was pointed out that silver alloys were of high academic interest because of the high number of low-melting-point compounds silver has. Many of these compounds belong to the Ag/AgCl reference electrode family, so abundant literature is available on their electrochemical behavior.[248] Besides, the fact that these silver compounds are sparingly soluble is an additional advantage, because their presence on the corroded surfaces can be confirmed by *ex situ* surface analysis.

The SCC susceptibility of silver alloys is well documented in the literature.[61,249-253] Unfortunately, for our present interest, the number of environments tested is limited, and no crack velocity measurements are available.

Duffó and Galvele[12,110,224] studied, in a more detailed way, the SCC behavior of Ag–15Pd and Ag–15Au alloys in various environments. SCC is frequently associated with dealloying,[186,254] and since the alloys tested in Duffó and Galvele's work showed different types of dissolution behavior, it was possible to check how such behavior correlated with SCC susceptibility.

Some authors associated selective dissolution with the presence of SCC.[254] Figure 74 shows the correlation between the c.p.r. and the polarization curve for an Ag–15Pd alloy in a $1.0M$ KCl solution. The first part of the polarization curve, with surface accumulation of Pd, is a typical "region (a)," according to Pickering's classification of alloy polarization

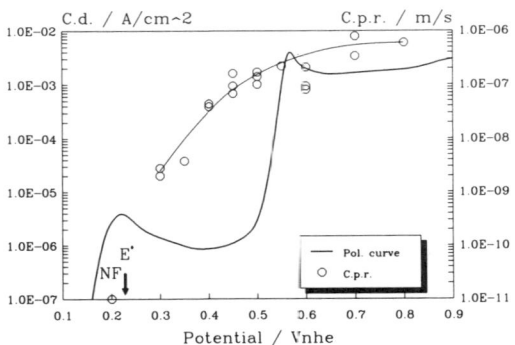

Figure 74. Correlation between the crack propagation rate (○) and the polarization curve (—) for an Ag–15Pd alloy in a 1.0M KCl solution. SCC was found before the current increase due to simultaneous dissolution [Pickering's region (c)] took place.[12]

curves.[254] The current increase observed at higher potentials was probably enhanced by a reaction of the type

$$Pd + 4Cl^- = (PdCl_4)^{2-} + 2e^-$$

and silver and palladium were simultaneously dissolved [i.e., Pickering's "region (c)"]. According to Pickering, the combination of region (a) plus region (c) leads to type II polarization curves. Type I polarization curves, that is, region (a) plus region (b) (selective dissolution), were found for an Ag–15Au alloy in a 1.0M KCl solution. According to Rambert and Landolt's work,[255] the transition from selective to simultaneous dissolution occurs at 10% of noble metal for Ag–Pd alloys and at above 20% for Ag–Au alloys. Then the current breakdown in Fig. 75 is due to selective dissolution of Ag, region (b) in Pickering's classification.

It is observed that the c.p.r. starts to increase as soon as the electrode potential surpasses the formation potential of AgCl and keeps increasing up to a maximum value at about 0.6 V. This range of potentials encompasses the region (a) of Pd surface enrichment and the region (c) of simultaneous Ag–Pd dissolution. From Fig. 74 there is no indication that the SCC process was affected by this change in the alloy dissolution behavior. It would seem that the presence of palladium on the alloy surface

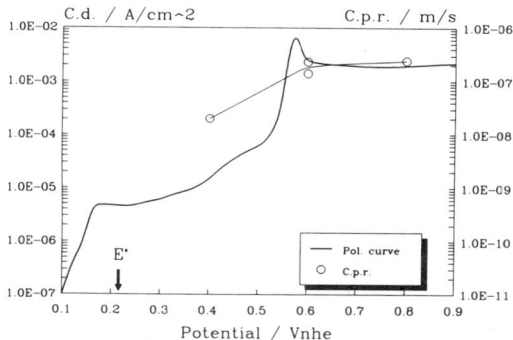

Figure 75. Correlation between the crack propagation rate (○) and the polarization curve (—) for an Ag–15Au alloy in a 1.0M KCl solution. SCC was found before the current increase due to selective dissolution occurred [Pickering's region (b)].[12]

had no effect on the c.p.r. Most probably, it favored the SCC process by reducing the anodic reaction in region (a).

A similar conclusion could be drawn for Ag–15Au in a 1.0M KCl solution (Fig. 75). In this case, as mentioned above, selective dissolution is present [region (b)]. Nevertheless, SCC was found at potentials lower than the breakdown potential. In these cases we are dealing with intergranular SCC, but a similar observation of SCC below the breakdown potential was reported by Lichter et al.[256] for transgranular SCC in Cu–Au single crystals.

Let us see now how the results in Duffó and Galvele's work[12] fit with the predictions of the surface-mobility SCC mechanism.[202] From a qualitative point of view, the predictions of this mechanism have been satisfied. According to Fig. 27, SCC was found only under those electrochemical conditions at which low-melting-point surface compounds, leading to high surface mobility, were formed on the alloy surface. In the case of the silver alloys, SCC susceptibility was not a specific property of the alloying element, since it was found in alloys with Pd and with Au.

The importance of the surface compound was confirmed through experiments in iodine-containing nonaqueous solutions. SCC was found

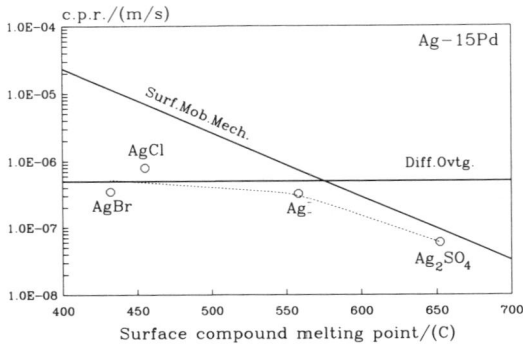

Figure 76. Crack propagation rates as a function of the surface compound melting point, for an Ag–15Pd alloy, $\sigma = 62.4$ MPa. Temperature = 25°C. Surf.Mob.Mech.: Values predicted by the surface-mobility SCC mechanism. Diff.Ovtg.: Maximum c.p.r. above which diffusion overvoltage becomes the rate-controlling step in the aqueous solutions used in the present experiments. The measured values follow the predicted values up to the point where diffusion overvoltage becomes the rate-controlling step.[12]

in all of them, with crack velocities close to the maximum values found in aqueous solutions. In conclusion, provided the low-melting-point compound was present on the alloy surface, no matter how it was produced, SCC took place.

For the demonstration of the validity of the surface-mobility SCC mechanism, it is important that the predictions should be fulfilled also at a quantitative level. According to the surface-mobility SCC mechanism,[202] the crack propagation rate is given by Eq. (46). All the c.p.r. measurements were performed at a very low strain rate, 2.6×10^{-6} s^{-1}. At the same time, the c.p.r. values found were usually very high. This combination of factors led to fractured samples with almost negligible plastic deformation. Consequently, it was considered adequate to take for σ the value $\sigma_{0.2} = 62.4$ MPa measured for the Ag–15Pd samples (see Fig. 4). The value taken for the atomic size was that for silver, $a = 2.88 \times 10^{-10}$ m. As for the D_s values induced by each silver compound, the following melting points were used in Eq. (49): AgBr, 432°C; AgCl, 455°C; AgI, 558°C; and Ag$_2$SO$_4$, 652°C.

With the above values and Eqs. (46) and (49), the c.p.r. values predicted by the surface-mobility SCC mechanism[202] were calculated and are compared in Fig. 76 with the maximum c.p.r. values measured in Duffó and Galvele's work.[12] The surface-mobility SCC mechanism predicts quantitatively the c.p.r. values for Ag–15Pd in $1.0M$ KI and $1.0M$ Na$_2$SO$_4$ solutions. For $1.0M$ KCl and $1.0M$ KBr solutions, where higher c.p.r. values were expected, the system seems to have reached a saturation limit. This saturation value is easily explained when the electrochemical reactions taking place inside the cracks are taken into account.

As mentioned above (Fig. 27) and as reported in a previous publication,[110] in each of the environments tested, SCC is only found above the formation potential of the respective silver compound. This proves that the presence of these low-melting-point silver compounds was a necessary condition for SCC. These compounds, in aqueous solutions, are formed by the following reaction:

$$n\text{Ag} + \text{X}^{n-} = \text{Ag}_n\text{X} + ne^- \tag{51}$$

Since these compounds must be present at the tip of the crack, two processes must take place inside the cracks. The first one, as indicated by Eq. (51), is a charge transfer reaction, which will start at the equilibrium potential E_o and will become faster, the higher the positive overpotential. The second is the transport, by diffusion and migration, of the X^{n-} ions from the bulk solution to the site inside the crack where the charge transfer reaction is taking place. This site does not need to be at the tip of the crack but a few atomic distances away, since the Ag$_n$X compound could move along the crack by surface diffusion.

The surface-mobility SCC mechanism[202] requires that at least a monolayer of Ag$_n$X should be present at the tip of the crack. In aqueous solutions the formation of such a monolayer involves the circulation of an anodic current and the transport of X^{n-} ions along the crack. As the crack propagates, it generates new metal surface that must be covered by the Ag$_n$X monolayer. The higher the c.p.r. is, the higher the current and the faster the transport of ions required. It is well known from electrochemical kinetics[257] that processes of this type increase within certain limits, because diffusion overvoltage becomes the rate-controlling step. In other words, in aqueous solutions, the c.p.r. in Fig. 76 will follow the theoretical values only up to a limit, above which the process will become limited by diffusion overvoltage.

The equation for diffusion overvoltage is[257]

$$\eta = \frac{RT}{nF} \ln\left(1 - \frac{i}{i_d}\right) \qquad (52)$$

where i_d is the limiting current, which is given by the following equation:

$$i_d = -Z_a\left(1 + \left|\frac{Z_a}{Z_b}\right|\right)\frac{F \cdot D_a \cdot C_a}{\delta} \qquad (53)$$

Z_a being the charge of the anion and Z_b that of the cation. Duffó and Galvele[12] assumed that $Z_a = -1$ and that $|Z_a/Z_b|$ is equal to 1. In these calculations, $n = 1$ and the following conservative assumptions were made to correlate the value of i with that of the c.p.r. It was assumed that only a monolayer of Ag_nX was being formed. Actually, thicker layers were formed on the crack surfaces; hence, higher i values were involved. The other conservative assumption made was that $\delta = 0.01$ cm. This is the thickness of the diffusion layer in a stagnant solution, and in the present case it was assumed that this value involved the diffusion layer plus the depth of the crack. Other values considered were $D_a = 10^{-9}$ m^2/s and $C_a = 1.0M$. With these values the maximum c.p.r. allowed by diffusion plus migration of X^{n-} ions was calculated and is compared in Fig. 6 with the c.p.r. values measured for Ag–15Pd in a $1.0M$ KBr solution. The results in Fig. 6 confirm that the limiting value observed in Fig. 76 is due to diffusion overvoltage.

Later, Duffó and Galvele extended their observations to an Ag–15Pd alloy in Na_3PO_4, $Na_4P_2O_7$, and NaOH solutions[258] and to an Ag–20Au alloy in $HClO_4$, $AgClO_4$, and KCl solutions.[200] In all these cases the surface-mobility SCC mechanism was confirmed.

These correlations are not restricted to silver alloys only. Recently, Rebak[259] has shown that the SCC of Alloy 600 (approximately 78% Ni, 15% Cr, 7% Fe) in high-temperature water could also be explained by the surface-mobility mechanism, assuming that NiO is formed on the alloy surface.

(iii) Effect of Temperature

It is well documented in the literature that SCC is affected by temperature. The higher the temperature is, the higher the crack velocity. The surface-mobility mechanism predicts that, provided there are no changes in the rate-controlling step during the temperature change, the crack velocity at any temperature can by calculated if the value at a given temperature is known. For this purpose, Eqs. (46) and (49) are used.

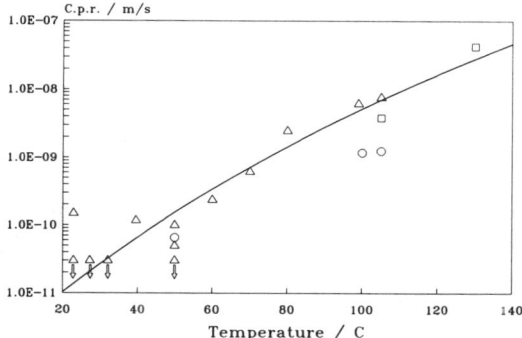

Figure 77. A comparison between theory and experiment: effect of temperature on the rate of SCC in austenitic stainless steels exposed to chloride solutions; experiments reported by Speidel.[260] △, 304 sensitized; ○, 304 annealed; □, 304L annealed. Experiment at 130°C in 42% MgCl$_2$; all the other experimental points are for experiments in an aerated 22% NaCl solution. For the predicted values the following parameters were used: Tm = 1100°C; σ = 240 MPa (yield strength of the samples used by Speidel).[202]

Speidel[260] reported a series of crack velocity measurements for 304 austenitic stainless steel in the presence of chloride solutions, at various temperatures. Figure 77 shows Speidel's experimental results compared with the values predicted by the theory.[202] For the calculations the following parameters were used: Tm = 1100°C; σ = 240 MPa (yield strength of the samples used by Speidel) (Fig. 4). The Tm value that gives the best fit (1100°C) is very close to the melting point of CrCl$_3$ (1150°C). SCC of austenitic stainless steels, in aqueous chloride solutions, takes place close to the pitting potential of the alloy, and CrCl$_3$ is the most probable species responsible for SCC in this system. Consequently, the Tm value gives further support to the surface-mobility SCC mechanism.

In agreement with the above observations, Duffó et al.[76] reported that the dependence of SCC on temperature for AISI 304 stainless steel in LiCl solutions was equal to that predicted by the surface-mobility mechanism.

(iv) Turbine Steels

Magdowski and Speidel[135] published an extensive work on the SCC behavior of steels in hot water. Their measurements are of particular interest because they give abundant information on the effects of yield strength (Y.S.) and temperature on the crack propagation rate for a wide variety of steels. The authors studied 13 different steels. Their study covered more than 30 different Y.S. values, ranging from 614 MPa up to 1420 MPa, and temperatures from below 0°C up to 288°C.

The work by Magdowski and Speidel was analyzed, from the point of view of the surface-mobility SCC mechanism, by Galvele.[261] From a first approach, it was assumed that clean iron surfaces were being exposed to water, and the surface mobility of iron was evaluated. Nevertheless, it is well known that magnetite is present when iron is exposed to high-temperature water. On the other hand, recent work[22] has shown that, as described below, the sources of vacancies have to be taken into account. Consequently, it was concluded that it was more appropriate to use, in the calculations, the melting point of magnetite (mp = 1597°C) instead of that of iron (mp = 1539°C). Because of the closeness of these two melting point values, the predictions made in the initial work[261] suffered no changes, and the initial conclusions remained valid. Nevertheless, the results shown here in Figs. 78–80 have been recalculated for the melting point of magnetite.

The SCC of iron in water was evaluated with Eqs. (46) and (49) using the melting point of magnetite (1597°C + 273) as T_m; $a = 2.5 \times 10^{-10}$ m, and T is the water temperature. The yield strength values are used for σ in Eq. (46) (Fig. 4).

Figure 78 shows the effect of temperature on the c.p.r. for 600- to 800-MPa and 1040-MPa steels. The lines show the values predicted by Eq. (46) for σ values of 700 MPa and 1040 MPa. Figure 79 shows the experimental and the theoretical values for steel with Y.S. = 1140 MPa. At temperatures below 60°C, a c.p.r. increase due to hydrogen is observed.

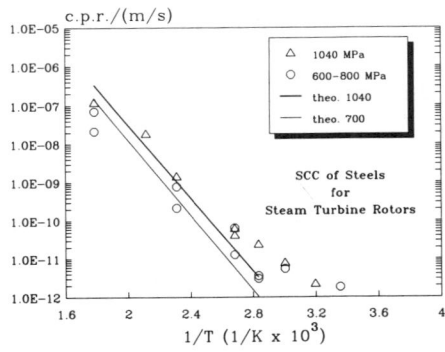

Figure 78. Effect of temperature on c.p.r. for 600- to 800-MPa and 1040-MPa steels. The points represent experimental results by Magdowski and Speidel,[135] and the lines represent the values predicted by Eq. (46), using yield strength values of 1040 MPa (upper line) and 700 MPa (lower line) for σ. Values are from Ref. 261 but have been recalculated for magnetite.

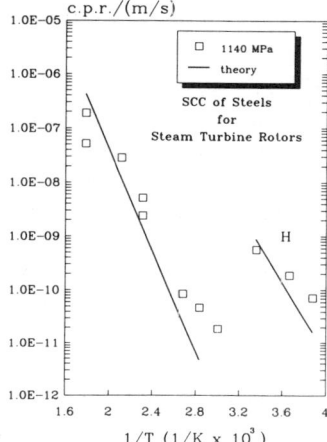

Figure 79. Effect of temperature on c.p.r. for 1140-MPa steel. The points represent experimental results by Magdowski and Speidel,[135] and the lines represent values predicted by Eq. (46). High c.p.r. values at low temperatures are due to hydrogen embrittlement. Values are from Ref. (261) but have been recalculated for magnetite.

As mentioned above, the effect of hydrogen is accounted for by the surface-mobility SCC mechanism with Eq. (50). In the case of Fig. 79 the hydrogen effect was calculated using a value of 0.53 for α.

Magdowski and Speidel found that there are two different effects of Y.S. on c.p.r.: a moderate one at low strength levels, and a strong one at high strength levels. The first one is attributed to SCC, while for the second the authors suggested an accelerating effect due to hydrogen. Figure 80 shows that Eq. (46) predicts very well the effect of Y.S. on c.p.r. for the

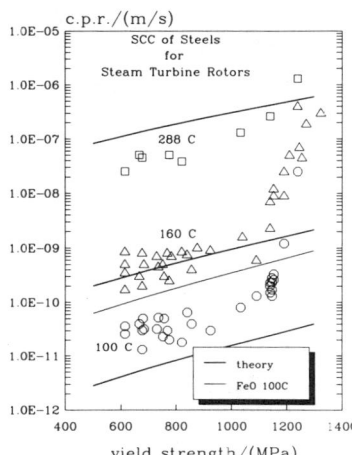

Figure 80. Effect of Y.S. on c.p.r. The points represent experimental results by Magdowski and Speidel,[135] and the lines represent the values predicted by Eq. (46). Values are from Ref. (261) but have been recalculated for magnetite. For 100°C calculations for FeO were also included.

first part of the curve. In all the cases studied, the values predicted for 288°C were higher than those measured. This is probably due to changes in the properties of the surface films, which could lead to D_S values lower than those predicted by Eq. (49). With regard to the data at 100°C, the nature of the surface contaminant should be confirmed. The experimental values fall between those predicted for Fe_3O_4 and those for FeO. The correlation between c.p.r. and yield strength shown in Fig. 80 is much better than that shown in Fig. 41.

Figures 78–80, therefore, prove that Eq. (46) gives a satisfactory prediction for the SCC of steel in hot water.

5. Sources of Vacancies

As mentioned above, the surface-mobility SCC mechanism[202] assumes that the stress concentration at the tip of a crack generates a very localized vacancy-deficient region. The capture of vacancies by the tip of the crack leads to the crack propagation process (Fig. 68). The systems studied so far indicate that the surface contaminant acts, not only by increasing the surface mobility, but also by supplying the vacancies required by the SCC process. Some examples of this process are described in the following sections.[22]

(i) Silver Alloys

The crack propagation rate was observed to be a function of the potential.[12,110,224] It was higher, the higher the anodic potential. It is well known that in the crystal structure of silver halides, cations in interstitial lattice sites and vacancies in the cation sublattice (Frenkel type defects) are predominant.[262] In this case only silver ions can move via interstitial lattice positions or vacancies (Fig. 81).

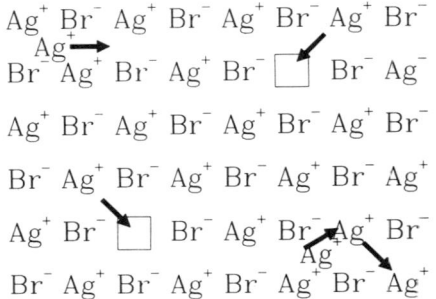

Figure 81. Frenkel type defects in an AgBr crystal.

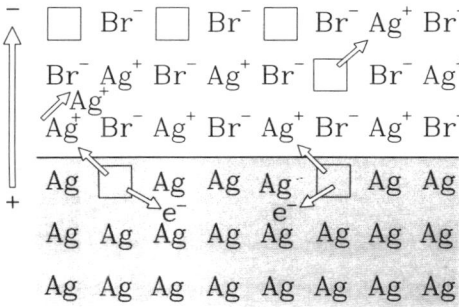

Figure 82. Crystal growth process in an AgBr crystal, accompanied by a production of metal vacancies at the metal surface.

When an electric field is applied, for example, to the AgBr crystal, the only migrating entities are those Ag^+ ions that move in the direction of the cathode via either interstitial lattice positions or vacancies.[262] The crystal growth process is accompanied by a production of metal vacancies on the metal surface (Fig. 82). At high temperature, these vacancies diffuse to the usual sinks, such as dislocations or grain boundaries. At the temperature at which SCC takes place, which, as mentioned above, is assumed to be below $0.5T_m$,[202] the vacancies are accumulated at the Ag/AgBr interface[22] (Fig. 83).

If the metal is subjected to a tensile stress, the vacancies will preferentially be captured by the tip of the crack, and the SCC process will take

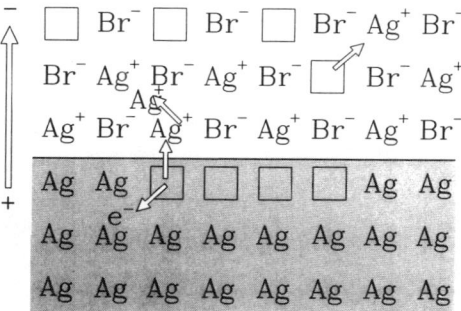

Figure 83. Accumulation of vacancies at the metal surface when volume diffusion in the metal is negligible.

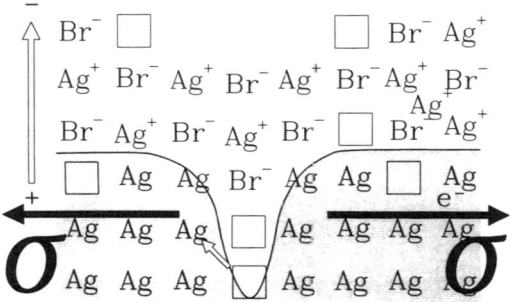

Figure 84. Capture of vacancies by the stressed lattice, at the tip of the crack, and crack propagation process in a silver alloy.

place[22] (Fig. 84). The supply of vacancies to the metal surface will be a function of the electric current circulating through the AgBr crystal, and the mobility of those vacancies along the metal surface will depend on the properties of the surface contaminant,[202] in this case AgBr.

(ii) Steels for Steam Turbine Rotors

It is well known that magnetite is formed on iron exposed to steam. As pointed out by Hauffe,[262] the diffusion through FeO and through Fe_3O_4 oxide films takes place by movement of iron-ion vacancies. Following the same line of thought used above for silver halide compounds, the crack

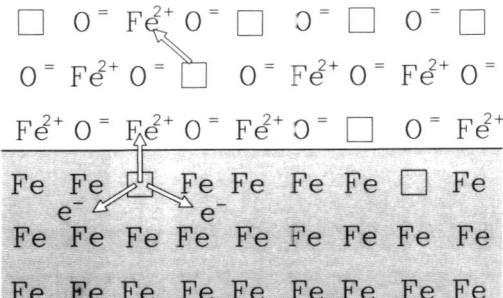

Figure 85. Crystal growth process in FeO or in Fe_3O_4 oxide films, by movement of iron-ion vacancies. The process is accompanied by the production of metal vacancies on the metal surface.

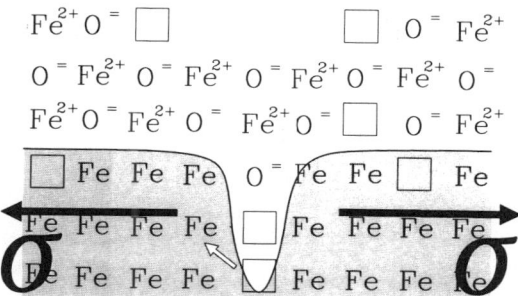

Figure 86. Capture of vacancies by the stressed lattice in iron, followed by crack growth.

propagation process for steels in water in the presence of tensile stresses can be described. The reaction of iron to steam, by producing magnetite, injects vacancies into the iron surface (Figs. 85 and 86). The surface mobility of those vacancies is that expected for the melting point of magnetite, and the experimental values measured by Magdowski and Speidel[135] as mentioned above, are in very good correlation with the predictions of the surface-mobility SCC mechanism.[261]

(iii) Liquid Metal Embrittlement

A very similar treatment can be applied to liquid metal embrittlement. Duffó and Galvele[263] have recently found that Ag–Cd alloys, while showing no cracking in halide solutions, show fast cracking in the presence of mercury. Surface contamination with a liquid metal is known to induce high surface mobility.[202] The source of vacancies, in this case, would be the liquid metal (Fig. 87).

(iv) Passive State

It is well known from the SCC literature[109] that cracking only appears outside the passive region. An explanation of this behavior could be found in the mechanism by which passive films grow.[22]

In high-temperature measurements, as pointed out by Hauffe,[262] there is a clear difference between the growth mechanism of oxides such as FeO or Fe_3O_4 and that of a typical component of passive films, such as Fe_2O_3. In the case of the first two oxide films, the growth process takes place by outward movement of iron ions. On the other hand, in the case of Fe_2O_3,

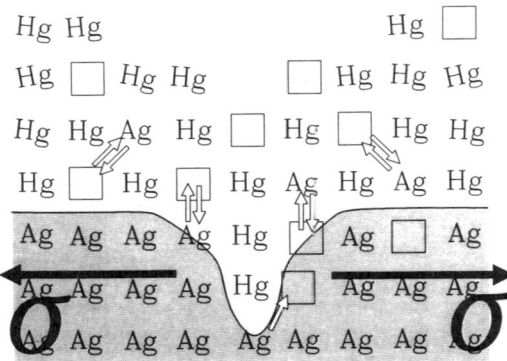

Figure 87. Crack propagation by surface mobility, in the presence of a liquid metal.

growth takes place by inward movement of oxygen ions. A similar situation is found in passive films at room temperature. This was discussed by Chao et al.[264] These authors concluded that the migration of oxygen anions or oxide-ion vacancies is essential for passive film growth to occur.

When the film growth mechanism involves inward movement of oxygen ions, as schematically shown in Fig. 88, the source of vacancies operative in the examples described above is either absent or restricted by

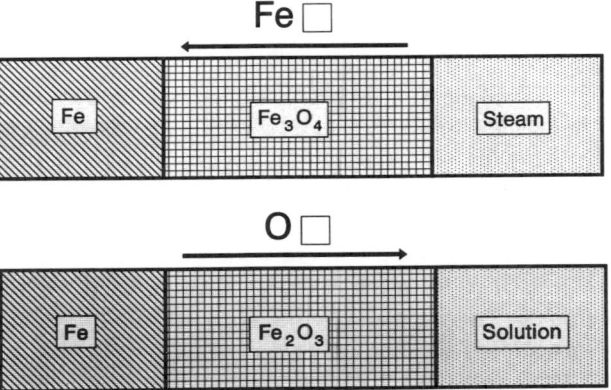

Figure 88. Growth mechanism of films leading to SCC (top) and films involving inward movement of oxygen ions, typical of passivity (bottom), where SCC by the surface-mobility mechanism is restricted.

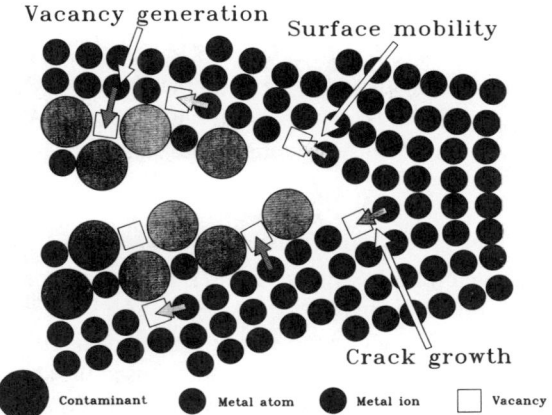

Figure 89. Description of the crack propagation process by a mechanism of surface mobility, in the presence of an ionic contaminant.[22]

the rate of passive film formation. Consequently, according to the surface-mobility SCC mechanism, without the supply of metal vacancies no SCC should be expected in the passive region.

To conclude, Fig. 89 depicts the crack propagation process by a mechanism of surface mobility, in the presence of an ionic contaminant.[22]

6. The Mechanism as a Predictive Tool

One of the most promising properties of the surface-mobility SCC mechanism is that it allows the prediction of crack velocities at any temperature value if the value at a given temperature is known. This is particularly important when crack velocities as low as 1×10^{-12} m/s or even lower are relevant, as happens in nuclear pressure vessels, nuclear repositories, prestressed concrete rebars, nuclear steam generators, etc. Those very low crack velocities are very difficult to measure with the techniques available at present. Nevertheless, by increasing the temperature, reliable crack velocities could be measured. If the crack velocity of the material is known at a given temperature, a D_S value can be found with Eq. (46). With Eq. (49) an estimate of Tm can be made. Then with Tm and Eqs. (46) and (49) the crack velocity for any given temperature can be calculated. Figures 77–79 are very promising in this respect.

It seems important to point out that the present mechanism allowed the prediction of crack velocities as low as 2×10^{-12} m/s, as was the case for steel contaminated with magnetite (steel in water at 84°C; Fig. 78),[261] or higher than 1×10^{-3} m/s, as was found with an Ag–20Au alloy in an AgClO$_4$ solution (at room temperature and under ultrafast-strain-rate conditions).[200]

7. Limitations of the Model

The surface-mobility SCC mechanism is a very recent development, and not enough time has elapsed for the researchers in the field to come out with experimental results contradicting it. The most critical work was that published by Kelly *et al.*[199] These authors used ultrafast-strain-rate techniques for straining an Ag–20Au alloy in an HClO$_4$ solution. These measurements included tests in liquid nitrogen, and the authors concluded that as their results could be explained neither by the anodic dissolution mechanism nor by the surface-mobility mechanism, they were a proof in favor of the cleavage mechanism. Duffó and Galvele[200] repeated the experiments and extended them to AgClO$_4$ and KCl solutions. The conclusions of Duffó and Galvele's work were that all the observations made could be explained by the surface-mobility mechanism. They also reported that the size of the cracks was a function of the electric charge circulated previous to the straining, that AgClO$_4$ specifically accelerated the SCC process, and that no cracks were observable when the corroded samples in a KCl solution were strained in liquid nitrogen. All these observations could be accounted for by the surface-mobility SCC mechanism, but had no explanation in terms of the cleavage mechanism.

One of the limitations of the surface-mobility SCC mechanism is the poor knowledge we have at present of the surface self-diffusion of contaminated metals. Another important limitation is found in the need for experimental demonstration of the validity of the fourth postulate: capture of vacancies by the stressed lattice at the tip of the crack. It is hoped that in the near future relevant information on this point can be obtained either through the use of computer simulation or by direct observation with the scanning tunneling microscope.

IX. GENERAL CONCLUSIONS

The general conclusions from this review are:

1. There are three competing SCC mechanisms available in the present literature: anodic dissolution, cleavage, and surface-mobility. The first one has serious theoretical limitations, which should be reconsidered if it is to be taken as an acceptable mechanism. The second, in spite of being as old as the previous one, is still in the stage of proving the credibility of its elementary stage: cleavage in ductile metals. It has not been developed yet as a full mechanism, since parameters such as stress, temperature, or environment composition have not yet been analyzed by its supporters. The third one is of very recent development. It has been successful in the description of numerous SCC cases but still requires more experimental and theoretical development.

2. The available experimental techniques for measuring the kinetics of SCC are still very poor. Numerous present-day applications require the measurement of crack velocities lower than 1×10^{-11} m/s. There are no techniques adequate for this purpose. Hence, there is an urgent need for the development of new crack propagation rate measurement techniques.

3. Among the available theories, a lack of predictive capability is observed. The anodic dissolution mechanism either requires previous experimental parameters or uses arbitrarily chosen limits. The cleavage mechanism needs more development before it can be used as a predictive tool. The surface-mobility SCC mechanism gives full predictive capabilities, but it is of recent development and requires further testing.

4. Further knowledge of the processes taking place, at an atomic level, at the tip of a crack is required. Computer simulation and the STM technique are very promising in this regard.

5. The electrochemical aspects of an elastically strained metal, particularly under the conditions of the crack tip, would require a more detailed analysis.

ACKNOWLEDGMENTS

The author wishes to express his thanks to his co-worker Dr. Gustavo S. Duffó for reading the manuscript and for his very useful discussions and comments. The author is particularly indebted to his wife, Aida, for correcting the manuscript and for her continuous support and encouragement.

The present work was done as a part of the Commission of the European Communities Contract No. CI1*-CT91-0915 and has been supported by the Consejo Nacional de Investigaciones Científicas y Técnicas, Argentina, and the Programa OEA-CNEA for Materials Degradation.

REFERENCES

[1] B. De Celis, A. S. Argon, and S. Yip, *J. Appl. Phys.* **54** (1983) 4864.
[2] J. K. McCoy and A. J. Markworth, in *Predictive Capabilities in Environmentally Assisted Cracking*, PVP-Vol. 99, Ed. by R. Rungta, The American Society of Mechanical Engineers, New York, 1985, p. 195.
[3] N. H. Macmillan, in *Atomistics of Fracture*, Ed. by R. M. Latanision and J. R. Pickens, Plenum Press, New York, 1983, p. 95.
[4] C. J. Smithells and E. A. Brandes, eds., *Metals Reference Book*, 5th ed., Butterworths, London, 1978.
[5] R. C. Newman, K. Sieradzki, and H. S. Isaacs, *Metall. Trans.* **13A** (1982) 2015.
[6] J. A. Feeney and M. J. Blackburn, in *The Theory of Stress Corrosion Cracking in Alloys*, Ed. by J. C. Scully, NATO, Scientific Affairs Division, Brussels, 1971, p. 355.
[7] T. R. Beck, in *The Theory of Stress Corrosion Cracking in Alloys*, Ed. by J. C. Scully, NATO, Scientific Affairs Division, Brussels, 1971, p. 64.
[8] J. R. Galvele, in *Proceedings of the 7th International Congress on Metallic Corrosion*, Vol. 1, ABRACO-Rio de Janeiro, 1978, p. 65.
[9] P. D. Williams, P. S. Pao, and R. P. Wei, in *Environment-Sensitive Fracture of Engineering Materials*, Ed. by Z. A. Foroulis, The Metallurgical Society of AIME, New York, 1979, p. 3.
[10] P. L. Lee and D. Tromans, in *Environment-Sensitive Fracture of Engineering Materials*, Ed. by Z. A. Foroulis, The Metallurgical Society of AIME, New York, 1979, p. 232.
[11] F. P. Ford, in *Corrosion Processes*, Ed. by R. N. Parkins, Applied Science Publishers, London, 1982, p. 271.
[12] G. S. Duffó and J. R. Galvele, *Corros. Sci.* **30** (1990) 249.
[13] M. O. Speidel, in *The Theory of Stress Corrosion Cracking in Alloys*, Ed. by J. C. Scully, NATO, Scientific Affairs Division, Brussels, 1971, p. 289.
[14] T. R. Beck, M. J. Blackburn, W. H. Smyrl, and M. O. Speidel, in *Advances in Corrosion Science and Technology*, Vol. 3, Ed. by M. G. Fontana and R. W. Staehle, Plenum Press, New York, 1973, p. 184.
[15] D. N. Fager and W. E. Spurr, *Corrosion-NACE* **26** (1970) 409.
[16] A. A. Griffith, *Phil. Trans. Roy. Soc. (London)*, **A221** (1920) 163.
[17] E. Orowan, *Fracture Strength of Solids, Report Progress in Physics*, Vol. 12, Physical Society of London, London (1949).
[18] R. M. Thomson, in *Physical Metallurgy*, 3rd ed., Ed. by R. W. Cahn and P. Haansen, North-Holland Physics Publishing, Amsterdam, 1983, p. 1487.
[19] R. N. Parkins, C. M. Rangel, and J. Yu, *Metall. Trans.* **16A** (1985) 1761.
[20] M. O. Speidel, in *Corrosion in Power-Generating Equipment*, Ed. by M. O. Speidel and A. Atrens, Plenum Press, New York, 1984, p. 85.
[21] M. O. Speidel, *Metall. Trans.* **12A** (1981) 779.
[22] J. R. Galvele, in *Parkins Symposium on Fundamental Aspects of Stress Corrosion Cracking*, Ed. by S. M. Bruemmer, E. I. Meletis, R. H. Jones, W. W. Gerberich, F. P. Ford, and R. W. Staehle, The Minerals, Metals and Materials Society, Warrendale, Pennsylvania, 1992, p. 85.
[23] H. L. Logan, *The Stress Corrosion of Metals*, John Wiley & Sons, New York, 1966, p. 156.

[24] G. M. Ugiansky and J. H. Payer, eds., *Stress Corrosion Cracking—The Slow Strain-Rate Technique*, American Society for Testing and Materials, STP 665, Philadelphia, 1979.
[25] B. F. Brown, *Stress Corrosion Cracking Control Measures*, National Bureau of Standards Monograph, Washington, D.C., 1977, p. 156.
[26] H. R. Smith and D. E. Piper, in *Stress-Corrosion Cracking in High Strength Steels and in Titanium and Aluminum Alloys*, Ed. by B. F. Brown, Naval Research Laboratory, U.S. Goverment Printing Office, Washington, D.C., 1972, p. 18.
[27] A. Paskin, B. Massoumzadeh, K. Shukla, K. Sieradzki, and G. J. Dienes, *Acta Metall.* **33** (1985) 1987.
[28] General discussion, International Conference on Corrosion–Deformation Interactions, Fontainebleau, France, October 5–7, 1992.
[29] Y. Ronen and P. Rozenak, The Development of the SCREW. An Early Metallurgy, Report, Faculty of Engineering, Ben-Gurion University, Beer Sheva, Israel, 1990.
[30] F. H. Keating, *Symposium on Internal Stresses in Metals and Alloys*, Institute for Metals, London, 1948, p. 311.
[31] Anonymous, *Sci. Am.* **244**(3) (1981) 11.
[32] W. C. Roberts-Austen, *Proc. Roy. Inst. G. Brit.* **11** (1886) 395.
[33] H. Leidheiser Jr., *The Corrosion of Copper, Tin and Their Alloys*, John Wiley & Sons, New York, 1971, p. 147.
[34] F. S. Spiers, ed., *The Failure of Metals under Internal and Prolonged Stress*, The Faraday Society, London, 1921.
[35] *Symposium on Stress-Corrosion Cracking of Metals*, American Society for Testing and Materials and American Institute of Mining, Metallurgical, and Petroleum Engineers, Philadelphia, 1944.
[36] W. Rosenhain and S. L. Archbutt, *Proc. Roy. Soc. (London)*, **A96** (1919) 55.
[37] O. B. Ellis, *Symposium on Stress-Corrosion Cracking of Metals*, American Society for Testing and Materials and American Institute of Mining, Metallurgical, and Petroleum Engineers, Philadelphia, 1944, p. 421.
[38] R. E. Johnson, in *The Science, Technology and Application of Titanium*, Ed. by R. I. Jaffee and N. E. Promisel, Pergamon Press, Oxford, 1970, p. 1175.
[39] B. Isecke, *Proceedings 3rd FIP Symposium*, FIP-Berkeley, Wexham Springs, Slough, U.K., 1981.
[40] J. S. Leak, *Proceedings 26th Conference National Association of Corrosion Engineers*, National Association of Corrosion Engineers, Houston, 1970, p. 497.
[41] M. O. Speidel, *J. Mater. Eng.* **9** (1987) 157.
[42] B. Cox, *J. Nucl. Mater.* **170** (1990) 1.
[43] J. R. Galvele, *Bol. Acad. Nac. Ciencias*, Córdoba, Argentina, **54** (1980) 79.
[44] H. J. Engell and A. Baumel, in *Physical Metallurgy of Stress-Corrosion Fracture*, Ed. by T. N. Rhodin, Interscience, New York, 1959, p. 341.
[45] C. Edeleanu, *J. Iron Steel Inst.* **173** (1953) 140.
[46] U. R. Evans, *The Corrosion and Oxidation of Metals*, Arnold, London, 1960, p. 689.
[47] R. A. Oriani, in *Stress-Corrosion Cracking and Hydrogen Embrittlement of Iron Base Alloys*, Ed. by R. W. Staehle, J. Hochmann, R. D. McCright and J. E. Slater, NACE-5, National Association of Corrosion Engineers, Houston, 1977, p. 351.
[48] R. A. Oriani and P. H. Josephic, *Acta Metall.* **22** (1974) 1065.
[49] W. A. Tiller and R. Schrieffer, *Scr. Metall.* **4** (1970) 57.
[50] J. C. Scully and D. T. Powell, *Corros. Sci.* **10** (1970) 719.
[51] H. H. Uhlig, in *Stress-Corrosion Cracking and Hydrogen Embrittlement of Iron Base Alloys*, Ed. by R. W. Staehle, J. Hochmann, R. D. McCright, and J. E. Slater, NACE-5, National Association of Corrosion Engineers, Houston, 1977, p. 174.
[52] E. G. Coleman, D. Weinstein, and W. Tostoke, *Acta Metall.* **9** (1961) 491.
[53] H. L. Logan, *J. Res. Nat. Bur. Stand.* **48** (1952) 99.

[54] T. P. Hoar and J. G. Hines, *J. Iron Steel Inst.* **182** (1956) 124.
[55] T. P. Hoar and J. M. West, *Proc. Roy. Soc., Ser. A* **268** (1962) 304.
[56] G. Taylor, in *The Sorby Centennial Symposium on the History of Metallurgy*, Ed. by C. S. Smith, Gordon and Breach, American Institute of Mining and Metallurgical Engineers, Cleveland, 1963, p. 355.
[57] E. Orowan, in *The Sorby Centennial Symposium on the History of Metallurgy*, Ed. by C. S. Smith, Gordon and Breach, American Institute of Mining and Metallurgical Engineers, Cleveland, 1963, p. 359.
[58] D. Tromans and J. Nutting, *Fracture of Solids*, Interscience, New York, 1962, p. 637.
[59] P. R. Swann, *Corrosion-NACE* **19** (1965) 161.
[60] D. L. Douglass, G. Thomas, and W. R. Roser, *Corrosion-NACE* **20** (1964) 15t.
[61] N. Ohtani and R. A. Dodd, *Corrosion-NACE* **21** (1965) 161.
[62] G. R. Irwin, *Metall. Rev.* **10** (1965) 223.
[63] B. F. Brown and C. D. Beachem, *Corros. Sci.* **5** (1965) 745.
[64] M. Pourbaix and F. Vandervelden, Les Méthodes Intensiostatiques et Potentiostatiques, CEBELCOR, Rapport Technique No. 89, 1961, p. 5.
[65] U. R. Evans and T. P. Hoar, *Proc. Roy. Soc., Ser. A* **137** (1932) 343.
[66] C. Wagner and W. Traud, *Z. Electrochem.* **44** (1940) 391.
[67] A. Hickling, *Trans. Faraday Soc.* **38** (1942) 27.
[68] C. Edeleanu, *Nature (London)* **173** (1954) 739.
[69] M. J. Humphries and R. N. Parkins, *Corros. Sci.* **7** (1967) 747.
[70] B. F. Brown, C. T. Fujii, and E. P. Dahlberg, *J. Electrochem. Soc.* **116** (1969) 218.
[71] R. N. Parkins, *Metall. Rev.* **9** (1964) 248.
[72] A. Turnbull, *Corros. Sci.* **23** (1983) 833.
[73] H. J. S. Sand, *Phil. Mag.* **1** (1900) 45; *Z. Phys. Chem.* **35** (1900) 641. (See K. J. Vetter, *Electrochemical Kinetics*, Academic Press, New York, 1967, p. 205.)
[74] T. R. Rosebrugh and W. Lash Miller, *J. Phys. Chem.* **14** (1910) 816.
[75] T. P. Hoar and J. R. Galvele, *Corros. Sci.* **10** (1970) 211.
[76] G. S. Duffó, I. A. Maier, and J. R. Galvele, *Corros. Sci.* **28** (1988) 1003.
[77] T. R. Beck, *J. Electrochem. Soc.* **116** (1969) 177.
[78] M. O. Speidel, *Metall. Trans.* **6A** (1975) 631.
[79] A. Turnbull, in *Embrittlement by the Localized Crack Environment*, Ed. by R. P. Gangloff, The Metallurgical Society of AIME, Warrendale, Pennsylvania, 1984, p. 3.
[80] J. R. Galvele, *Corros. Sci.* **21** (1981) 551.
[81] C. F. Baes and R. E. Mesmer, *The Hydrolysis of Cations*, John Wiley & Sons, New York, 1976.
[82] H. Wendt, *Chimia* **27** (1973) 575.
[83] J. P. Hunt, *Metal Ions in Aqueous Solutions*, Benjamin, New York, 1963.
[84] K. J. Vetter, *Electrochemical Kinetics*, Academic Press, New York, 1967, p. 180.
[85] J. N. Butler, *Ionic Equilibrium, A Mathematical Approach*, Addison-Wesley, Reading, Massachusetts, 1964.
[86] L. G. Sillen and A. E. Martell, *Stability Constants of Metal–Ion Complexes*, Special Publication No. 17, The Chemical Society, London, 1964.
[87] L. G. Sillen and A. E. Martell, *Stability Constants of Metal–Ion Complexes*, Supplement No. 1, Special Publication No. 25, The Chemical Society, London, 1970.
[88] S. M. Gravano and J. R. Galvele, *Corros. Sci.* **24** (1984) p. 517.
[89] R. R. Shuck and J. L. Swedlow, in *Localized Corrosion*, Ed. by R. W. Staehle, B. F. Brown, J. Kruger, and A. Agrawal, National Association of Corrosion Engineers, Houston, 1974, pp. 190 and 208.
[90] W. H. Smyrl and J. Newman, *J. Electrochem. Soc.* **121** (1974) 1000.
[91] R. B. Rebak, R. M. Carranza, and J. R. Galvele, *Corros. Sci.* **28** (1988) p. 1089.

[92] J. C. Scully, in *The Theory of Stress Corrosion Cracking in Alloys*, Ed. by J. C. Scully, NATO, Scientific Affairs Division, Brussels, 1971, p. 126.
[93] C. Manfredi, I. A. Maier, and J. R. Galvele, *Corros. Sci.* **27** (1987) 887.
[94] I. L. Muller and J. R. Galvele, *Corros. Sci.* **17** (1977) 179.
[95] J. R. Galvele and S. M. de De Micheli, *Corros. Sci.* **10** (1970) 795.
[96] J. R. Galvele, S. M. de De Micheli, I. L. Muller, S. B. de Wexler and I. L. Alanis, in *Localized Corrosion*, Ed. by R. W. Staehle, B. F. Brown, J. Kruger, and A. Agrawal, NACE-3, National Association of Corrosion Engineers, Houston, 1974, p. 580.
[97] S. M. de De Micheli and J. R. Galvele, in *Aluminum Transformation Technology and Applications—1981*, American Society for Metals, Metals Park, Ohio, 1982, p. 521.
[98] S. M. de De Micheli, *Rev. Coatings Corrosion* **2** (1977) 73.
[99] G. Alvarez and J. R. Galvele, *Corrosion-NACE* **32** (1976) 285.
[100] V. P. Batrakov, *Third International Congress on Metallic Corrosion* **Vol. I**, 1966, Moscow (1969), p. 313. Distributor: Swets-Zeitlinger, Amsterdam.
[101] R. E. Hanneman, P. Rao, and J. C. Danko, in *Environment-Sensitive Fracture of Engineering Materials*, Ed. by Z. A. Foroulis, The Metallurgical Society of AIME, New York, 1979, p. 153.
[102] M. Kowaka, T. Kudo, and K. Ota, in *Environment-Sensitive Fracture of Engineering Materials*, Ed. by Z. A. Foroulis, The Metallurgical Society of AIME, New York, 1979, p. 178.
[103] G. S. Was and V. B. Rajan, *Metall. Trans.* **18A** (1987) 1313.
[104] G. S. Duffó, M. Giordano, and J. R. Galvele, unpublished work, 1993.
[105] S. M. Payne and P. McIntyre, *Corrosion-NACE* **44** (1988) 314.
[106] E. D. Hondros and M. P. Seah, *Interfacial and Surface Microchemistry* in *Physical Metallurgy*, 3rd ed., Ed. by R. W. Cahn and P. Haansen, North-Holland Physics Publishing, Amsterdam, 1983, p. 856.
[107] C. Lea and E. D. Hondros, *Proc. Roy. Soc., Ser. A* **377** (1982) 477.
[108] R. N. Parkins, in *Environment-Induced Cracking of Metals*, Ed. by R. P. Gangloff and M. B. Ives, National Association of Corrosion Engineers, Houston, 1990, p. 1.
[109] R. W. Staehle, in *The Theory of Stress Corrosion Cracking in Alloys*, Ed. by J. C. Scully, NATO, Scientific Affairs Division, Brussels, 1971, p. 223.
[110] G. S. Duffó and J. R. Galvele, *Corros. Sci.* **28** (1988) 207.
[111] R. M. Carranza and J. R. Galvele, *Corros. Sci.* **28** (1988) 233.
[112] R. M. Carranza and J. R. Galvele, *Corros. Sci.* **28** (1988) 851.
[113] R. Oltra, *Métaux et Techniques*, Numéro special, Corrosion sous contrainte, **76**, (September) (1988) 17.
[114] R. W. Staehle, in *Environment-Induced Cracking of Metals*, Ed. by R. P. Gangloff and M. B. Ives, National Association of Corrosion Engineers, Houston, 1990, p. 561.
[115] R. N. Parkins, *Journal of Minerals, Metals, and Materials Society* (JOM) **44** (December) (1992) 12.
[116] P. Doig and P. E. J. Flewitt, *Metall. Trans. A* **12** (1981) 923.
[117] P. Doig and P. E. J. Flewitt, in *Corrosion in Power-Generating Equipment*, Ed. by M. O. Speidel and A. Atrens, Plenum Press, New York, 1984, p. 139.
[118] P. Doig and P. E. J. Flewitt, in *Embrittlement by the Localized Crack Environment*, Ed. by R. P. Gangloff, The Metallurgical Society of AIME, Warrendale, Pennsylvania, 1984, p. 305.
[119] A. R. Despic, R. G. Raicheff, and J. O'M. Bockris, *J. Chem. Phys.* **49** (1968) 926.
[120] K. Nobe and S. Tan, *Corrosion-NACE* **18** (1962) 391t.
[121] S. Tan and K. Nobe, *Can. J. Chem.* **41** (1963) 495.
[122] J. O'M. Bockris, in *Modern Aspects of Electrochemistry*, Ed. by J. O'M. Bockris and B. E. Conway, Butterworths, London, 1954, p. 180.
[123] E. A. Flood, *Can. J. Chem.* **36** (1958) 1332.

[124] J. J. Harwood, *Corrosion-NACE* **6** (1950) 249.
[125] J. R. Galvele, *J. Electrochem. Soc.* **133** (1986) 953.
[126] R. W. Staehle, J. J. Royuela, T. L. Raredon, E. Sarrate, C. R. Morin, and R. V. Farrar, *Corrosion-NACE* **26** (1970) 451.
[127] B. E. Wilde, *J. Electrochem. Soc.* **118** (1971) 1717.
[128] A. Windfeldt, *Electrochim. Acta* **9** (1964) 1139, 1295.
[129] S. Haruyama and S. Asawa, *Corros. Sci.* **13** (1973) 395.
[130] J. Newman, *Electrochemical Systems*, Prentice-Hall, Englewood Cliffs, New Jersey, 1973.
[131] J. R. Galvele, *J. Electrochem. Soc.* **123** (1976) 464.
[132] J. M. West, *Nature (London)* **185** (1960) 92.
[133] P. Doig and P. E. J. Flewitt, *Proc. Roy. Soc. London, Ser. A* **357** (1977) 439.
[134] D. Singbeil and D. Tromans, *Metall. Trans. A* **13** (1982) 1091.
[135] R. Magdowski and M. O. Speidel, *Metall. Trans. A* **19** (1988) 1583.
[136] M. O. Speidel and H. V. Hyatt, in *Advances in Corrosion Science and Technology*, Ed. by M. G. Fontana and R. W. Staehle, Plenum Press, New York, 1972, p. 115.
[137] J. R. Galvele, in *Environment-Induced Cracking of Metals*, Ed. by R. P. Gangloff and M. B. Ives, National Association of Corrosion Engineers, Houston, 1990, p. 163.
[138] H. Gerischer, in *The Surface Chemistry of Metals and Semiconductors*, Ed. by H. C. Gatos, John Wiley & Sons, New York, 1960, p. 177.
[139] H. Kaesche, *Metallic Corrosion*, National Association of Corrosion Engineers, Houston, 1985, p. 127.
[140] N. S. Stoloff, in *Atomistics of Fracture*, Ed. by R. M. Latanision and J. R. Pickens, Plenum Press, New York, 1983, p. 921.
[141] R. M. Latanision, in *Atomistics of Fracture*, Ed. by R. M. Latanision and J. R. Pickens, Plenum Press, New York, 1983, p. 3.
[142] S. P. Lynch, *Scr. Metall.* **13** (1979) 1051.
[143] H. L. Logan, *The Stress Corrosion of Metals*, John Wiley & Sons, New York, 1966, p. 16.
[144] H. Uhlig, K. Gupta, and W. Liang, *J. Electrochem. Soc.* **122** (1975) 343.
[145] R. N. Parkins, *Corros. Sci.* **20** (1980) 147.
[146] J. R. Galvele, in *Predictive Capabilities in Environmentally Assisted Cracking*, PVP-Vol. 99, Ed. by R. Rungta, The American Society of Mechanical Engineers, New York, 1985, p. 273.
[147] F. P. Ford and P. L. Andresen, in *Parkins Symposium on Fundamental Aspects of Stress Corrosion Cracking*, Ed. by S. M. Bruemmer, E. I. Meletis, R. H. Jones, W. W. Gerberich, F. P. Ford, and R. W. Staehle, The Minerals, Metals and Materials Society, Warrendale, Pennsylvania, 1992, p. 43.
[148] F. P. Ford, in *Environment-Induced Cracking of Metals*, Ed. by R. P. Gangloff and M. B. Ives, National Association of Corrosion Engineers, Houston, 1990, p. 139.
[149] R. N. Parkins and N. J. H. Holroyd, *Corrosion* **38** (1982) 245.
[150] R. W. Staehle, in *Passivity and Its Breakdown on Iron and Iron Base Alloys in Aqueous Environments*, Ed. by R. W. Staehle and H. Okada, National Association of Corrosion Engineers, Houston, 1976, p. 155.
[151] F. P. Ford, Mechanisms of Environmental Cracking in Systems Peculiar to the Power Generation Industry, Final Contract Report, EPRI report NP 2589, September 1982.
[152] T. P. Hoar and J. M. West, *Nature (London)* **181** (1958) 835.
[153] T. P. Hoar and J. C. Scully, *J. Electrochem. Soc.* **111** (1964) 348.
[154] J. C. Scully and T. P. Hoar, *Proceedings of the 2nd International Congress on Metallic Corrosion*, National Association of Corrosion Engineers, Houston, 1966, p. 184.
[155] J. R. Galvele, S. B. de Wexler and I. Gardiazabal, *Corrosion-NACE* **31** (1975) 352.

[156] J. R. Galvele and I. Maier, in *Passivity and Its Breakdown on Iron and Iron Base Alloys in Aqueous Environments*, Ed. by R. W. Staehle and H. Okada, National Association of Corrosion Engineers, Houston, 1976, p. 178.
[157] I. Maier and J. R. Galvele, *Corrosion-NACE* **36** (1980) 60.
[158] I. A. Maier, E. Lopez Perez, and J. R. Galvele, *Corros. Sci.* **22** (1982) 537.
[159] T. P. Hoar and R. W. Jones, *Corros. Sci.* **13** (1973) 725.
[160] I. A. Maier, C. Manfredi, and J. R. Galvele, *Corros. Sci.* **25** (1985) 15.
[161] Y. S. Park, J. R. Galvele, A. K. Agrawal, and R. W. Staehle, *Corrosion-NACE* **34** (1976) 413.
[162] M. G. Alvarez, C. Manfredi, M. Giordano, and J. R. Galvele, *Corros. Sci.* **24** (1984) 769.
[163] M. G. Alvarez, C. Manfredi, M. Giordano, and J. R. Galvele, *Corrosion-NACE* **46** (1990) 717.
[164] S. F. Bubar and D. A. Vermilyea, *J. Electrochem. Soc.* **113** (1966) 892.
[165] J. R. Galvele, Anodic Behaviour of Mild Steel during Yielding, Ph.D. Dissertation, Department of Metallurgy, University of Cambridge, 1966.
[166] S. B. de Wexler and J. R. Galvele, *J. Electrochem. Soc.* **121** (1974) 1271.
[167] M. C. Andrade, private communication, 1992.
[168] R. N. Parkins, in *Parkins Symposium on Fundamental Aspects of Stress Corrosion Cracking*, Ed. by S. M. Bruemmer, E. I. Meletis, R. H. Jones, W. W. Gerberich, F. P. Ford, and R. W. Staehle, The Minerals, Metals and Materials Society, Warrendale, Pennsylvania, 1992, p. 3.
[169] J. C. Scully, *Corros. Sci.* **7** (1967) 197.
[170] J. C. Scully, *Corros. Sci.* **8** (1968) 513.
[171] T. J. Smith and R. W. Staehle, *Corrosion-NACE* **23** (1967) 117.
[172] J. R. Galvele, Paper presented at the International Conference on Advances in Corrosion and Protection, University of Manchester, 1992; *Corros. Sci.* **35** (1993) 419.
[173] J. M. Silcock and P. R. Swann, in *Environment-Sensitive Fracture of Engineering Materials*, Ed. by Z. A. Foroulis, The Metallurgical Society of AIME, Warrendale, Pennsylvania, 1979, p. 133.
[174] H. W. Pickering, F. H. Beck, and M. S. Fontana, *Corrosion-NACE* **18** (1962) 230t.
[175] N. A. Nielsen, *Physical Metallurgy of Stress Corrosion Fracture*, Metallurgical Society Conferences, AIME, Vol. 4, Interscience, New York, 1969, p. 121.
[176] J. R. Ambrose, in *Treatise on Materials Science and Technology*. Vol. 23. *Corrosion: Aqueous Processes and Passive Films*, Ed. by J. C. Scully, Academic Press, London, 1983, pp. 175–204.
[177] T. R. Beck, *J. Electrochem. Soc.* **129** (1982) 2500.
[178] M. Barbosa and J. C. Scully, in *Environment-Sensitive Fracture of Engineering Materials*, Ed. by Z. A. Foroulis, The Metallurgical Society of AIME, Warrendale, Pennsylvania, 1979, p. 91.
[179] J. R. Galvele, R. M. Torresi, and R. M. Carranza, *Corros. Sci.* **31** (1990) 536.
[180] M. M. Schultze, M. M. Lohrengel, and D. Ross, *Electrochim. Acta* **28** (1983) 973.
[181] T. E. Evans, in *Passivity of Metals*, Ed. by R. P. Frankenthal and J. Kruger, The Electrochemical Society, Princeton, New Jersey, 1978, p. 410.
[182] N. Cabrera and N. Mott, *Rep. Prog. Phys.* **12** (1949) 163.
[183] S. Szklarska-Smialowska, in *Localized Corrosion*, Ed. by R. W. Staehle, B. F. Brown, J. Kruger, and A. Agrawal, National Association of Corrosion Engineers, Houston, 1974, p. 312.
[184] A. R. C. Westwood, in *Fracture of Solids*, Vol. 20, American Institute of Mining, Metallurgical and Petroleum Engineers, Cleveland, 1963, p. 553.
[185] E. N. Pugh, in *Atomistics of Fracture*, Ed. by E. M. Latanision and J. R. Pickens, Plenum Press, New York, 1983, p. 997.
[186] K. Sieradzki and R. C. Newman, *Phil. Mag. A*, **51** (1985) 95.

[187] C. Edeleanu and A. J. Forty, *Phil. Mag.* **5** (1960) 1029.
[188] A. J. Forty, *Teknisk-Vetenskaplig Forskning (Royal Swedish Academy of Engineering Sciences)* **32** (1961) 104.
[189] W. W. Gerberich and S. Chen, in *Environment-Induced Cracking of Metals*, Ed. by R. P. Gangloff and M. B. Ives, National Association of Corrosion Engineers, Houston, 1990, p. 167.
[190] F. A. Champion, in *Symposium on Internal Stresses in Metals and Alloys*, Institute of Metals, London, England, 1948, p. 468.
[191] A. J. McEvily, Jr. and A. P. Bond, *J. Electrochem. Soc.* **112** (1965) 131.
[192] E. N. Pugh, in *The Theory of Stress Corrosion Cracking in Alloys*, Ed. by J. C. Scully, NATO, Brussels, 1971, p. 418.
[193] T. R. Pinchback, S. P. Clough, and L. A. Heldt, *Corrosion-NACE* **32** (1976) 469.
[194] K. Sieradzki and R. C. Newman, *J. Phys. Chem. Solids* **48** (1987) 1101.
[195] M. Henthorne and R. N. Parkins, *Corros. Sci.* **6** (1966) 357.
[196] J. M. Silcock, *Br. Corros. J.* **16** (1981) 78.
[197] J. D. Fritz, B. W. Parks, and H. W. Pickering, *Scr. Metall.* **22** (1988) 1063.
[198] B. D. Lichter, R. M. Bhatkal, and W. F. Flanagan, in *Parkins Symposium on Fundamental Aspects of Stress-Corrosion Cracking*, Ed. by S. M. Bruemmer, E. I. Meletis, R. H. Jones, W. W. Gerberich, F. P. Ford, and R. W. Staehle, The Minerals, Metals and Materials Society, Warrendale, Pennsylvania, 1992, p. 279.
[199] R. G. Kelly, A. J. Frost, T. Shahrabi, and R. C. Newman, *Metall. Trans. A.* **22** (1991) 531.
[200] G. S. Duffó and J. R. Galvele, *Metall. Trans. A.* **24** (1993) 425.
[201] I. Maier, S. Fernández, and J. R. Galvele, *Corros. Sci.* **37** (1995) 1.
[202] J. R. Galvele, *Corros. Sci.* **27** (1987) 1.
[203] G. E. Rhead, *Surf. Sci.* **15** (1969) 353.
[204] G. E. Rhead, *Surf. Sci.* **22** (1970) 223.
[205] F. Delamare and G. E. Rhead, *Surf. Sci.* **28** (1971) 267.
[206] O. Oda and G. E. Rhead, *Scr. Metall.* **13** (1979) 985.
[207] J. Henrion and G. E. Rhead, *Surf. Sci.* **29** (1972) 20.
[208] F. Delamare and G. E. Rhead, *Surf. Sci.* **35** (1973) 172, 185.
[209] M. Pichaud and M. Drechsler, *Surf. Sci.* **36** (1973) 813.
[210] R. A. Oriani, in *Environment-Induced Cracking of Metals*, Ed. by R. Gangloff and B. Ives, NACE-10, National Association of Corrosion Engineers, Houston, 1990, pp. 263 and 264.
[211] A. C. Cocks and M. F. Ashby, *Prog. Mat. Sci.* **27** (1982) 189.
[212] J. P. Hirth and W. D. Nix, *Acta Metall.* **33** (1985) 359.
[213] A. Seeger, *J. Less-Common Met.* **28** (1972) 387.
[214] A. H. Cottrell, *An Introduction to Metallurgy*, St. Martin's Press, New York, 1967, p. 356.
[215] N. A. Gjostein, in *Surfaces and Interfaces—I*, Ed. by J. J. Burke, N. L. Reed, and V. Weiss, Syracuse University Press, Syracuse, New York, 1967, p. 271.
[216] G. Neumann and G. M. Neumann, Surface Self-Diffusion of Metals, DMS-1, Diffusion Information Center, Riehen, Switzerland, 1972.
[217] G. Ehrlich and K. Stolt, *Annu. Rev. Phys. Chem.* **31** (1980) 603.
[218] G. E. Rhead, *Int. Mater. Rev.* **34** (1989) 261.
[219] G. Ehrlich, *J. Vac. Sci. Technol.* **17** (1980) 9.
[220] C. Alonso, R. C. Salvarezza, J. M. Vara, and A. J. Arvia, *Electrochim. Acta* **35** (1990) 1331.
[221] I. C. Oppenheim, C. E. D. Chidsey, D. J. Trevor, and K. Sieradzki, Corrosion Research in Progress Symposium, National Association of Corrosion Engineers, Las Vegas, Nevada, April 23–24, 1990, Extended Abstracts, p. 3.
[222] M. P. García, M. M. Gómez, R. C. Salvarezza, and A. J. Arvia, 43rd ISE Meeting, Córdoba, Argentina, September 20–25, 1992.

[223] P. Wynblatt and N. A. Gjostein, *Surf. Sci.* **12** (1968) 109.
[224] G. S. Duffó and J. R. Galvele, in *Environment-Induced Cracking of Metals*, Ed. by R. Gangloff and B. Ives, NACE-10, National Association of Corrosion Engineers, Houston, 1990, pp. 261–264.
[225] G. Grimvall and S. Sjödin, *Phys. Scr. (Sweden)*, **10** (1974) 340.
[226] J. R. Galvele, CEFRACOR-SF2M Meeting, Ecole de Bombannes, Paris, June 26, 1992.
[227] H. K. Birnbaum, in *Environment-Induced Cracking of Metals*, Ed. by R. Gangloff and B. Ives, NACE-10, National Association of Corrosion Engineers, Houston, 1990, p. 21.
[228] S. P. Lynch, *Acta Metall.* **32** (1984) 79.
[229] A. W. Thompson and I. M. Bernstein, in *Advances in Corrosion Science and Technology*, Ed. by M. G. Fontana and R. W. Staehle, Vol. 7, Plenum Press, New York, 1980, p. 53.
[230] R. M. Latanision, O. H. Gastine, and C. R. Compeau, in *Environment-Sensitive Fracture of Engineering Materials*, Ed. by Z. A. Foroulis, The Metallurgical Society of AIME, New York, 1979, p. 48.
[231] J. K. Nørskov, F. Besenbacher, J. Bøttiger, B. B. Nielsen, and A. A. Pisarev, *Phys. Rev. Lett.* **49** (1982) 1420.
[232] K. B. Kim and S. I. Pyun, *Arch. Eisenhüttenwes.* **53** (1982) 397.
[233] R. Gibala, in *Stress Corrosion Cracking and Hydrogen Embrittlement of Iron Base Alloys*, Ed. by R. W. Staehle, J. Hochmann, R. D. McCright, and J. E. Slater, National Association of Corrosion Engineers, Houston, 1977, p. 244.
[234] H. K. Birnbaum, in *Environment-Sensitive Fracture of Engineering Materials*, Ed. by Z. A. Foroulis, The Metallurgical Society of AIME, New York, 1979, p. 326.
[235] R. A. Oriani, *Annu. Rev. Mater. Sci.* **8** (1978) 327.
[236] H. K. Birnbaum, in *Atomistics of Fracture*, Ed. by R. M. Latanision and J. R. Pickens, Plenum Press, New York, 1983, p. 733.
[237] G. E. Kerns and R. W. Staehle, *Scr. Metall.* **6** (1972) 1189.
[238] K. Sieradzki and P. Ficalora, *Scr. Metall.* **13** (1979) 535.
[239] K. Sieradzki, *Acta Metall.* **30** (1982) 973.
[240] B. Cox and R. Haddad, *J. Nucl. Mater.* **137** (1986) 115.
[241] B. Cox and R. Haddad, in *Zirconium in the Nuclear Industry, 7th International Symposium*, Ed. by R. B. Adamson and L. F. P. Van Swam, ASTM STP 939, American Society for Testing and Materials, Philadelphia, 1987, p. 717.
[242] I. Schuster and C. Lemaignan, *J. Nucl. Mater.* **189** (1992) 157.
[243] M. Peehs, H. Stehle, and E. Steinberg, in ASTM Zirconium Conference, Stratford-upon-Avon, England, 1978.
[244] M. Peehs, F. Garzarolli, R. Hahn, and E. Steinberg, *J. Nucl. Mater.* **87** (1979) 274.
[245] G. L. Bianchi and J. R. Galvele, *Corros. Sci.* **34** (1993) 1411 and **36** (1994) 611.
[246] F. Delamare, *Scr. Metall.* **8** (1974) 991.
[247] G. L. Bianchi and J. R. Galvele, *Corros. Sci.* **27** (1987) 631.
[248] D. J. G. Ives and G. J. Janz, Eds., *Reference Electrodes. Theory and Practice*, Academic Press, New York, 1961, p. 179.
[249] L. Graf, in *Proceedings of the Conference on Fundamental Aspects of Stress Corrosion Cracking*, Ed. by R. W. Staehle and A. J. Forty, National Association of Corrosion Engineers, Houston, 1959, p. 187.
[250] L. Graf, in *Proceedings of 2nd International Congress on Metallic Corrosion*, National Association of Corrosion Engineers, Houston, 1966, p. 89.
[251] L. Graf and J. Budke, *Z. Metallkd.* **46** (1955) 378.
[252] L. Graf and N. Wieling, *Z. Metallkd.* **63** (1972) 379.
[253] L. Graf, *Z. Metallkd.* **66** (1975) 751.
[254] H. W. Pickering, *Corros. Sci.* **23** (1983) 1107.
[255] S. Rambert and D. Landolt, *Electrochim. Acta* **31** (1986) 1433.

[256] B. D. Lichter, W. F. Flanagan, J. B. Lee, and M. Zhu, in *Conference on Environment Induced Cracking of Metals*, Ed. by R. P. Gangloff and M. B. Ives, NACE-10, National Association of Corrosion Engineers, Houston, 1990, p. 251.
[257] K. Vetter, *Electrochemical Kinetics*, Academic Press, New York, 1961, p. 170.
[258] G. S. Duffó and J. R. Galvele, *Corros. Sci.* **34** (1993) 79.
[259] R. B. Rebak, Ph. D. Dissertation, The Ohio State University, 1993.
[260] M. O. Speidel, *Metall. Trans.* **12A** (1981) 779.
[261] J. R. Galvele, *Corros. Sci.* **30** (1990) 955.
[262] K. Hauffe, *Oxidation of Metals*, Plenum Press, New York, 1965, pp. 4, 9, 152, 280, 282, and 285.
[263] G. S. Duffó and J. R. Galvele, research in progress.
[264] C. Y. Chao, L. F. Lin, and D. D. Macdonald, *J. Electrochem. Soc.* **128** (1981) 1187.

4

Metal Hydride Electrodes

Thomas F. Fuller

International Fuel Cells, South Windsor, Connecticut 06074

John Newman

Department of Chemical Engineering, University of California, and Materials Sciences Division, Lawrence Berkeley Laboratory, Berkeley, California 94720

I. INTRODUCTION

The energy shock of the seventies and, more recently, cold fusion spurred interest in metal hydrides. Because of their ability to store large quantities of hydrogen, they have been considered for a variety of energy-conversion applications.[1,2] Although some applications have not been realized on a large scale, and some not at all, several companies are aggressively pursuing the commercialization of the NiOOH–metal hydride battery.[3,4] As the demand for and the production of portable appliances grow, environmental concerns about the disposal of cadmium, a toxic component of the nickel/cadmium (Ni/Cd) cell, mount. Many anticipate that the NiOOH–metal hydride battery will replace the traditional Ni/Cd battery.[5] Moreover, the U.S. Advanced Battery Consortium (USABC)[6] and others[7] are evaluating the NiOOH–metal hydride cell for electric-vehicle applications. Ovshinsky et al.[8] recently discussed Ovonics's efforts to develop a metal hydride battery. Our objectives are to review the fundamentals of operation of the metal hydride electrode, to compare critically the NiOOH–metal hydride cell with competing technologies, and to highlight the remaining technical challenges in its development.

Modern Aspects of Electrochemistry, Number 27, edited by Ralph E. White *et al.* Plenum Press, New York, 1995.

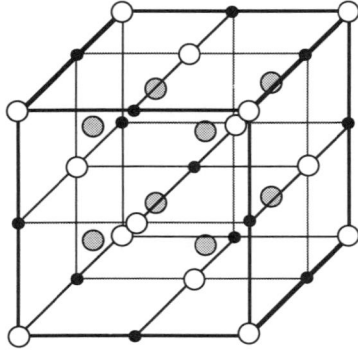

Figure 1. Face-centered cubic (fcc) hydride structure. Open circles are the metal atoms, small solid circles represent hydrogen atoms in octahedral sites, and large shaded circles represent hydrogen in tetrahedral sites.

Hydrides are expected to form when hydrogen is combined with an element less electronegative than itself. For convenience, hydrides are categorized by the nature of the chemical bonding as covalent, ionic, or metallic hydrides. Hydrogen reacts with many transition metals or their alloys to give compounds called metal hydrides, even though there may be no "hydrides" present. The electronic structure of metal hydrides has been elucidated only recently.[9] In general, these alloys are complex, with multiple phases and wide divergences from stoichiometry.

Metal hydrides are a class of solid-state systems called interstitial alloys wherein hydrogen is incorporated into empty sites of a simple metal lattice. Figure 1 shows the insertion of hydrogen in a typical alloy; hydrogen is found in tetrahedral or octahedral interstices of a face-centered-cubic metal structure, for example. Often, only partial occupancy is thermodynamically favored. Nonetheless, high hydrogen densities can be obtained in these metals, leading to their application for hydrogen storage and reactor moderator materials.[10] The insertion of hydrogen into interstices only slightly distorts the crystal lattice, and hence the reactions can be highly reversible. The absorption of hydrogen in alloys has many similarities with topochemical reactions of lithium in transition-metal oxides and sulfides.[11]

For practical devices, metal hydrides are sought with the following properties: high capacity to store hydrogen, high rates of absorption and desorption, fast kinetics, low cost, light weight, and good stability. This combination is not possible with elemental hydrides. Hydrides formed from pure metals are too stable for use as reversible electrodes.

About 200 intermetallic alloys that form hydrides are known and are sometimes grouped according to their structure. The major groups are

Table 1
Major Classes of Intermetallic Alloys

Group	Example	Structure
AB_5	$LaNi_5$, $MmNi_5$	Hexagonal or Orthorhombic
AB_2	$ZrMn_2$	Laves phases
AB	TiFe	CsCl
A_2B	Ti_2Ni	Complex

listed in Table 1; the minor classes of intermetallic hydrides are given by Buschow et al.[12] and Ivey and Northwood.[13] More detailed information can also be found in Refs. 14 and 15. In Table 1, A is an alkaline earth, transition metal, rare-earth, or actinide; B is a transition metal of the iron group. Mm, or misch metal, is a combination of lanthanum and other rare-earth elements. $LaNi_5$ is often referred to as the conventional material. It should be noted that the classifications are only generalizations; typical alloys contain multiple phases and complex structures.

The study of hydrides, in particular hydrogen in metals, is a broad interdisciplinary research topic. Metal hydrides have been the subject of several international symposia.[14–16] We concentrate on the electrochemical applications of metal hydrides, where our expertise lies. Originally, these alloys were used to reduce the pressure of stored hydrogen in the Ni/H_2 cell, a well-established energy-storage system for satellites.[17] More recently, the metal hydride electrode has been considered as a replacement for the cadmium negative electrode in the nickel/cadmium cell. Figure 2 depicts a typical cell; the potential of the cell is about 1.2 V, similar to that of the Ni/Cd cell. The first reversible metal hydride electrode was reported by Justi et al.[18] These cells had poor performance; dramatic improvement in the cycle life and durability were obtained by Willems.[19] The NiOOH–metal hydride cell will be discussed in detail later in this chapter. The status of research and development of the metal hydride electrode has been addressed in earlier reviews.[19,20] We will focus on the recent research and development activities on intermetallic hydrides and the NiOOH–metal hydride battery.

Before analyzing the NiOOH–metal hydride cell, we will discuss the principles of operation of metal hydride electrodes. The fundamentals of thermodynamics, kinetics, and transport will provide us with the tools to

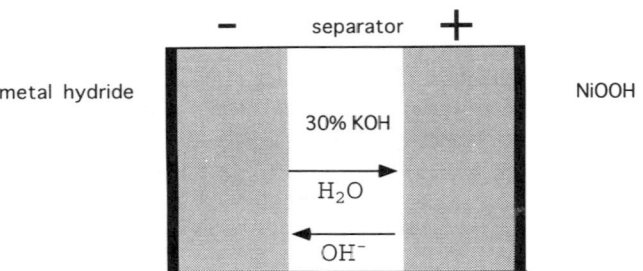

Figure 2. NiOOH–metal hydride cell. The cell potential is about 1.2 V. There is no net production or consumption of water. During discharge, hydroxyl ions are transported from the positive to the negative electrode, and water is transported from negative to positive, resulting in a shuttle of hydrogen between electrodes.

identify the keys to improving the performance of metal hydride electrodes. In addition, the equations presented would be the basis for a detailed mathematical model, which could ultimately be used to design and scale up battery systems.

II. FUNDAMENTALS

1. Thermodynamics

First, let us consider the open-circuit potential of the cell shown below. Here, we measure the potential of the metal hydride electrode with a mercury–mercuric oxide electrode.

α	β	δ	ε	β'	α'
Pt(s)	M(s) H_2	KOH H_2O	HgO(s)	Hg(l)	Pt(s)

The cell potential is

$$FU = -F(\Phi^\alpha - \Phi^{\alpha'}) = \mu_e^\alpha - \mu_e^{\alpha'} \quad (1)$$

At equilibrium, we can equate the electrochemical potential of species in adjacent phases. For the mercury–mercuric oxide electrode,

$$Hg + 2OH^- \Leftrightarrow HgO + H_2O + 2e^- \quad (2)$$

Metal Hydride Electrodes

Thus,

$$\mu_{Hg}^{\beta'} + 2\mu_{OH^-}^{\delta} = \mu_{HgO}^{\varepsilon} + \mu_{H_2O}^{\delta} + 2\mu_e^{\alpha'} \qquad (3)$$

For the metal hydride, two reactions are possible:

$$M + H_2O + e^- \Leftrightarrow M \cdot H_{ad} + OH^- \qquad (4)$$

and

$$2M \cdot H_{ad} \Leftrightarrow H_2 + 2M \qquad (5)$$

If reactions (4) and (5) are equilibrated,

$$\mu_e^{\alpha} = \frac{1}{2}\mu_{H_2}^{\beta} + \mu_{OH^-}^{\delta} - \mu_{H_2O}^{\delta} \qquad (6)$$

In effect, the metal hydride electrode functions as a hydrogen electrode in an alkaline medium. We can express the cell potential as

$$U = U^\theta + \frac{RT}{2F} \ln \frac{p_{H_2}^{\beta}}{a_{H_2O}^{\delta}} \qquad (7)$$

where

$$U^\theta = \frac{1}{2F}\left(\mu_{H_2}^* + \mu_{HgO}^o - \mu_{H_2O}^o - \mu_{Hg}^o\right) \approx 0.926 \text{V} \qquad (8)$$

The potential of the cell is related to the fugacity of hydrogen through Eq. (7). These two quantities are plotted against the stoichiometry of a hypothetical hydride in Fig. 3, the shape of the plot being typical of the absorption isotherm for a single-phase system. In the simplest analysis, we would relate the fugacity of hydrogen in the gas phase to a concentration in the metal using a Langmuir or similar isotherm based on Eq. (5). Often, phase changes accompany the absorption of hydrogen, and the isotherms are complex. For a typical two-phase system, a small amount of hydrogen dissolves into the metal, occupying interstitial sites in the metal lattice. This is referred to as the α-phase. Further absorption of hydrogen causes a phase transformation. The new hydrogen-rich phase is called the β-phase. Such a phase transformation results in a plateau in the absorption isotherm. After the phase transformation is complete, the hydrogen pressure increases with increasing hydrogen concentration. The monograph by Lewis,[21] detailing the palladium–hydrogen system, con-

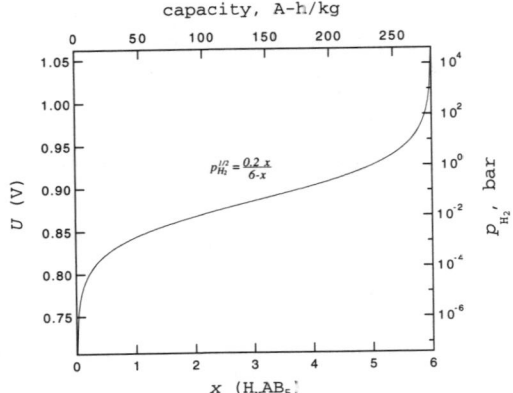

Figure 3. Open-circuit potential of a typical metal hydride electrode. The potential is that of the Hg/HgO reference electrode relative to the metal hydride electrode, and x is related to the capacity, assuming a molecular weight of 590 g/mol for AB_5. The composition in the metal is related to the fugacity of hydrogen in the gas phase, assuming a Langmuir type isotherm.

tains a wealth of information, applicable to most metal hydride systems, and provides a historical perspective of the field.

The adsorption of hydrogen is an exothermic reaction. The effect of temperature on the equilibrium, quantified by the Gibbs–Helmholtz relationship, has been exploited in heat engines using metal hydrides. The effect of temperature on the absorption of hydrogen has been measured in detail for the palladium system.[22] With increasing temperature, the maximum solubility of hydrogen in the α-phase increases, and at the critical point the two phases are miscible. At constant composition, the equilibrium pressure of hydrogen increases with increasing temperature.

One major consideration for the utility of a metal hydride electrode is the hydrogen storage capacity of the alloy. Hydrogen can be stored in gaseous, liquid, or solid forms. Table 2 compares the capacity of some metal hydride electrodes with that of pure hydrogen. The electrochemical capacity is calculated from the molecular weight of the compound for the stoichiometry given in Table 2. The parameter c_t is the maximum concentration of hydrogen in the compound. Although these alloys are capable of storing hydrogen in high concentrations, the capacity per mass is low

Table 2
Hydrogen Storage Capacity

Compound	c_t (mol/dm^3)	Theoretical capacity[a] (A·h/g)
H$_2$ gas, 150 bar	11.6	26.8
Liquid H$_2$, 20 K	66.41	26.8
Liquid CH$_4$	99.6	6.7
LaNi$_5$H$_6$	91.3	0.372
TiFeH$_{1.93}$	100.2	0.501
ZrMn$_2$H$_{3.6}$		0.480

[a] Theoretical capacity calculated from the molecular weight of each compound for the stoichiometry given.

compared to that of pure forms of hydrogen. The capacities listed do not include the mass of the containers used to store the hydrogen.

2. Electrode Kinetics

Because the rates of reaction can limit the rate of charge or discharge of the metal hydride electrode, the surface or kinetic properties of the alloys are important. The main reactions for metal hydride electrodes are:

$$M + H_2O + e^- \Leftrightarrow M \cdot H_{ad} + OH^- \text{ (Volmer)} \quad (9)$$

$$M \cdot H_{ad} + H_2O + e^- \Leftrightarrow M + H_2 + OH^- \text{ (Heyrovský)} \quad (10)$$

$$2M \cdot H_{ad} \Leftrightarrow H_2 + 2M \text{ (Tafel)} \quad (11)$$

and

$$M \cdot H_{ad} \Leftrightarrow M \cdot H_{ab} \quad (12)$$

The last reaction represents the diffusion of hydrogen from the surface of the particle into the metal lattice. These reactions are fundamental to the hydrogen electrode and have been studied extensively.[23,24] For the metal hydride electrode, along with the above reactions, there are self-discharge, recombination, over-discharge, and corrosion reactions; these are important to the operation of the cell and will be considered in more detail in the analysis of the NiOOH–metal hydride battery. The overall reaction at the electrode is

$$M + H_2O + e^- \underset{\text{discharge}}{\overset{\text{charge}}{\rightleftharpoons}} MH + OH^- \qquad (13)$$

The rates of electron transfer reactions are commonly described with the Butler–Volmer equation:

$$i = i_o\left\{\exp\left[\frac{(1-\beta)nF}{RT}\eta_s\right] - \exp\left[\frac{-\beta nF}{RT}\eta_s\right]\right\} \qquad (14)$$

The kinetics of the reaction may be limiting if it is slow compared to the ohmic resistance or when compared to diffusion of hydrogen in the particles.

The catalytic properties of the alloy are critical to the battery designer. Some reactions that are thermodynamically favorable may proceed at such a slow rate as to be negligible, with desirable or undesirable effects. For example, if the kinetics for recombination of atomic hydrogen (Eq. 11) is poor, then substantial amounts of hydrogen may be electrochemically inserted into the lattice without significantly raising the cell pressure (a favorable situation), assuming the cell is discharged shortly thereafter. This also suggests that if hydrogen is generated on the metal hydride electrode during overcharge, it will be slow to be absorbed into the metal hydride electrode. The area of the electrode is important for heterogeneous reactions. Often these alloys are activated; that is, the alloy is cycled several times to break the metal into smaller fragments. The finely divided alloys have a high specific area, reducing the kinetic resistance. Thus, they are termed "activated" because of their ability to absorb and desorb hydrogen rapidly.

Schlapbach[25] has reviewed surface properties of metals with respect to the formation of metal hydrides. Flanagan[26] pointed out the difficulties of kinetic measurements and interpretation of results for two-phase alloys. Notten and Einerhand measured the exchange current densities for several AB_5 materials.[27] Because the alloys are in a strongly oxidizing environment, the surface may be passivated. A metal oxide film could form, inhibiting the adsorption of hydrogen and catalyzing other reactions. The rate of mass transfer can also limit the performance of the electrode and is discussed below.

3. Transport Phenomena

Transport phenomena in the metal hydride particles as well as in the electrolyte can affect the performance of the metal hydride electrode. The metal hydride electrode is similar to the lithium ion insertion systems. The interested reader is referred to the work of Doyle and co-workers[28,29] or of Newman.[30]

Transport of hydrogen in the alloy is typically analyzed with the diffusion equation

$$\frac{\partial c_s}{\partial t} = -\nabla \cdot N \tag{15}$$

The relevant physical property for the solid particle is the diffusion coefficient of hydrogen in the metal lattice. Conway and Wojtowicz[31] investigated the maximum rate of charge or discharge in systems limited by solid-state diffusion in a single-phase system. Züchner et al.[32] used the current-pulse method to measure the diffusion coefficient in the α-phase of H in LaNi$_5$. They found that the diffusion coefficient depended on the crystal orientation. There are often phase changes associated with insertion, and the alloys have complex structures, suggesting that simple diffusion may be a severe approximation. Li et al.[33] presented a relationship between discharge capacity and current density for two-phase AB$_5$ materials, assuming diffusion of hydrogen is rate limiting. Sakamoto et al.[34] modeled diffusion in the metal particles, treating diffusion in the two phases of the palladium system, and compared the calculated potentials of the electrode with experimental values. Pollard and Newman[35] developed a shrinking core model for a two-phase alloy. Reilly et al.[36] used a shrinking core model to describe the isothermal decomposition of LaNi$_5$, assuming that the phase change was rate limiting.

For the LaNi$_5$ electrode, Willems[19] compared the time of discharge to the time of diffusion in the particles and found no transport limitations for his system. In battery applications, the metal hydride electrode would have a porous structure, and this ratio of time constants can be generalized:

$$S_s = \frac{R_s^2 I}{D_s n F (1 - \varepsilon) c_t \delta} \tag{16}$$

Tiedemann and Newman[37] developed a similar parameter. The time constant for diffusion in the solid is given by $\tau = R_s^2/D_s$. This assumes spherical particles, where R_s is the particle radius. The time for a galvanostatic charge or discharge is calculated from the amount of hydrogen that can be absorbed into the electrode and the current density. I is the separator current density, and ε is the porosity of the electrode. For small particles ($S_s \ll 1$), the time for diffusion is small compared to the time of discharge, and diffusion would not be limiting.

The typical electrolyte is concentrated potassium hydroxide. For such a binary electrolyte, the important transport properties are the electrical conductivity (κ), the transference number of hydroxyl ions (t_-^0), and the diffusion coefficient of the salt (D). Potassium hydroxide is widely used in other cells and is reasonably well characterized. The fluxes can be represented by

$$N_+ = c_+ v^\square - \varepsilon D \nabla c_+ + \frac{(1-t_-^0)}{z_+ F} i_2 \tag{17}$$

and

$$N_0 = c_0 v^\square - \varepsilon D \nabla c_0 \tag{18}$$

These equations are for a concentrated binary electrolyte, using a volume-average velocity.[38,39] At steady state, the flux of cations is zero, and a concentration gradient develops even though there is no net consumption of electrolyte. Although steady states, other than $I = 0$, are not obtainable in battery systems, concentration gradients will develop during charge and discharge.

There are two important issues regarding electrolyte transport. First, we can compare the time of discharge with the diffusion time for the electrolyte as we did above for the particles. For a cell of thickness L,

$$S_e = \frac{L^2 I}{D n F (1-\varepsilon) c_t \delta} \tag{19}$$

This expression is based on the time to discharge the cell completely and tells us whether there is sufficient time for concentration gradients to be established. We can also examine a short pulse of high current. Although a cell may not be able to sustain a large current density over the full discharge, it may be able to deliver a short pulse of power without significant concentration polarization. Second, we can estimate the magnitude of the concentration difference. This will be determined by the transference number of OH$^-$, the diffusion coefficient of potassium hydroxide in the solvent, and the current density:

$$\Delta c \approx \frac{(1-t_-^0) L I}{z_+ F D} \tag{20}$$

This approximation, obtained from Eq. (17) with $N_+ = 0$, suggests that the larger the anion transference number, the smaller the concentration gradients.

The potential Φ_2 in the electrolyte, measured with an Hg–HgO electrode in solution, is related to the superficial current density in the pore system by

$$i_2 = -\kappa \nabla \Phi_2 - \frac{\kappa}{F}\left(t_+^0 + \frac{c}{2c_o}\right)\nabla \mu_e \qquad (21)$$

This is a modified form of Ohm's law, accounting for concentration polarization. Similarly, for the electronically conducting solid phase,

$$i_1 = -\sigma \nabla \Phi_1 \qquad (22)$$

One advantage of these metal electrodes is that the electrodes have high electronic conductivity. This is in contrast to the NiOOH electrode, which forms a poorly conducting hydroxide.

Vitanen[40] developed a mathematical model of the metal hydride electrode but did not examine the performance of the electrode in detail. White and co-workers[41,42] have modeled the nickel/hydrogen cell. Heat transfer in metal hydride beds has been studied extensively. Thermal effects are frequently neglected for electrochemical applications, but undoubtedly they play an important role and should be studied in detail. Bernardi et al.[43] presented a general energy balance for battery systems.

III. ANALYSIS

1. The NiOOH–Metal Hydride Cell

Clearly, one of the most promising applications of metal hydrides is the NiOOH–MH cell. The potential of the cell is about the same as that of the Ni/Cd cell, but the metal hydride electrode has a greater energy density. The major advantage, however, is that the battery does not contain toxic materials. Other advantages of this cell are that no dendrites are formed, the main reactions do not involve precipitation or dissolution, and there is no net consumption or production of water. The NiOOH electrode has been studied extensively[44,45]; we will review its operation only in connection with the NIOOH–metal hydride cell. The negative metal hydride electrode is typically made by compressing the metal hydride powder onto a nickel grid. The negative and positive electrodes and separator are then impregnated with aqueous potassium hydroxide. As was mentioned in the preceding section, this cell has many similarities to the dual lithium ion insertion cells. Here hydrogen is inserted and deinserted from the metal electrodes. In the NiOOH–metal hydride cell, in contrast to the lithium systems, hydrogen ions are not available for transport in the alkaline medium. On discharge, for example, hydrogen from the metal combines with hydroxyl ions to produce water (Eq. 13). The net effect is that the hydroxyl ions move from the positive (NiOOH) to the negative (metal hydride) electrode, and water is transported from negative to positive, resulting in a shuttle of hydrogen between the two electrodes.

The success of the metal hydride electrode depends on a few critical issues:

- Capacity to absorb hydrogen reversibly
- Cycle life and mechanical stability of the electrode
- Nominal operating pressure
- Cost

The amount of hydrogen absorbed by the hydride determines the specific energy of the electrode. The absorption must be rapid and reversible. Although there are numerous factors (discussed in more detail below) that affect the characteristics listed above, there are two avenues to enhance the performance: modifying the intermetallic alloys or improving the cell design and control of the operating conditions.

2. Optimization of Alloys

The composition of the alloy is the most important factor in the performance of the cell, and optimization of the alloy compositions is an active area of research.[46–48] There have been two approaches to improving the characteristics of metal hydride materials: modification of existing alloys and development of new alloys. Most current work centers on the conventional AB_5 materials and AB_2 compounds. The goals of alloy modification are:

- Increasing the capacity to absorb hydrogen reversibly
- Reducing the rate of corrosion
- Controlling the catalytic/surface properties
- Reducing the cost of raw materials and processing
- Lowering the equilibrium hydrogen pressure

For the conventional lanthanum-based materials, numerous combinations of metals have been tried, trade-offs being made between the desired properties: high capacity for hydrogen absorption, minimal expansion of the metal lattice, resistance to corrosion, and highly catalytic surfaces. $LaNi_5$ is a conventional metal hydride electrode and has good hydrogen absorption capacity but poor stability. Attempts to improve the stability generally result in lower capacity and poor kinetics. Small amounts of Al and Si were found to improve the stability dramatically[49]; replacing half the nickel with cobalt improved the cycle life.[20] The cobalt reduces the expansion of the lattice, and the Al and Si help to form a more protective layer on the surface of the alloy. Misch metals are commonly used in place of pure elements. These are a mixture of naturally occurring rare-earth elements (Ce, La, Nd, Pr), often added as they are found in nature because of the difficulty and expense of separation. Tadokoro et al.[50] varied the value of x between 3.85 and 5.56 in $Mm(Ni-Co-Mn-Al)_x$ to optimize the hydrogen absorption capacity. They found that with lower values of x the equilibrium pressure was lower, but the capacity diminished. At the same time they observed minimal loss of capacity on cycling. Alloys with Mo, B, Ta, W, and Zr improved the discharge characteristics because of increased surface area and better catalytic properties. Notten and Einerhand[27] reported improved catalytic activity with the addition of palladium to AB_5 materials and with double-phase alloys.

An alternative approach is to use the laves phase (AB_2) alloys; there are several strategies that may be employed to improve the characteristics of these intermetallic alloys. Moriwaki et al.[51] attempted to optimize the composition of $ZrMn_{0.6}Cr_{0.2}Ni_{1.2}$ by adjusting the A:B ratio and the Mn:Cr ratio. Fetcenko et al.[52] compared and contrasted the AB_5 and AB_2 materials and outlined their approach of introducing compositional and structural disorder to identify improved electrode materials. By adding different elements, the capacity to store hydrogen, bond strength, catalytic properties, and corrosion resistance of the alloys can be modified.

Sakai et al.[53] found that heat treatment was effective in prolonging cycle life, suggesting that the microstructure of the alloy is important. Thin films of alloys were found to behave differently depending on how the films were prepared; however, Sakai et al.[54] concluded that the thin films were unsuitable for common applications. Other methods to improve the electrodes include preactivation of the alloy, surface pretreatment, and the use of composite electrodes. Petrov et al.[55] added binder and carbon to the electrode, attempting to improve the performance.

Table 3
Theoretical Energy Densities of Selected Batteries

Battery	Theoretical energy density (W·h/kg)
Pb–PbO$_2$	161
Cd–NiOOH	209
LaNi$_5$H$_6$–NiOOH	216
H$_2$–NiOOH	378
LaNi$_5$H$_6$–air	458
Li–Li$_y$Mn$_2$O$_4$	478 (0.2 < y < 1)
LiC$_6$–Li$_y$CoO$_2$	644 (0 < y > 1)
Na/S	758

For the NiOOH–metal hydride cell, increasing the capacity of the alloy (the negative electrode) results in progressively smaller increases in the specific energy of the cell when the mass of other battery components is considered. Table 3 gives theoretical energy densities for several systems. Ovshinsky et al.[8] reported a theoretical capacity of 400 A·h/kg for current alloys and projected much larger capacities in the near future. The capacity of the NiOOH electrode is about 300 A·h/kg. This, of course, limits the maximum energy density obtainable for the NiOOH–metal hydride battery. Furthermore, the equilibrium hydrogen pressure must be considered. The more hydrogen that is inserted, the greater the equilibrium pressure; consequently not all the capacity may be accessible without raising the cell pressure excessively.

3. Stability

The importance of the corrosion resistance of the alloys cannot be overstated. The stability of the alloys directly affects the cycle life of the electrode. The goal of the USABC is to obtain 500 to 1000 cycles. It is well known that many of the hydride phases are not thermodynamically stable. Nevertheless, the kinetics are slow at ambient temperatures, and thus this is not a problem for electrochemical applications. Repeated cycling stresses the alloy and tends to break the particles into small fragments (pulverization). The smaller particles have a greater surface area available for adsorption, there could be loss of contact between particles, and the rate of corrosion may increase as well.

Notten and Hokkeling[56] gave the corrosion stability of several AB_5 systems in strongly alkaline solutions and discussed the kinetics of the electrochemical reactions. Willems[19] showed that the capacity of $LaNi_5$ (without additives) decreased rapidly with cycling and that $La(OH)_3$ was formed. Other concerns with corrosion of the alloys are loss of electrolyte through irreversible reactions and changing of the balance of the electrodes (positive-to-negative capacity ratio). Oxygen generated at the nickel electrode, a mechanism of self-discharge, may oxidize the metal hydride electrode. Markin et al.[56a] have suggested that the alloy can be protected by coating the surface with Pt black, catalyzing the recombination reaction given in Eq. (25) below. Sakai et al.[57] microencapsulated the alloy in porous copper or nickel, protecting the alloy from attack by oxygen. The copper increases the effective electronic conductivity of the solid matrix, acting as a microcurrent collector, and the microencapsulation also increases the thermal conductivity of the electrodes.

4. Cell Design

The battery designer has substantial control over the performance of the cell. Although a better design does not increase the theoretical capacity to absorb hydrogen or reduce the equilibrium pressure of hydrogen, the operating conditions and design of the battery can be modified to increase the utilization of active material and to reduce the maximum pressure of the cell during charge, for example. Some of the variables that should be considered are:

- Particle size
- Use of partially electrolyte-starved electrodes
- Thickness of electrodes and separator
- Positive-to-negative-electrode capacity ratio
- Temperature of operation and thermal control
- Rates of charge/discharge and charge termination
- Porosity of the electrode

5. Cell Pressure

Commercial success depends on the amount of hydrogen that can be reversibly cycled without loss of capacity or excessive buildup in pressure of hydrogen. A completely sealed system is required for most applications, and a gauge pressure above about 1 bar will require a metal casing, adding significant weight to the system. If excess pressure is vented, water is lost from the cell, and the relative states of charge of the electrodes change.

When the metal hydride is coupled with a NiOOH positive electrode, the following reactions must also be considered. If the cell is overcharged, oxygen can be generated at the NiOOH electrode,

$$4OH^- \rightarrow O_2 + 2H_2O + 4e^- \quad (0.401 \text{ V}) \tag{23}$$

and hydrogen at the metal hydride electrode,

$$H_2O + e^- \rightarrow \frac{1}{2} H_2 + OH^- \quad (-0.828 \text{ V}) \tag{24}$$

There are also recombination reactions, in an oxygen atmosphere, at the metal hydride electrode,

$$\frac{1}{2} O_2 + 2M \cdot H_{ad} \rightarrow H_2O + 2M \tag{25}$$

and for a hydrogen atmosphere, at the positive electrode,

$$\frac{1}{2} H_2 + NiOOH \rightarrow Ni(OH)_2 \tag{26}$$

For recombination to occur, the gases must have access to the electrode surface, and the kinetics should be facile. The critical issues are the rate of mass transfer of H_2 or O_2, the solubility of the gases in the electrolyte for immersed electrodes, and the catalytic properties of the surface. Hydrogen pressure can build up when the rate of transport in the particles or the kinetics is slow or when the cell is overcharged. If the rate of the reaction given by Eq. (9) is greater than the rate of diffusion into the alloy, i.e., $S_s > 1$, the concentration of hydrogen at the surface of the particle will increase. When the maximum surface concentration is reached, hydrogen gas will be evolved (Eq. 24). An analogous situation occurs at the NiOOH electrode, generating oxygen. If the rate of oxygen or hydrogen generation is greater than the rate of recombination, the cell pressure will rise.

Overcharge mechanisms are important to the operation of the cell, particularly for cells in series. It is generally recognized that the recombination of hydrogen at the nickel electrode has poor kinetics and that hydrogen generation is not a practical approach for overcharge protection. Consequently, one must rely on oxygen recombination (generated at the NiOOH electrode) for an overcharge mechanism. Making the capacity of the metal hydride electrode larger than that of the positive electrode will promote oxygen generation on overcharge. This also suggests that electrolyte-starved electrodes may be desirable, allowing the oxygen access to the metal hydride electrode if significant overcharge is anticipated. Because of the poor kinetics of hydrogen recombination, any hydrogen produced on overcharge must be reabsorbed into the alloy. This may require limiting the range over which the electrode is cycled. Making the negative electrode larger than the positive electrode should reduce the hydrogen pressure in the cell. Microencapsulation, mentioned above, was also reported to reduce the operating pressure of the cell, presumably because of more effective oxygen reduction and possibly the prevention of hydrogen evolution.

On overdischarge it is possible to form hydrogen at the NiOOH electrode and oxygen at the metal hydride electrode, although these reactions are unlikely in an alkaline environment unless the cell is severely overdischarged, a consideration for multiple cells in series.

6. Self-Discharge

The rate of self-discharge is an important characteristic of the cell. Oxygen and hydrogen can be generated from the following reactions:

$$NiOOH + \frac{1}{2}H_2O \Leftrightarrow Ni(OH)_2 + \frac{1}{4}O_2 \quad (27)$$

$$M \cdot H \Leftrightarrow M + \frac{1}{2}H_2 \quad (28)$$

Iwakura et al.[58] and Sakai et al.[53] reported the self-discharge rates for metal hydride electrodes. They distinguished between reversible and irreversible capacity loss. The reversible losses are from the reactions given above and are larger than the rate of self-discharge for the Ni/Cd cell. As expected, they found that for an alloy with a lower equilibrium hydrogen pressure, the rate of self-discharge is reduced. For a sealed system, hydrogen stored in the alloy may desorb from the metal according to Eq. (28), be transported to the NiOOH electrode, and recombine according to Eq. (26), but the kinetics is slow. Oxygen generated (Eq. 27) can also diffuse to the metal hydride electrode and recombine (Eq. 25), resulting in transport of hydrogen from the negative to the positive electrode.

There are several strategies for reducing the rate of self-discharge. First, a material with a low equilibrium pressure of hydrogen for the desired range over which x (H_xAB_5) is cycled can be selected. When the partial pressure of hydrogen in the cell header is equal to the equilibrium pressure, desorption of hydrogen will stop at equilibrium. Second, a barrier can be placed between the electrodes, preventing the transport of hydrogen to the positive electrode and transport of oxygen to the negative electrode. This will, however, increase the operating pressure of the cell, and is not considered a practical approach.

Fetcenko et al.[59] were able to reduce the rate of self-discharge by modifying the composition of the alloy, using a more inert separator, and using NiOOH electrodes without nitrate impurities. Their results suggest that shuttle mechanisms for self-discharge are present. Nitrate impurities in NiOOH electrodes are known to increase the rate of self-discharge in the Ni/Cd cell, and the nitrates, as well as other impurities, may contribute to the rate of self-discharge. We are not aware of a detailed analysis of shuttle mechanisms in the literature.

7. Competing Technologies

The absence of toxic materials alone will ensure greater acceptance of the NiOOH–metal hydride cell over Ni/Cd cells. Nonetheless, it is useful to compare the advantages and disadvantages of the metal hydride cells by comparison with other batteries. The Ni/Cd cell has a slightly lower specific energy but has a lower rate of self-discharge, greater tolerance for overcharging, better temperature range, and a greater cycle life.[5] The NiOOH–metal hydride cells do not exhibit the "memory effect" seen in Ni/Cd cells (attributed to the cadmium electrode). Metal hydride batteries, however, cannot be charged with the "peak-detect fast chargers" currently used with Ni/Cd cells. Attempts are under way to standardize the size and capacity of the NiOOH–metal hydride cells and to make them compatible with Ni/Cd chargers.[60] Today NiOOH–metal hydride cells cost about twice as much as comparable Ni/Cd cells.

The lead acid cell is a widely used and well-understood secondary battery. The specific energy of the cell is low, and lead is toxic. The sodium sulfur cell has exceptional specific energy but is a high-temperature battery. As a consequence, there are problems with materials corrosion, sealing, and freeze-thaw cycles. Cycle life is short for cells in series, largely attributed to failure of the β''-alumina separators. The H_2–NiOOH cell has a comparable specific energy, but the required supply of pressurized hydrogen presents a safety issue. A sealed system is required to allow for the recombination of hydrogen and oxygen, and the H_2–NiOOH cell is not practical for traction applications. Characteristics of and prospects for other battery systems can be found in the literature.[61]

The NiOOH–metal hydride cell has a much lower specific energy than lithium cells (see Table 3). Although there are some secondary lithium cells on the market, they are not widely used. The lithium cells have two serious drawbacks. First, pure lithium is unstable and reacts irreversibly with the electrolyte. This has been circumvented to some extent by replacing the lithium foil negative electrode with an intercalation material, such as carbon. This concept (dual lithium ion insertion or "rocking-chair" cells) has been used in commercial cells for portable appliances.[62] Second, there is no practical overcharge or overdischarge mechanism in these cells, making control of charging and operation of series of cells more difficult. The lithium rechargeable cell is an active area of research, and improvements can be expected.

8. Assessment of Research Needs

Numerous technical challenges remain in the development of the NiOOH–metal hydride cell, particularly for electric-vehicle applications, where low cost and high performance are required. Some possible future research areas are:

- Detailed mathematical modeling
- Design and scale-up of large batteries
- Analysis of thermal effects
- Optimization of alloy composition
- Charging/discharging multiple cells in series
- Fundamental studies of kinetic mechanisms including overcharge/overdischarge mechanisms
- Development of a positive electrode with a larger specific energy
- Fundamental studies of transport in intermetallic alloys
- Fundamental studies of the state of hydrogen in candidate metal hydride electrodes

Detailed mathematical modeling and design and scale-up considerations are important to the development of metal hydride cells. Electric vehicles, for example, require much larger batteries than commercially available NiOOH–metal hydride cells. Matsumoto et al.[63] identified several problems with the scale-up of NiOOH–metal hydride cells. In particular, they saw high temperatures and high pressures and lower utilization of active material in larger batteries. They postulated that the low utilization was due to the distribution of electrolyte in the cell. Little work has been published on modeling and design of metal hydride electrodes.

Thermal management is critical for the NiOOH–metal hydride cell. Achieving a large number of cycles, low cell pressure, and high specific energy will depend on a good understanding of temperature distributions in large cells. As we have seen, the equilibrium pressure of hydrogen increases with temperature. Physicochemical properties are strongly dependent on temperature. In addition, the charge acceptance of the NiOOH electrode is poor above 35°C. A thorough analysis of thermal effects on the performance of these cells is needed.

Some researchers have investigated solid-state metal hydride batteries requiring a polymer or other solid electrolyte.[64] The following advantages are usually cited: no corrosive solvent is present, the cell can easily be configured into a variety of shapes, the cell is easy to manufacture, and the functions of separator and electrolyte are combined. Some of the possible problems are the low conductivity of polymer electrolytes at room temperature, loss of contact with the electrolyte, and poor overcharge and overdischarge capabilities.

Optimization of the composition of the alloys can be expected to continue as an active area of research. Klein and Salkind[65] estimated the costs of AB_5 alloys. Although not critical for space applications and less critical for small appliances, cost is a key consideration for electric-vehicle applications. They pointed out that major improvement is needed in this area. Research would focus on development of alloys made from raw materials of lower cost and reducing the manufacturing costs.

Because the specific energy of the NiOOH–metal hydride cell is limited by the positive electrode, this cell should be viewed as a candidate for meeting only the midterm goals of the USABC. To meet the long-range goals, the positive electrode must be improved or replaced to achieve a significantly larger specific energy in metal hydride cells.

Individual cells connected in series will inevitably be at different states of charge. Although it is possible to monitor the potential of single cells, it may not be practical to monitor individual cells in a large battery or to provide for charging of individual cells. Overcharge and overdischarge mechanisms provide a means to pass current without altering the state of charge of the cell. A theoretical study of the effect of operating multiple cells in series with and without overcharge mechanisms is warranted. This analysis would also apply to batteries other than the NiOOH–metal hydride cell.

ACKNOWLEDGMENTS

This work was supported by the Assistant Secretary for Energy Efficiency and Renewable Energy, Office of Transportation Technologies, Electric and Hybrid Propulsion Division of the U.S. Department of Energy under Contract No. DE-AC03-76SF00098.

LIST OF SYMBOLS

c_i Concentration of species i, mol/m^3
D Diffusion coefficient, m^2/s
F Faraday's constant, 96,487 C/mol
i Current density, A/m^2
I Superficial current density, A/m^2
L Width of cell, m
n Number of electrons transferred
N_i Flux of species i, mol/(m^2·s)
p Pressure or fugacity, bar
R_s Radius of solid particles, m
S Dimensionless ratio defined in Eqs. (16) and (19)
t_i^o Transference number of species i
U Open-circuit potential, V
v Velocity, m/s
z_+ Charge number
δ Thickness of electrode, m
Φ electrical potential, V
κ Electrical conductivity, S/m
μ_i Electrochemical potential of species i, J/mol
ε Porosity of electrode

Subscripts

s Solid phase
t Maximum concentration
$+$ Cation
$-$ Anion

o   Solvent
e   Electrolyte
1   Solid phase
2   Electrolyte phase

Superscripts

θ Standard cell potential
*   Secondary reference state
o   Pure component
□   Volume-average quantity

REFERENCES

[1] A. F. Andresen and A. J. Maeland, eds., *Hydrides for Energy Storage*, Proceeding of an International Symposium, International Association for Hydrogen Energy, Pergamon Press, Oxford, 1978.
[2] F. E. Lynch, *J. Less-Common Met.* **172–174** (1991) 943.
[3] D. Malinik, *Electronics Design* **1992** (July 23) 18.
[4] *PC Magazine* **1991** (May 28) 36.
[5] C. H. Small, *Electronic Design News* **1992** (December 10) 156.
[6] E. J. Cairns, *Interface* **1** (1992) 38.
[7] T. Sakai, H. Miyamura, N. Kuriyama, I. Uehara, M. Muta, A. Takagi, U. Kajiyama, K. Kinoshita, and F. Isogai, *J. Alloys Compounds* **192** (1993) 159.
[8] S. R. Ovshinsky, M. A. Fetcenko, and J. Ross, *Science* **260** (1993) 176.
[9] A. C. Switendick, in *Transition Metal Hydrides*, Ed. by R. Bau, Advances in Chemistry Series, Vol. 167, American Chemical Society, Washington, D.C., 1978, pp. 264–282.
[10] W. M. Mueller, J. P. Blackledge, and G. G. Libowitz, *Metal Hydrides*, Academic Press, New York, 1968.
[11] M. S. Whittingham, *J. Electrochem. Soc.* **123** (1976) 315.
[12] K. H. J. Buschow, P. C. P. Bouten, and A. R. Miedema, *Rep. Prog. Phys.* **45** (1982) 937.
[13] D. G. Ivey and D. O. Northwood, *J. Mater. Sci.* **18** (1983) 321.
[14] L. Schlapbach, ed., *Hydrogen in Intermetallic Compounds II*, Topics in Applied Physics, Vol. 67, Springer-Verlag, Berlin, 1992.
[15] D. A. Corrigan and S. Srinivasan, eds., *Proceedings of the Symposium on Hydrogen Storage Materials, Batteries, and Electrochemistry*, The Electrochemical Society, Pennington, New Jersey, 1992.
[16] Metal-Hydrogen Systems. Fundamentals and Applications, Proceedings of the First International Symposium Combining "Hydrogen in Metals" and "Metal Hydrides," *Z. Phys. Chem. N. F.* **164** (1989) 747.
[17] Z. P. Zagrodnik and K. R. Jones, *J. Power Sources* **36** (1991) 375.
[18] E. W. Justi, H. H. Ewe, A. W. Kalberlah, and N. M. Saridakis, *Energy Conv.* **10** (1970) 183.
[19] J. J. G. Willems, *Philips J. Res.* **39** (Suppl. No. 1) (1984) 1.
[20] H. F. Bittner and C. C. Badcock, *J. Electrochem. Soc.* **130** (1983) 193C.
[21] F. A. Lewis, *The Palladium Hydrogen System*, Academic Press, London, 1967.
[22] H. Frieske and E. Wicke, *Ber. Bunsenges. Phys. Chem.* **77** (1973) 48.

[23] P. K. Subramanyan, in *Comprehensive Treatise of Electrochemistry*, Vol. 4, J. O'M Bockris, B. E. Conway, E. Yeager, and R. E. White, eds., Plenum Press, New York, 1981.

[24] A. N. Frumkin, in *Advances in Electrochemistry and Electrochemical Engineering*, Vol. 3, C. W. Tobias and P. Delahay, eds., Wiley-Interscience, New York, 1963, pp. 287–391.

[25] L. Schlapbach, in *Hydrogen in Intermetallic Compounds II*, Topics in Applied Physics, Vol. 67, L. Schlapbach, ed., Springer-Verlag, Berlin, 1992, pp. 15–95.

[26] T. B. Flanagan, in *Hydrides for Energy Storage*, A. F. Andresen and A. J. Maeland, eds., Proceeding of an International Symposium, International Association for Hydrogen Energy, Pergamon Press, Oxford, 1978.

[27] P. H. L. Notten and R. E. F. Einerhand, *Adv. Mater.* **3** (1991) 343.

[28] M. Doyle, T. F. Fuller, and J. Newman, *J. Electrochem. Soc.* **140** (1993) 1526.

[29] T. F. Fuller, M. Doyle, and J. Newman, *J. Electrochem. Soc.* **141** (1994) 1.

[30] J. Newman, *Electrochemical Systems*, Prentice-Hall, Englewood Cliffs, New Jersey, 1991.

[31] B. E. Conway and J. Wojtowicz, *J. Electroanal. Chem.* **326** (1993) 277.

[32] H. Züchner, T. Rauf, and R. Hempelmann, *J. Less-Common Met.* **172–174** (1991) 611.

[33] Z. Li, Y. Lei, C. Chen, J. Wu, and Q. Wang, *J. Less-Common Met.* **172–174** (1991) 1260.

[34] Y. Sakamoto, K. Kuruma, and Y. Naritoma, *Ber. Bunsenges. Phys. Chem.* **96** (1992) 1813.

[35] R. Pollard and J. Newman, *J. Electrochem. Soc.* **128** (1981) 491.

[36] J. J. Reilly, Y. Josephy, and J. R. Johnson, *Z. Phys. Chem. N. F.* **164** (1989) 1241.

[37] W. Tiedemann and J. Newman, *J. Electrochem. Soc.* **122** (1975) 1482.

[38] J. Newman and T. W. Chapman, *AIChE J.* **19** (1973) 343.

[39] J. Newman and W. Tiedemann, *AIChE J.* **21** (1975) 25.

[40] M. Vitanen, *J. Electrochem. Soc.* **140** (1993) 936.

[41] J. B. Kim, T. V. Nguyen, and R. E. White, *J. Electrochem. Soc.* **139** (1992) 2781.

[42] D. Y. Fan and R. E. White, *J. Electrochem. Soc.* **138** (1991) 2952.

[43] D. Bernardi, E. Pawlikowski, and J. Newman, *J. Electrochem. Soc.* **132** (1985) 5.

[44] J. McBreen, in *Modern Aspects of Electrochemistry*, No. 21, R. E. White, J. O'M Bockris, and B. E. Conway, eds., Plenum Press, New York, 1991.

[45] D. A. Corrigan and A. H. Zimmerman, eds., *Proceedings of the Symposium on Nickel Hydroxide Electrodes*, The Electrochemical Society, Pennington, New Jersey, 1990.

[46] T. Sakai, T. Hazama, H. Miyamura, N. Kuriyama, A. Kato, and H. Ishikawa, *J. Less-Common Met.* **172–174** (1991) 1175.

[47] M. Tadokoro, M. Nogami, Y. Chihano, M. Kimoto, T. Ise, K. Nishio, and N. Furukawa, *J. Alloys Compounds* **192** (1993) 179.

[48] T. Sakai, H. Miyamura, N. Kuriyama, A. Kato, K. Oguro, and H. Ishikawa, *J. Electrochem. Soc.* **137** (1990) 795.

[49] T. Sakai, H. Miyamura, N. Kuriyama, A. Kato, K. Oguro, H. Ishikawa, and C. Iwakura, *J. Less-Common Met.* **159** (1990) 127.

[50] M. Tadokoro, K. Moriwaki, K. Nishio, M. Nogami, N. Inoue, Y. Chihano, M. Kimoto, T. Ise, R. Maeda, F. Mizutaki, Mi Takee, and N. Furukawa, *Proceedings of the Symposium on Hydrogen Storage Materials, Batteries, and Electrochemistry*, D. A.. Corrigan and S. Srinivasan, eds., The Electrochemical Society, Pennington, New Jersey, 1992.

[51] Y. Moriwaki, T. Gamo, H. Seri, and T. Iwaki, *J. Less-Common Met.* **172–174** (1991) 1211.

[52] M. A. Fetcenko, S. Venkatesan, and S. R. Ovshinsky, *Proceedings of the Symposium on Hydrogen Storage Materials, Batteries, and Electrochemistry*, D. A., Corrigan and S. Srinivasan, eds., The Electrochemical Society, Pennington, New Jersey, 1992.

[53] T. Sakai, H. Miyamura, N. Kuriyama, H. Ishikawa, and I. Uehara, *J. Alloys Compounds* **192** (1993) 155.

[54] T. Sakai, H. Ishikawa, H. Miyamura, and N. Kuriyama, *J. Electrochem. Soc.* **138**, 908–915 (1990).

[55] K. Petrov, A. Visitin, S. Srinivasan, and J. Appleby, 183rd Meeting of the Electrochemical Society, Honolulu, Hawaii, 1993, Abstract No. 27.
[56] P. H. L. Notten and P. Hokkeling, *J. Electrochem. Soc.* **138** (1991) 1877.
[56a] T. L. Markin and R. M. Dell, *J. Electroanal Chem.* **118** (1981) 217.
[57] T. Sakai, H. Ishikawa, K. Oguro, C. Iwakura, and H. Yoneyama, *J. Electrochem. Soc.* **134** (1987) 558.
[58] C. Iwakura, Y. Kajiya, H. Yoneyama, T. Sakai, K. Oguro, and H. Ishikawa, *J. Electrochem. Soc.* **136** (1989) 1351.
[59] M. A. Fetcenko, S. Venkatesan, and S. R. Ovshinsky, 183rd Meeting of the Electrochemical Society, Honolulu, Hawaii, 1993, Abstract No. 51.
[60] T. Albano, *PC Magazine* **1993** (January 12) 29.
[61] C. W. Tobias, Chairman, Assessment of Research Needs for Advanced Battery Systems, National Materials Advisory Board Commission on Engineering and Technical Systems, Publication NMAB-390, Washington, D.C., 1982.
[62] *JEC Press.* **1988** (February) 26.
[63] I. Matsumoto, M. Ikoma, A. Ohota, H. Matsuda, and Y. Matsushita, 183rd Meeting of the Electrochemical Society, Honolulu, Hawaii, 1993, Abstract No. 26.
[64] N. Kuriyama, T. Sakai, H. Miyamura, A. Kato, and H. Ishikawa, *J. Electrochem. Soc.* **137** (1990) 355.
[65] M. Klein and A. Salkind, 183rd Meeting of the Electrochemical Society, Honolulu, Hawaii, 1993, Abstract No. 25.

5

Recent Advances in the Study of the Dynamics of Electrode Processes

T. Z. Fahidy and Z. H. Gu

Department of Chemical Engineering, University of Waterloo, Waterloo, Ontario, Canada N2L 3G1

I. INTRODUCTION

Electrochemical systems are an abundant source of dynamic phenomena. Although the first periodic dissolution/deposition of silver metal on iron in an acidic silver nitrate solution was reported as early as 1812,[1] a chapter by Wojtowicz in *Modern Aspects of Electrochemistry* in 1972[2] is the first comprehensive review of early research in this area, to the authors' knowledge. Because of the rapid development of instruments and computer technology, various current/potential patterns ranging from regular oscillations to high chaos have been found and studied in recent years. A great deal of attention has been paid to metal dissolution, particularly dissolution of copper, iron, cobalt, and nickel.[3] Metal deposition processes have also been studied in detail.[4] Dynamic behavior has also been demonstrated in various organic electrochemical systems[5] and gaseous reactions.[6]

While the exact mechanistic nature of electrochemical dynamics is not fully understood, it is believed that physical and chemical characteristics of the electrode surface and films formed on it play an essential role. In studying surface phenomena, classical techniques, for example, scanning electron microscopy (SEM) and X rays, are increasingly complemented by modern methods, such as direct visualization associated

Modern Aspects of Electrochemistry, Number 27, edited by Ralph E. White *et al.* Plenum Press, New York, 1995.

with image processing and analysis[7] and Raman spectroscopy.[8] Very promising for dynamic studies is the considerable progress achieved in experimental data acquisition and analysis. Current/potential variations have been recorded at sampling rates as high as 50 kHz,[9] and oscillation patterns are routinely analyzed by means of modern nonlinear techniques, such as phase portraits and return maps[3] and fractional Brownian motion (FBM) theory.[10]

This chapter is intended to present an overview of various experimental and theoretical explorations of dynamic phenomena in electrochemical systems, with emphasis on publications in the last five years. The paper entitled "Electrochemical Reaction Dynamics: A Review" in Vol. 49, no. 10, of *Chemical Engineering Science* by T. T. Tsotsis and J. Hudson is a particularly important reference to acknowledge in this respect. The existence of the accepted but then still unpublished manuscript was brought to the authors' attention at the very final stage of preparing this chapter.

II. ANODIC PROCESSES

1. The Oxidation of Copper

(i) In Acidic NaCl

Anodic dissolution of copper in acidic NaCl electrolyte has been studied over a long time period.[3] The apparent mechanism involves initial precipitation of a porous CuCl layer on the electrode surface. The resulting locally high current densities and Cu^+ concentration trigger a strong migration of H^+ ions, and hence an increase in pH, which leads to Cu_2O formation. The formation of Cu_2O decreases the local pH and causes partial dissolution of the precipitate on the surface. This cyclic formation/deformation of surface films produces periodic current. The kinetics, mechanism, and mathematical formulation of copper dissolution processes have been the subject of studies by Nobe and co-workers.[11–14] Various oscillatory patterns and instabilities have been found experimentally in a single dissolution reaction.[15] The existence of the sequence 1-band chaos, 2-band chaos, double torus, torus, limit cycle has been demonstrated, and instability has been attributed to the slowly varying thickness of the surface film. The effect of NaCl concentration (0.1, 0.3 and $0.5M$), the speed of electrode rotation (200 and 1000 rpm), and imposed anode potential (100–1100 mV vs. SCE) on the oscillatory behavior have been examined.[16,17] Morphological aspects of the electrode surfaces were studied using SEM, and it appears that surface morphology and film thickness, composition, and structure are closely associated with particular oscillation patterns. The pH of the electrolyte significantly influences the dynamic behavior; no oscillation in neutral NaCl solutions has been observed by these authors.

(ii) In NaCl/KSCN

While no oscillation has been reported in concentrated neutral NaCl electrolytes, the addition of small amounts of KSCN (0.3 to 3mM) to such solutions can generate current oscillations at certain potentials at pH 7.[18] These oscillations exist in the transition region between apparent charge transfer and mass transport control. Oscillatory characteristics depend on the potential and imposed magnetic field, and, in an appropriate combination, externally imposed magnetic fields can rapidly destroy an oscillation. Direct flow visualization techniques indicate the appearance of a white CuSCN precipitate in the vicinity of the copper anode during oscillations, implying that saturation of CuSCN in the neighborhood of the anode is an essential condition for oscillation. In the absence of magnetic fields, the oscillation patterns are relatively regular and can be described via a truncated Fourier expansion. However, the imposition of a magnetic field greatly increases structure irregularity.[19] Over a short time period, various combinations of the electric and magnetic fields generate five typical oscillatory patterns. Over a long time period, the oscillations in magnetic fields show much less intercorrelation and higher nonstationarity.[10,20] The effect of the potential on oscillation has been systematically examined using a potential step as small as 5 mV in the range of –60 mV to –10 mV (vs. SCE).[8] The oscillation structure has been characterized by Hurst exponents based on FBM theory (reviewed in Section IV). Surface-enhanced Raman spectroscopy (SERS) indicates the existence of a black film without and with gray-white particles. The former contains only Cu_2O and some CuO, whereas the latter also contains KSCN. Oscillation occurs only in the presence of gray-white particles in the film, and hence the presence of KSCN in the anodic film is a necessary condition for oscillation in this system. The effects of the chemical composition of the electrolyte (e.g., SCN^-, Cl^-) and the pH on the oscillation behavior have also been investigated.[21]

Direct visualization combined with image processing and analysis demonstrates a high correlation between the anodic film growth and the pattern of current oscillation[7,22,23]: the onset of oscillation coincides with the time instant at which the black film fully covers the anode. The onset of oscillation can be substantially delayed by an external magnetic field that causes premature "stripping" of the anode film. The growth of the anode film can be closely modeled by the analogy of first-order reaction kinetics.

There is also experimental evidence for the occurrence of oscillatory anodic copper dissolution in *acidic* NaCl/KSCN solutions within an appropriate range of SCN^- ion concentration.[23] The mechanism of such oscillation in this system seems to be more complicated than in SCN^--free NaCl solutions. This remains an interesting subject matter to be explored.

(iii) In H_3PO_3

Copper dissolving anodically into H_3PO_3 solution is of practical importance in the metal polishing industry. Using a rotating copper disk anode in a solution containing 85% H_3PO_3, Albahadily and co-workers[24,25] found that potentiostatically induced current oscillations pass through a Hopf bifurcation or exhibit dual (mixed large- and small-amplitude) pattern and that the set of mixed-mode oscillations constitutes a Farey sequence. (A Farey sequence is a periodic sequence for which a one-to-one correspondence exists with an ordered sequence of rational noise.) A chemical disproportionation step is considered as part of a feedback mechanism for dynamic behavior.

Using an *in situ* ellipsometric technique associated with potentiostatic measurements, Tsitsopoulos *et al.*[26] monitored the copper surface and electrochemical reactions at various stages of anodization. In the oscillatory region, both thickness and optical properties of the anodic film were found to oscillate in phase with current density. Other techniques [e.g., Auger electron spectroscopy (AES), photoelectron spectroscopy (XPS), and SEM/energy dispersive analysis of X rays (EDAX)] are currently being employed in this work.

(iv) In Other Media

The oscillatory anodic dissolution of copper in acetate buffer solution was studied by Dewald *et al.*[27] in terms of surface reaction, film formation/dissolution, and mass transfer effects. The amplitude of the oscillation was found to increase with increasing potentials, but the frequency decreased with increasing potentials and increased with an increase in the rotation rate. Periodic and chaotic current oscillations were also observed in a pH 3.5 sodium acetate/acetic acid buffer solution.[28] The oscillations arise after the formation and dissolution of an acetate salt film preceding oxide passivation. The nature and composition of the surface film were studied via SEM and X-ray powder diffraction.

Uniform and sustained oscillatory anodic potentials were observed in the galvanostatic dissolution of copper in chlorate electrolyte[29] in the current density range of 0.3–150 A/cm^2, and the effect of current density, electrolyte temperature, acidity, and flow rate on the frequencies and amplitudes were analyzed. The rise and fall of the potential were attributed to the alternate growth and destruction of an adherent cuprous oxide surface layer.

2. The Oxidation of Iron

(i) In H_2SO_4

The dynamic dissolution of iron in H_2SO_4 has been extensively studied. Using rotating disk and hemispherical electrodes, current oscillations were observed along the limiting current plateau.[30] The frequency of the oscillations is proportional to the square root of rotation speed for the disk electrode, but in the case of a hemispherical electrode the frequency of oscillations is less dependent on the rotation speed. A mathematical model of sustained oscillations[31] assumes that the oscillations are caused by a continuous cycling of a portion of the electrode between active and passive states of the electrode surface covered by a porous $FeSO_4$ film. Transient changes in the potential and the concentration profile in the pores of the salt film and in the diffusion layer are responsible for continuous cycling.

In the experimental study of iron dissolution from a disk electrode rotating at 900 rpm in 1M H_2SO_4, Diem and Hudson[32] observed and analyzed chaotic current in terms of space plots, Poincaré sections, and correlation dimension, which varied smoothly between 2.4 and 6.0. The effect of the electrode surface area on chaotic attractor dimensions was studied by Wang and Hudson.[9] The experiments were carried out under potentiostatic conditions by setting constant values of the potential and varying the active area of the electrode. As the electrode area increases, periodic oscillation changes to low-order chaos and then to a high-order oscillation, and the dimension of the attractor increases. In the case of large electrodes at more positive potentials, a secondary higher frequency oscillation was also observed. The effect of coupled electrodes (one, two, and three electrodes embedded in the end of a rotating disk) on the oscillatory patterns has also been demonstrated.[33] The coupling of periodic oscillators leads to chaos, and that of chaotic oscillators to higher chaos. The Franck–FitzHugh (F-F) model (developed for large-amplitude oscillations occurring potentiostatically between active and passive states[34]) was also applied,[35] and results were compared with those of a similar study.[36] The F-F model was recently extended by Pagitsas and Sazou,[37] and a logarithmic (instead of a linear) dependence of the Flade potential (where the current shows a very sharp change in the polarization curve; see Ref. 2) on H^+ concentration was considered in this work. The effect of the transference number of the H^+ ions was also considered to account for charge transport by Fe^{2+} and SO_4^{-2}. In a further effort, a three-dimensional F-F model was established through the introduction of an external periodic-modulated potential, for example, a sinusoidal forced potential.[38] Phase plane, stroboscopic maps, time-delay reconstructions, and Liapunov characteristic exponents have been used for the interpretation of model predictions.

The oscillation of iron is also affected by electrode wetting.[39] A mechanism for passivating film stripping was proposed based on the Marangoni effect.

Impedance techniques have been used in the study of iron corrosion and passivation phenomena on hemispherical Fe electrodes.[40] The experiments include potentiodynamic anodic polarization, current oscillation occurring on limit current plateau at constant potential, and the measurement of AC impedance using a frequency-response analyzer. Experimental results do not compare well with theoretical predictions of the behavior of the $Fe–H_2SO_4$ system.

(ii) In NaCl

Oscillatory anodic dissolution of iron in NaCl has been widely studied by Nobe and co-workers, using rotating ring-disk electrodes.[41,42] The dependence of oscillation characteristics on chloride concentration, potential and current density, and hydrodynamic conditions was obtained from linear potential-sweep and potential/current transient measurements. Current oscillations were observed in both the current plateau and postplateau region, and potential oscillations were observed at current densities higher than the limiting current density. Both oscillations demonstrate various waveforms and are attributed to an alternate buildup and breakdown of a porous, nonprotective $FeCl_2$ and $Fe(OH)_2$ salt film on the Fe surface. The chaotic nature of oscillations was analyzed in terms of power spectral density, phase portrait, Poincaré section, and Liapunov exponent.[43,44] A mathematical model for potential oscillation was also developed. Recently, the effect of benzotriazole (an organic additive) on potential oscillation was studied and explained in terms of a "duplex" inner nonporous layer–outer porous salt film model.[45]

(iii) In $(NH_4)_2SO_4$

The anodic dissolution of iron in ammonium sulfate has been studied by Moina and Posadas.[46,47] Part of the transient behavior can be interpreted via a crystal growth model.

3. The Oxidation of Cobalt

(i) In H_2SO_4

The anodic dissolution of cobalt in the electrolytic system containing $1.58M$ H_2SO_4 and a certain amount of nitrate ions was studied under potentiostatic conditions by Sazou et al.[48] When the potential is fixed within a certain range of the limiting current plateau region, sustained current oscillations are observed. The width of the potential range for oscillation depends on the experimental conditions including disk rotation rate, nitrate and sulfuric acid concentration, and temperature. The onset potential for current oscillation varies linearly with the logarithm of nitrate concentration. The frequency and amplitude of oscillations depend on the potential and rotation speed. A proposed mechanism is based on the oxidizing action of the nitrate ions. The dissolution of cobalt in the absence and presence of chloride ions was also examined.[49] In the presence of chloride ions, the plateau region is extended in the direction of the oxygen evolution potential. Current oscillations appear within the potential range of the limiting current region, and they are related apparently to a "pitting" type corrosion.[50] In the $[Cl^-] = 0.03–0.09M$ region, the current exhibits various waveforms, for example, relaxation, monoperiodic, mixed mode, aperiodic, and chaotic, depending on the potential and rotation speed, which are the major factors affecting the formation and deformation of the electrode film. In the $[Cl^-] = 0.09–0.7M$ range, rotation speed still is important but the potential effect becomes weaker. Iodine ions also exert an influence on current oscillations associated with pitting and corrosion.[51] The nature of the current oscillation depends on iodide ion concentration, the rotation speed of the disk, and the applied potential. Oscillation and pitting corrosion of the partially passive cobalt surface are related.

(ii) In H_3PO_3

Cobalt dissolution in H_3PO_3 systems is also accompanied by oscillation.[52,53] Bistability, hysteresis, and current oscillations of relaxation type have been observed during transition between passive and active states of the Co electrode, and a hysteresis loop is formed by jump transitions between these two states. Current oscillations are attributed to competition between growth and dissolution of oxide films. Oscillatory wave shapes were observed under various electrolytic concentration, rotation speed, and ohmic potential-drop conditions. Current oscillations of the relaxation type can be "reproduced" by the solution of a system of differential equations having two independent parameters with widely different time scales.

(iii) In HCl/H_2CrO_4

Oscillatory cobalt dissolution is caused by the oxidation of cobalt and the reduction of chromic acid.[3] Autonomous potential oscillations have been studied by Franck and Meunier,[54] and the dynamics under potentiostatic conditions have been investigated by Hudson and co-workers[55,56] on a rotating electrode using potential-sweep and potentiostatic experiments. The range of potentials in potentiostatic experiments over which oscillations occur extends to much higher values than in sweep experiments. The effects of rotation speed and coupled electrodes embedded in the disk were also examined. The coupling of electrodes generates a more chaotic behavior.

(iv) In Co^{2+}/Co^{3+} Mixtures

Both biamperometric and bipotentiometric systems with Pt electrodes immersed in solutions of $NaClO_4$, Co^{2+}/Co^{3+}, H_2O, and acetic acid exhibit oscillatory behavior over extended time periods.[57,58] The frequency and amplitude of oscillations depend on the composition of the solution, temperature, and speed of agitation. Reactions occurring at electrodes and in the solutions, which are most probably responsible for oscillations, have also been identified.

4. The Oxidation of Nickel

(i) In H_2SO_4

The dynamics of nickel dissolution in acidic media have been extensively analyzed by Lev and his co-workers in terms of two-dimensional diagrams of complex behavior under galvanostatic conditions.[59,60] Periodic, quasi-periodic, and chaotic states were identified, characterized, and mapped in the plane of current versus acid concentration. At low current, transition to simple periodicity through quasi-periodicity is evident via power spectra and return maps. Chaos is characterized by a high correlation dimension. The domain of simple periodicity is bounded by Hopf and saddle-loop bifurcations. A set of micro-reference electrodes situated close to the nickel wire were employed to explore time-dependent local activity.[61] The electrolytic cell was controlled by a galvanostat and a potentiostat set in parallel. Two types of spatial motion were observed: galvanostatic potential oscillations with standing "activity waves" and almost potentiostatic current oscillations with traveling pulses. A mathematical model was also offered for the interpretation of the observed spatial patterns. A systematic approach for modeling and parameter estimation was also presented, and good qualitative agreement between observed and predicted bifurcation maps was demonstrated.[62] Sustained oscillations were also observed with nickel electrodes containing sulfur.[63]

(ii) In Other Media

Electrodissolution of nickel in the presence of chloride and iodide ions was investigated using low-sweep voltammetry (<2 mV/s).[64] The rates of dissolution are related to the concentration of chloride ions, and the addition of iodide ions modifies dynamical behavior: oscillation occurs over a wide range of potentials at a low iodide concentration, but it is suppressed as the iodide concentration is increased.

5. The Oxidation of Silicon

(i) In Fluoride Electrolyte

Current oscillations accompanying anodic dissolution of silicon in fluoride electrolyte have been investigated by Ozanam et al.[65] These oscillations occur only upon a perturbation in the electrode potential and are associated with the fluctuation of the thickness of an oxide film on the electrode surface. The surface may be envisaged as a juxtaposition of small self-oscillating areas in an appropriate range of potentials. The oscillations are not interrelated at a steady potential, but a perturbation in the potential has a "synthesizing" effect, giving rise ultimately to macroscopic oscillations. Silicon dissolution was also studied via impedance measurements and cyclic voltammetry.[66,67] Under nonresonance conditions, typical characteristics of metal corrosion are observed: (1) at a low positive potential, and past the first local current maximum (region of porous Si generation), impedance is essentially inductive and is due to the roughening of the surface; (2) at more positive potentials, corresponding to the first current plateau (electropolishing region), the oxide layer formed on the Si surface is the main contributor to impedance. At potentials beyond a second local maximum in the current, a steady-state current is reached, but it exhibits "resonance" in the sense that there is no spontaneous oscillation but the current is prone to oscillate if a small potential perturbation is applied. Oscillations decay and the steady-state current is recovered upon the removal of a perturbation. A model that predicts closely the value of the impedance can be constructed from the experimental data. The domain of self-oscillation, hidden oscillation, and synchronization impedance have also been discussed theoretically.[68]

In a study of the anodic dissolution of p-silicon in ammonium fluoride based on ring-disk voltammetry and ellipsometry techniques, the Si(II)/Si(IV) branching ratio was related to the surface condition of the electrode and the overall dissolution rate.[69] The periodic buildup and decay of space charge within the surface oxide layer was suggested as the predominant physical process. Hydrogen production, changes in optical properties of different regimes, and current oscillations observed at high potentials were interrelated. Light-induced oscillating reactions of n-silicon in ammonium fluoride were studied by Lewerenz and Schlichthoerl.[70] Profiles of the simultaneous photocurrent and the excess microwave reflectivity were recorded during light-induced oscillation. At small positive potentials, oscillations were stationary, but at higher positive potentials, irregular oscillations were found. In the initial time period, electronic interface parameters can be tentatively determined via simultaneous measurement of photocurrent and excess microwave reflectivity. At large times, oscillation is related to surface recombination and charge transfer processes.

(ii) In Other Media

The kinetics of growth of an anodic oxide film on silicon in aqueous solutions containing sulfuric, phosphoric, and oxalic acid were analyzed by Parkhutik.[71] Oscillations in the region of galvanostatic anodization and potentiodynamic and potentiostatic kinetics were explained in terms of simultaneous oxide growth and dissolution of the porous oxide. Both n-type and p-type silicon were used as anode, and the chemical composition of the oxide was analyzed by X-ray photoelectron spectroscopy. The observed kinetic behavior was explained via a model based on the porous structure formation of the oxide: the porosity of the oxide is responsible for the penetration of the acid anions into the oxide/electrode interface. The process of porous oxide growth starts with pore nucleation followed by barrier-oxide growth at the bottom of pores. This phenomenon is attributed to an electric-field-dependent intensity of oxidation/dissolution of the pores.

6. The Oxidation of Hydrogen

The oxidation of hydrogen in various acidic solutions containing small amounts of Ag^+, Cu^{2+}, and Pb^{2+} ions is accompanied by sustained potential oscillations in a certain range of the anodic current.[72] The maximum values of oscillating potentials are related linearly to the logarithm of the anodic current but are virtually independent of the concentration of the metal ion. The smaller the bubbling rate of H_2 is, the higher the potential maximum and the longer the time period of oscillation. The coupling of the anodic oxidation of H_2 with Ag deposition and dissolution has been described by a kinetic model.[73] Attractive interaction between Ag adatoms on the Pt electrode can also produce limit cycle behavior.[74] Similar experimental results were found in hydrogen oxidation associated with CO oxidation.[75]

The dynamics of hydrogen oxidation on a platinum electrode was extensively studied in the presence of Cu^{2+} and Cl^-.[6,76] The transition from a steady-state to small-amplitude and then to mixed-mode oscillation depends on the current density and Cu^{2+} concentration. Various bifurcation diagrams were constituted with emphasis on transition from small chaotic attractor-based to mixed-mode oscillations. The effect of the electrode structure on the dynamics was also examined[77,78] via experiments performed on three low-index plane surfaces of platinum. No oscillation was observed with Pt(110) and simple oscillation was observed with Pt(111), but complicated patterns including simple periodic oscillation, period doubling, chaos, and mixed-mode behavior were demonstrated with Pt(100). Investigation of potential dynamics at various levels of current density and Cu^{2+} concentrations indicated three broad regions:

at low current densities, a steady state exists; Small-amplitude oscillations then appear as the current density is increased; and at high current density, typical mixed-mode oscillations were noticed. Depending on Cu^{2+} ion concentration, the transition from the small-amplitude to mixed-mode region originates from a simple-periodic, period-doubled, or chaotic attractor. Various properties of the bifurcation diagrams have also been described.

7. Organic Oxidations

With a steadily expanding analytical knowledge of oscillations, many researchers began to explore the dynamic behavior of organic electrochemical systems in recent years. A survey of the oxidation of small electroactive organic molecules on different catalytic metal electrodes was published in 1990,[79] facilitating the search for new oscillating systems, especially in alkaline solutions. The HCHO/Rh system in alkaline medium, with large and easily reproducible current oscillations, was the subject of intensive studies in which the influence of various experimental parameters on oscillation structure was established. Particularly interesting is the work of Schell and co-workers on the oxidation of formaldehyde and formate/formic acid.[80-82] In their investigation of the oxidation of HCHO at a rotating Pt disk electrode, experiments were carried out under galvanostatic conditions using sulfate acid solutions. As the current was increased slowly, the system moved through a sequence of oscillation states. Most of the states demonstrate temporal patterns made up of large and small oscillations. The existence of peak values in measured waveforms is consistent with the assumption that a strongly absorbed intermediate reacts with hydroxyl radicals. The oxidation process exhibits bistability: high- and low-potential states were found within the same range of current values. Models based on formulated kinetic steps and rate laws predict that potential oscillations accompany the oxidation process. Similar bistability was also demonstrated in the oxidation of formaldehyde, along with period-doubling bifurcations, leading to chaos. The existence of various types of oscillation suggests that the oxidation of formaldehyde is accompanied by several stages of oxide layer formation on the platinum electrode.

The structural effects in the electrocatalysis of formic acid oxidation were investigated by Tripkovic et al.[83] The kinetics of oxidation of HCOOH strongly depend on the crystallographic orientation of the electrode surface. Oscillations were found at the Pt(100) and vicinal Pt(11,1,1) and Pt(610) faces. A rotating ring-disk assembly was used to generate forced oscillation, caused by a stream of hydrogen pulsed from the disk to the ring electrode during the oxidation of formaldehyde on a polycrystal platinum ring.[84] A mechanistic study of the potential oscillation and induction period in the oxidation of formic acid on platinum was conducted by Okamoto.[85] Current oscillations were observed during formic acid oxidation at the Pt(100) electrode plane in a potential range where a simultaneously produced surface-poisoning species is oxidized by absorbed OH.[86] Current oscillations are of the relaxation type, and the system is essentially in one of two quasi-stationary states.[87] Using *in situ* UV-visible reflectance spectroscopic techniques, Hachkar et al.[5] studied formaldehyde oxidation on a rhodium electrode in order to determine whether Rh oxides are involved in the oscillation process. The experimental results suggest that a cyclic buildup/removal of film oxides on the surface is involved in the mechanism. *In situ* Fourier transform infrared (FTIR) spectroscopy suggests an electrocatalytic mechanism in the oxidation of formic acid on a modified Pt electrode.[88]

8. Other Oxidation Systems

(i) Ag in $HClO_4$

Corrosion of silver electrodes in $1M$ $HClO_4$ and in an artificial pit configuration under potentiostatic control is characterized by periodic and chaotic oscillations within specific regimes of applied potential and pit depth.[89] Oscillations result from the instability of a porous salt film formed on the metal surface.

(ii) Nb in HF

Kinetic instability has been observed in the case of the electrochemical dissolution of Nb in an HF solution[90] at elevated temperatures and concentrations. Current oscillation (under constant potential) and potential oscillation (under constant current) are explained by the formation of an oxide and its subsequent dissolution.[91,92]

(iii) Al in Alkaline Solutions

Potential oscillations were found as single Al crystals spontaneously dissolve in KOH or NaOH solutions.[93,94] Optimal conditions for oscillations and changes in potential were established on oriented sections of single crystals.

(iv) Alloys

Oscillations are generated by the dissolution of alloy electrodes, for example, Fe–Ni alloy in concentrated chloride solutions.[95] Potential oscillations result from the anodic film formation and dissolution and have been analyzed in terms of power spectrum densities, phase portrait, Poincaré section, and correlation dimensions. The irregularity of the oscillatory structure is attributed to the surface heterogeneity or pitting conditions.

III. CATHODIC PROCESSES

1. The Reduction of Zinc Ions

(i) In $ZnSO_4$

The electrochemical deposition of Zn^{2+} ions can generate a wide spectrum of patterns, ranging from random fractals to orderly dendrites. In the dendritic regime, the cathodic current undergoes periodic oscillations at a constant potential,[96] and the oscillation frequency is proportional to the applied voltage and the concentration of $ZnSO_4$. Periodic and nonperiodic potential oscillations were also found at constant current in the regime between dendritic growth and the formation of diffusion-limited metallic clusters.[97] The analysis of data in terms of phase portrait, Poincaré maps, and one-dimensional maps indicates that the deposition process is characterized by low-dimensional deterministic chaos dynamics. The nature of the chaotic behavior is homoclinic,[4] and its physical origin is traced to the competition between the electrochemical reduction of Zn^{2+} ions and transport processes (migration, diffusion, and convection). Video recording of the deposition patterns shows that oscillations in the growth rate are in phase with cell potential oscillations across the entire cell width.

(ii) In Alkaline Solution

The mechanism of cathodic potential oscillations of a Zn electrode in KOH solution was studied by St. Pierre and Piron.[98] The oscillation period may be divided into three successive intervals, corresponding to depletion, supersaturation, and recovery. A controlling mechanism for electrochemical reaction and complex transport was also described. A model for potential oscillations was developed by measuring the oscillation period as function of (constant and pulsed) current.[99] Oscillations were observed only as a superposition over a threshold value of the mean current.

(iii) In Acidic Solution

In a study of cathodic potential oscillations of the zinc electrode in $ZnCl-KCl-H_3BO_3$ solutions, two types of potential oscillations under galvanodynamic conditions were found.[100] They differ in frequency, amplitude, and the threshold potential for cyclic behavior.

2. The Reduction of Indium(III) Ions

Indium(III)/thiocyanate is a classical electrochemical system for generating multiperiodic "mixed-mode" and aperiodic oscillations. In recent years, Koper and co-workers[101–105] developed a three-variable model to describe the nonlinear behavior of this system observed with a hanging mercury drop electrode and a rotating disk electrode. The authors ascribed dynamic instabilities to the coupling of diffusion with surface processes in the region of negative faradaic impedance. Relaxation delay in the diffusion process, neglected in many previous models for oscillations in electrochemical systems, was taken into account as an essential variable. Its role is increasing the phase-space dimension and causing mixed-mode oscillations as well as chaos. Very good agreement with experimental observations in In/SCN⁻ and Cu/H_3PO_3 systems provides direct support for this complex model and suggests that the effect of diffusion on oscillation patterns may be more complex, in general, than what has been proposed in the literature.

3. The Reduction of Hydrogen Peroxide

The galvanostatic reduction of H_2O_2 on a rotating Pt disk electrode in acidic electrolyte was studied by Fetner and Hudson.[106] The oscillatory patterns of potential are determined by H_2O_2 concentration, disk rotation speed, and applied current. Current oscillations were also observed in O_2 and H_2O_2 reduction processes at the (100) plane of the gold electrode. The role of adsorption of hydroxide ions and its effect on the oscillation behavior have also been discussed.[107]

When $CuFeS_2$ was used as a cathode,[108,109] the frequency and the amplitude of current oscillations were found to depend on ohmic resistance and mass transport. A small-amplitude potential modulation may trigger oscillations with the same frequency. XPS experiments show that oscillation phenomena are coupled with corrosion of the electrode surface.

When p-CuInSe$_2$ is used as a photocathode, current oscillations (at constant potential) and potential oscillations (at constant current) may occur in the presence of a suitable external resistance.[110,111] Particularly interesting is the finding that oscillation with a preset frequency can be obtained by synchronization to a periodic potential or a light triggering action. The cell can function as a light sensor, responding with individual current/potential pulses to short photon interruptions. A qualitative analysis of circuit stability was also included in this work.

4. Other Reduction Systems

(i) Sn(II)

Cathodic potential oscillations on an Sn electrode in an electrolyte containing 0.23M SnO and 6.25M KOH were observed at current densities above 300 A/cm^2.[112] Potential oscillations were sustained without hydrogen evolution. The oscillation period increased with time, but the amplitude decreased gradually until oscillations vanished completely.

(ii) Pb(II)

Current oscillations were observed in the electrochemical reduction of PbCl$_2$ on a liquid lead electrode of renewable surface immersed in an NaCl–KCl eutectic melt.[113] It was suggested that the Pb(II) reduction occurs in a single reversible step involving the transfer of two electrons. Active ion diffusion was considered to be the rate-determining step.

(iii) Cu(II) and Cd(II)

In a study[114] of potential oscillations during Cu(II) reduction on a copper wire electrode in 1M H$_2$SO$_4$ with varying CuSO$_4$ concentrations, the periods of oscillations were found to be correlated with the lifetime of hydrogen bubbles. Current oscillations were also observed in Cu(II) and Cd(II) reduction systems with surfactant in the solution.[115] The oscillations were attributed to the adsorption or desorption of the surfactant.

(iv) Fe(CN)$_6^{3-}$

Reduction of Fe(CN)$_6^{3-}$ at a glassy carbon electrode was shown to proceed via chaotic oscillations by using low-amplitude (40 mV) potential sine waves.[116] Oscillatory behavior was attributed to nonlinearities in diffusion and chemical reactions.

(v) Transition Metals

Karnaukhov et al.[117] considered that the cathodic reductions of transition (polyvalent) metal ions are accompanied by film formation on the cathode. The reduction processes occur at the electrode/film and film/electrolyte interfaces. A mathematical model, based on the assumption of multiple steps in polyvalent-metal-ion discharge processes and a nonuniform electric field in the film along the normal to the electrodes, was proposed to explain the experimental dynamic behavior.

IV. MATHEMATICAL TECHNIQUES APPLIED TO THE ANALYSIS OF DYNAMIC PHENOMENA: DATA INTERPRETATION AND MODELING

The survey presented in Sections II and III illustrates the wealth of information available on the dynamic behavior of electrochemical systems. Because imposed potential and current can be closely controlled (in the 273 PAR potentiostat, for example, the measurement of the applied potential is accurate to within 0.2% of the reading ±2mV, and that of the current is accurate to within 0.2% of full-scale current), measured signals carry very little noise and can be rapidly fed into a microcomputer at sampling speeds as high as 50 kHz. Hence, electrochemical systems are particularly attractive for the investigation of complex dynamics. In recent years, a variety of techniques have been developed to interpret and model experimental data. In the authors' opinion, the most important progress has hitherto been achieved in the extensive application of the time-series technique, where data are collected at equally spaced time instants. In electrochemical experiments, the imposed potential (or current) usually remains constant, and the measured response current (or potential) is recorded as a function of time: time-series analysis can be applied to such records directly and without *a priori* data manipulation. A survey of modern approaches to the analysis of dynamic behavior based on the time-series technique is presented below.

1. Geometrical Techniques— Construction of the Phase Portraits and Return Maps

The origin of modern geometrical techniques can be traced to the pioneering work of Poincaré leading to the (by now routine) use of phase portraits and return maps, which allow the establishment of the deterministic extent of (apparent) chaotic behavior. Fundamental concepts of the approach have been described in a large number of texts on nonlinear dynamics, for example, those by Thompson and Stewart[118] and Drazin.[119] In general, the "history" of a real system can be traced by a time series or as a trajectory in a geometrical state space. The set of phase-space trajectories constitutes a phase portrait of the system. The m-dimensional phase portrait can be constructed from one-dimensional time series through the time-delay method: $\Psi(t) = I(t), I(t + \tau), \ldots, I[t + (m - 1)\tau]$.[120,121] The choice of time delay τ is relatively arbitrary, and its effect on the phase portrait effect has been discussed in detail.[121] A Poincaré section is established by the intersection of "positive directed" orbits of an m-dimensional phase portrait with an $(m - 1)$-dimensional hypersurface. The map defined by $F:R^2 \to R^2$ such that $X_{n+1} = f(X_n)$ is called a Poincaré map or first return map (in R^2 space). If the X_n versus X_{n+1} plot is a smooth and single curve, the system is considered to be deterministic because X_{n+1} is fully determined by X_n: as an example, the return map of an essentially linear phase portrait is quasiquadratic.[122] Chaotic appearance may be caused by the superposition of two incommensurate harmonic motions. For an irrational ratio of the component frequencies, the Poincaré map is a closed elliptical curve, which indicates the deterministic nature of the harmonic motion.[123]

In the area of electrochemical dynamics, the technique may now be considered as a "standard" approach for the description of low-order chaotic behavior and the determination of bifurcation points; the efforts of Hudson, Schell, and Swinney (see the reference list) played a seminal role in this respect. Figure 1 illustrates a typical chaos occurring during the dissolution of copper in an acidic NaCl system.[124] Figure 2 shows a chaotic process during the cathodic deposition of Zn^{2+} ions.[4]

2. Fractional Brownian Motion (FBM) Theory

Fractional Brownian motion theory is a formal extension of a statistical method called rescaled range analysis (R/S analysis) invented by Hurst[125] in the experimental study of reservoirs. The experimental Hurst exponent H characterizes the time record of a natural phenomenon. The trace of the record is a curve with a fractal dimension of $2 - H$.[126]

FBM was defined originally by Mandelbrot and Van Ness[127] in terms of a random function of time $B_H(t)$ as

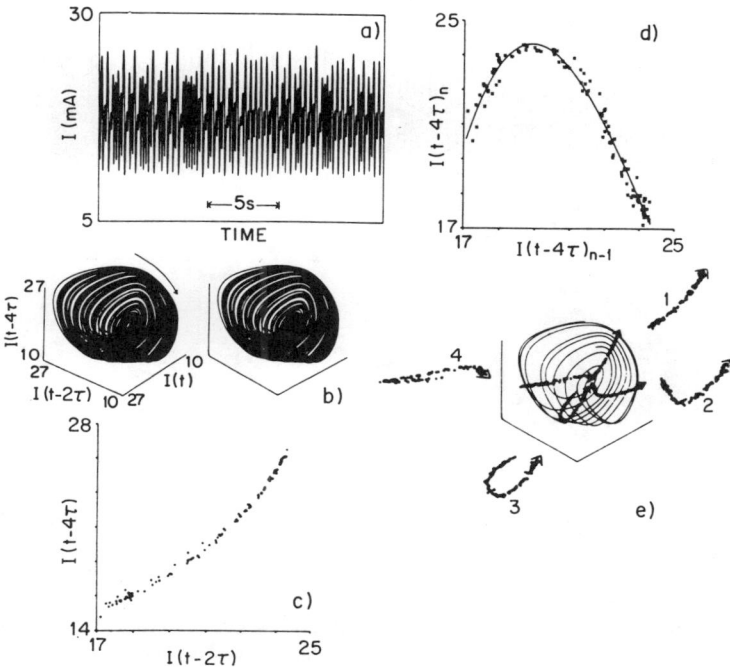

Figure 1. Chaos. (a) Time series. (b) Attractor (split fit). (c) Poincaré section taken at $I(t) = 17.5$ mA; increasing $I(t - 4\tau)$. (d) Return map constructed from the Poincaré section taken at $I(t - 2\tau) = 17.5$ mA with increasing $I(t - 4\tau)$; using variable $I(t - 4\tau)$. (e) Series of Poincaré section showing the flow of trajectories around the attractor. Sections 1 and 3 were taken at $I(t - 2\tau) = 17.5$ mA, and Sections 2 and 4 were taken at $I(t - 4\tau) = 7.5$ mA. After Ref. 124 (by permission of J. L. Hudson).

$$B_H(t) = \frac{1}{\Gamma(H + 0.5)} \int_{-\infty}^{t} (t - \lambda)^{H - \frac{1}{2}} dB(\lambda) \tag{1}$$

in generalizing the one-dimensional random walk model of ordinary Brownian motion. In this context, the Hurst exponent acquires the following significance as a trend descriptor: $0 < H < 0.5$ signals that an increasing trend in the time series is followed by a decreasing trend, whereas $0.5 < H < 1.0$ signals that an increasing trend is followed by an increasing trend. A value of H close to 1 indicates essentially no randomness, and $H = 0.5$

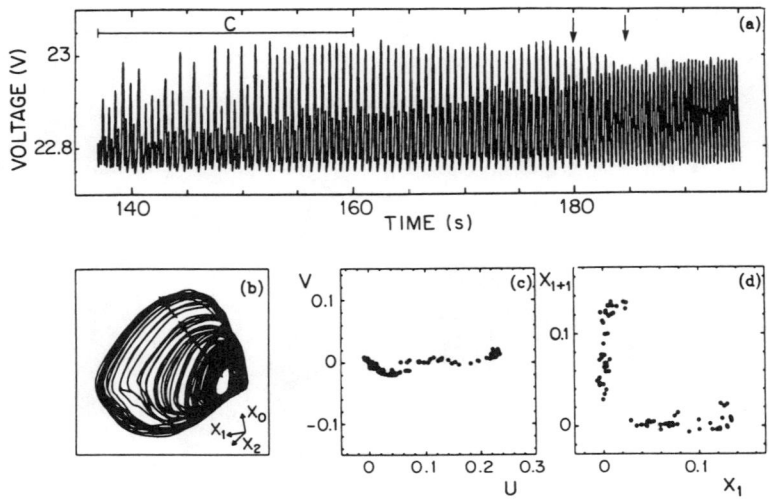

Figure 2. (a) A Voltage time series exhibiting a transition from a chaotic regime [130 s, 160 s] to a periodic regime (above 190 s) The reverse period-doubling bifurcation to the periodic regime is bracketed by two vertical arrows. (b) Phase space portrait of the chaotic regime C: $X_0 = \Phi(t)$, $X_1 = \Phi(t+\tau)$, $X_2 = \Phi(t+2\tau)$, where $\tau = 36$ ms. (c) Poincaré section constructed from trajectories crossing the plane (in the positive direction) normal to the page through the dashed line in (b). (d) One-dimensional map obtained by plotting ordered pairs (X_i, X_{i+1}), where $X_i = U_i \cos\theta + V_i \sin\theta$ and $\theta = 70°$. Experimental conditions: $[ZnSO_4]$ = $0.2M$, $I = 0.41$ mA ($j = 27.3$ mA/cm^2), $W = 5$ mm, $L = 110$ mm, $b = 0.3$ mm, $\Phi_{t=0} = 22.6$ V. After Ref. 4 (by permission of H. L. Swinney).

represents pure Brownian motion.[128,129] The numerical value of H can be determined from experimental Pox diagrams,[126,130,131] where the logarithm of the rescaled range R_r is plotted against the logarithm of the time lag s:

$$\log R_r = \text{const.} + H \log s \quad (2)$$

The rescaled range is defined as

$$R_r = s \frac{\max_{(i)} c_T(i) - \min_{(i)} c_T(i)}{\left[s \sum_i x_i^2 - (\sum_i x_i)^2 \right]^{1/2}} \quad (3)$$

where $x_i = y_{i+1} - y_i$, if y_i are observations at t_i uniformly spaced time instants within a time interval spanned from an arbitrary time instant to lag time s.

The range parameter in Eq. (3) is defined as

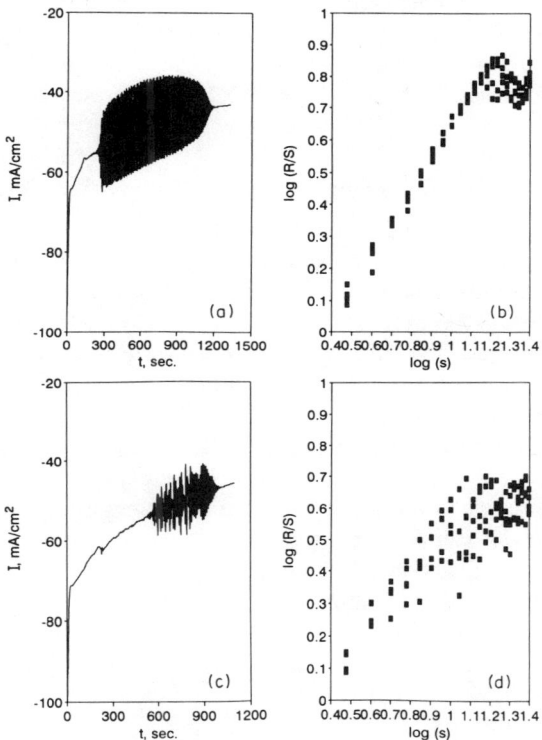

Figure 3. (a) Current time series, $B = 0$. (b) Pox diagram for (a). (c) Current time series, $B = 133$ mT. (d) Pox diagram for (c). Experimental conditions: [NaCl]= $4M$, [KSCN] = $4mM$. After Refs. 7 and 8.

$$c_T(i) \equiv \sum_{j=1}^{i} (x_j - \bar{x}_T); \quad i = 1,...,s \qquad (4)$$

where

$$\bar{x} \equiv \sum_i x_i/s$$

FBM theory has been extensively applied to detect the effect of potential, electrolyte composition, and the imposition of external magnetic fields on the oscillation structure of current in Cu/NaCl/KSCN systems. As shown in Fig. 3, the destabilizing effect of the magnetic field is clearly portrayed by the decrease in the numerical value of the Hurst exponent from essentially unity to about 0.72 in the presence of a magnetic field.[8,10,20–22]

3. The Time Differencing Technique: Box–Jenkins Approach

Although certain individual processes in dynamic electrochemical systems may be considered to be in an approximately stationary state, at least over a limited time interval, overall electrochemical processes can be nonstationary because the surface of the electrodes, the interface between electrodes and electrolyte, and the chemical composition of the solution may continuously change with time.

To circumvent difficulties in analysis based on the stationary-state assumption, the differencing technique of Box and Jenkins[132] is useful for the transformation of the original time series into a stationary series. The technique consists of creating successive autoregressive series until stationary behavior can be demonstrated. Let I_j be the current observed at time instant t_j. Successive differencing (up to the nth difference) is obtained by generating functions

$$\nabla^n I_j = \sum_{k=0}^{n} (-1)^k \binom{n}{k} I_{j-k} \tag{5}$$

To ascertain nonstationarity for the nth difference, the autocorrelation function

$$r_s = \frac{\sum_{j=s+1}^{n} (I_j - \bar{I})(I_{j-s} - \bar{I})}{\sum_{j=1}^{n} (I_j - \bar{I})^2} \tag{6}$$

is plotted against s. The r_s versus s plots associated with increasing values of n indicate at what particular value of n the autocorrelation function begins to exhibit the characteristics of a stationary process.[132,133] The utility of this technique in the analysis of oscillatory copper dissolution has recently been demonstrated.[10] It is instructive to note that the stationary series obtained after the nth differencing step may be used to establish a forecasting model through the stages of model specification, parameter estimation, and diagnostic checking, applied to iterative model construction.

V. SUMMARY AND CONCLUSIONS

Over the past five years, much progress has been made in the understanding of the dynamics of electrochemical phenomena. An already large and fast-growing literature attests to the incessant interest that this relatively new field enjoys. While this review cannot possibly pay tribute to all contributions to this field, the examples presented here indicate the width of achievements as an important component of the theory and the data base of modern dynamics. Many basic questions are still unanswered, however. Why does a single copper dissolution process possess several bifurcating points? Why do oscillation patterns remain highly regular even at a frequency as high as 30 Hz when iron dissolves in H_2SO_4? Why does an external magnetic field destabilize the oscillation structure in a specific electric/magnetic field configuration? What is the exact role of "impurities" in the electrolyte (e.g., the role of thiocyanate ions in copper dissolution into chloride solutions)? And so forth. It appears logical to presume that the discovery of the exact links between oscillation structure and the chemical and physical nature of processes that generate oscillations will require the crossing of conventional borders between chemical kinetics, surface science, mass transport, statistics, and mathematics. May this challenge encourage increasing collaboration among chemists, physicists, mathematicians, and engineers.

ACKNOWLEDGMENTS

The authors are indebted to Professors J. L. Hudson (University of Virginia), H. L. Swinney (The University of Texas at Austin), and T. T. Tsotsis (University of Southern California) for their help in preparing this manuscript. Research in the authors' laboratory has been supported by the Natural Sciences and Engineering Research Council of Canada (NSERC).

REFERENCES

[1] G. Th. Fechner, *Schweigg. J. Chemie Physik* **53** (1828) 129.
[2] J. Wojtowicz, in *Modern Aspects of Electrochemistry*, No. 8, Ed. by J. O'M. Bockris and B. E. Conway, Plenum Press, New York, 1972, p. 47.
[3] J. L. Hudson and M. R. Bassett, in *Reviews in Chemical Engineering*, Ed. by D. Luss and N. R. Amundson, Freund, London, 1991.
[4] F. Argoul, J. Huth, P. Merzeau, A. Arneodo, and L. Swinney, *Physica D* **62** (1993) 170.
[5] M. Hachkar, M. Martinez, A. Rakotondrainibe, B. Beden, and C. Lamy, *J. Electroanal. Chem. Interfacial Electrochem.* **302** (1991) 173.
[6] K. Krischer, M. Luebke, M. Eiswirth, W. Wolf, and J. L. Hudson, *Physica D* **62** (1993) 123.
[7] Z. H. Gu, J. Chen, A. Olivier, and T. Z. Fahidy, *J. Electrochem. Soc.* **140** (1993) 408.
[8] Z. H. Gu, J. Chen, and T. Z. Fahidy, *Electrochim. Acta* **37** (1992) 2637.
[9] Y. Wang and J. L. Hudson, *AIChE J.* **37** (1991) 1833.
[10] Z. H. Gu, T. Z. Fahidy, and J. P. Chopart, *Electrochim. Acta* **37** (1992) 97.
[11] H. P. Lee and K. Nobe, *J. Electrochem. Soc.* **131** (1984) 131.
[12] H. P. Lee, K. Nobe, and A. J. Pearlstein, *J. Electrochem. Soc.* **132** (1985) 1031.
[13] A. J. Pearlstein, H. P. Lee, and K. Nobe, *J. Electrochem. Soc.* **132** (1985) 2159.
[14] H. P. Lee and K. Nobe, *J. Electrochem. Soc.* **133** (1986) 2035.
[15] M. R. Bassett and J. L. Hudson, *Physica D* **35** (1989) 289.
[16] M. R. Bassett and J. L. Hudson, *J. Electrochem. Soc.* **137** (1990) 922.
[17] M. R. Bassett and J. L. Hudson, *J. Electrochem. Soc.* **137** (1990) 1815.
[18] Z. H. Gu, A. Olivier, and T. Z. Fahidy, *Electrochim. Acta* **35** (1990) 933.
[19] Z. H. Gu and T. Z. Fahidy, *Can. J. Chem. Eng.* **70** (1992) 127.
[20] P. Fricoteaux, Z. H. Gu, and T. Z. Fahidy, *J. Electroanal. Chem.* **324** (1992) 161.
[21] Z. H. Gu, J. Chen, T. Z. Fahidy, and A. Olivier, *J. Electroanal. Chem.* **367** (1994) 7.
[22] Z. H. Gu, J. Chen, and T. Z. Fahidy, *Electrochim. Acta* **38** (1993) 2631.
[23] Z. H. Gu and T. Z. Fahidy, An Experimental Study of Anode Dynamics During The Dissolution of Copper in Acidic NaCl/KSCN Electrolyte, Paper presented at the AIChE Annual Meeting, St. Louis, Missouri, November 7–12, 1993.
[24] F. N. Albahadily and M. Schell, *J. Chem. Phys.* **88** (1988) 4312.
[25] F. N. Albahadily, J. Ringland, and M. Schell, *J. Chem. Phys.* **90** (1989) 813.
[26] L. T. Tsitsopoulos, T. T. Tsotsis, and I. A. Webster, *Surf. Sci.* **191** (1987) 225.
[27] H. D. Dewald, P. Parmananda, and R. W. Rollins, *J. Electroanal. Chem. Interfacial Electrochem.* **306** (1991) 297.
[28] H. D. Dewald, P. Parmananda, and R. W. Rollins, *J. Electrochem. Soc.* **140** (1993) 1969.
[29] J. F. Cooper, R. H. Muller, and C. W. Tobias, *J. Electrochem. Soc.* **127** (1980) 1733.
[30] P. Russell and J. Newman, *J. Electrochem. Soc.* **133** (1986) 2093.
[31] P. Russell and J. Newman, *J. Electrochem. Soc.* **133** (1987) 1051.
[32] C. B. Diem and J. L. Hudson, *AIChE J.* **33** (1987) 218.
[33] Y. Wang and J. L. Hudson, *J. Phys. Chem.* **21** (1992) 8667.
[34] U. F. Franck and R. FitzHugh, *Z. Electrochem.* **65** (1961) 156.
[35] Y. Wang, J. L. Hudson, and N. I. Jaeger, *J. Electrochem. Soc.* **137** (1990) 485.
[36] A. J. Pearlstein and J. A. Johnson, *J. Electrochem. Soc.* **136** (1989) 1290.
[37] M. Pagitsas and D. Sazou, *Electrochim. Acta* **36** (1991) 1301.
[38] M. Pagitsas and D. Sazou, *Electrochim. Acta* **37** (1992) 1047.
[39] O. Teschke, F. Galembeck, and M. A. Tenan, *Langmuir* **3** (1987) 400.
[40] C. C. Haili, Report LBL-23776; *Energy Res. Abstr.* **12** (1987) abstract 47827 (order no. DE87014679).
[41] W. Li, X. Wang, and K. Nobe, *J. Electrochem. Soc.* **137** (1990) 1184.

[42] X. Wang and K. Nobe, *Gaodeng Xuexiao Huaxue Xuebao* **9** (1988) 301.
[43] W. Li, K. Nobe, and A. J. Pearlstein, *Corros. Sci.* **31** (1990) 615.
[44] W. Li, K. Nobe, and A. J. Pearlstein, *J. Electrochem. Soc.* **140** (1993) 721.
[45] W. Li and K. Nobe, *J. Electrochem. Soc.* **140** (1993) 1642.
[46] C. Moina and D. Posadas, *Electrochim. Acta* **32** (1987) 1203.
[47] C. Moina and D. Posadas, *Electrochim. Acta* **34** (1989) 789.
[48] D. Sazou, M. Pagitsas, and G. Kokkinidis, *J. Electroanal. Chem. Interfacial Electrochem.* **289** (1990) 217.
[49] D. Sazou and M. Pagitsas, *J. Electroanal. Chem. Interfacial Electrochem.* **304** (1991) 171.
[50] D. Sazou and M. Pagitsas, *J. Electroanal. Chem. Interfacial Electrochem.* **312** (1991) 185.
[51] D. Sazou and M. Pagitsas, *Electrochim. Acta* **38** (1993) 835.
[52] D. Sazou and M. Pagitsas, *J. Electroanal. Chem.* **323** (1992) 247.
[53] M. Pagitsas and D. Sazou, *J. Electroanal. Chem.* **334** (1992) 81.
[54] U. F. Franck and L. Meunier, *Z. Naturforsch B* **8** (1953) 396.
[55] J. L. Hudson, J. Bell, and N. I. Jaeger, *Ber. Bunsen-Ges. Phys. Chem.* **92** (1988) 1383.
[56] J. Bell, N. I. Jaeger, and J. L. Hudson, *Phys. Chem.* **96** (1992) 8671.
[57] T. J. Pastor and M. M. Pastor, *J. Electroanal. Chem. Interfacial Electrochem.* **316** (1991) 335.
[58] T. J. Pastor, G. Petkovic, and D. D. Manojlovic, *Acta Chim. Hung.* **129** (1992) 421.
[59] O. Lev, A. Wolffberg, M. Sheintuch, and L. M. Pisman, *Chem. Eng. Sci.* **43** (1988) 1339.
[60] O. Lev, A. Wolffberg, L. M. Pismen, and M. Sheintuch, *J. Phys. Chem.* **93** (1989) 1661.
[61] O. Lev, M. Sheintuch, H. Yarnitsky, and L. M. Pismen, *Chem. Eng. Sci.* **45** (1990) 839.
[62] D. Haim, O. Lev, L. M. Pismen, and M. Sheintuch, *J. Phys. Chem.* **96** (1992) 2676.
[63] S. Sternberg, I. V. Cotarta, and D. Constantinescu, *Rev. Roum. Chim.* **34** (1989) 623.
[64] B. Li and O. Vittori, *Can. J. Chem.* **66** (1988) 1525.
[65] F. Ozanam, J. N. Chazalviel, A. Radi, and M. Etman, *Ber. Bunsen-Ges. Phys. Chem.* **95** (1991) 98.
[66] F. Ozanam, J. N. Chazalviel, A. Radi, and M. Etman, *J. Electrochem. Soc.* **139** (1992) 2491.
[67] M. Etman, J. N. Chazalviel, and F. Ozanam, *Proc.—Indian Acad. Sci., Chem. Sci.* **104** (1992) 299.
[68] J. N. Chazalviel and F. Ozanam, *J. Electrochem. Soc.* **139** (1992) 2501.
[69] J. Stumper, R. Greef, and L. M. Peter, *J. Electroanal. Chem. Interfacial Electrochem.* **310** (1991) 445.
[70] H. J. Lewerenz and G. Schlichthoerl, *J. Electroanal. Chem.* **327** (1992) 85.
[71] V. P. Parkhutik, *Electrochim. Acta* **36** (1991) 1611.
[72] T. Yamazaki and T. Kodera, *Electrochim. Acta* **34** (1989) 969.
[73] T. Yamazaki and T. Kodera, *Electrochim. Acta* **35** (1990) 431.
[74] T. Kodera, T. Yamazaki, M. Masuda, and R. Ohnishi, *Electrochim. Acta* **33** (1988) 537.
[75] T. Yamazaki and T. Kodera, *Electrochim. Acta* **36** (1991) 639.
[76] K. Krischer, M. Luebke, W. Wolf, M. Eiswirth, and G. Ertl, *Ber. Bunsen-Ges. Phys. Chem.* **95** (1991) 820.
[77] M. Eiswirth, M. Luebke, K. Krischer, W. Wolf, J. L. Hudson, and G. Ertl, *Chem. Phys. Lett.* **192** (1992) 254.
[78] K. Krischer, Ph.D Thesis, Free University of Berlin, 1990.
[79] M. Hachkar, B. Beden, and C. Lamy, *J. Electroanal. Chem. Interfacial Electrochem.* **287** (1990) 81.
[80] M. Schell, F. N. Albahadily, J. Safar, and Y. Xu, *J. Phys. Chem.* **93** (1989) 4806.
[81] Y. Xu and M. Schell, *J. Phys. Chem.* **94** (1990) 7137.

[82] F. N. Albahadily and M. Schell, *J. Electroanal. Chem. Interfacial Electrochem.* **308** (1991) 151.
[83] A. Tripkovic, K. Popovic, and R.R. Adzic, *J. Chim. Phys. Phys.-Chim. Biol.* **88** (1991) 1635.
[84] S. Nakabayashi and K. Akira, *J. Phys. Chem.* **96** (1992) 1021.
[85] H. Okamoto, *Electrochim. Acta* **37** (1992) 37.
[86] F. Raspel, R. J. Nichols, and D. M. Kolb, *J. Electroanal. Chem. Interfacial Electrochem.* **286** (1990) 279.
[87] P. De. Kepper and J. Boissonade, in *Oscillations and Travelling Waves in Chemical Systems*, Ed. by R. J. Field and M. Burger, John Wiley & Sons, New York, 1985, p. 223.
[88] S. Sun and Y. Lin, *Prog. Nat. Sci.* **4** (1992) 351.
[89] S. G. Corcoran and K. Sieradzki, *J. Electrochem. Soc.* **139** (1992) 1568.
[90] M. I. Eidel'berg, D. B. Sandulov, and V. N. Ustimenko, *Zh. Prikl. Khim.* **64** (1991) 665.
[91] V. Yu. Novichkov and V. M. Orlov, *Elektrokhimiya* **26** (1990) 771.
[92] M. I. Eidel'berg, D. B. Sandulov, and V. N. Ustimenko, *Elektrokhimiya* **26** (1990) 272.
[93] M. Miadokova and J. Siska, *Collect. Czech. Chem. Commun.* **52** (1987) 1461.
[94] I. Halasa and M. Miadokova, *Collect. Czech. Chem. Commun.* **55** (1990) 345.
[95] D. H. Shen, W. Li, and K. Nobe, *Proc. Electrochem. Soc.* **92-9** (1992) 511.
[96] R. M. Suter and P. Z. Wong, *Phys. Rev. B: Condens. Matter* **39** (1989) 4536.
[97] F. Argoul and A. Arneodo, *J. Phys.* **51** (1990) 2477.
[98] J. St. Pierre and D. L. Piron, *J. Electrochem. Soc.* **137** (1990) 2491.
[99] J. St. Pierre and D. L. Piron, *J. Electrochem. Soc.* **134** (1987) 1689.
[100] J. A. Weber, *Powloki Ochr* **18** (1990) 4.
[101] M. T. M. Koper and P. Gaspard, *J. Phys. Chem.* **95** (1991) 4945.
[102] M. T. M. Koper and J. H. Sluyters, *J. Electroanal. Chem. Interfacial Electrochem.* **303** (1991) 65.
[103] M. T. M. Koper and J. H. Sluyters, *J. Electroanal. Chem. Interfacial Electrochem.* **303** (1991) 73.
[104] M. T. M. Koper and P. Gaspard, *J. Chem. Phys.* **96** (1992) 7797.
[105] M. T. M. Koper, P. Gaspard, and J. H. Sluyters, *J. Phys. Chem.* **96** (1992) 5674.
[106] N. Fetner and J. L. Hudson, *J. Phys. Chem.* **94** (1990) 6505.
[107] S. Strbac and R. R. Adzic, *J. Electroanal. Chem.* **337** (1992) 355.
[108] S. Cattarin and H. Tributsch, *J. Electrochem. Soc.* **137** (1990) 3475.
[109] S. Cattarin, S. Flechter, C. Pettenkofer, and H. Tributsch, *J. Electrochem. Soc.* **137** (1990) 3484.
[110] S. Cattarin and H. Tributsch, *J. Electrochem. Soc.* **139** (1992) 1328.
[111] S. Cattarin and H. Tributsch, *Electrochim. Acta* **38** (1993) 115.
[112] D. L. Piron, I. Nagatsugawa, and C. Fan, *J. Electrochem. Soc.* **138** (1991) 3296.
[113] M. E. de Almeidi Lima, J. Bouteillon, and J. P. Diard, *J. Appl. Electrochem.* **22** (1992) 577.
[114] Z. H. Gu and M. Pritzker, A Visual Study of the Relation between the Electrode Response and Dynamics of Hydrogen Evolution during Copper Electrodeposition, Paper presented at the AIChE Annual Meeting, Los Angeles, November 17–22, 1991.
[115] H. D. Doerfler, *Nova Acta Leopold* **61** (1989) 25.
[116] V. S. Varma and P. K. Upadhyay, *J. Electroanal. Chem Interfacial Electrochem.* **271** (1989) 345.
[117] I. N. Karnaukhov, A. I. Karasevskii, N. D. Ivanova, A. V. Gorodyskii, and Y. I. Boldyrev, *J. Electroanal. Chem. Interfacial Electrochem.* **288** (1990) 35.
[118] J. M. T. Thompson and H. B. Stewart, *Nonlinear Dynamics and Chaos*, John Wiley & Sons, New York, 1986.
[119] P. G. Drazin, *Nonlinear Systems*, Cambridge Texts in Applied Mathematics, Cambridge University Press, Cambridge, 1992.

[120] N. H. Packard, J. P. Crutchfield, J. D. Farmer, and R. S. Swar, *Phys. Rev. Lett.* **45** (1980) 712.
[121] J. C. Roux, R. H. Simoyi, and H. L. Swinney, *Physica D* **8** (1983) 257.
[122] K. Tomita, in *Chaos*, Ed. by A. V. Holden, Princeton University Press, Princeton, New Jersey, 1986.
[123] F. C. Moon, *Chaotic Vibrations*, John Wiley & Sons, New York, 1987.
[124] J. L. Hudson, in *Patterns and Dynamics in Reactive Media*, Ed. by R. Aris, D. G. Aronson, and H. L. Swinney, Springer-Verlag, New York, 1991, p. 89.
[125] H. E. Hurst, *Trans Am. Soc. Civ. Eng.* **116** (1951) 770.
[126] J. Feder, *Fractals*, Plenum Press, New York, 1988.
[127] B. B. Mandelbrot and J. W. Van Ness, *SIAM Rev.* **10** (1968) 422.
[128] D. Saupe, *Fractals and Chaos*, Ed. by R. A. Crilly, R. A. Earnshaw, and M. Jone, Springer-Verlag, New York, 1991.
[129] L. T. Fan, D. Neogi, M. Yashima, and R. Nassar, *AIChE J.* **36** (1990) 1529.
[130] B. B. Mandelbrot and J. R. Wallis, *Water Resour. Res.* **5** (1969) 321.
[131] B. B. Mandelbrot and J. R. Wallis, *Water Resour. Res.* **5** (1969) 967.
[132] G. E. P. Box and G. M. Jenkins, *Time Series Analysis, Forecasting and Control*, Holden Day, San Francisco, 1976.
[133] B. Abraham and J. Ledolter, *Statistical Methods for Forecasting*, John Wiley & Sons, New York, 1983.

6

WATER ELECTROLYSIS AND SOLAR HYDROGEN DEMONSTRATION PROJECTS

Gerd Sandstede

Battelle Europe, Battelle Institut e.V., D-60486 Frankfurt, Germany

Reinhold Wurster

Ludwig-Bölkow-Systemtechnik GmbH, D-85521 Ottobrunn, Germany

I. INTRODUCTION

In this chapter, nearly all conventional and newly developed processes for water electrolysis will be considered, and an overview of demonstration projects in which hydrogen is electrolytically produced from solar, wind, and hydroenergy will be given. After a brief historical description of hydrogen, water electrolysis, and solar hydrogen economy, a systematic classification of all possible water electrolysis processes will be presented. In order to help the reader understand the development opportunities of the processes, the physicochemical fundamentals and the problems of engineering for water electrolysis will be discussed, including considerations of thermodynamics, electrocatalysis, and the reaction mechanisms of hydrogen and oxygen evolution.

After the delineation—mainly in tabular form—of the material technologies for the components of water electrolysis cells and plants, a compilation of the characteristic data for the technical state of the art of nearly all types of water electrolysis will be presented, and many conventional and advanced electrolyzers will be described.

Modern Aspects of Electrochemistry, Number 27, edited by Ralph E. White *et al.* Plenum Press, New York, 1995.

Subsequently, the relevant renewable energies will be considered briefly, and then the coupling of a water electrolyzer to an electricity generation system will be discussed. Demonstration projects on renewable hydrogen, national and international ones, not only in Germany but also in other European countries as well as in other parts of the world, will be considered. In the discussion of these projects, the significance, applications, benefits, and economy of hydrogen will be included.

II. TECHNICAL BACKGROUND FOR THE DEVELOPMENT OF WATER ELECTROLYSIS FOR THE ENERGETIC APPLICATION OF HYDROGEN

When we use the term solar hydrogen in this chapter, we mean hydrogen that is produced by water electrolysis using electricity generated from renewable energy; this may be not only direct solar energy but also includes all indirect possibilities—from hydro power to biomass. Of course, direct photolysis of water or certain thermal methods might also be counted as processes producing solar hydrogen, but these will not be considered in this chapter. Electrolysis of water is the central process, and the techniques and new developments for it will be dealt with in the following. In order to be complete, it may be mentioned that electrical power from fossil or nuclear energy, of course, leads to the same electrochemical result, namely, hydrogen and oxygen, which means that the purely technical aspects are the same—but not the systems aspects. On the other hand, the production of hydrogen in a purely chemical way, with no emissions to the environment, is currently being discussed (but not in this chapter). At any rate, hydrogen technologies can be used for the development of sectorial or partial hydrogen energy systems.

Hydrogen is a nearly ideal energy carrier (in the sense that, generally speaking, a really ideal system does not exist—some disadvantages always remain); it is universally applicable, its oxidation (combustion) product is only water (vapor) (and, only under certain conditions, relatively little NO_x), it is the lightest element, etc., but it has to be produced from other sources—it is not a raw material as such. So far, hydrogen is mainly produced from fossil raw materials—natural gas, mineral oil fractions, or even coal—with the result of corresponding generation of carbon dioxide (and possibly other pollutants).

There are a number of well-established processes, including electrolytic ones, for the generation of hydrogen,[1–17] because hydrogen plays a

Table 1
Reasons for the Increase in the Electrolytic Production of Hydrogen

A. Technological reasons
- Electrolytic technologies are making great progress.
- The number of hydro power stations has increased substantially.
- Nuclear energy and other base-load electricity can be used in off-peak hours.

B. Increased demand for hydrogen
- The immediate power reserve of fossil power stations can be substituted by hydrogen/oxygen burners,
- Peak power can be obtained from hydrogen generated during the night.
- Demand for hydrogen as a chemical and petrochemical feedstock has grown.
- Hydrogen is used increasingly in other industrial processes, e.g., in the semiconductor industry.
- Because of the extension of air traffic, liquid hydrogen (LH_2) will be used as fuel for large aircraft.

C. Environmental aspects and restrictions
- Hydrogen production from fossil raw materials should be diminished for the purpose of CO_2 reduction.
- Geothermal power, wind energy, and photovoltaics are slowly emerging as power source.
- Partial and regional hydrogen energy economy systems will be possible soon.
- Hydrogen propulsion of automobiles might become necessary for legal/environmental reasons—all the more as storage technologies and fuel cell technologies make progress.

---

central role in the chemical industry; that is why one can start into a solar energy economy from known technologies for the production of hydrogen. However, one cannot be content with the current processes because they do not satisfy the necessary demand either in technical or in economic respects and especially not in environmental respects. Therefore, on the one hand, improvements are under way and, on the other hand, new processes are being developed that use a nonfossil basis. Water electrolysis for the generation of hydrogen (and oxygen) in big plants is carried out in only a few places in the world, at sites where cheap electricity, mostly hydro power, is available. For some time, substantial progress in electrolytic technologies has been made as a result of remarkable improvements and new developments. These and other reasons for the increase in the electrolytic generation of hydrogen are compiled in Table 1.

Moving beyond its present uses, namely, as feedstock for the chemical industry and, in addition, as a fuel in space applications (LH_2), hydrogen is about to develop as a secondary energy carrier. In addition to its application as energy raw material and fuel, hydrogen might also become a means for energy storage and an energy transport medium in the energy-supply economy of tomorrow. For this purpose, it has to be produced on a large scale by electrochemical means. Water electrolysis is looked upon as part of the large field of electrochemical energy technologies, for the following reasons:

- Hydrogen in general is a fuel.
- Hydrogen produced by electrolysis can be used in the field of energy technologies and propulsion technologies.
- The electrolysis technology is the reverse of fuel cell technology.
- The combination of water electrolysis and an electrochemical fuel cell renders possible a storage device for electrical energy.
- The construction of a reversible cell, which can be alternately operated in the fuel cell mode and in the electrolysis mode, is possible.

The application of hydrogen as a secondary energy carrier will lead to partial hydrogen energy systems. In order to investigate more closely these possibilities in terms of scale and limits and, in particular, to gain experience, a number of demonstration projects have been started, and their state of development will be described in more detail in the second part of this chapter. These projects already have truly pioneered the introduction of hydrogen. Especially important is the investigation of the conditions of coupling the electrical energy from various sources to the electrolysis units. An outline of the energy sources for the production of hydrogen is shown in Table 2. At present, photolysis and thermolysis do not play a part; pyrolysis is of minor importance, whereas water vapor reforming (and partial oxidation) is presently the main process. The range of primary energy sources for electricity generation is broad; however, using fossil power stations for water electrolysis does not make much sense because of the CO_2 problem. The other nonpolluting energy conversion processes will not be dealt with here. With the exception of geothermal and tidal energy, all other renewable energies can be traced back to solar energy. It can be imagined that the number of biomass power stations, including those using bigas, will increase, particularly in warm regions. In these zones, however, purely thermal solar power stations are also possible. Electricity from ocean energy—thermal or from waves—is still in the research stage, but may be realized in the future. Tidal power stations

Table 2
Energy Sources for the Generation of Hydrogen by Various Processes

Process	Energy source
Electrolysis	Hydro power
	Wind power
	Geothermal energy
	Photovoltaic
	Ocean energy
	Solar power stations
	Nuclear power stations
	Biomass power stations
	Fossil power stations
Photolysis	Solar radiation
Thermolysis	Water vapor plus high-temperature reactor or high-temperature solar energy
	Hydrogen sulfide
Pyrolysis	Biomass
	Hydrocarbons
Water vapor reforming	Biomass
	Fossil raw materials

are already in operation, and wind and hydro power have been known for a long time. In addition, wind towers may be built in the future.

III. HISTORICAL REMARKS ABOUT THE GENERATION OF HYDROGEN, WATER ELECTROLYSIS, AND SOLAR HYDROGEN

At first, hydrogen was generated from dissolution of metal in acid, which is a reaction of electrochemical nature and, therefore, a forerunner of electrolysis. The first application of hydrogen, however, was not of a chemical but of a physical nature, although the characteristic property of hydrogen, its combustibility, led to its discovery. As early as the 17th century, Boyle reported that "combustible air" is developed when iron is dissolved in sulfuric acid or hydrochloric acid. In the 1760s, Cavendish made quantitative experiments, which led to the discovery that water is formed from burning "combustible air."[18] After Cavendish had measured the density of "combustible air" and in 1781 Cavallo in Italy and later

Charles in Paris had demonstrated its low density by means of soap bubbles, the Montgolfier brothers had the idea that "combustible air" might render possible the flight of a balloon. They failed, however, because of the leakage of the bags used, made from paper and silk, and, immediately, invented the "hot air" balloon, which they first raised into the air in the beginning of 1783. At the same time, Charles, Faujas, and the Robert brothers were about to construct a hydrogen balloon. The balloon was filled with 40 m^3 of hydrogen, which was produced from iron and sulfuric acid. On August 27, 1783, only three months after the invention of the "hot air" balloon, the first flight of the hydrogen balloon took place; the hydrogen balloon was called Charlière. Within one year Charles, as a kind of project manager, had carried out the whole project, from the research stage to the development and erection of the production plant for hydrogen and, in addition, the construction of the first hydrogen balloon up to the second successful start on December 1, 1783. At the same time he has conceived essential elements of chemical technology and of the gas industry, such as equipment for gas washing, cleaning, and drying as well as the heat exchanger and constructed them on a technical scale (1000 pounds of iron!).

In 1784, Lavoisier produced hydrogen from water vapor and iron.[19] In 1785, after quantitative experiments regarding the formation of water from hydrogen and oxygen, he named the two gases hydrogenium and oxygenium.[6]

In 1800 a multicell primary battery—the first technically usable electrical energy source—was invented and introduced to the public by Volta. This means that the electrical age started with electrochemistry, since, after the invention of the electrostatic machine in 1650 by Otto von Guericke, there had been practically no technical progress for centuries. In the same year (1800), this source of electrical energy was used for the decomposition of water; thus, water electrolysis was the first electrochemical process, apart from earlier galvanic experiments for metal deposition. First Carlisle and Nicholson in London and, a few months later, Ritter in Jena carried out experiments on electrochemical hydrogen evolution using Volta's pile.[20,21] Strictly speaking, the inventors of electric water splitting were—after preliminary experiments by van Marum—the Dutchmen van Troostwijk and Diemann, who in 1789 decomposed water into "combustible air" (hydrogen) and "life air" (oxygen) using a sparking current.[20] The apparatus used by Carlisle and Nicholson is shown in Fig. 1; the electrolysis tube next to Volta's pile looks insignificant. As they used

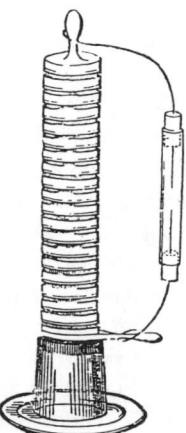

Figure 1. Volta's battery pile used for water electrolysis (electrolysis tube on the right) by Carlisle and Nicholson in London in 1800.

electrodes made from brass, they only achieved hydrogen evolution at first. There was no oxygen evolution but rather oxide formation, which they assumed to be a lime deposit. A few months later, Ritter constructed the apparatus shown in Fig. 2 using gold electrodes and achieved also oxygen evolution, and thus the first complete water electrolysis. Progress was made after Faraday explained the principles of electrolysis in London in 1820, and thus he is called the father of electrochemistry.

The next step forward in the development of water electrolysis processes took place only at the end of the 19th century after the electro-

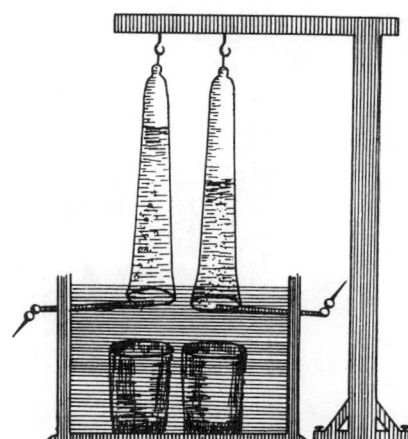

Figure 2. Water electrolyzer constructed by Ritter in Jena in 1800.

dynamic principle was invented and technically applied. In a similar way, the recent availability of new primary energy sources has advanced the development. In the beginning, an acid electrolyte was employed, but a switch to an alkaline electrolyte was made in order to use cheaper materials for the technical process. In 1890, Renard constructed a water electrolysis plant in order to produce hydrogen for the airships of the French army.[22] Progress, however, was made only slowly because the chemical technologies of reforming of hydrocarbons were, and still are, much less expensive.

Another important step followed exactly 100 years after the discovery of water electrolysis: in 1900 Schmidt in Zurich introduced his electrolyzer, which was constructed according to the bipolar filter press principle (see Fig. 3); the cell voltage amounted to 2.5 V.[20] Schmidt's apparatus was the first industrial electrolyzer to be produced by the company Oerlikon.

The development of a hydrogen economy, which started at the end of last century, was based mainly on chemical processes; after all, the chemical industry, and petrochemistry as well, is based on a chemical hydrogen economy. Furthermore, around the turn of the century the spreading of town gas, produced in coking plants, started. Large pipeline networks were established in Europe, the United States, and elsewhere, consisting of hydrogen pipelines, because town gas contained 60% hydrogen. Only in the 1950s did the substitution of this hydrogen gas by natural

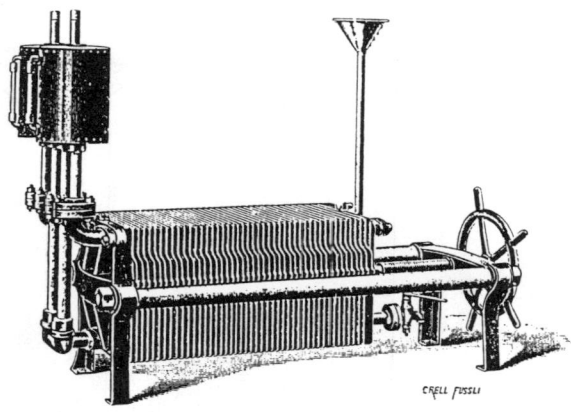

Figure 3. Bipolar water electrolyzer constructed by Schmidt in Zurich in 1900.

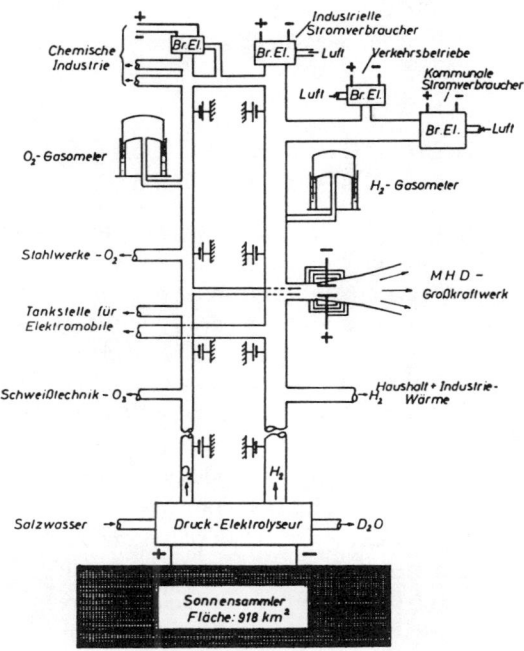

Figure 4. First diagram of a solar hydrogen energy system by Justi in Braunschweig, published in 1964 (Br.El. = fuel cell).

gas begin. Now, as before, however, there are chemical hydrogen pipelines, which have a length of several hundred kilometers and a pressure of 5 MPa (= 50 bar), in Germany, France, and the United States. Supplementary to the use of hydrogen as fuel, hydrogen has been discussed as a general energy carrier only since the beginning of the 1970s. Scientists in Germany, especially Justi, and in the United States, particularly Bockris, have developed concepts of a solar hydrogen economy; in addition, Appleby, Bölkow, Gregory, Marchetti, Veziroglu, and Winter may be mentioned as pioneers. The first published diagram of a solar hydrogen economy originated from Justi as early as 1964 (see Fig. 4). Jules Vernes, however, must not be forgotten; over 100 years ago he wrote *L'Isle Mystérieuse*, in which his hero gave the following answer to the question, What will happen when coal is exhausted? We will secure the energy supply by hydrogen, which is made by electrolysis of water. About 10 years ago demonstration projects for solar hydrogen started, with Hysolar being the first big one.

IV. SYSTEMATIC CLASSIFICATION OF WATER ELECTROLYSIS PROCESSES

Although all large-scale plants for water electrolysis exclusively use alkaline electrolytes, the electrolytic production of hydrogen is achieved by a broad variety of processes. This stems in particular from the fact that the aim to produce hydrogen less expensively, with less energy and less harm to the environment, as well as more safely, and to develop processes for various applications in practical use. As a basis for classification of the possible processes, the kind of electrolyte and the temperature can be employed. Depending on the electrolyte employed, different electrode reactions occur, as shown in Table 3. Hydrogen is formed at the cathode from water molecules in alkaline solution, and oxygen is formed from hydroxide ions at the anode, whereas half of the amount of water used up at the cathode is reformed at the anode [see Eq. (1) in Table 3]. The reverse is true in acid electrolytes [see Eq. (2) in Table 3]: Three moles of water react at the anode to give oxygen and hydroxonium ions, which migrate to the cathode, where they are discharged to give hydrogen. Equations (2) are only approximate because the proton may be hydrated to a greater or lesser extent than that corresponding to the hydroxonium ion. These equations for acid electrolytes are also valid for the proton-exchange-membrane cell (see below). The third case in Table 3—Eqs. (3)—only holds for a purely proton-conducting electrolyte, which exists especially at higher temperatures, such as a ceramic electrolyte; however, electrolytes of this type have not yet been used in technical cells. The fourth case

Table 3
Electrode Reactions of the Electrolysis of Water

Cathode	Anode	Electrolyte	Equation
$2H_2O + 2e^- = H_2 + 2OH^-$	$2OH^- = 1/2\, O_2 + H_2O + 2e^-$	Alkaline	(1)
$2H_3O^+ + 2e^- = H_2 + 2H_2O$	$3H_2O = 1/2\, O_2 + 2H_3O^+ + 2e^-$	Acid	(2)
$2H^+ + 2e^- = H_2$	$H_2O = 1/2\, O_2 + 2H^+ + 2e^-$	H^+-conducting	(3)
$H_2O + 2e^- = H_2 + O^{2-}$	$O^{2-} = 1/2\, O_2 + 2e^-$	O^{2-}-conducting	(4)

Table 4
Electrolytic Processes for the Generation of Hydrogen

I. Direct water electrolysis
   1. Water electrolysis with liquid electrolytes
      a. Alkaline (KOH, NaOH; bipolar, mon polar) (AWE)
      b. Acid (PWE)
   2. Water electrolysis with immobilized liquid electrolytes
      a. Alkaline (IAWE)
      b. Acid (IPWE)
   3. Water electrolysis with polymer membrane electrolyte
      a. Anion-exchange membrane (AMWE)
      b. Proton-exchange membrane (PMWE)
II. High-temperature water vapor electrolysis (HTWE)
   1. Water electrolysis with molten electrolyte (MEWE)
      a. Alkaline (MAWE)
      b. Acid (MPWE)
      c. Hydride (MHWE)
   2. Water electrolysis with ceramic electrolyte
      a. Ceramic proton conductor water electrolysis (CPWE)
      b. Ceramic oxygen ion conductor water electrolysis = solid oxide water electrolysis) (SOWE)
III. Electrolysis with thermochemical cycles
IV. Electrolysis with chemical support (anodic depolarization)
V. Electrolysis with biochemical support
VI. Electrolysis with photochemical support
VII. Photoelectrochemical water electrolysis
VIII. Electrolysis generating hydrogen as a by-product

concerns the high-temperature water vapor electrolysis presently being developed, which employs a purely oxygen-conducting ceramic, namely, doped zirconium oxide [see Eqs. (4) in Table 3].

A categorization of electrolytic processes for the generation of hydrogen, including those which are still in the research stage, on the basis of the electrolytes employed and a number of additional criteria is presented in Table 4. The state of development of these processes will be briefly explained later on—it will be seen that processes III–VII are unimportant at present, although they may gain significance in the future. Process VIII is not a water electrolysis but an electrolysis generating hydrogen as a by-product. In particular, chlorine electrolysis, which is the largest electrolytic hydrogen generator of today, belongs in this category.

V. THERMODYNAMICS, CONDUCTIVITY PROBLEMS, AND ELECTROCATALYSIS OF WATER ELECTROLYSIS

From the overall reaction of water electrolysis, the decomposition (splitting) of water into hydrogen and oxygen,

$$H_2O = H_2 + \frac{1}{2} O_2 \tag{5}$$

it follows that the reaction enthalpy for water splitting, $\Delta_d H_{H_2O}$ (d is short for decomposition) equals the negative enthalpy of formation of water, $-\Delta_f H_{H_2O}$, and this, in turn, equals the negative enthalpy (heat) of combustion of hydrogen, $-\Delta_c H_{H_2}$:

$$\Delta_d H_{H_2O} = -\Delta_f H_{H_2O} = -\Delta_c H_{H_2} \tag{6}$$

Some thermodynamic data are compiled in Table 5, including also the values for 100°C and 1000°C. The corresponding free enthalpies of reaction are also listed in Table 5. These values have been used to calculate the reversible cell voltages of the hydrogen/oxygen cell according to the well-known equation

$$U_{rv} = \frac{\Delta G}{nF} \tag{7}$$

In addition, the values for the thermoneutral voltage are also listed in Table 5, the thermoneutral voltage is defined by

Table 5
Thermodynamic Data of Water Splitting (Heat of Combustion of Hydrogen $\Delta_c H_{H_2}$ and that of Formation of Water $\Delta_f H_{H_2O}$) As Well As the Thermoneutral Voltage U_{tn} and the Reversible Voltage U_{rv}

	298 K	298 K (gas)	373 K (as)	1273 (gas)
$\Delta_f H_{H_2O}$ (kJ/mol)	285.9	241.8	242.6	249.4
$\Delta_f G_{H_2O}$ (kJ/mol)	237.2	228.6	225.1	177.1
U_{tn} (V)	1.48	1.25	1.26	1.29
U_{rv} (V)	1.23	1.18	1.17	0.92
$\Delta_c H_{H_2}$ (kJ/kg)	141,800	120,000	120,300	123,700
$\Delta_c H_{H_2}$ (kJ/m^3)	12,750	10,800	10,800	11,100
$\Delta_c H_{H_2}$ (kWh/kg)	39.4	33.3	33.4	34.4
$\Delta_c H_{H_2}$ (kWh/m^3)	3.55	3.0	3.0	3.1

$$U_{tn} = \frac{\Delta H}{nF} \qquad (8)$$

It is not a physical quantity, but only a formal quantity that indicates the cell voltage at which the electrical energy equals the heat of combustion of hydrogen (enthalpy of formation of water).

Assuming complete conversion of the electrical energy into hydrogen and oxygen, the energetic degree of efficiency can be calculated from the voltage values; it is defined by

$$\eta = \frac{\Delta H}{\text{El.energy}} = \frac{\Delta H}{nFU} = \frac{U_{tn}}{U} \qquad (9)$$

U is the cell voltage under operation. If the thermoneutral voltage is referred to the standard reaction enthalpy, that is, to 25°C, the electrolysis would run at a cell voltage of 1.48 V and a degree of efficiency of 100%—which is not yet technically possible at room temperature. If the electrolysis is carried out at the thermoneutral voltage of 1.29 V at 1000°C, which has been already done, the degree of efficiency would amount to 115%.

Using $U_{tn} = 1.48$ V in the calculations, the degree of efficiency of the electrolysis is referred to the higher heating value, which corresponds to the heat of combustion $\Delta_c H_{H_2}^{298}$ forming liquid water as combustion product at 298.15 K and 1 bar (0.1 MPa); this means the same as using liquid water as educt for the electrolysis. For gaseous water at 298.15 K and 0.1 MPa, the following relationship is obtained:

$$\Delta_{dg} H_{H_2O}^{298} = -\Delta_{fg} H_{H_2O}^{298} = -\Delta_{cg} H_{H_2}^{298} \qquad (10)$$

where $\Delta_{cg} H_{H_2}^{298}$ equals the lower heating value of hydrogen (g is short for gaseous).

If one wants to refer the degree of efficiency of water electrolysis to the lower heating value, which means water vapor at 25°C as educt, the thermoneutral voltage $U_{tn} = 1.25$ V has to be applied. In that case an electrolysis with a degree of efficiency of $\eta = 87\%$ would only have a degree of efficiency of $\eta_g = 75\%$ which value seems to be less favorable.

Using the well-known equation

$$T\Delta S = \Delta H - \Delta G \qquad (11)$$

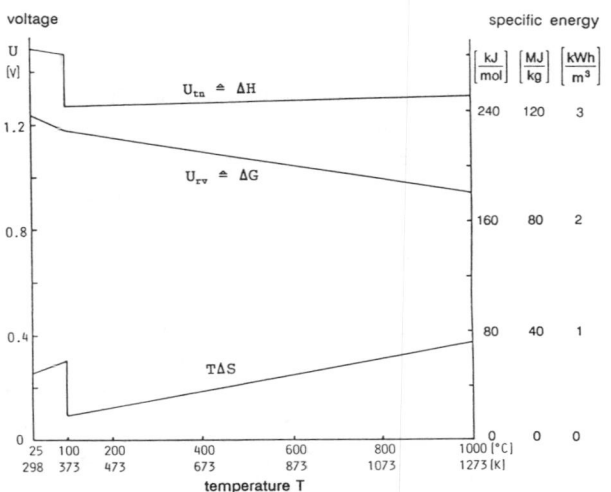

Figure 5. Temperature dependence of thermodynamic quantities and cell voltages of water electrolysis and hydrogen combustion.

the entropy part $T\Delta S$ of the reaction enthalpy can be calculated to be 0.9 kWh/m^3 of hydrogen at 1000°C. Up to this amount of thermal energy might be theoretically fed into the reaction at 1000°C (cf. also Fig. 5).

Figure 5 evidently shows that the minimum electrical energy for the splitting of water, ΔG, which is proportional to U_{rv}, decreases substantially with increasing temperature. The angle in the curve at $U_{rv} \propto \Delta G$ at 100°C, which corresponds to the step in the $T\Delta S$ curve, is due to the heat of vaporization of water.

In order to improve the water electrolysis process, in particular with respect to energy consumption, the components of the cell, the module, and the whole plant have to be investigated one by one. First, the increase of the cell voltage, which results in a decrease in the degree of efficiency, will be considered. In particular, the ohmic resistance of the cell (cathode, electrolyte, separator, electrolyte, anode) and module must be as small as possible. The connection of several cells can be executed either in a monopolar or in a bipolar arrangement. The bipolar arrangement, which is most frequent, is illustrated in Fig. 6.

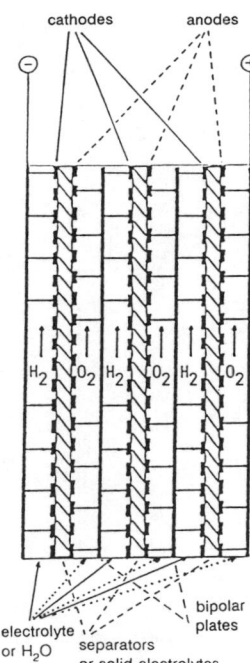

Figure 6. Bipolar cell arrangement for water electrolysis.

Obviously, an especially compact design is obtained if the back of the anode of one cell is combined with the back of the cathode of the next cell using an electrically conductive spacer and an electrically conducting but gastight separating plate (bipolar plate) and again an electrically conducting spacer in between. This leads to the electrical series connection, whereas parallel connection is used for the monopolar arrangement. The cells in such a module are robust and easily exchangeable, but a high electrical current is needed. It is possible to construct double cells or to immerse a number of cells, electrically connected in parallel, in one tank, which gives the so-called tank arrangement. In contrast, the total voltage is high and the current is relatively small in the bipolar arrangement. In addition, the ohmic drop in the electrodes and the connections is smaller than in the monopolar design because of the short distances, so that thinner cells and electrodes with larger areas can be constructed.

Besides the electrical resistance R of the cell, and hence the ohmic drop iR, the polarization of the electrodes also plays a major role in

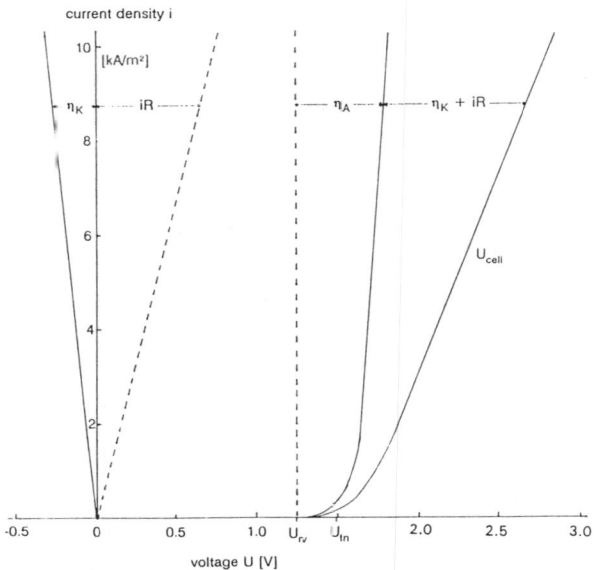

Figure 7. Current–voltage curve of a water electrolysis cell as well as the polarization curves for the anode and the cathode and the ohmic loss.

determining the cell voltage and hence the degree of efficiency. A typical current–voltage curve, together with the partition into the polarization curves, for alkaline water electrolysis is shown in Fig. 7.

The cell voltage corresponding to the current density i can be written as

$$U = U_{rv} + \eta_{Cathode} + \eta_{Anode} + iR \tag{12}$$

where $\eta_{Cathode}$ is the cathode polarization (also called hydrogen overvoltage), and η_{Anode} is the anode polarization, the oxygen overvoltage. In order to reduce the operational voltage at a given current and hence to increase the energetic degree of efficiency, smaller overvoltages have to be attained, which is achieved principally by applying electrocatalysts to the electrodes. Up to now, iron cathodes and stainless steel anodes have been in use in technical electrolysis. Electrodes made from nickel would be better but more expensive. On the other hand, the electrodes could be essentially improved by generating iron and nickel in finely divided form on their surface.

As the surface of the anode is always oxidized, a great number of mixed oxides have been investigated in order to decrease the oxygen overvoltage. The possible electrocatalysts will be described in Section VI.

As electrocatalysis essentially determines the reaction mechanism, the mechanisms of the electrode reactions will be summarized here in a unified general description based on the literature. A redox mechanism is assumed nowadays for the anodic oxygen evolution, especially due to the investigations of Trasatti.[23] This should be true also with metallic electrodes, because all of these are covered by an oxide layer at high potentials. In order to formulate the mechanism in a general way, a surface metal atom $\underline{M}$ is taken, to which oxygen is bound. The underlineing means that the surface metal atom (ion) belongs to the crystal structure. This metal ion undergoes a change of the charge (valence state) from II to III or from III to IV or from IV to V or even from I to II. (In the latter case, and possibly in the I/II case, OH may be on the surface instead of O.) The reaction sequence would then be for acid electrolyte:

$$\underline{M}O = \underline{M}^+O + e^- \tag{13}$$

$$\underline{M}^+O + 2H_2O = \underline{M}OOH + H_3O^+ \tag{14}$$

$$\underline{M}O + 2H_2O = \underline{M}OOH + H_3O^+ + e^- \tag{15}$$

For such a redox reaction, at first a (at least partial change of the charge (valence state) has to occur due to the change of potential of the anode. After the oxidation reaction (13), the saturation of the charge by H_2O (reaction with H_2O to form a hydrated proton) follows [see Eq. (14)]. If this consecutive reaction is very fast, Eq. (15) can take place directly. In order to achieve the second electron transfer, one can assume the same oxidation reaction (Eq. 13) on a neighboring site. After that, the $\underline{M}OOH$ formed previously according to Eq. (14) can react with the newly formed $\underline{M}^+O$ in the neighborhood to form an oxygen bridge (Eq. 16). Finally, saturation of the charge by H_2O occurs again (Eq. 17). If this reaction is very fast, the reaction sequence may be possible in one step [see Eq. (18)]. The sequence is as follows:

$$\underline{M}O = \underline{M}^+O + e^- \qquad (13)$$

$$\underline{M}^+O + \underline{M}OOH = \underline{M}^+OO\underline{M}OH \qquad (16)$$

$$\underline{M}^+OO\underline{M}OH + H_2O = \underline{M}OO\underline{M}O + H_3O^+ \qquad (17)$$

$$\underline{M}O + \underline{M}OOH + H_2O = \underline{M}OO\underline{M}O + H_3O^+ + e^- \qquad (18)$$

Alternatively, the bridge complex might be formed by elimination of H_2O from two neighboring $\underline{M}OOH$ groups:

$$2\underline{M}OOH = \underline{M}OO\underline{M}O + H_2O \qquad (19)$$

When finally two neighboring oxygen bridge complexes have been formed, they can decompose with formation of an oxygen molecule:

$$2\underline{M}OO\underline{M}O = 4\underline{M}O + O_2 \qquad (20)$$

It can also be imagined that, instead of Eq. (16) or (18), a second electron transfer takes place at the same surface ion $\underline{M}^+$ according to Eq. (13) or (15). This means a total change of two charges at the same surface metal atom (ion) $\underline{M}$, namely, I/III, II/IV, III/V, or IV/VI, which would again be possible in two steps:

$$\underline{M}OOH = \underline{M}^+OOH + e^- \qquad (21)$$

$$\underline{M}^+OOH + H_2O = \underline{M}O_2 + H_3O^+ \qquad (22)$$

$$\underline{M}OOH + H_2O = \underline{M}O_2 + H_3O^+ + e^- \qquad (23)$$

or in one step according to reaction (23).

An oxygen molecule may be split off from two neighboring $\underline{M}O_2$ complexes:

$$2\underline{M}O_2 = 2\underline{M}O + O_2 \qquad (24)$$

Alternatively, an oxygen bridge complex may also be involved. Furthermore, even more complicated surface structures may occur due to the possibility of electron transfer reactions taking place in parallel and consecutively.

As the mechanisms have mostly been derived for alkaline electrolyte up to now, the reaction sequences will also be formulated for alkaline solution:

$$\underline{M}O = \underline{M}^+O + e^- \quad (13)$$

$$\underline{\underline{M}^+O + OH^- = \underline{M}OOH} \quad (25)$$

$$\underline{M}^+O + OH^- = \underline{M}OOH + e^- \quad (26)$$

$$\underline{M}O + = \underline{M}^+O + e^- \quad (13)$$

$$\underline{M}^+O + + \underline{M}OOH = \underline{M}^+OO\underline{M}OH \quad (16)$$

$$\underline{\underline{M}^+OO\underline{M}OH + OH^- = \underline{M}OO\underline{M}O + H_2O} \quad (27)$$

$$\underline{M}O + \underline{M}OOH + OH^- = \underline{M}OO\underline{M}O + H_2O + e^- \quad (28)$$

After that comes Eq. (20) for the splitting off of the oxygen molecule. The alternative pathway, namely, the second electron transfer at the same surface ion $\underline{M}$ (Eqs. 21–23) is written as follows for alkaline electrolyte:

$$\underline{M}OOH = \underline{M}^+OOH + e^- \quad (21)$$

$$\underline{\underline{M}^+OOH + OH^- = \underline{M}O_2 + H_2O} \quad (29)$$

$$\underline{M}OOH + OH^- = \underline{M}O_2 + H_2O + e^- \quad (30)$$

This is followed by Eq. (24) for the evolution of oxygen.

If one takes the reaction sequence of Eqs. (26), (30), and (24) and substitutes the metal M for the oxidized surface $\underline{M}O$, then Bockris's electrochemical path is obtained.[24]

The mechanism of the cathodic reaction, the hydrogen evolution, at metallic electrodes was explained long ago. It takes place according to the chemisorption mechanism at the surface of the metal M, starting with the Volmer reaction; for acid electrolyte, this reaction is

$$M + e^- + H_3O^+ = MH_{ad} + H_2O \quad (31)$$

This electrosorption reaction is followed by the Tafel reaction, namely, the desorption of hydrogen:

$$2MH_{ad} = 2M + H_2 \quad (32)$$

Instead, a second electron transfer reaction may occur, which is called Heyrovský reaction:

$$MH_{ad} + e^- + H_3O^+ = M + H_2 + H_2O \tag{33}$$

For alkaline medium, the reactions corresponding to Eqs. (31) and (33) are:

$$M + e^- + H_2O = MH_{ad} + OH^- \tag{34}$$

$$MH_{ad} + e^- + H_2O = M + H_2 + OH^- \tag{35}$$

whereas Eq. (32) remains the same. Lately, oxides have also been used as electrocatalysts for the cathodic reaction, but the question has not yet been settled as to whether the chemisorption mechanism (Volmer–Tafel or Volmer–Heyrovský) or a redox mechanism takes place at the metal oxide surface not being reduced. In the redox case the surface metal ion $\underline{M}$ of the oxide would undergo a change in valence state, either II/I or III/II or IV/III. If one designates the surface of the metal oxide by $\underline{M}O$, the red x mechanism could be formulated as follows:

$$\underline{M}O + e^- = \underline{M}^-O \tag{36}$$

$$\underline{M}^-O + H_3O^+ = \underline{M}OH + H_2O \tag{37}$$

$$\underline{M}O + H_3O^+ + e^- = \underline{M}OH + H_2O \tag{38}$$

The electron transfer is either followed by the chemical reaction (37) with the hydrated proton or occurs at the same time as reaction (37) to give reaction (38).

Hydrogen may be evolved chemically from two neighboring groups, in a similar manner as in the Tafel reaction:

$$2\underline{M}OH = 2\underline{M}O + H_2 \tag{39}$$

or with the involvement of a second electron transfer in a similar manner as in the Heyrovský reaction:

$$\underline{M}OH + e^- = \underline{M}^-OH \tag{40}$$

$$\underline{M}^-OH + H_3O^+ = \underline{M}O + H_2 + H_2O \tag{41}$$

$$\underline{M}OH + H_3O^+ + e^- = \underline{M}O + H_2 + H_2O \tag{42}$$

For alkaline electrolyte the equations would read:

$$\underline{M}O + e^- = \underline{M}^-O \qquad (36)$$

$$\underline{M}^-O + H_2O = \underline{M}OH + OH^- \qquad (43)$$

$$\underline{M}^-O + H_2O = \underline{M}OH + OH^- \qquad (44)$$

$$\underline{M}OH + e^- = \underline{M}^-OH \qquad (40)$$

$$\underline{M}^-OH + H_2O = \underline{M}O + H_2 + OH^- \qquad (45)$$

$$\underline{M}OH + H_2O + e^- = \underline{M}O + H_2 + OH^- \qquad (46)$$

One could imagine that such a redox mechanism would take place or at least contribute in the case of mixed catalysts such as nickel with molybdenum oxide or the platinum metal oxides, especially with RuO_2.

Unfortunately, a further discussion is not possible for reasons of space (cf. Refs. 23–27). The nature of the electrocatalysts for the cathode as well as for the anode must be such that the Tafel slopes obtained are as low as possible over a wide range of current density. This would mean that the first electron transfer would be fast and the desorption would be fast as well, but the coverage would not be too high.

Quite another mechanism can occur in the medium temperature range, at least at the anode. In a hydroxide melt the anodic reaction leads to the superoxide anion as intermediate:

$$4OH^- = O_2^- + 2H_2O + 3e^- \qquad (47)$$

$$O_2^- + \frac{1}{2} H_2O + \frac{3}{4} O_2 + OH^- \qquad (48)$$

An unfavorable side reaction can occur, because the superoxide anion can diffuse to the cathode, where it will be reduced according to the reverse of reaction (47).[28]

At still higher temperature, from about 800°C upwards, the reactions are substantially accelerated in such a way that intermediates practically do not occur or do not affect the rate. The following reaction mechanism can be formulated for a proton-conducting solid electrolyte [cf. Eqs. (3) and (4)]:

Cathode: $2H^+ + 2e^- = H_2$ (49)

Anode: $H_2O = H^+ + OH_{ad} + e^-$ (50)

$OH_{ad} = H^+ + O_{ad} + e^-$ (51)

$2O_{ad} = O_2$ (52)

For an oxygen-conducting solid electrolyte, we obtain:

Cathode: $H_2O + e^- = H_{ad} + OH^-_{ad}$ (53)

$OH^-_{ad} + e^- = H_{ad} + O^-$ (54)

$2H^-_{ad} = H_2$ (55)

Anode: $2O^- = O_2 + 4e^-$ (56)

Table 6
Parameters for the Increase of the Degree of Efficiency of Water Electrolysis

A. Reduction of ohmic losses by
- Changing the electrolyte composition
- Increasing the temperature
- Decreasing the thickness of cell components
- Decreasing the distance, e.g., zero gap
- Employing more conductive structure materials
- Leading the gas bubble stream to the back of the electrodes
- Improving the separator

B. Decrease of the overvoltages by
- Employing improved electrocatalysts
- Increasing the temperature (thermal activation)
- Increasing the pressure

C. Lowering of the reversible voltage by
- Increasing the temperature

D. Increase of the current density (and optimization of the current density together with the voltage) by
- Improving electrocatalysis as well as increasing the temperature and pressure (as in B above)

The uptake and removal of an oxygen ion onto and from the surface of the solid electrolyte, for example, doped zirconium oxide, take place via a vacancy mechanism.

The considerations presented in this section about thermodynamics, electrical conductivity, and the electrocatalysis/reaction mechanism of water electrolysis constitute the basis for establishing ways in which the effectiveness of hydrogen and oxygen evolution may be increased. The possible measures that may be taken to increase the degree of efficiency, including considerations related to the cell components, are summarized in Table 6.

VI. MATERIAL TECHNOLOGIES FOR WATER ELECTROLYSIS, INCLUDING THE ELECTROCATALYSTS

Material technologies play an ever increasing role not only in improving water electrolysis processes but also in reducing their costs. Improved materials are needed to achieve better performance, for example, higher efficiency. Higher efficiency, in turn, means lower cell voltage, which saves electrical energy. On the other hand, higher current density enables a more compact construction, which saves material and, hence, reduces investment costs. Therefore, material technologies are decisive for progress. The most important material technologies are compiled in Table 7. This table should give an idea of what has to be taken into account. Unfortunately, a detailed consideration of all items is not possible; instead,

Table 7
Material Technologies for Water Electrolysis

Parts and systems	Components and properties
Electrode	Mechanical rigidity
	Corrosion stability
	Transport pore structure
	High internal surface area
	Electrocatalysts
	Activity/aging
	Insensitivity to poisoning
	Electrical conductivity
	Heat conductivity

(continued)

Table 7 (continued)

Parts and systems	Components and properties
Cell	Structure materials
	Frame
	Electrode support materials
	Current conductors
	Gaskets
	Gas removal
	Electrolytes (liquid/solid)
	Separator/diaphragm
	Spacer
	High electrical conductivity (electronic/ionic)
	Large size
Module	Electrical connections
	End plates
	Bipolar plates
	Electrolyte channels
	Electrolyte tanks
	Corrosion stability
	Electrolyte transport
	Gas transport
	Shunt avoidance
	Heat insulation
	Cooling
	Thermal cycling tolerance
Plant	Container
	Tubes
	Pipes
	Pumps
	Gas vessels
	Heat exchanger
	Pressure stability
	Temperature stability
	Size
	Lifetime
Additional conditions	Safety measures
	Environmental neutrality
	Service friendliness
	Manufacturing technology
	Economy

Table 8
Electrolytes for Water Electrolysis

Aqueous KOH
Aqueous NaOH
Aqueous H_2SO_4
Polymer membrane
KOH melt
Molten acid
Molten hydrid
Proton-conducting ceramic
Oxygen-ion-conducting ceramic

only electrolytes, electrocatalysts, and separators will be dealt with briefly here.

The electrolytes that have been investigated so far fall into the groups listed in Table 8. Highly concentrated KOH or NaOH solutions have their conductivity maximum at about 160°C; this is why a temperature as high as possible is desired. Because of the water vapor pressure, a pressure of about 3 MPa has to be applied. For the time being, aqueous acids are not used, but they have been investigated (see below). However, a quasi-acid electrolyte is used, namely, a cation-exchange membrane, also called a solid polymer membrane. It contains bound sulfonic acid groups and, therefore, conducts hydroxonium ions, which are hydrated protons. The first membrane of this kind that was sufficiently stable was Du Pont's Nafion, made from PTFE. Since its introduction, other manufacturers have produced membranes of different compositions; recently, the Paul Scherrer Institute has published some interesting results.[29] At higher temperatures, corrosion is still problematic with molten KOH and especially with the molten acids, although a lot of know-how has been gained from work with the phosphoric acid fuel cell, but the oxygen anode in an electrolysis cell using pyrophosphoric acid as electrolyte is not stable enough. Molten metal hydrides are being investigated.[30]

Electrolysis cells with a proton-conducting solid electrolyte would be interesting, because they can potentially be operated at temperatures above 100°C, up to about 800°C. Initial results have been promising.[31] At present, such electrolytes are systematically being investigated by many groups in Germany (including groups at Battelle Frankfurt, KfK, the Max

Table 9
Proton-Conducting Inorganic Solids

	Conductivity (Ω^{-1} cm^{-1})	
β-Alumina(H)	~10^{-5}	(25°C)
Nasicon	~10^{-4}	(25°C)
Zirconium phosphate	~10^{-4}	(25°C)
Uranyl phosphate	~10^{-4}	(25°C)
Antimonic acid	~10^{-5}	(25°C)
Tantalum oxide	~10^{-5}	(25°C)
Molybdo phosphoric acid	~10^{-2}	(25°C)
montmorillonite	~10^{-4}	(25°C)
Sn-mordenite	~10^{-2}	(120°C)
Perovskite SrCe(Y,Yb)O$_3$	~10^{-2}	(955°C)
Perovskite BaCe(Gd,Nd)O$_3$	~10^{-3}	(700°C)
CaY-phosphate	~4×10^{-4}	(800°C)

Planck Institute,[32,33] TH Darmstadt, and ZSW), and encouraging results have also come from Denmark,[34,35] England,[36,37] Japan, the United States, and other countries. A systematic review of the publications on this subject is beyond the scope of this chapter, but the compilation of data is presented in Table 9.

High-temperature water electrolysis with oxygen-ion-conducting solid electrolytes will be dealt with later on. The ceramic material zirconium oxide is one example of a pure oxygen conductor. If it is doped with yttrium oxide, its conductivity amounts to about 0.1 Ω^{-1} cm^{-1} at 1000°C. Doping with ytterbium or scandium leads to even higher conductivity but at a higher price, too. Cerium-containing electrolytes, which also have a higher conductivity, are under development. If the electronic part of their conductivity can be optimized, then the operating temperature of the cell may be lowered to about 900°C.

Next, we will consider the electrocatalysts, which are used normally at temperatures of about 50°C to about 100°C or a little higher. Tables 10 and 11 show the electrocatalysts that have been investigated in detail for alkaline cell and acid cells, respectively.

As Raney nickel (Ra Ni) is widely used nowadays, it will be examined more closely here. Raney alloys were introduced in electrochemistry by Justi and Winsel, and already in 1957 Winsel developed a Raney nickel

Table 10
Electrocatalysts for Water Electrolysis in Alkaline Electrolytes (AWE)

Cathode, H_2	Anode, O_2
Fe, Fe/Ni, Ni/S, Raney Ni, NiB_2, Ni/Zn, Pt/Ni, Ni/Cd, Ni/Mo, Si–W–O, amorphous metals	Steel, Ni, spinels (e.g., $NiCo_2O_4$, Co_3O_4), perovskites (e.g., $LaNiO_3$, $La(Sr)Co(Ni)O_3$), Raney Ni, amorphous metals

cathode.[38] Later on, Vielstich[39] as well as the group of one of the present authors[40] applied Raney nickel electrodes. Raney electrodes for technical applications were first constructed by Müller, Lohrberg, and Wüllenweber at Lurgi.[41] In order to form such electrodes, a Raney nickel/aluminum alloy is mixed with nickel carbonyl powder, and the mixture is hot pressed onto a metal sheet and finally sintered. Figure 8 shows a schematic drawing of the type of electrode.

In the last step, aluminum will be dissolved by a concentrated alkaline solution. Likewise, the nickel–aluminum alloy or nickel–zinc alloy can be formed from the metals directly onto the electrode sheet (see Plzak and Wendt[42,43]). The Raney alloy can also be deposited on the metal sheet with the use of a plasma spraying technique (see Schnurnberger *et al.*[44,45]). In addition, a nickel–zinc alloy can be formed by cathodic deposition (see Divisek *et al.*[46–49]). The catalytic activity of nickel can be increased by alloying with cobalt and molybdenum.[50] A Ni–Mo alloy catalyst is used for a 300-kW electrolysis plant by the French company ACB = Alsthom.[51]

With respect to the decrease of the polarization of the anode, a large number of mixed oxides have been investigated. Good results have been achieved with spinels, namely, cobalt spinel (see, e.g., the DLR group[52,53]) and nickel cobalt spinel, as well as with perovskites. Lanthanum nickel perovskite and lanthanum strontium cobalt perovskite have been most

Table 11
Electrocatalysts for Water Electrolysis in Acid Electrolytes (PWE)

Cathode, H_2	Anode, O_2
Pt, Pt metals, RuO_2, IrO_2, Ni–Mo, WC, WO_2/TiO_2	Pt, RuO_2, IrO_2, Rh, Pt/TiO_2, GeS_2, Fe/Ta_2O_5

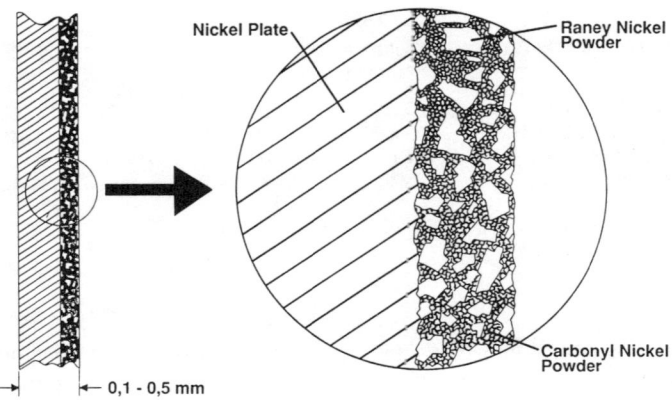

Figure 8. Lurgi rolled plate Raney nickel electrode.

effective (see Divisek and co-workers[54,55]). In addition, lithiated nickel oxide is used as an electrocatalyst for oxygen evolution. Raney nickel, however, is also best at the oxygen anode,[41,48,56] based on many years of experience at Lurgi. New possibilities for electrocatalysis might result from the application of vitreous amorphous metals.[57] The section of materials for water electrolysis in acid electrolytes is much smaller than in the case of alkaline electrolytes. As the development of acid cells is quite promising owing to the performance of the electrodes, there have recently been attempts to investigate non-noble-metal catalysts besides platinum metal electrodes. Water electrolysis in liquid acid electrolytes (PWE), however, is only carried out in the laboratory.

On the other hand, cells with a solid acid electrolyte—a proton-exchange membrane or polymer membrane (PMWE)—are very interesting. With this type of cells, the cathode chamber contains only pure water (cf. Table 3 and Fig. 6). The first cells of this kind have been developed by General Electric Co.[5] and are called SPE (solid polymer electrolyte) cells. The catalyst is mostly platinum, which is applied in a very thin layer either onto the electrode carriers or directly onto both sides of the membrane. In this case the porous electrode carriers are pressed against the catalyst layers on both sides of the membrane for electrical conduction. Mixed ruthenium oxide/iridium oxide on titanium oxide is used as the oxygen anode catalyst. Surprisingly, this mixed oxide is also effective as

Table 12
Separators for Alkaline Electrolytes

(Asbestos)
Polysulfone/Sb_2O_5 (PAM)
Polyacrylethersulfone (Udal)
Polyphenylsulfone (Radel)
Polyethersulfone (Victrex)
Polyphenylene sulfide (PPS, Ryton)
Polybenzimidazole (PBI)/K_2TiO_3
PTFE
$BaTiO_3$, $CaTiO_3$
NiO

a cathode electrocatalyst and is less sensitive, for instance, with respect to iron deposition.[58]

A complete list of the electrocatalysts for hydrogen evolution has been compiled by Trasatti[26] both for alkaline and for acid electrolytes. A compilation of electrocatalysts for oxygen evolution can be found in a book by Kinoshita.[27]

The separator, also called the diaphragm, is an essential part of the cell. A selection of diaphragm materials is compiled in Table 12. Asbestos has been widely used but has had to be replaced for environmental reasons. The new separators are more expensive but have a higher temperature stability with a relatively low resistance, from 0.1 to 0.2 Ω cm^2. Ceramic titanate separators have been developed at TH Darmstadt,[59] and nickel oxide separators at KFA Jülich.[50] In addition, ion-exchange membranes have been developed in addition to nonconducting separators, also for alkaline electrolytes (see, e.g., Rapp et al. [60]).

The structural materials, spacers, conduction materials, sealing materials, etc., listed in Table 7 as well as the engineering and other conditions cannot be dealt with in this chapter. The reader is referred to an article by Wendt[61] for a discussion of these topics. At his point, it may be mentioned that much progress in water electrolysis has been achieved within a recently terminated program of the European Community, coordinated by Wendt.[62] Furthermore, the International Energy Agency (IEA) has a cooperative program for hydrogen generation including water electrolysis, which has been coordinated for many years by Struck (see the proceedings of the last workshop[63]).

VII. TECHNICAL STATE OF THE ART OF WATER ELECTROLYSIS

All large-scale industrial water electrolysis plants operate on alkaline electrolytes, about 30% potassium hydroxide solution. The largest plant was built by BBC (now ABB) at the Aswan coffer dam in Egypt. It has a hydrogen production rate of 33,000 m^3/h, corresponding to an electrical power of 150 MW. The yearly production of hydrogen by water electrolysis is about ca. 2 billion m^3 in the world. This may be compared with chlorine brine electrolysis, which generates about 20 billion m^3 of hydrogen per year as a side product. The total amount of hydrogen mainly produced by thermal methods comes to about 500 billion m^3/yr in the world. To get an idea what this means energetically, this figure may be recalculated with the theoretical energy consumption of 3.55 kWh/m^3 of

Table
Characteristic Data of

Manufacturer	Cell type	Operating temperature (°C)	Operating pressure (bar)
ABB	Proton-exchange membrane, bipolar	80	1.5
GE	Proton-exchange membrane, bipolar	80	40
ABB	Alkaline, bipolar	80	1
Davy Bamag	Alkaline, bipolar	85	1
De Nora	Alkaline, bipolar	80	1
GHW	Alkaline, bipolar	130	30
Hydrogen systems	Alkaline, bipolar	110	5
Hydrotechnik	Alkaline, bipolar	80	1
KFA	Alkaline, bipolar	100	1.1
Krebskosmo	Alkaline, bipolar	90	1.1
Lurgi	Alkaline, bipolar	90	32
Norsk Hydro	Alkaline, bipolar	80	1
Teledyne	Alkaline, bipolar	80	7
Electrolyzer Corp.	Alkaline, monopolar	80	1

hydrogen: The worldwide hydrogen production corresponds to about 1800 billion kWh, which means nearly 17% of the world electricity generation. The total production of hydrogen in Germany amounts to about 25 billion m^3.

Characteristic data for most of the water electrolyzers (in the commercial or pilot stage) are summarized in Table 13. Some of the types listed will be examined more closely as examples.

Not the largest total plants but the largest blocks of cells are from Lurgi. The Lurgi pressure electrolyzer, which goes back to the Zdansky–Lonza cell, produces hydrogen and oxygen under a pressure of 3 MPa = 30 bar.[64] The pressure electrolyzer depicted in Fig. 9 has a capacity of 740 m^3 of hydrogen per hour, which corresponds to an electrical power of about 3400 kW. It is approximately 10 m long and consists of 560 cells of a diameter of 1.60 m. The zero-gap technology, which has been applied for years, is now being modified, and the electrodes with the rolled-on Raney alloys are being improved. Such an electrode is shown in Fig. 10. It goes without saying that the diaphragms are of the same size. At present,

13
Water Electrolyzers

Current density (kA/m^2)	Cell voltage (V)	Specific energy (kWh/m^3)	Degree of efficiency (%)
10	1.7	4.1	85
10	1.8	4.2	81
2	2.1	4.9	72
2.5	1.9	4.5	77
1.5	1.9	4.6	80
5	1.7	4.1	85
5	1.7	4.1	85
2.5	1.9	4.6	77
5	1.7	4.1	87
5	1.8	4.3	82
4	1.7	4.1	85
3	1.8	4.3	82
4	1.9	4.5	80
2.5	1.8	4.3	81

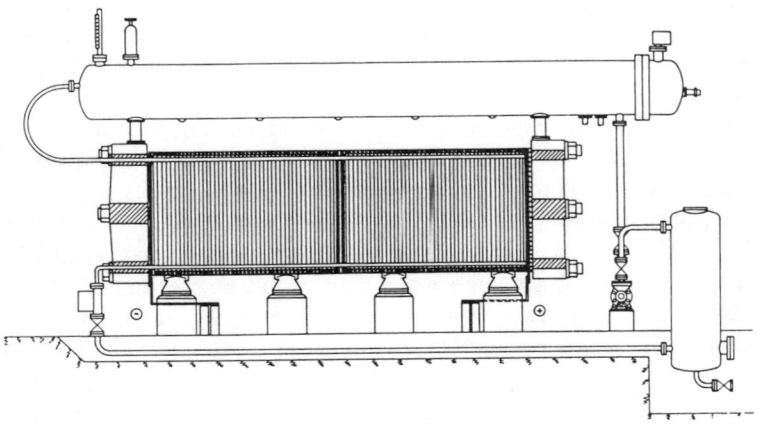

Figure 9. Diagram of a Lurgi pressure electrolyzer.

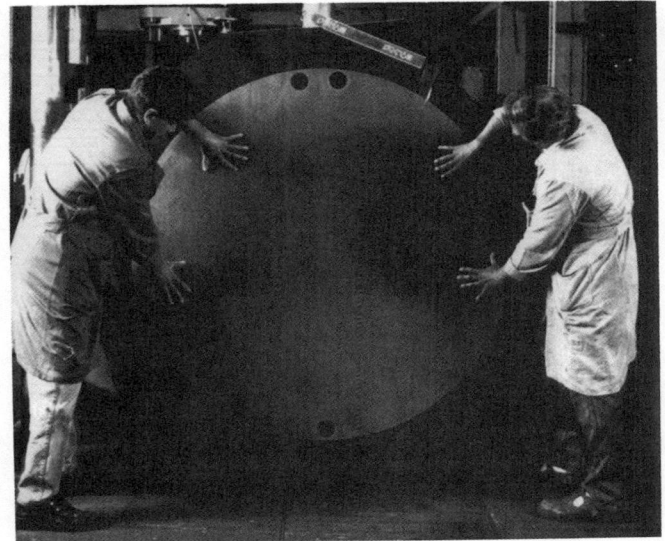

Figure 10. New electrode for the Lurgi pressure electrolyzer.

Figure 11. Cell stack of the new Lurgi pressure electrolyzer.

the asbestos diaphragm is replaced by a nickel oxide diaphragm, which has been developed by KFA. Now the production technology is being worked out by Lurgi.[65] The cell stack is shown in Fig. 11. The new performance data are contained in Table 13.

KFA Jülich has demonstrated the performance of their Raney electrodes and newly developed nickel oxide diaphragm in the form of a stack of cells with a diameter of 50 cm.[66–68] Their 10-kW pilot plant has been successfully tested in the Hysolar project in Stuttgart. Presently, new electrodes with a small amount of platinum catalyst (0.1 mg/cm^2) are being developed; these electrodes tolerate a lengthly exposure to air without any further production measures.[69] Winsel and co-workers developed the eloflux technique (slow passing of electrolyte through the electrodes) further in Kassel, Germany.[70,71] Besides fuel cell blocks, electrolysis blocks of a size of about 1 l have been constructed. The current–voltage curves are presented in Fig. 12. Work on this technique is

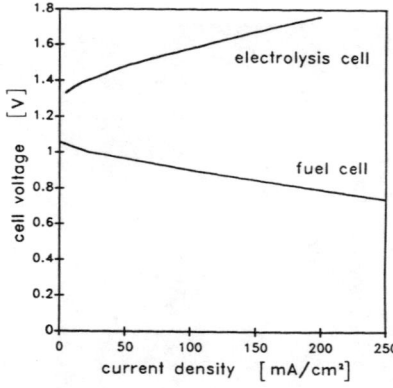

Figure 12. Current–voltage curves of an electrolysis cell and a fuel cell in the corresponding eloflux cell blocks at 80°C and 0.1 MPa.

being continued at the Institute for Solar Energy Technology (ISET) in Kassel.

In Belgium Vandenborre and co-workers[72,73] have commercialized their electrolyzer developed by application of separators made from polysulfone with antimonic oxide as a filler (called IMET). The company is named Hydrogen Systems (see Table 13). The TH Darmstadt group has developed the first oxide ceramic separators[74–76] and, in addition, a compound unit, which consists of a cathode, a diaphragm, and an anode. The porous nickel electrodes are firmly connected to the ceramic, made of calcium titanate. This unit is called EDE and is only 0.8 mm thick; a cross section of this unit is shown in Fig. 13. Its technical development is being continued at GHW, Munich, funded by Daimler-Benz/DASA, Hamburger Electricitäts Werke, and Linde.[77,78] The fabrication of a 1-m^2 compound unit is shown in Fig. 14. These electrodes are constructed for a temperature of 130–150°C at a pressure of about 3 MPa (see Fig. 15) and are to be tested in the Solar Hydrogen Bavaria (SWB) project.

For a long time Krebskosmo has manufactured electrolyzers, which go back to the Demag design.[79,80,5] Recently, this company developed a new electrolyzer with a power uptake of 110 kW, which is used in the SWB project[81] (see Fig. 16). The separator material employed is polysulfone. Unfortunately, Krebskosmo went out of business in 1992.

Other alkaline electrolyzers contained in Table 13 are those of ABB (formerly Oerlikon), Davy Bamag (now belonging to Lurgi), Hydrotech-

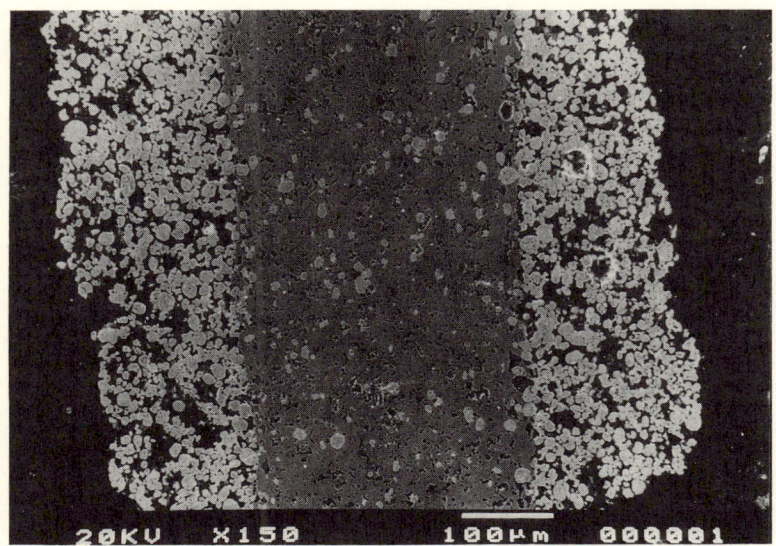

Figure 13. Cross section of the EDE compound unit of GHW.

Figure 14. EDE compound unit fabricated by GHW. The size of the unit is 1 m^2.

Figure 15. The 100-kW water electrolyzer constructed by GHW. The operating temperature is 130–150°C, and the pressure is about 3 MPa.

nik (based on the Demag technology),[82] Norsk Hydro,[5] and Teledyne[5] in the United States.

The last alkaline electrolyzer included in Table 13 is one manufactured in Canada by Electrolyser Corp. It is the only commercial electrolyzer that operates according to the monopolar principle.[5] It has recently been improved by application of Ryton as a separator.[83]

There are other alkaline systems that are not mentioned in Table 13. One of them has been manufactured by the company CJB in England[3] for a long time. The American company Life Systems has built an alkaline electrolyzer of high performance (6.5 kA/m^2 at 1.84 V).[84] Recently, the Swiss firm Metkon/Alyzer Division has introduced several electrolyzers, which are of the alkaline type with Raney nickel electrodes.[85] A plant with a power uptake of 350 kW has been produced for the Hysolar project.

Another alkaline electrolysis cell is that under development by DASA-Dornier/Daimler-Benz. It consists of a matrix containing the immobilized electrolyte, so that the cathode chamber is only filled with

Figure 16. The 110-kW Krebskosmo electrolyzer based on new design.

hydrogen. Water is vaporized, and the vapor diffuses through the hydrogen.[86] A similar arrangement has been developed by Life Systems.[50] Such electrolysis units have been designed for the life support systems on board a spacecraft. They would, however, be suitable also for other smaller applications.

Other concepts that can be used for larger applications may be mentioned. The company Hoechst has developed a so-called "falling film cell," which can be used both for electrolysis, for example, water electrolysis or brine electrolysis, and as a fuel cell.[87] The advantages are a thin electrolyte film and, consequently, a short diffusion length and application of high pressure. The combination of water electrolysis with the fuel cell is envisaged to arrive at a hydrogen/oxygen storage and energy system. In addition, the "capillary electrode cell" of Matschiner and Wenske[88] should

Figure 17. The 1-kW alkaline water electrolysis module developed by DLR, Stuttgart, with 250-cm^2 electrodes, fabricated by the plasma spraying technique.

be mentioned. This cell, which can be used for various electrochemical processes, has recently been further developed, particularly with respect to alkaline water electrolysis. Due to the special flow arrangement for the electrolyte, high current densities can be obtained easily. Finally, the work of DLR, Stuttgart, may be mentioned again. This institute has now fabricated electrodes with a larger area by the plasma spraying technique using Raney alloys or mixed oxides. Figure 17 shows the bipolar module with Raney-Ni–Mo cathodes and Co_3O_4 anodes with an area of 250 cm^2.[53,89]

Acid water electrolysis cells with a cation-exchange membrane in the protonated state have been in the pilot stage for some time. Cells based on this proton-exchange membrane (or polymer membrane) water electrolysis (PMWE) were first introduced by General Electric, who named them SPE (solid polymer electrolyte) cells. ABB (formerly BBC) has also developed such an electrolyzer under the name Membrel cell.[90–93] Other companies now also offer membrane cells; these include the Belgian company Hydrogen Appliances and the Japanese company GIRI. The first 100-kW Membrel electrolyzer was produced by ABB in 1987. In 1991 another 100-kW electrolyzer was delivered to the SWB project.[94] The company then discontinued production of the Membrel electrolyzer, as GE had done before with the SPE cells. In spite of this, the Paul-Scherrer Institute in Zurich has recently begun experiments with the Membrel electrolyzer.[95] Cheaper membranes have become available, which may lead to renewed interest. The Fraunhofer society in Freiburg (FhG-ISE) has developed membrane cells, too, using new methods for catalyst preparation.[96] So far, only small electrolyzers (1 m^3 of hydrogen per hour) are available for solar operation (Fig. 18). In principle, such a membrane cell could be operated alternately as an electrolysis cell and as a fuel cell, thus being a reversible cell. However, this is only possible if the electrodes can operate as bifunctional electrodes. Such electrodes have been developed by Ledjeff and Heinzel.[97] The best results were obtained when the bifunctional electrodes were always operated in the same mode: one always as anode and the other one always as cathode (i.e., either anodic hydrogen oxidation and cathodic oxygen reduction or anodic oxygen evolution and cathodic hydrogen evolution). Aging was higher if the electrodes were switched between anodic and cathodic operation. Such cells, together with a gas container for hydrogen and oxygen, might be used as an energy storage device.

In addition to low-temperature cells (up to ca. 150°C) high-temperature water vapor electrolysis using an oxygen-conducting solid electrolyte (SOWE) is in the technical development stage. The research stage of cells with molten electrolyte (MEWE) and with ceramic proton-conducting electrolyte (CPWE) has already been reported. As already mentioned, the best known solid electrolyte is zirconium oxide stabilized by calcium or yttrium; it is a pure oxygen-ion conductor at high temperature (not at very low oxygen partial pressure) and has been introduced in fuel cell technology. It has been used for the measurement of the oxygen partial pressure in automobile exhaust with the help of the lambda sensor. The reversal of

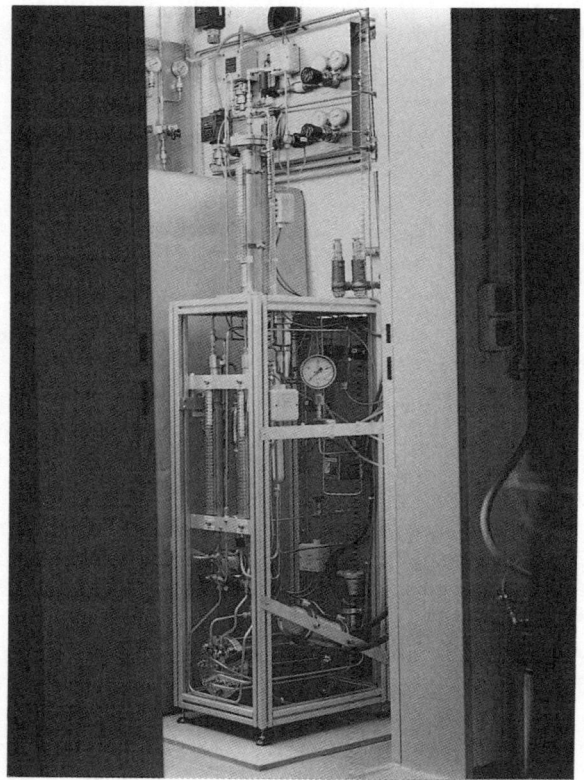

Figure 18. Small membrane electrolyzer developed by FhG-ISE, Freiburg.

the high-temperature fuel cell principle led to water vapor electrolysis,[5,98–101] which was investigated first by Battelle, BBC, and Westinghouse and later on by others. A cell based on water vapor electrolysis is being developed in Germany by Dornier/Daimler-Benz in cooperation with Lurgi and Badenwerk. This cell is called Hot Elly.[102,103]

In the initial work on high-temperature water vapor electrolysis, conical cells have been developed by Battelle and by BBI for material technology reasons, as outlined in Fig. 19 (Ref. 99). This figure shows

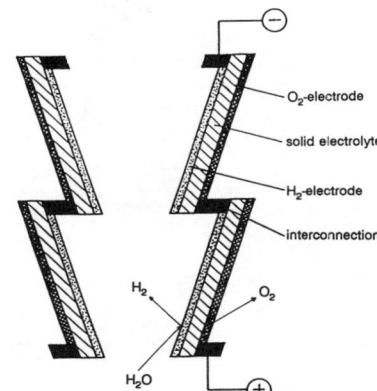

Figure 19. Outline of the bipolar arrangement of conical cells for high-temperature water vapor electrolysis (early design).

that five types of materials are required for module construction; they are compiled in Table 14.

The bipolar electrically connecting material between the cells of the cell stack must meet special requirements: it has to be a good electrical conductor, it has to be gastight, it has to be chemically stable in a reducing as well as an oxidizing atmosphere, and its thermal expansion must be similar to that of the electrolyte. Doped perovskites are used for this purpose (see Table 14). As can be seen from the results of earlier work at Battelle (Fig. 20), there are some perovskites that have a high conductivity both at low and at high oxygen partial pressure. Others have a high conductivity only at high oxygen partial pressure. These perovskites can also be used as oxygen electrodes; the best material so far is lanthanum

Table 14
Materials for High-Temperature Water Electrolysis (SOWE)

Component	Material
H_2 electrode	Ni, Co, Pt (cermet)
O_2 electrode	$La(Ca,Sr)MnO_3$, Pt, $In(Sn)O$
Interconnect	$La(Ca,Sr)CrO_3$
Electrolyte	$Zr(Ca,Y,Yb,Sc)O_2$
Nonconductor	$CaZrO_3$

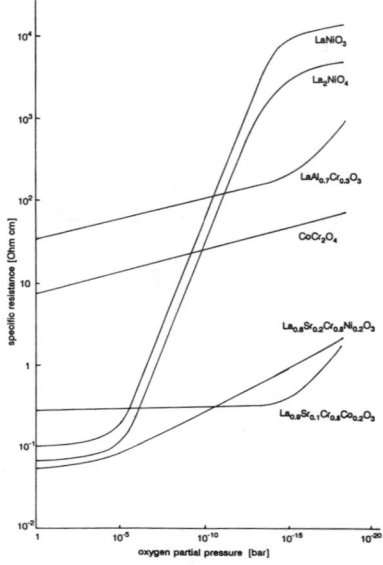

Figure 20. Specific resistance of mixed oxides at 1000°C. From results obtained by the Battelle Institute, Frankfurt.

manganite doped with strontium or calcium. As the hydrogen electrode, nickel is used, which is fabricated together with $Zr(Y)O_2$ as a cermet.

The development of the SOWE is being pursued in conjunction with the development of the solid oxide fuel cell (SOFC) in the United States as well as in Japan. At present, Mitsubishi and other Japanese companies and Lurgi + Dornier are investigating how the high-temperature cell can be operated as a reversible cell, that is to say, how the same cell can be operated alternately as SOWE and SOFC.[104,105] Meanwhile, the ceramic material technology has made great progress; this is why quite a number of groups in the world are trying to develop flat cells in larger dimensions (dm^2). This concept leads to a more compact cell stack, which can be produced more cheaply; see, for instance, the concept of Lurgi + Dornier in Fig. 21.[106]

It is to be expected that SOWE plants will be put into practice only after sufficient experience has been gained with the SOFCs. On the other hand, the situation with the PMWE plants and the new alkaline water electrolysis (AWE) (see Table 4) plants is different: The development is progressing speedily in order to achieve investment costs, which are lower

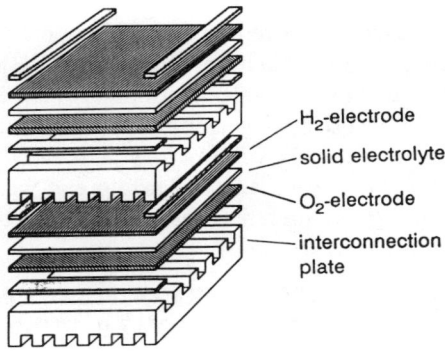

Figure 21. Planar concept for SOWE stacks by Lurgi + Dornier.

than with conventional plants, in addition to having a higher degree of efficiency.

Typically, the investment costs for large commercial electrolysis plants amount to ca. 1000 DM/kW = ca. 4000 DM/(m^3/h), if one takes 4 kWh/m^3 of hydrogen as a basis for the energy consumption in new plants. The price of electrolytic hydrogen is largely determined by the cost of the current; it consists of the investment costs, operating cost, and cost of electricity. The investment part amounts to 0.08 DM/m^3 of hydrogen, if one takes an annuity of 15% and an 85% employment of the capacity (corresponding to 7500 hours of operating time per year). For the time being, the operating cost could be about 0.05 DM/m^3, a figure that may decrease owing to larger service intervals in the future. Taking the cost of electricity as 0.20 DM/m^3 of hydrogen (at a price of 0.05 DM/kWh), the production cost amounts to 0.33 DM/m^3. If the price of electricity were 0.10 DM/kWh, the production cost would be 0.53 DM/m^3. The DECHEMA study on intercontinental hydrogen reported 0.27 DM/m^3 as the lowest value[107,108]; similar values can be found in the VEBA company study.[109] Slightly higher values found in the very detailed Euro-Québec feasibility study[156] (which was a more detailed extension of the DECHEMA study). Further cost reductions can be foreseen in the future.

In the meantime new pilot plants are being built in large demonstration systems[110] in order to gather test experience, which will be described in more detail in the following. An important aspect is the coupling of the fluctuating energy generation to the current requirements of the electrolysis. Special problems are produced by intermittent operation, that is to say, when the cell is left idling for some length of time. Then, the potential of the cathode can rise, because hydrogen diffuses away, and the catalyst can corrode. Furthermore, the anode catalyst may be reduced by hydrogen, which diffuses into the anode chamber. Therefore, a weak electrolysis current is applied in order to prevent damage to the electrodes.

VIII. CHARACTERISTICS FOR COUPLING OF WATER ELECTROLYSIS WITH HYDRO POWER, WIND ENERGY, AND PHOTOVOLTAICS

1. Nonfluctuating Energy Sources

Hydro power is supplied continuously over the entire year—several thousand hours per year at design capacity. Therefore, hydro power does not fluctuate in such intervals that electrode activation coatings would be affected. From this point of view, and considering the low electricity generation costs, advanced electrolysis systems are not necessary and therefore should not be considered before their investment costs have dropped significantly, thus improving the cost/benefit ratio (efficiency/cost ratio).

2. Fluctuating Energy Sources

Mainly wind and direct solar energy conversion typically suffer from sharp and rapid changes in the power levels provided. In general, the dynamic characteristics of photovoltaic/electrolysis coupling are excellent [as long as fluctuations in solar insolation are not greater than 250 $W/(m^2 \cdot s)$]. By the use of active electronic control systems, short mismatches caused by passing clouds or by temperature changes in the electrolysis cells and the resulting electrocatalytic degradation effects can be compensated for very well. Advanced electrolysis technologies are more susceptible to energy changes owing to the electrocatalytic coating materials on the electrodes.

Conventional high-efficiency bipolar electrolysis systems, because of the tendency of the catalytic coatings at the electrodes to degrade owing to reversion of electrochemical mechanisms during longer shutoff times, are not well suited to cope with periods when no energy at all is generated (during the night in the case of solar energy or at dead calm in the case of wind energy). R&D is under way in order to understand these effects better and to take appropriate countermeasures.

3. Resulting Requirements for the Coupling of Solar Electric Sources with Electrolyzer Technology and Existing Coupling Concepts

Since the coupling of hydro power and electrolysis is a state-of-the-art procedure that has been carried out for more than half a century, only the more challenging concepts for the coupling of fluctuating solar sources with water electrolysis are considered here. If bulk hydrogen production from cheap hydro power is the objective, the major goal for improvement of the relevant electrolysis technology is the identification of low-cost materials.

The major goals for the improvement of the coupling of photovoltaics and electrolysis are:

- Reduction of overvoltages at high current densities
- Zero-gap diaphragm and cell concepts
- Improved stationary behavior under varying operating conditions
- Avoiding large current fluctuations and thus hot spots in electrolysis diaphragms/membranes
- Improved overload capability
- Improved startup and shutdown characteristics
- Mitigating overnight depolarization effects
- Mitigating corrosion and degradation effects
- Improved fluid dynamics
- Advanced control dynamics and improved efficiencies of DC–DC conversion
- Improved interaction of system components in continuous and transient operation
- Minimization of parasitic and intrinsic losses
- Intelligent optimized automatic control and operating strategies

The three main concepts for the coupling of photovoltaics and electrolysis are:

- Direct coupling of a photovoltaic (PV) generator and electrolysis at selected voltage levels
- DC/DC bypass concept, involving the connection in parallel of a PV main field (which is directly coupled to the electrolysis) and a bypass

field (which is decoupled and tracked at its maximum power point (MPP) by a converter)
- DC/DC connection of a PV generator and electrolysis via converters that track the PV generator output (i.e., its voltage over a wide range) at any instant at MPP

Although the direct coupling concept is the most efficient one on a system level,[111] the DC/DC-MPP direct connection concept yields the highest annual energy output. It best accepts possible mismatches and flaws in design. It can be operated over a wide voltage range. In comparison to the direct coupling concept, the bypass concept has no significant advantage. To achieve the long-term goal of high renewable hydrogen yields, the DC/DC-MPP direct connection concept is the most favorable one. Nevertheless, the direct coupling concept also is not very sensitive to MPP mismatching and seems to be suited for future large-scale plants as long as the characteristics of the photovoltaic plant and the electrolysis plant coincide.

4. Concepts of Selected Advanced Water Electrolysis Systems and Presently Ongoing Developments into Commercialization

(i) Gesellschaft für Hochleistungswasserelektrolyseure mbH (GHW)

GHW plans a car fueling system to be implemented in existing fuel stations which is based on an advanced pressurized electrolyzer concept operating at up to 150°C and more than 3 MPa pressure.[114] The bipolar filter-press stack contains so-called highly porous metal-oxide electrode–diaphragm–electrode (EDE) cells of a screen-printed, tape-casted cermet (calcium titanate) type. The EDE is a virtual zero-gap arrangement of diaphragm and electrodes of only 0.8-mm thickness (see Fig. 22). The operating range of the electrolyzer system lies between 20% and 120% of design load, and its current density is 10-kA/m^2 at an efficiency of greater than 82% at the higher heating value (hhv) of hydrogen, thus representing an energy consumption of about 4.1 kWh/Nm^3 of H_2. The cell stack is completely enclosed in a pressure vessel. The system will also be suited for fluctuating solar electricity. The state of the art is a 100-kW prototype of a GHW filling station electrolyzer. The design concept of the EDE and the electrolyzer system will permit a scaling up into the megawatt range; a 2.5-MW unit for load leveling applications is presently being considered.

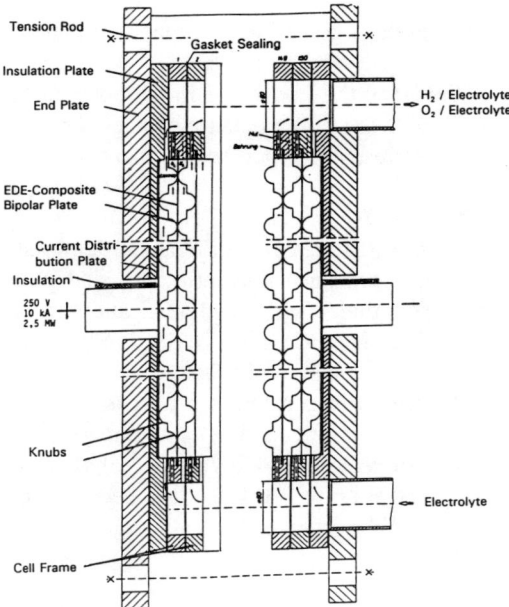

Figure 22. Schematic of EDE-composite cell block by GHW.[114]

(ii) Forschungszentrum Jülich—KFA

The KFA concept for an advanced electrolyzer has already been implemented in some smaller units; for example, 10-kW units have been developed for the Hysolar project. At 80°C, ambient pressure, and a current density of 2.5 kA/m^2, the specific energy consumption is as low as 3.84 kWh/Nm^3 of H$_2$. The electrolyzer consists of 25 cells with an area of 0.05 m^2, a zero-gap arrangement, activated Raney Ni electrodes, and ceramic NiO diaphragms. The cathodic catalyst is composed of a Raney nickel (Mo) alloy, whereas the anodic catalyst is composed of a mixture of Raney Ni and iron powder. The Mo applied to the Ni structure improves the depolarization stability slightly.[126] Design criteria for this concept are low specific energy consumption, high gas purity, good current efficiency at partial load, good dynamic response times under conditions of intermittent solar electricity supply, automatic limiting of gas production at 2% hydrogen in oxygen and vice versa, maximum temperature of 100°C to limit the risk of electrode degradation, and maximum operating pressure of 0.1–0.5 MPa.

The state of the art of this concept is in the range of several tens of kilowatts; 100-kW units are in the planning stage. New electrodes with traces of platinum are presently under development.

(iii) Hydrogen Systems N.V.

Hydrogen Systems features the Inorganic Membrane Electrolyte Technology (IMET), based on a polyantimonic acid polysulfone ion-exchange membrane combined with a cathodic catalyst of nickel sulfide, an anodic catalyst of Ni or Ni/Co spinel oxide, and structural materials such as nickel, nickeliferous steel, and polysulfone. This bipolar electrolysis design can produce hydrogen under a pressure of 0.3–4.0 MPa at current densities of 5–10 kA/m^2 and at temperatures of up to 120°C. Electricity consumption as low as 4 kWh_e/Nm^3 of H_2 can be achieved. Prototype electrolyzers of up to 100 kW_e have been tested successfully. The commercial state of the art is compact automatic hydrogen generator units of about 5–10 kW_e.

(iv) Alyzer S.A.

Alyzer fabricates 2500-mm^2, 10,000-mm^2, 0.06-m^2, 0.25-m^2, and 0.5-m^2 cells which are configured to electrolyzer units with production capacities of 1–5, 5–50, and 50–100 Nm^3/h and which can be operated, depending on the cell frame system used (metal or plastic), at pressures of 1, 2.5, and 4 MPa. Current production employs pure nickel bipolar plates and front and back electrodes. Electrode activation is performed via plasma-spray metal deposition. The diaphragms are of polyphenylene sulfide fabrics. The frame reinforcing technique allows operating pressures of up to 4 MPa and operating temperatures as high as 120°C. The present state of the art comprises various units for solar hydrogen production in the range of 1 to 32 kW_e and extends to a 350-kW_e unit provided for the Hysolar project in Saudi Arabia.

An advanced setup for an alkaline electrolysis unit may comprise operating pressures exceeding 10 MPa suited for pressurized H_2 storage, operating temperatures not exceeding 120°C, NiO diaphragms, and activated electrodes based on metal oxide catalysts, all accommodated in a reinforced frame suited to maintain the pressure level in the cell stack configured in a filter press arrangement. The operating range of such a set up is feasible within 6% to 100% and will avoid cell protecting polarization. The problems related to the necessary tightness to be achieved safely under commercial conditions certainly will need significant time and efforts for solution.

As of the summer of 1994, Alyzer S.A. terminated its electrolysis business.

(v) GEC Alsthom Group/Electricité de France

Electricité de France (EDF) and ACB, a subsidiary of Alsthom, are developing a large-scale, low-cost electrolyzer for the utilization of off-peak nuclear electricity.[207] The system will operate at 130°C and 3 MPa and achieve a current density of 10 kW/m^2. The energy consumption is estimated at 5 kWh/Nm^3 of hydrogen. The electrolyzer plant unit size will reach 20 MW and will be composed of eight modules in a pressure containment, each module consisting of 120 square cells stacked in a filter-press arrangement and reaching 2.4-MW capacity. Nickel frames proved reliable whereas composite materials frames did not show sufficient reliability in 30-kW laboratory tests and had to undergo further evaluation. The main problem seemed to be internal tightness between the different cells. The capital cost goal to be reached was set at 12,500 FF per Nm^3/h.

For the polymer material cell frames, some prerequisites have to be fulfilled such as tightening of the cell stack in hot conditions in order to equalize stresses occurring under operating conditions, optimized yield rate of polymers with respect to differential stain to surrounding metallic structural components, and sufficiently high creep flow of polymer to improve cell tightness.

For the diaphragms polyphenylene sulfide was selected as appropriate owing to its low corrosion in potash, its high bubble pressure, its high liquid permeability, and its low equivalent electric resistance. The hydrophilic properties of the material still have to be improved. The differences between the electrolyte flow rates in the different cells and also in the different modules are less than ±5%.

(vi) Norsk Hydro Electrolyzers

The advanced concept of Norsk Hydro consists of a diaphragm made of a woven polyphenylene sulfide multifill material called Ryton. The ohmic losses are improved in comparison to those obtained with previous asbestos diaphragms. Energy consumption as low as 4 and 4.3 kWh/Nm^3 of hydrogen is achieved at currents of 4 and 6 kA, respectively.

Norsk Hydro electrolyzer units are produced in capacity ranges of between 100 and 400 Nm³ H_2/h at a maximum of 80°C on a commercial scale with diaphragm diameters of 1.78 m.

(vii) Electrolyser, Inc.

Electrolyser employs the uni- or monopolar electrolysis design with an activated electrode system. Owing to the unipolar design, power interruptions do not cause loss of activity because there are no current leakage paths which would permit the buildup of uncontrolled potentials. Therefore, this type of electrolyzer is not affected by plant shutdown, unlike electrolyzers based on the bipolar design. The operating range of this electrolysis concept is from 0 to 120% of design capacity. A major development goal is to reduce the investment costs of the technology below its already very low investment costs, by comparison with those of other existing technologies, as well as to improve the performance, for example, by the use of Ryton as a new separator material. The state of the art is several tens of megawatts of capacity installed in over 700 plants in 85 countries worldwide.

(viii) Life Systems Technology

Life Systems Technology[127] employs a type of low-temperature steam electrolysis where the evaporated internal source water diffuses through the water feed membrane into the cathodic hydrogen compartment and there to the porous cathode. Because only water vapor contacts the catalytically coated porous nickel electrodes and is electrolyzed there, no impurities are brought into contact with the electrodes. These are only in external contact with gases. A similar concept is being pursued by DASA-Dornier.[86]

(ix) Teledyne Energy Systems

In the Teledyne concept, the electrolyte is circulated in the anode compartment. The cathode is reached only through the moistened diaphragm. The cathode compartment contains only hydrogen and water vapor. The gas pressure on the hydrogen side is higher and thus prevents the electrolyte from leaving the anodic compartment and entering into the cathodic one. Advanced diaphragm materials suitable for up to 125°C are under investigation, such as polybenzimidazole (PBI)-bonded K_2TiO_3. The preferred cathode catalyst materials are Ni–Mo alloys or platinized nickel screen structures, whereas the best suited anode catalyst seems to be Ni powder/sintered Ni structures.[127]

(x) Asea Brown Boveri (ABB)

The perfluorinated ion-exchange membrane of the so-called ABB Membrel water electrolyzer is covered on both sides by thin layers of electrocatalysts which are in contact with a porous gas-permeable layer conducting the generated electrons, thus serving as both a current collector and a 'gas separator' permitting the gas generated at the contact zone of membrane and electrolyte to pass to the gas-collecting channels. The electrodes in contact with the membrane consist of noble metals such as platinum (cathode) and a mixed oxide $Ru_xIr_{1-x}O_2$ (anode). The porous current collectors are made of metallized graphite–PTFE compounds (cathode) and of a Pt-coated porous titanium plate (anode). The bipolar plates are graphite structures.[127]

The state of the art is units of 100 kW_e, operated at 80°C, at a pressure of 0.15 MPa and at current densities of 10.4 kA/m^2, yielding an energy consumption of 4.5 kWh/Nm^3 of H_2. The operating range lies between 8% and 100%. Fabrication of these units was abandoned by ABB in 1991.

(xi) DASA-Dornier

The concept of a tubular solid-oxide, hot-water-vapor electrolysis (Hot Elly) has reached laboratory/pilot plant status in the 10-kW range. Theoretically, very low electrical energy consumption levels can be achieved, such as 2.6 kW_e/Nm^3 of H_2 when high-temperature heat (0.5 kW_e /Nm^3 of H_2) or low-temperature heat (0.6 kW_e/Nm^3 of H_2), respectively, is fed into the system (allothermal operating mode) or 3.2 kW_e/Nm^3 of H_2 when low-temperature vapor (0.6 kW_e/Nm^3 of H_2) is fed into the system (autothermal operating mode). For the autothermal operating mode, the current density is at 3.3 kA/m^2 at the design voltage of 133 V. The current efficiency reaches 100% and the overall efficiency based on the lower heating value of hydrogen is 95% for autothermal operation and 115% for allothermal operation.[128] The most widely used solid-oxide electrolyte is a calcium- and yttrium-stabilized zircon oxide. The tubular segments are connected with bipolar connecting elements made of perovskites (e.g., $LaCrO_3$).

An advantage of the Hot Elly concept is its capability to make use of high-temperature water vapor from concentrating solar energy systems or from high-temperature reactors as an energy source, partially replacing electrical energy. Design studies were carried out for medium-pressure (0.5 MPa) electrolysis versions.

There is presently no practical application of the system. Much more advanced is the solid-oxide fuel cell system (basically the reverse of the Hot Elly concept), especially in its planar sandwich design. With respect to technical realization on a large scale, an optimization of the cell concept used and of the module arrangement thereof configured has to be achieved in order to arrive at a simple and cost-effective production process.

(xii) Bamag/Lurgi

The improved commercial Lurgi pressure electrolysis makes use of rolled plate electrodes which consist of a solid nickel foil covered with a layer of a mixture of nickel carbonyl and Raney Ni powder in combination with a catalyst of modified Raney Ni in a nickel carbonyl matrix (see Fig. 8). The diaphragm developed by KFA Jülich, is 0.5 mm thick and consists of a thin nickel wire mesh covered on both sides with a ceramic porous layer of nickel oxide. The cell design features electrodes with a 2.5-mm spacing to the diaphragm, maintained by ceramic spacers. On the active side of gas evolution at the electrodes, the gas bubbles can move freely, supported by the enhanced flow of electrolyte, and then pass through the perforated electrodes and rise to the top in the space between the electrode and the cell wall. Because of the improved electrolyte flow, electrolyte temperature and concentration are equalized much better than in conventional zero-gap designs, thus reducing corrosion and improving long-term reliability.[193]

A commercial electrolyzer of such design with a cell diameter of 1.6 m and unit capacities of up to almost 1000 Nm^3 of H_2 per hour is typically operated at 3.2 MPa, a current density of 4 kA/m^2, a temperature level of 90°C, and an electrolyte concentration of 25% KOH and reaches a cell voltage of 1.74 V and thus an energy consumption of 4.2 kWh/Nm^3 of H_2. Lower cell voltages and thus higher energy efficiencies could be achieved by the use of higher operating temperatures and higher electrolyte concentrations.

As of the date of September 30, 1994, Bamag quit the electrolysis business.

IX. HYDRO POWER/ HYDROGEN DEMONSTRATION PROJECTS

1. Rationale

Hydro power utilization worldwide amounted to a total of about 2000 TWh of electricity in 1988. The technically exploitable hydro power potential is around 17,000 TWh/yr[142] (estimates of the technically exploitable hydropower potential range between roughly 15,000 and 19,000 TWh/yr), which is about eight times the present electricity generation from hydro power and almost one-quarter of the world primary energy consumption in 1987. Owing to the possible environmental impact of large-scale hydro power utilization, these estimates probably have to be reduced by about 35%[172] in the long run, yielding exploitable potentials of slightly above 10,000 TWh/yr. Other estimates of the utilizable potential have been derived from the rate applicable to industrialized countries and amount to 6000–9000 TWh/yr.[182] Other reasons why existing potentials might not be exploited are scarcity of funding and the lack of nearby markets for the electricity.[177] From the present hydroelectricity production of 2000 TWh/yr, an increase of more than 50% to slightly over 3000 TWh/yr will take place by the year 2000. An estimate of the world hydro power potential is given in Fig. 23.

Many locations with high renewable potentials for electricity generation are far away from the major centers of energy consumption.[131] In the

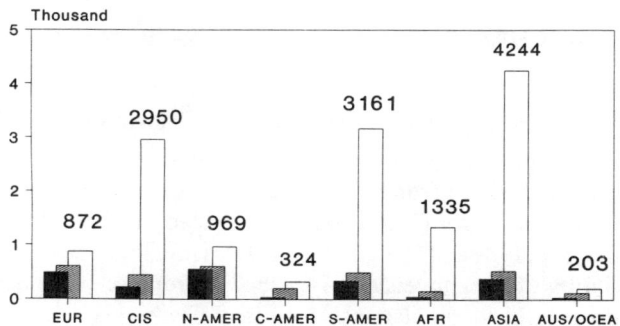

Figure 23. Hydro power potential worldwide. EUR, Europe; CIS, Commonwealth of Independent States; N-, C-, S-AMER, North, Central, South America; AFR, Africa; AUS/OCEA, Australia/Oceania. After Ref. 129.

case of hydro power potential, this applies, for example, to Zaire, Indonesia, Iceland, Greenland, Brazil, and some other locations in Africa, Asia, and North and South America. Several long-distance energy transportation technologies are available, such as gas pipelines, high-voltage direct current electricity transmission lines, and liquid-gas tanker transportation. For intercontinental (maritime) linkage of production and consumption centers, only large seagoing tankers seem technically feasible in the medium term. At present, practical experience is limited to containerized cryogenic transport of liquid hydrogen. In this context, hydrogen can play a role as a "facilitating" or "enabling" technology.

The remaining global potential for energy trade (excluding possible scenarios in which 100% of the electricity demand is covered) is considered to be in the range between 6500 and 10500 TWh/yr (in 2010), which is equivalent to a capacity in the range between 1400 and 2300 GW. The remaining potential calculated for the year 2040 is smaller than that for 2010 because electricity demand will increase steadily. It has been estimated to be about 4800–9500 TWh/yr (1070–2100 GW).[145]

The utilization of these potentials would mean a manyfold increase of generation capacities—existing ones and ones under construction. An environmentally compatible development of these potentials certainly will not be possible in all cases. Hydro power therefore can provide a first possibility for the step into international hydrogen trade. In the long-term perspective, besides the hydrological cycle, larger renewable energy potentials will have to be developed by utilizing direct and indirect solar radiation.

In the following some selected hydro-hydrogen projects are described in more detail.

2. Euro-Québec Hydro-Hydrogen Pilot Project (EQHHPP)

The concept of a hydro-hydrogen-based clean energy system, conceived by the Joint Research Centre Ispra of the Commission of the European Communities (CEC), was investigated on a 100-MW$_e$ scale during the years 1989 and 1990. The goal of this investigation was to prove the feasibility of the conversion of Québec hydropower via electrolysis into hydrogen, the maritime transport of liquid hydrogen (LH$_2$) (see Fig. 24) or methylcyclohexane (MCH) to Europe, and its storage, distribution, and end use there (see Fig. 25). The end-use technologies comprise vehicle and aviation propulsion, electricity/heat cogeneration, and hydrogen enrichment of natural gas for use in industry and households.[156,157]

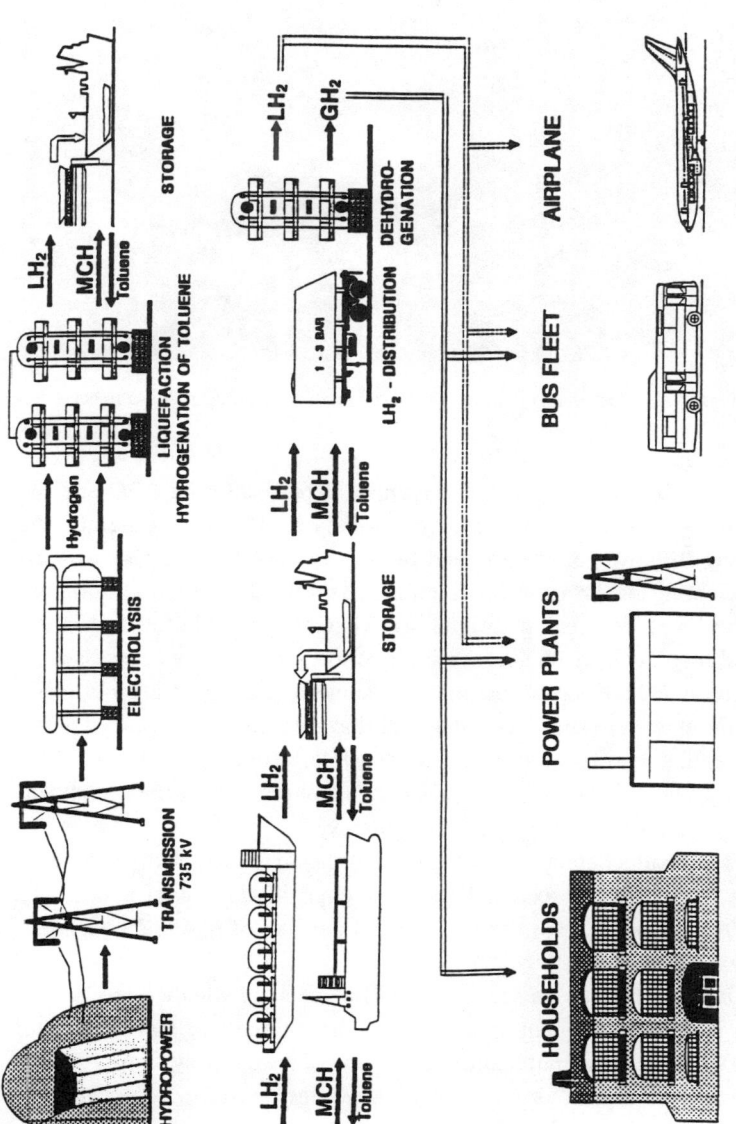

Figure 24. EQHHPP barge carrier for translantic LH2 transport (Thyssen Nordseewerke).

Figure 25. Block diagram for the Euro-Québec Hydro-Hydrogen Pilot Project.

Organized by the Joint Research Center Ispra of the CEC and the Québec government (QG), managed jointly by Hydro-Québec and the Ludwig-Bölkow Foundation, and funded by the CEC, the QG, and the participating industries and institutions, a 3 million ECU feasibility study,[†] on the basis of existing technology wherever possible, evaluated the operability and the technical feasibility of the overall system, assessed its near-term and long-term potential for technological improvements, clarified the relevant aspects of safety and hazards, environmental impacts, and regulatory licensing as far as possible, promoted and stimulated research and development, and produced a cost estimate on the investment and operation costs of ±15 % cost accuracy.

The results obtained in the course of this feasibility study (phase II) came to a delivery cost of liquid hydrogen in Europe of about 0.15 ECU/kWh$_{th}$ or approximately 1.75 ECU per liter of gasoline equivalent (untaxed).

If financing can be secured, it will be decided whether and how to enter into the subsequent phases III (detailed engineering and blueprints) and IV (hardware manufacture and erection of the entire system). Already before this decision is made, liquid hydrogen can be exported from

[†] 1 ECU ≅ U.S. $1.20.

existing liquefaction plants operated with hydroelectricity in Québec via containerized transport.

The EQHHPP is structured in the following project phases, which have either been completed or are presently being carried out:

Phase I	Assessment	1986/04/01–1987/03/31
Phase II	Detailed system definition	1989/01/01–1991/03/31
Phase II Suppl. Task Program	Additional R&D programs in Europe	1990–1992
Phase III.0-1	Preapproval activities in Europe	1991
Phase III.0-2	Hydrogen demonstration program	1992–1994
Phase III.0-3	Hydrogen demonstration program (Europe)	1993–1998
Phase III	Detailed engineering & specifications	Duration 1–2 years
Phase IV	Construction	Duration 4–5 years

Since early in 1992, demonstration activities have been started by the CEC and the Québec government within Phases III.0-2 and III.0-3, focusing on the following technologies and concepts[187,188]:

- Hydrogen Systems city bus with internal combustion engine and LH_2 hydrogen storage
- M.A.N. city bus with internal combustion engine and LH_2 storage (see Fig. 26)
- Ansaldo Ricerche city bus with membrane fuel cell drive and LH_2 storage
- Three hythane city buses in Montreal (hythane is typically 20% H_2 and 80% methane)
- Operational management system for the integration of H_2 buses in public operators' networks
- Passenger transport boat with electric drive powered by a membrane fuel cell and LH_2 storage
- Safety tests for LH_2 vehicle tanks (approximately 125 liters) and for a down-scaled maritime barge transport container (approximately 180 m^3)
- Emission testing on sectors of LH_2-adapted combustors of Airbus jet engines for low NO_x emissions
- H_2 direct reduction of iron ore fines in plasma arc furnaces, optionally also with biofuel
- Cogeneration plant with a piston engine
- Cogeneration plant with a phosphoric acid fuel cell

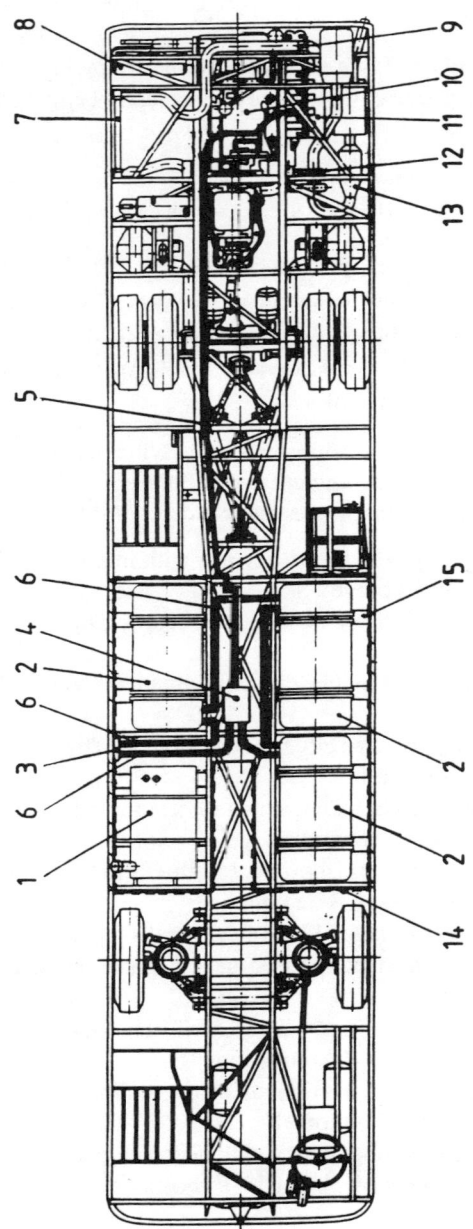

Figure 26. M.A.N. city bus with liquid hydrogen storage.[188]

- Study on rules and regulations, safety, and acceptability of hydrogen
- Study on socioeconomic effects of reduced emissions by hydrogen buses and airplanes compared with conventional technology investigated for three metropolitan areas
- Study activities on LH_2 transport containers
- Several smaller study activities

Some hardware demonstrations extend until 1996 or even 1998.

One of the driving forces of the EQHHPP is the chance to develop technology that allows the tapping of vast, still unused hydro power potentials worldwide, especially those which are far away from the centers of consumption and cannot be utilized otherwise. Very important additional aspects are the possible diversification of primary energy sources, the probable competitiveness of liquid hydro-hydrogen from Canada in an increasing European market in the late 1990s, and environmental concerns, such as local pollution in cities and the increasing greenhouse effect.

Important factors supporting early application of hydro-hydrogen systems are that most of the components involved already exist, virtually no engineering breakthroughs are required, and real-scale pilot/demonstration projects can be built and operated already today or in the foreseeable future in order to gain valuable systems experience.

3. *N*orwegian *H*ydro-*E*nergy for *G*ermany (NHEG)

An investigation of a European hydrogen energy cycle with hydrogen being produced and liquefied in Norway from hydroelectricity and transported to Germany for end use was proposed by Norsk Hydro (NH) and Ludwig-Bölkow-Systemtechnik (LBST) and carried out between 1990 and 1992. The funding was provided by the German Ministry for Research and Technology (BMFT), the Commission of the European Communities (CEC), the Norwegian Government, Norsk Hydro, and Ludwig-Bölkow-Systemtechnik.[189] The objective of the case study was to draw an overall picture and to evaluate alternative technologies and energy transport vectors with respect to minimum energy losses, economic optimization, and least influence on the environment.

The main topics discussed were hydrogen production by water electrolysis (existing NH installations at the Glomfjord), purification and liquefaction of gaseous hydrogen for storage in Norway, and transport by sea to Germany. Two alternative routes (see Fig. 27) for transportation of liquid hydrogen were evaluated: a special gas tanker with permanently installed cryogenic tanks, and LH$_2$ transport vessels within the frame of ISO 40-ft containers.

Other aspects discussed were regional storage and distribution of hydrogen for later use in Germany. Three end-user centers were chosen to facilitate a comparison of distribution vectors representing various technical alternatives, based on rail, road, and river transportation. State-of-the-art LH$_2$ transport vessels, such as standard ISO 40-ft or jumbo containers, as well as road trailers were considered.

The NHEG investigation demonstrated that LH$_2$ can be supplied at competitive market prices. Unfortunately, the market does not yet absorb all of the LH$_2$ produced in conventional liquefaction plants in Europe. On the other hand, the environmental bonus of clean, renewably produced hydrogen is also not yet valued highly by the market/consumer. Therefore, NHEG has not yet reached a stage of hardware realization.

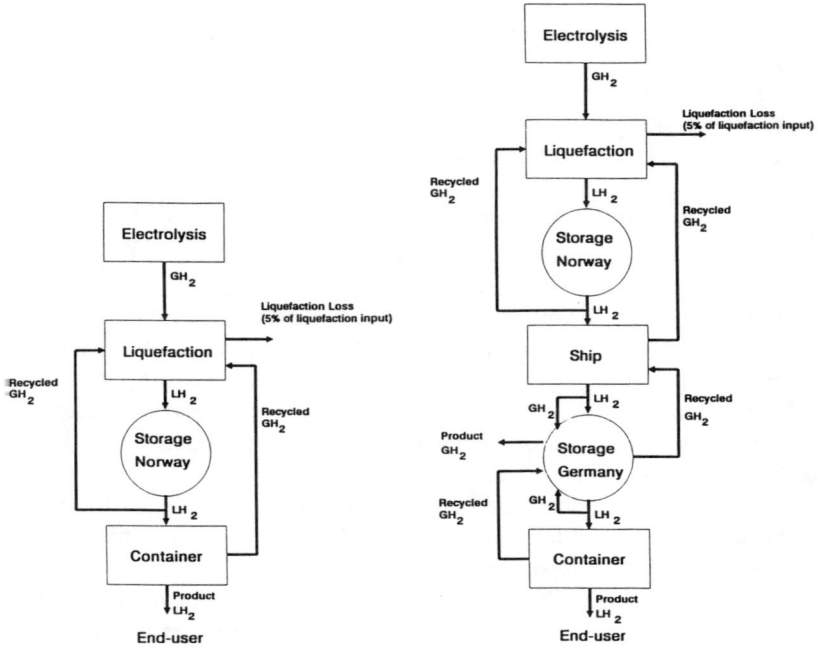

Figure 27. NHEG mass flow charts for 20-MW case (left) and 100-MW case (right).[189]

As in the EQHHPP, a simulation model was used, providing a basis for further variation analysis of energy and mass balances and cost evaluations.

4. Hydrogen-Powered Automobiles Using Seasonal and Weekly Surplus of Electricity (HYPASSE)

In various alpine countries, excess hydroelectricity could be made available in late spring or early summer.[134] Also, in some of these countries nuclear off-peak electricity may be available over weekends. For climatic reasons and also because of local pollution in cities, there is interest in storing this clean excess energy.

Within the framework of the Swiss–German project Hypasse, carried out by Daimler-Benz and the Paul-Scherrer Institute, it is intended to convert this seasonal and periodic surplus electricity via electrolysis into hydrogen. A 400-kW electrolyzer will be operated 3200 hours per year. The hydrogen from seasonal surplus electricity will be made available in chemically bound form as methylcyclohexane (MCH) throughout the year. Periodic excess electricity will be stored as pressurized hydrogen (CGH_2) at 30 MPa in a weekly cycle. The gaseous hydrogen generated will be used in public bus transport five days a week and 300 km per day, 70% as MCH in order to operate two MCH buses and 30% as CGH_2 in order to operate one CGH_2 bus.[190]

Within the framework of Hypasse, Daimler-Benz will equip one bus with a pressurized hydrogen storage system made of composite materials, whereas the Paul-Scherrer Institute will equip one bus with an on-board integrated dehydrogenation unit.[133] Both buses will be operated with a modified spark-ignited diesel engine with internal high-pressure mixture formation. The first four-year project phase (which started in 1990) will lead to test-bed qualified components. The second four-year project phase will comprise demonstration of buses and necessary infrastructure (H_2 production, fueling system, stationary dehydrogenation) as well as the testing of the entire system.

5. Concepts for Utilization/Export of Electricity Generated from Hydro Power/Geothermal Energy in Iceland and Greenland

(i) Iceland

Besides significant geothermal resources, Iceland possesses remarkable, still untapped hydro power potential. The existing hydro electric generating capacity amounts to almost 700 MW and delivers approximately 4 TWh$_e$/yr in safe and reliable energy with a capacity factor of 66%.[169]

The technical potential for hydro power generation is estimated at 64 TWh/yr, of which approximately 31 TWh/yr can be utilized under economically and ecologically acceptable conditions. This means that only 13% of this potential is presently utilized. The potential for geothermal electricity generation is estimated at 20 TWh/yr, of which only 220 GWh/yr or some 1% is utilized.

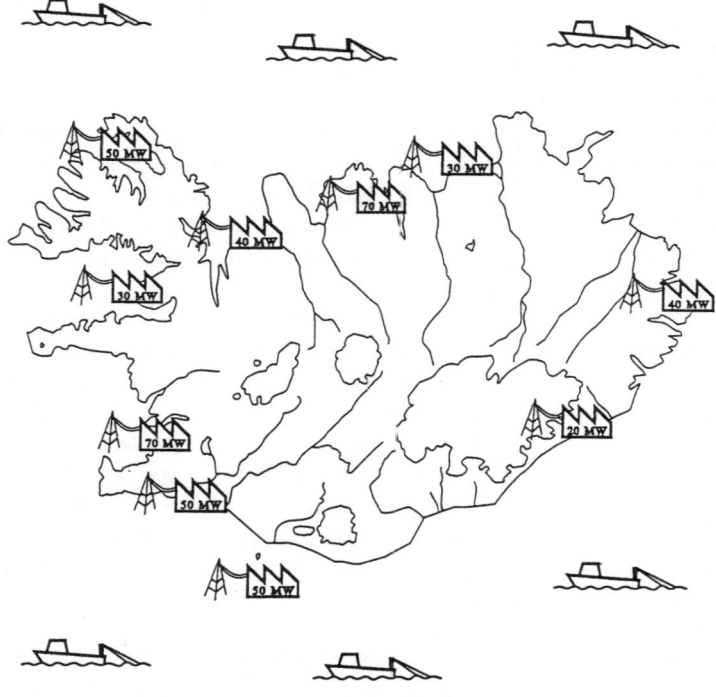

Figure 28. Icelandic fishing fleet conversion.[185] A proposed future scenario for a hydrogen-driven Icelandic fishing fleet is shown, with hydrogen plants and fuel depots situated within major harbor areas. Power amounts (indicated in megawatts of power) have been adjusted to 1991 fuel consumption data.

A first pragmatic approach proposes the conversion of Iceland's fishing fleet to operation with hydrogen fuel produced electrolytically from hydro- and geothermal electricity (see Fig. 28). This would replace almost 30% of the fossil fuel imported. About 450 MW_e generating capacity, distributed over 10 production sites of between 20 and 70 MW_e around the Icelandic coastline, would be required in order to provide a total of 70,000 t of hydrogen annually. An advantage of this decentralized concept would be the possibility for a stepwise buildup of the entire fuel supply system. The cheap gaseous hydrogen would be stored in magnesium hydride stores on board fishing ships. Upon refueling of the ships, the heat released by the exothermic recharging process of the hydride stores could be fed into the local space heating system. The heat input needed for the endothermic process of liberating hydrogen gas from the metal hydride stores would mainly come from the ship engines' exhaust heat.[185]

(ii) Greenland

In southern Greenland during a three- to four-month period in the summer the large water masses from snow melting can be utilized. Power plants could be installed in the massive rock formations at the coastal edges of the island, which rise some 2000 m. The water from the snow melting would be collected in reservoirs and then led to turbines installed in the power plants via underground tubes.

If we assume 20 glacier power plants with about 40,000 km^2 of reservoir surface area each, a hydraulic head of 1000 m, and the utilization of 1 m in reservoir level difference (= annual precipitation), about 2000 TWh can be produced annually. If the water liberated by the annual glacier ice melting of about 1 m were utilized, another 2000 TWh could be produced annually.[140,141] Even if only one-tenth of this total 4000 TWh were utilized, 30% more hydroelectricity would be produced than is presently generated, and this production would amount to 70% of the entire Canadian hydroelectricity potential.

In a first pragmatic approach for Greenland, a similar concept as for Iceland might be useful, making use of the existing conventional hydro power resources. Hydrogen could be produced and used as fuel for the entire transport sector and an energy commodity in remotely located settlements, replacing conventional fuels imported from far away.

X. WIND POWER/ HYDROGEN DEMONSTRATION PROJECTS

1. Rationale

In order to make technical use of wind energy on a large scale, an average wind speed of approximately 5m/s is necessary. The relationship between wind speed and the power obtained from it is a cubic one. This relationship is best illustrated by comparing good, excellent (good + 86%), and outstanding (good + 212%) wind sites with average wind speeds of 5.8, 7.2, and 8.5 m/s, respectively.[146] Areas with average wind speed high enough for the technical use of wind energy can be found in many coastal countries. Large-scale use of wind energy may be feasible primarily in northern Canada, northwestern Europe (mainly the United Kingdom), eastern China, the Antarctic, and coastal areas of northwest Africa as well as of South and North America.

Wind resources vary in speed, direction, and availability. Identifying truly exploitable resources is a highly complicated and complex process, and therefore parallel resource assessments for identical areas may lead to a wide range of results. In a study carried out by the International Institute for Applied Systems Analysis (IIASA), the global technical potential of wind energy has been estimated to be approximately 3 TW or 8000 to 12,000 TWh/yr. These figures show that the technical potential of wind energy could almost cover the current global consumption of electricity (1987).

Most of the regional resource assessments regarding wind energy have been carried out in Canada, the United States, and the European countries. Conservative estimates show that, in the United States, accessible resources of wind energy have the potential of providing more than ten times the electric power currently consumed in the United States. A study for Europe shows that the European wind energy potential amounts to three times the electric power consumption of Europe.

More recent investigations estimate the theoretical wind energy potential for electricity production at about 500,000 TWh/yr (excluding offshore applications, most islands, Antarctica, and Greenland). If all environmental and other known restrictions are imposed, an exploitable potential of 53,000 TWh/yr remains (Table 15).

Worldwide, some 17,800 wind turbines with a capacity of approximately 1700 MW have been installed up to now, 1500 MW of this capacity being in California and more than 100 MW in Denmark.[182] In Europe and

Table 15
World Wind Electricity Potentials[a]

Region	Gross electricity potential (TWh/yr)	Estimated second-order electricity potential (TWh/yr)
Africa	106,000	10,600
Australia	30,000	3,000
North America	139,000	14,000
Latin America	54,000	5,400
Western Europe	31,400	4,800
Eastern Europe and former USSR	106,000	10,600
Rest of Asia	32,000	4,900
Total	498,400	53,300

[a]Ref. 182.

in the United States, most investments in wind energy production have been subsidized by governments in order to surmount the break-even point compared to competing electricity production. Since 1985, 622 MW of wind converter capacity have been installed in California without the benefit of tax incentives, in total representing a capital investment of almost U.S. $1 billion.

Wind energy has not only proven to be apart from hydro power—the most cost-competitive (indirect) solar energy technology, but it can also be made available for the bulk power market. The production of electricity by wind turbines is at present concentrated in industrialized countries. There exists, however, an increasing interest in developing countries in using wind energy for both the operation of water pumps and the production of electricity.[147]

The development of large-sized turbines is more cost-intensive and difficult than the development of smaller turbines. Nevertheless, major efforts regarding the technical development and construction of large-sized turbines have been made in several countries. This is reflected in the fact that the European countries, Japan, and the United States make major subsidies available in order to promote wind energy R&D. An agreement for cooperation regarding the development of large-scale wind conversion systems has been worked out under the auspices of the International Energy Agency.[148]

In areas with an average annual wind speed of 8 to 9 m/s, large-sized turbines lead to generation costs of U.S. $0.07–0.10/kWh (rates for consolidated technologies are U.S. $0.04–0.07/kWh) and a cost reduction to U.S. $0.05/kWh is presently considered achievable. In the long term (30 years), costs could be reduced to as low as U.S. $0.03/kWh.

In order to produce hydrogen on the basis of wind energy, large-sized turbines are required, and these have to be grouped and interconnected in high numbers. These wind farms may be situated in remote areas. However, the choice of suitable sites may be restricted because the wind power produced in industrialized countries will generally be fed into the electricity grid. Potential sites of hydrogen production on the basis of wind power are located in West Africa, Greenland, Iceland, and the southern part of Latin America (if transcontinental electricity transport tends to be too costly or inefficient).

2. Hawaii Wind Energy Storage Test Facility

The main components of the renewable electricity generating system conceived by the Hawaii Natural Energy Institute (HNEI) are wind energy converters of, in total, 130-kW capacity, which operate at average wind speeds of 9.8 m/s with capacity factors of higher than 60%.[158] This generating capacity is supported by 2.2 kW of crystalline photovoltaic modules.

For the storage of the energy, three methods are applied: battery storage of 130 kWh, pumped hydro-storage of 200 kWh, and hydrogen gas storage—both pressurized storage of some 150 kWh_{th} and low-temperature metal hydride storage of 160 kW_{th}.

The hydrogen is produced in a conventional alkaline electrolyzer (from Harbing, China) of 11-kW electric capacity at a rate of 2 Nm^3/h and under 1 MPa of pressure. Fifty cubic meters of hydrogen can be stored in a 5-m^3 storage tank. Later on, a solid polymer electrolyte electrolyzer (by Mitsubishi Heavy Industries) of 0.5 Nm^3/h capacity will be tested in the system. The hydrogen can be converted into electricity by using one of five spark-ignited diesel engines (135 kW, 60 kW, 2 × 10 kW, 5 kW, 3 kW) integrated into the system.

Additional hydrogen applications are envisaged for a future guest house, such as cooking, heating, and power generation. In a modified pickup truck with electric traction and battery storage, hydrogen will serve as a range extender, fueling either a fuel cell drive or a generator.

According to HNEI, the above-described project is the first of its type in the world and will demonstrate renewable hydrogen applications that are close to being economic. HNE1 sees the above-described system concept also as a cost-effective approach for the development of rural areas in developing countries and furthermore as an export opportunity for the U.S. industry.

A 2.5-MW_e hydrogen demonstration project is being considered for the north of the island of Hawaii, in the Kohala region, making use of the existing Hawaii hydro power plant. Step by step, the hydrogen production capacity (consisting of Chinese elctrolyzers) can be built up and the pipeline grid for the distribution of hydrogen to local home, office, industrial, and commercial consumers can be established. Making use of cheap nighttime off-peak hours, the electrolytically produced hydrogen and oxygen would be compressed and stored in pressurized storage. The hydrogen would be fed into the pipeline system and sold as an energy commodity or raw product to consumers, whereas the oxygen would be sold in pressure bottles to medical users.

3. Wind–Hydrogen Test Installation at the Fachhochschule Wiesbaden, Germany

Since 1988, a small test installation has been operated by the Fachhochschule Wiesbaden on the Feldberg/Taunus mountain in Germany (Fig. 29). It consists of a 20-kW wind energy converter, an alkaline pressurized electrolyzer of 20 kW (at 3 MPa, 50 cells at 125 V and 160 A), pressurized gas storage in bottles (of 2.4-m^3 volume), and two hydrogen-operated gas motor generators (8 kW and 4 kW) with external mixture formation built into one generator unit with individual fuel conditioning equipment for each engine.[159] The overall primary energy efficiency of the "wind energy converter–electrolysis–combustion engine–electric generator" system is approximately 15%.

The wind energy converter is designed for stand-alone operation and feeds its pulsating direct current in the first place into the electrolyzer. Additionally, the electricity is used for heating purposes in the staff shelters of the test installation. Also, a battery is fed which provides electricity for the measuring and monitoring equipment in case wind is not available. An electronic power conditioning unit optimizes the electricity supply to the various consumers.

The goal of this project is to optimize the "wind energy converter–electricity supply–pressurized electrolyzer–storage–generator" system, as

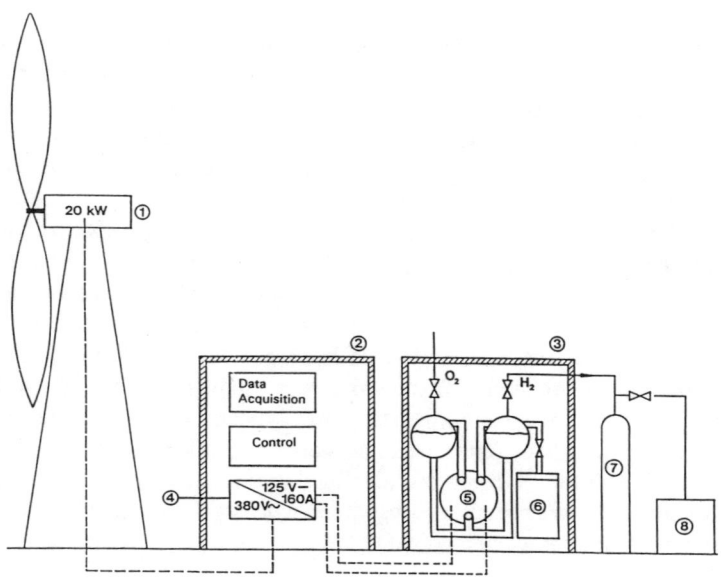

Figure 29. Wind energy–hydrogen test site on the Feldberg/Taunus mountain.[159] 1) Wind energy converter; 2) building housing electronics; 3) building housing electrolyzer; 4) power electronics; 5) pressurized electrolyzer; 6) water storage; 7) hydrogen storage; 8) Otto Cycle internal combustion engine.

well as each component itself. The energy production costs calculated over a 10-year depreciation period are estimated to be 1.6 DM/kWh$_e$.

4. Photovoltaic/Wind-Hydrogen Test Installation at the University of Oldenburg, Germany

After four years of preparation time, an installation for the testing of a hydrogen energy storage system for small self-sufficient renewable energy systems was inaugurated in the energy laboratory of the University of Oldenburg in August 1990.[160] The system comprises 6.2 kW of photovoltaic modules, a 5-kW wind energy converter, an alkaline pressurized electrolyzer, a lead–acid battery with 300-Ah storage capacity, pressurized storage for 30 Nm^3 of H$_2$ and 15 Nm^3 of O$_2$, a 5-MPa compressor, an alkaline fuel cell of 0.6 kW, and a gas motor generator of 12-kW capacity.

The photovoltaic and wind energy electric generator system provides electricity to the internal grid of the energy laboratory. A battery system equalizes fluctuations in the supply/demand curve. Excess electricity is converted into hydrogen and oxygen. In cases of low or missing renewable electricity supply, hydrogen is converted into electricity with the fuel cell and/or gas motor generator systems. A future optimized system will not need a gas motor generator anymore.

Already, a small, efficient battery storage system can cope with most of the mismatch of renewable electricity production and load, thus leading to hydrogen systems (electrolysis, storage, fuel cell) drastically reduced in size and capacity. The present system has an overall efficiency of about 35% and can be improved to up to 50% by making use of future advanced technologies. The monitoring of the energy flow mainly focuses on losses, leaks, and auxiliary energy consumption. A mathematical modeling of the entire system with the results obtained in the tests will allow simulations of possible new system configurations.

5. Wind/Photovoltaic/Heat/Hot Water/Hydrogen Test System at the Technical University of Ilmenau, Germany

In the final stage of the project at the Technical University of Ilmenau, a 30-kW_e horizontal-axis wind energy converter, photovoltaic modules of in total 30 $kW_{p,e}$, a solar–thermal low-temperature collector system, a pressurized electrolyzer with storage, and refueling stations for electric vehicles and for hydrogen-powered vehicles are planned.[191] Presently, in preparation for the realization of the final project stage, small-scale and laboratory experiments for all envisaged technological concepts are under way.

XI. SOLAR HYDROGEN TEST AND DEMONSTRATION PROJECTS

1. Rationale

Worldwide, about 1.9 million km^2 of vast stony deserts with a global irradiance of 2000 $kWh/(m^2 \cdot yr)$ exist, representing about 1.3% of the global surface area.[135,136,144,148] Depending on the level of technology applied (e.g., solar–thermal power plants or photovoltaic power plants and their respective efficiencies for the conversion of solar radiation to electricity, hydrogen transportation via pipelines in gaseous form or containerized in liquid form, transportation distance, end-use distribution, availability of entire system) and on the assumed land use factor, between 105,833 and 138,397 TWh/yr of solar hydrogen could be generated on it.

To cover the present world end energy consumption of roughly 70,000 TWh/yr, a suitable area in North Africa of only some 850,000 km^2 (representing approximately 10% of the Sahara or little more than one-third of the acreage of Algeria) would be required. Unused deserts and wastelands exist in many other areas in the world (in Spain alone, some 20,000 km^2).

Assuming a photovoltaic module efficiency of only 10%, the following theoretical solar electricity production potentials are obtained[150]:

Northern Europe: 88×10^{12} kWh/yr
Southern Europe: 195×10^{12} kWh/yr
Germany: 25×10^{12} kWh/yr
Algeria: 476×10^{12} kWh/yr

In all these cases, the present electricity consumption can be covered at least 50 times by the theoretically existing potential. Vast areas of land in Australia, in Brazil, in North Africa, in the United States, and in the Commonwealth of Independent States can be used for large-scale solar energy conversion. All these considerations show that land availability itself does not represent any obstacle to extensive solar hydrogen production in many areas all over the world.

The energy production costs for the two most promising technologies—solar–thermal parabolic trough electricity generating systems and photovoltaic (pv) electricity generating systems—amount to about U.S. $0.08–0.12/kWh in California (1990). The costs can be expected to be reduced to about U.S. $0.06 /kWh in the mid-1990s for the solar–thermal systems. For utility-scale PV systems, the present costs under Californian insolation conditions have to be assumed realistically to be still on the order of U.S. $0.50/kWh (1990) with the potential to drop to U.S. $0.20/kWh by the year 2000 and finally to about U.S. $0.04–0.07/kWh by 2030.[22-26]

2. Solar Hydrogen Project in Bavaria [Solar Wasserstoff Bayern GmbH (SWB)]

Already in 1986, Ludwig Bölkow convinced the utility Bayernwerk AG (BAG), Munich, to start a demonstration project covering the whole solar hydrogen concept including production of solar electricity, conversion to hydrogen, hydrogen storage, and hydrogen application technologies. A company was formed [Solar Wasserstoffbayern GmbH (SWB)] with the participation of Siemens, Linde, MBB (now DASA), and BMW, each with 10% capital stock, with BAG holding 60%. The plant site is located at Neunburg vorm Wald in Bavaria, and the first stage of the project was put into operation in 1987[161] and completed at the end of 1991. The photovoltaic generators have been operative in the grid connection mode since 1989. The cost of phase I is on the order of 64 million DM, half of it borne by the federal and Bavarian governments and half by SWB.

Phase I (1987–1991) for 64 million DM consisted of the following plant components:

- 135-kW_p monocrystalline Si modules (Siemens) and 132.8-kW_p polycrystalline Si modules (DASA-TST) = 267.8-kW_p solar generator peak capacity)
- DC/DC converters to DC-bus level: 3 × 54 kW, 200–300 V_{DC} to 300–435 V_{DC} (Siemens AG) and 2 × 70 kW, 195–295 V_{DC} to 300–435 V_{DC} (DASA-TST)
- Electrolysis power supplies (111 kW and 100 kW) (Siemens AG)
- Three self-commutated inverters of 160 kVA each (Siemens AG)
- 80-kW DC/AC inverter (Siemens AG)
- 111-kW_{el} alkaline electrolysis [Cadwel 1000 at 0.008 MPa (Krebskosmo)] and 100-kW_{el} membrane electrolysis [Membrel MHK-04-120 at 0.15 MPa (ABB)]
- H_2 and O_2 treatment (120 Nm^3 of H_2 and 60 Nm^3 of O_2 per hour) and 3.3-MPa pressure storage (5000 Nm^3 of H_2 and 500 Nm^3 of O_2) (Linde)
- Two boilers of 20 kW_{th} each (Buderus GmbH/ Kromschröder AG)
- 6.4-$kW_{p,e}$ alkaline fuel cell at 0.1 MPa and 80°C (Siemens) and 79.3 $kW_{p,e}$ phosphoric acid fuel cell at an overpressure of 0.004 MPa and 190°C (KTI B.V./Fuji Electric Co., Ltd.)
- LH_2 car filling station with < 3000-liter LH_2 storage capacity and 30 liters of LH_2/min filling speed (Linde) (separately financed without government contribution)
- Auxiliary systems, main process-control system, gas alarm device

Phase II (1992–1996) for approximately 60 million DM will incorporate the following components[192]:

- 25-kW_p amorphous Si modules [Siemens Solar GmbH (SSG)], 25-kW_p amorphous Si modules (Phototronics PST), 10-kW_p polycrystalline Si modules with AS design (DASA), 10-kW$_p$ monocrystalline high-performance Si modules (SSG), and 9.7-kW$_p$ MIS Si modules (Nukem)[†]

- 100-kW$_{el}$ alkaline high-pressure electrolysis (Metkon S.A./KFA) and 100-kW$_{el}$ alkaline high-pressure electrolysis (GHW)
- 10-kW$_{th}$ catalytic heater, natural gas and H$_2$/natural gas fueled, air as oxidant (KFA)
- 17.5-kW$_{th}$ catalytically heated absorption refrigeration unit, H$_2$ fueled with air as oxidant, for cold water production (FHG-ISE)
- 10-kW$_{el}$ PEM fuel cell fork lift drive with hydride storage (Siemens)
- Automatic LH$_2$ car filling station with drastically increased filling speed (Messer Griesheim and Linde) (separately financed without government contribution)
- New main process control system and integration of existing phase I and new phase II subsystems (Siemens)

The technological concepts of phases I and II are depicted in Fig. 30.

3. Hysolar Project in Germany and Saudi Arabia

The Hysolar project is being carried out by the Deutsche Forschungsanstalt für Luftund Raumfahrt (DLR) and the University of Stuttgart in cooperation with three Saudi universities (King Saud University, Riyadh; King Abdulaziz University, Jeddah; King Fahad University of Petroleum and Minerals, Dharan) and the King Abdulaziz City for Science and Technology (KACST), Riyadh. Phase I of the program, which started in 1985 and lasted until the end of 1989, had a total financial commitment of about 40 million DM, brought up by the Saudi and German partners (KACST 50%, Bundesministerium für Forschung und Technologie Ministerium für Wiessendraft und Forschung (BMFT) 25%, Bundesministerium für Forschung und Technologie Ministerium für Wiessendraft und Forschung (MWI) 25%).[162] To complete some of the goals that could not be fulfilled within project phase I and to prepare a subsequent phase II, a supplementary phase Ib covered the period 1990–91 with an additional budget of 8 million DM. Seven Hysolar teams with about 60 scientific and technical collaborators in total were active.

Phase I consisted of the following activities:

[†]These 80 kWp seem to be the feasible technological options for the PV applications within phase II. Neigher the buried contact Si modules (DASA) nor the copper indium diselenide (CIS) modules (SSG) could be made available for this phase.

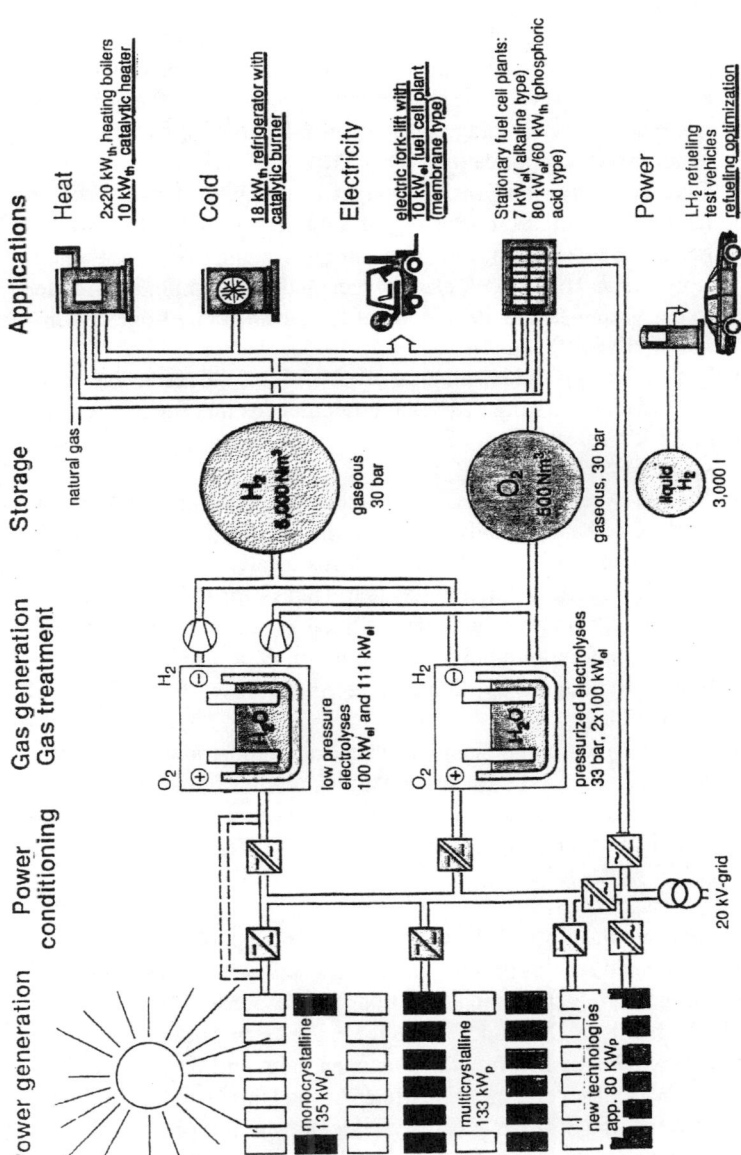

Figure 30. Block diagram of the photovoltaic–hydrogen project of Solar-Wasserstoff-Bayern GmbH (SWB).[192]

- Task 1: Design and installation of a 350-kW demonstration plant in the "Solar Village" near Riyadh, consisting of a concentrating photovoltaic power system, an advanced electrolyzer system with a power supply system, a grid-operated rectifier, and the necessary gas handling and storage system in order to gain experience in technical-scale application (HYS 350)
- Task 2: Design and installation of a 10-kW test and research facility in Stuttgart, consisting of a multicrystalline photovoltaic generator system, a power conditioning system, two 10-kW$_e$ electrolyzers and one 2-kW$_e$ electrolyzer, and a PV simulator, in order to carry out systems development for advanced hydrogen equipment (HYS 10)
- Task 3: Design and installation of a 2-kW test and research facility in Jeddah, consisting of similar equipment as that for task 2 (only one 2-kW$_e$ electrolyzer), but especially suitable as an experimental setup (HYS 2)
- Task 4: Accomplishment of fundamental research program, including research on advanced alkaline electrolysis and catalytic mechanisms in fuel cells and gas diffusion electrodes (HYS-FUP)
- Task 5: Accomplishment of system studies for the assessment of the Hysolar program and of a utilization program for the evaluation of safety, reliability, and environmental aspects of the selected hydrogen application technologies (HYS-SYS and HYS-UTIL)
- Task 6: Accomplishment of educational and training program, comprising periodic review meetings, regular scientific and program review seminars, annual summer schools for students, and technical training for Saudi scientists, engineers, and technicians (HYS-EDU)

Phase II will last from 1992 to 1995, and a budget of 28 million DM has been allocated for this period. The major focus within this phase will be on hydrogen production and utilization. The responsibility, as in phase I, lies in the hands of DLR and KACST. On the German side, additional participants will be the Zentrum für Sonnenenergie- und Wasserstoff-Forschung (ZSW) and several other institutes of the University of Stuttgart.

In phase II, the 350-kW electrolysis demonstration plant in Riyadh will be put into continuous solar-connected operation as soon as the new cell block has been installed. From the long-term experience to be accumulated, performance data will be obtained in order to optimize and scale up the system. In the 10-kW research and test center in Stuttgart, testing will be continued, dynamic analyses of pressurized electrolyzers will be carried out, and simplified control concepts for the stand-alone operation mode will be developed.[111-113]

In the field of hydrogen utilization technologies, basic theoretical and laboratory research will be performed on alkaline fuel cell concepts (e.g., characterization of gas diffusion electrodes) as well as on catalytic burners (reaction kinetics of H_2/air mixtures). Experimental investigation of non-stationary combustion phenomena will be performed. Practical tests will be carried out on internal combustion engines, also of the compression ignition type (diesel).

4. Energy-Self-Sufficient Solar Houses in Europe

(i) *Solar House of the Institute for Solar Energy Systems (ISE) of the Fraunhofer Society in Germany*

At the Fraunhofer Institute for Solar Energy Systems, the concept of the "Self-Sufficient Solar House 2000" was developed over several years,[136,164] and, between September and October 1992, a solar house was built in Freiburg, Germany. Its design encompasses solar architecture principles, transparent insulation, thermal collectors, photovoltaics, battery storage, hydrogen storage, and an electricity and hydrogen application system (Fig. 31). All energy needs are provided by solar radiation.

The energy supply strategy of this concept supposes direct electricity use during sunshine. Excess energy is stored in a lead–acid battery system. As soon as this system is charged completely, hydrogen and oxygen are produced in a membrane pressure electrolysis unit at a level of 3 MPa and stored in gas cylinders. The hydrogen storage is seasonal, with a cycling period of one year. During wintertime, the hydrogen is converted to electricity by an alkaline fuel cell system or to high-temperature heat in a catalytic cooker with four diffusion burners and in a catalytic air heater with a two-step diffusion burner. The battery buffers electricity peak demand that cannot be covered by the fuel cell. The catalytic cooker replaces the electric one when the sun is not shining. The oven is operated electrically. The daily energy consumption is assumed to be on the order of 3 kWh, and the overall efficiency of the storage system is calculated to be about 70%.

The solar hydrogen system consists of the following components:

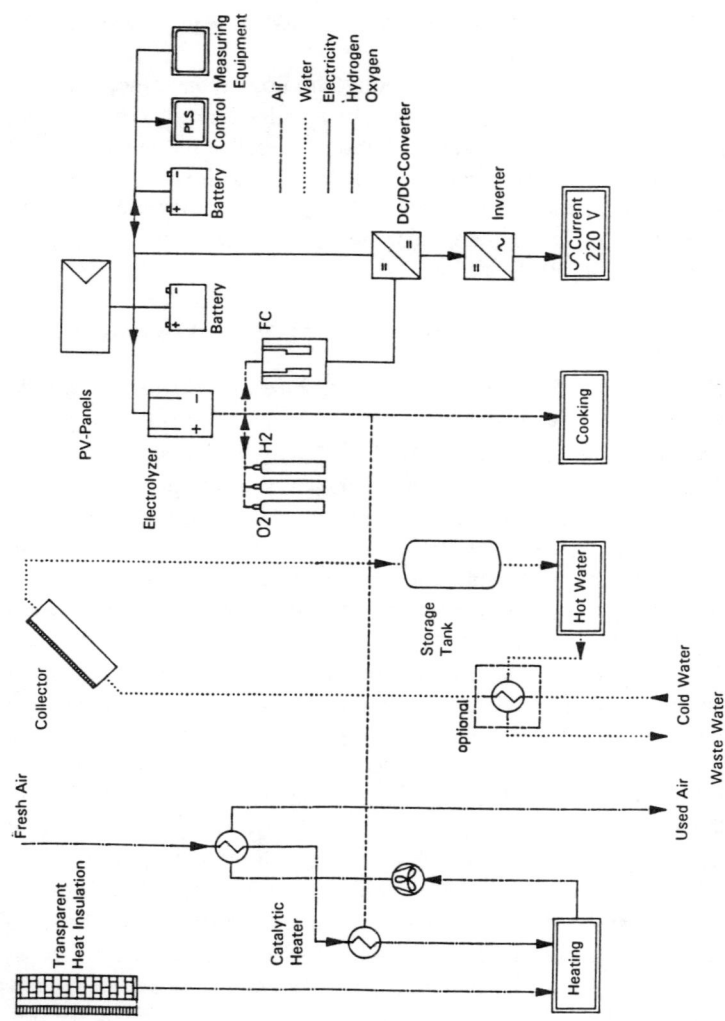

Figure 31. Energy self-sufficient solar house at Freiburg, Germany.[164]

- Solar generator: 84 Siemens M 50 modules with 36 monocrystalline silicon cells, each covering 30 m^2, with a total output of 4.2 kW$_p$
- Battery bank: 48 lead–acid batteries (200 Ah at 2 V each), delivering a system voltage of 48 V and a design capacity of 19.2 kWh
- Auxiliary batteries: 11 lead–acid batteries (75 Ah at 2 V each), delivering a system voltage of 22 V and a design capacity of 1.65 kWh (for the process control system)
- Inverter: Sinus type with a design rating of 3 kW
- Electrolyzer: 30-cell membrane electrolyzer by FhG-ISE with a design capacity of 2 kW, an operating pressure of up to 3 MPa at a current of up to 90 A, an operating temperature of 80°C, a hydrogen production capacity of 1.2 m^3/h, and an oxygen production capacity of 0.6 m^3/h
- Fuel cell system: 14-cell alkaline fuel cell unit operated with 30% KOH, with Ni gas diffusion electrodes, operating temperature of 70°C, and design output of 0.5 kW
- Gas storage: Hydrogen, 15 m^3, 3 MPa, 1436-kWh energy content; oxygen, 7.5 m^3, 3 MPa
- Cooking stove: Four catalytic diffusion burners with 2.6-, 1.7-, 1.7-, and 1.0-kW design output (FhG-ISE development); electric baking oven with 230 VAC supply
- Air heater: Two-step catalytic diffusion burner of 0.5/1.5 kW power rating built into the air supply duct; automatic ignition (PhG-ISE development)

The solar thermal collector system for the production of hot water consists of selective absorbers, type "BEIKO," with a cermet cover achieving an absorption degree of 0.93 and an emissivity of 0.09, a 1000-liter water layer storage, a flat-plate heat exchanger for the collector circuit and the fuel cell excess heat circuit, and 24-V circulation pumps with variable flow rates.

The air ventilation system has a design capacity of 200 m^3/h, providing an exchange rate of 0.7. The fresh air enters into a system of tubes buried 4 m deep in the earth; by passing through the tubes the air is preheated, then it passes a heat exchanger (recovery factor 0.85) recovering the heat contained in the used air, and finally it is heated to its required final temperature by the hydrogen-powered air heater.

(ii) Solar House in the Swiss Emmental

In Switzerland, the architect Markus Friedli has supplied the energy needs of his highly insulated residential house in off-grid operation mode with a solar–hydrogen energy supply system since 1990.[165] Solar thermal collectors provide hot water, and solar electricity is supplied by 6-kW$_e$ photovoltaic modules. Not directly used electricity is stored in batteries or converted into hydrogen by a pressurized electrolyzer (Hydrogen Systems) at a rate of 2 m^3 per hour. The hydrogen is stored in a 100-liter metal-hydride hydrogen storage with a capacity of 19 m^3 of hydrogen. A stove and the heating of a washing machine are fueled by H$_2$. The total investment costs for the solar and the hydrogen components amounted to some 500,000 Swiss francs (83,000 SFr were from public funds).

In 1991, Mr. Friedli modified a Toyota Hiace 4WD van for multifuel operation (hydrogen, propane, gasoline). The solar hydrogen produced via photovoltaics and electrolysis is stored in a metal-hydride storage providing between 100 and 150 km of operating range at a maximum velocity of 90 km/h.

5. Photovoltaics/Hydrogen Project at the Agricultural College of Triesdorf in Germany

In the Agricultural College of Triesdorf, various electricity consumers are supplied with photovoltaic electricity.[193] Among these consumers are the educational workshop for agricultural equipment, the seed cultivation, cattle breeding, and pig breeding programs, the fish rearing pond, and the domestic science department. The last two departments of the college have interfaces with the hydrogen/oxygen production and storage system. Their photovoltaic generators—the mobile one of the fish rearing pond (2 kW$_p$) as well as the stationary one of the domestic science department (5 kW$_p$)—can feed their electricity into the electrolysis module when electricity is not used directly by other consumers and when the battery (48 V/420 Ah) is charged. The stored electricity, via a 5-kW inverter, feeds electric domestic appliances in the college's domestic science department (see Fig. 32).

The hydrogen generated is used for cooking in the domestic science department, whereas the oxygen generated is used for aeration of the fish pond. This permits the parallel use of the product gases from the electrolysis. The alkaline electrolyzer has 3.15 kW$_{el}$ and is of an Alyzer model 0100, operating at 0.6-MPa pressure and at a maximum temperature of 75°C.

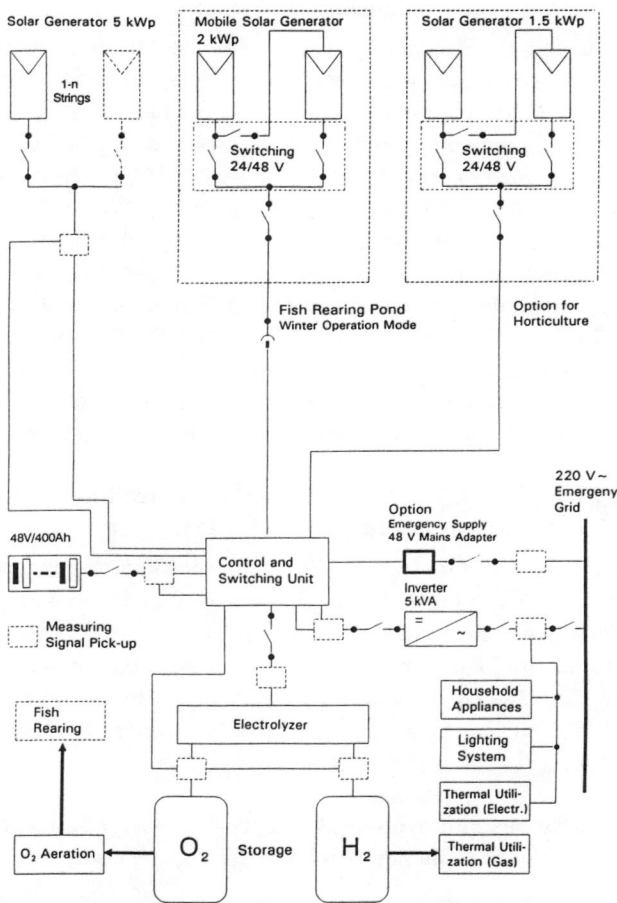

Figure 32. Block Diagram for the photovoltaic hydrogen project at Triesdorf.[193]

The system was conceived by Ludwig-Bölkow-Systemtechnik GmbH for the Agricultural College of Triesdorf and the local utility Fränkisches Überlandwerk. The main contractor for the photovoltaic electricity supply and storage system was IBC Solartechnik.

6. Photovoltaics/Hydrogen Project at the Helsinki University of Technology in Finland

In Finland, the Helsinki University of Technology is carrying out an experimental program on a seasonal energy storage system consisting of a monocrystalline PV generator, lead–acid buffer battery storage, alkaline electrolysis, H_2 pressure storage, and a phosphoric acid fuel cell.[194] The system is designed to provide between 1 and 2 kMw of electricity daily. Extensive simulations led to a system layout of 1.3 kW_p for the PV array, 0.8 kW_e for the electrolysis, 0.5 kW_e for the fuel cell unit, and up to 0.1 kW_e for a resistive load. Measurements showed very good compliance with earlier simulations. The round-trip efficiency of the H_2 storage is 25%. The overall system efficiency is 3.8% ("load/insolation"). The power consumption of the system controls is still rather high, amounting to 65% of the system load. For 100% self-sufficiency at 60° latitude, the PV array has to be designed to produce three times the load energy.

7. Photovoltaics/Hydrogen Project at the Instituto Nacional Técnica Aerospacial (INTA) in Spain

In the Spanish province of Huelva, the Instituto Nacional Técnica Aerospacial (INTA) has been carrying out an experimental program on solar hydrogen production and utilization since 1990.[195] A photovoltaic generator of 144 BP modules (BP260S) generates a peak power output of 8.5 kW_p. The 26-cell, 7-kW_e alkaline electrolyzer by Alyzer operates at 0.6 MPa and 80°C and can be coupled flexibly with the PV generator via a 7.4-kVA AC/DC converter.

8. Autonomous Photovoltaic/Hydrogen/Electricity Supply System of the Library at KFA in Jülich, Germany

At the KFA Forschungszentrum Jülich GmbH, the self-sufficient supply of the library building with electricity over the whole year is planned.[196] The system will include a photovoltaic generator, a DC/DC converter, a lead–acid battery system, alkaline electrolysis, and an alkaline fuel cell system. The entire system envisaged was simulated in the JULSIM simulation program in order to ensure a successful functioning of all control devices when the hardware system was put into operation at the end of 1993.

The system conceived has the following design data and technical concepts:

- PV generator: Four facade and rooftop integrated generators consisting of monocrystalline modules with an active area of 312 m^2 and a peak power output of 43 kW$_p$ (1000 W/m^2, 25°C, AM 1.5).
- DC/DC converter: Each PV away has two converters of 5 kW each, operated in a master/slave mode and controlled with a pulse frequency of 20 kHz, adjusting the voltage to the level given by the DC grid, which has the actual voltage level of the battery system.
- Pb battery system: Design current of the DC grid is 220 V (200... 260 V); design capacity of 250 kWh and 1380 Ah over 10 h; batteries are of the OPzS OCSM type with electrolyte recirculation.
- Electrolysis system: Bipolar 21-cell electrolyzer with an active cell area of 2500 cm^2, a current of 750 A, and a current density of 3 kA/m^2, with 30% KOH solution, 80°C operating temperature, and 90% efficiency in design load operation at a design power rating of 26 kW.
- H$_2$/O$_2$ storage system: The product gases hydrogen (6.5 Nm^3/h) and oxygen (3.25 Nm^3/h) leaving the electrolyzer at 0.6 MPa are com pressed to 15 MPa and stored in pressure bottles of 1.4 m^3 each and a total geometrical volume of 20 m^3 for H$_2$ and 10 m^3 for O$_2$ sufficient for the entire seasonal storage requirements.
- Fuel cell system: Alkaline KOH gas diffusion electrode fuel cell of the Siemens BZA 4-2 type; design power output of 6.5 kW at 48 V and 135 A; system efficiency at design load is 63% (lower heating value of H$_2$) and 70% at 30% partial load.

The overall efficiency of the electrolysis/storage/fuel cell/DC-converter chain is 45%. The requirements for self-sufficient operation impose new considerations concerning the design, process engineering, and setup of the involved components such as fuel cells, electrolyses, and hydrogen storage systems. Advanced systems such as proton-conducting membrane fuel cells (PEM), reversible alkaline electrolysis/fuel cell system, and Stirling motor with AC generator will be investigated and tested.

9. Decentralized Solar Hydrogen Energy Supply System at the Fachhochschule Wiesbaden, Germany

Since 1991 a solar hydrogen energy supply system has been in operation at the Fachhochschule Wiesbaden. A small PV generator with an output of 1.5 kW$_{e,p}$ feeds its energy into a battery bank (350 Ah). This battery supplies alternating current (220 V, 50 Hz) to an independent local grid via an inverter. Excess energy that cannot be fed into the battery system produces pressurized hydrogen in a 2-kW$_e$ pressurized electrolyzer (2 MPa). For long-term storage, pressurized hydrogen and oxygen are stored in bottles for pressurized gases. In an alkaline fuel cell system of 1.2-kW$_e$ capacity, hydrogen and oxygen can be converted into electricity with an average efficiency (hhv of H$_2$) of 50–55%. The overall electrolysis/fuel cell efficiency amounts to 40%. The electrolyzer and the fuel cell have only one electrolyte/water circuit, thus avoiding feed water treatment.[197]

Within the framework of the hydrogen activities of the Fachhochschule Wiesbaden, various hydrogen application technologies that are not yet available on the market have been developed, some derived from natural gas or propane systems. For the later introduction of hydrogen,* consumer equipment/appliances will have to be available. Among the hydrogen appliances developed or adapted are an absorption refrigerator, a gas lamp, a camping cooker, a hydrogen heater, internal combustion engines suited for hydrogen, a cogeneration unit, a motor pump, a seawater desalination plant, a cooling device, welding equipment, an H$_2$/O$_2$ generator, a catalytic burner, a gas purification unit, and gas sensors.

10. Photovoltaic–Hydrogen Project at ENEA in Italy

Ente per le nuove Technologie, l'Energia e l'Ambiente (ENEA) operates a small photovoltaic–hydrogen generation plant in its research center near Rome. The system is in the 10-kW range and includes a 2-MPa pressurized alkaline electrolyzer fabricated by Alyzer. Storage of hydrogen is accomplished in metal hydrides. With the hydrogen generated, a small (2 kW) membrane fuel cell unit by Ballard Power Systems is operated. Furthermore, on the basis of a Fiat Ducato, a small passenger bus with pressurized hydrogen storage in fiber composite containers and with an internal combustion engine has been conceived.

---

*Into decentralized hydrogen-based energy systems.

XII. OUTLOOK ON HYDROGEN ACTIVITIES WORLDWIDE

1. Japanese Activities on International Clean Energy Network Using Hydrogen Conversion [World Energy Network (WE-NET)] and Others

The Japanese clean energy advancement program focuses on the supply of clean renewable energy from resources abroad in the form of hydrogen or hydrogen compounds (Fig. 33). The time frame of the project is 1993 to 2030, and the project budget is 300 billion yen (approximately U.S. $3 billion). The WE-NET project[198] is included in the overall framework of the New Sunshine Project with an overall budget of 1550 billion yen (approximately U.S. $15 billion).[199]

The 28-year WE-NET hydrogen project (see Fig. 33) covers different steps from component development to pilot and demonstration activities to popularization and commercialization of the entire system. The technologies considered important are hydrogen technologies for production (improved water electrolysis using electricity from hydro power, PV, and wind energy), conditioning (liquefaction or hydrogenation to methanol or cyclohexane), large-scale maritime transportation (containerized), decentralized storage and transportation, power generation (turbines, fuel cells), mobile applications (car, aviation), CO_2 fixation, and others.

At first, hydro power, which is still the cheapest renewable clean energy source, will be used for WE-NET. Japan is already in negotiations with Canada on the purchase of 4000 MW_e of hydroelectricity. Later on, solar energy, for example, from Saudi Arabia will provide the primary energy source. This clean, renewably generated electricity will be converted into hydrogen, which will be delivered to Japan either in liquefied form or chemically bound, for example, as methanol.

A very important feature of the New Sunshine Program[198] is the transfer of appropriate energy technologies and hydrogen to less developed countries and the development of appropriate technologies for the growing markets in least developed countries (LDCs), making use of cheaper infrastructures there. Besides the reduction of global CO_2 emissions, the buildup of global partnerships and the creation of new industries based on hydrogen technologies are expected to be major effects of the WE-NET project.

Very soon—already in the second half of the 1990s—the Japanese automotive company Mazda intends to introduce its rotary engine hydrogen car HR-X via an extended demonstration activity in Japan. Hydrogen

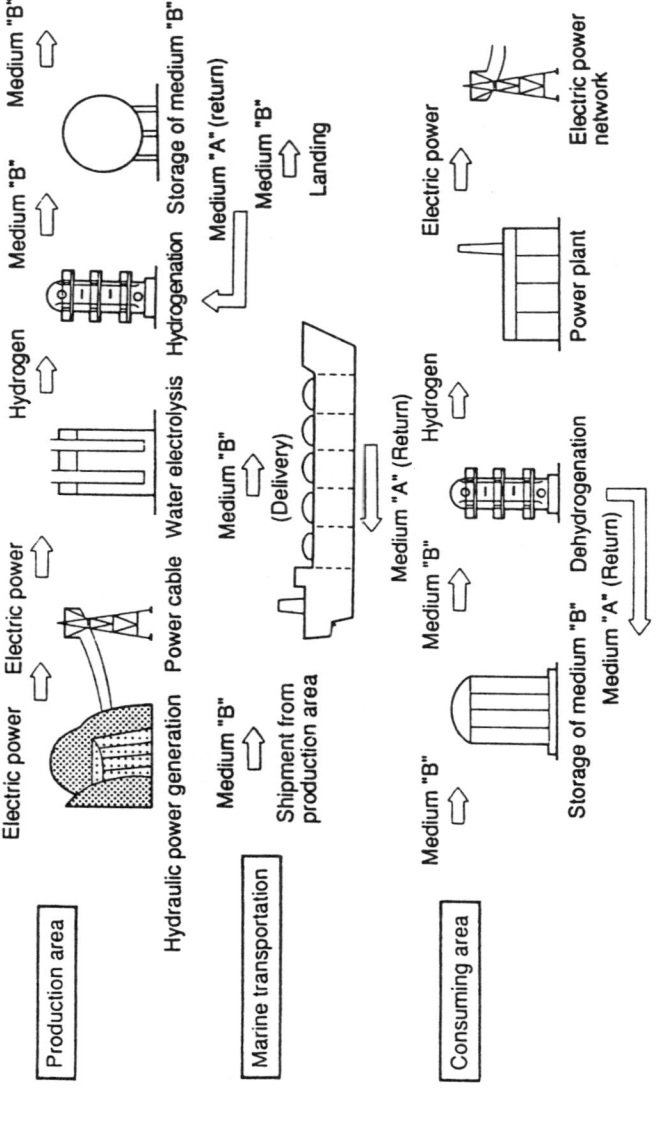

Figure 33. Block Diagram for the WE-NET concept.[199]

will be produced electrolytically in a decentralized way at each fueling station and distributed in pressurized form to the metal-hydride stores of the hydrogen vehicles.[200]

2. Other Hydrogen-Related Activities in the United States and Canada—Summary

At the Riverside City campus of the Riverside Community College, a 4-kW$_{p,e}$ photovoltaic generator and an electrolyzer unit will be coupled for the daytime production of solar hydrogen. During nighttime, the electrolyzer is connected to the grid in order to use off-peak electricity, thus permitting almost 24-h operation.[201] A Dodge D-50 truck, which has been being operated with hydrogen for 12 years, will be fueled by the hydrogen produced in this solar hydrogen plant. The project is being carried out by a consortium consisting of California's South Coast Air Quality Management District, Riverside Community College, American Lung Association of Inland Counties, Electrolyser Inc., Canada, and the Ministry of Energy of Ontario, Canada. The goal of this U.S. $1 million project is to demonstrate the zero-emission production of solar hydrogen and the benefits of its application as a clean, alternative automotive fuel, as well as to develop a student training program for hydrogen fuel technology. The system has been operative since the summer of 1993.

Within a Department of Energy (DoE) and Department of Transport (DoT) project, the concept for a 12-m city bus driven by a phosphoric acid fuel cell (PAFC) fueled by hydrogen obtained from methanol via on-board catalytic steam reforming is being developed in cooperation with Georgetown University in Washington, D.C. Three 9-m test bed buses were scheduled to be built and tested by a consortium consisting of H Power Corporation, Booz Allen & Hamilton, Bus Manufacturing U.S.A. Inc., Fuji Electric, Soleq Corporation, and Transportation Manufacturing Corporation in September of 1994 and end of early 1994. One tested 37-passenger bus will be delivered to the South Coast Air Quality Management District. The bus is of the high-floor design, has a single-speed differential, regenerative braking, either lead–acid or nickel–cadmium batteries for peak energy supply, and a 55-kW Fuji PAFC (175 cells, 0.66 V DC each, 2.4 A/m^2, 0.2-m^2 active area, 190°C operating temperature, 38% overall efficiency). The project is sponsored by DoE, DoT, and Georgetown University.[202] The first batch of the 20–30 buses will roll off the assembly line in 1997.

Sacramento Municipal Utility District (SMUD) plans to realize a completely carbon-free bus system in a U.S. $7.3 million three-year project with several partners (tentatively, Aerojet, Rockwell, DoE, University of California at Davis, Sacramento Regional Transit District). A 200-kW photovoltaic solar hydrogen production/storage/refueling facility will supply hydrogen to a fuel cell bus using the same fuel cell technology (PAFC) as the DoE/Georgetown University bus, but without a reformer. The mode of hydrogen storage (metal hydrides, pressurized storage, or direct iron reduction storage system) has not yet been decided. The bus will be operated on a line linking Sacramento's Municipal Airport, the city of Sacramento, and the Northern Sacramento County.[203]

In the Canadian province of British Columbia, Ballard Power Systems developed a 20-passenger fuel cell city bus with a membrane fuel cell (PEM) system (24 stacks of 5 kW each) delivering 75 kW power to the wheels and with a compressed hydrogen storage, providing 150-km operating range. Battery storage for peak power is not required. The bus was put into demonstration service at British Columbia Transit in early summer 1993.[204]

In the United States, there have been proposals to shift funds from DoE's nuclear defense budget to sustainable energy, energy conservation, and hydrogen at significant rates, reaching almost U.S. $400 million annually for hydrogen in the year 2000. A significant share will be allocated to fuel cells in transportation applications. DoE presently intends to commercialize methanol- and hydrogen-driven ultralow- and zero-emission fuel cell vehicles by 2010.[205]

For reasons of diversification, the province of Québec in Canada is considering or already practices the production of energy-intensive products for export. Among these products, besides aluminum and manganese, is hydrogen. In the North American market the average transport distance for hydrogen is more than 1000 km. Therefore, hydrogen is transported in liquefied form, thus saving a lot of transport fuel. Electricity in Québec is very reasonably priced, and therefore LH_2 is produced at very competitive prices and completely free of CO_2 (whereas gaseous hydrogen from steam reforming would generate 0.5 m^3 of CO_2 per cubic meter of H_2 produced); an LH_2 production facility has been in operation at Bécancour since 1987.[206] Furthermore, Québec imports almost all its automotive fuel, whereas it produces most of its electric energy from huge hydro power resources; hydroelectricity's share in primary energy increased by a factor of two in the period 1972–92. The remaining resources for development are estimated to be on the order of 50,000 MW_e.[167,168]

3. Other Hydrogen-Related Activities in Europe—Summary

Besides activities on the production of hydrogen, various activities on the application of hydrogen and hydrogen-rich gases are under way in Europe. Some of them are highlighted in the following.

In a 130 million DM joint Danish/German project, carried out by an industrial consortium consisting of DASA, RWE, Ruhrgas, Elkraft, and Haldor Topsöe, molten carbonate fuel cell (MCFC) technology will be developed to a commercial state by 1996. Three 100-kW MCFC units will be built and tested.[208] These units will be operated with hydrogen-rich gases obtained from hard coal, lignite, and natural gas via reforming (external or internal).

In a development program from 1991 to 1993, Ansaldo developed and built a 1.34-MW_{DC}/1.28-MW_{AC} phosphoric acid fuel cell (PAFC) power plant in Italy to be installed in Milan. The system will be operated with steam-reformed hydrogen from natural gas (reformer technology by Haldor Topsöe).[209]

Within the EQHHPP (see Section IX.2) various hydrogen bus demonstration projects on the basis of internal combustion engines (M.A.N., Hydrogen Systems), an external combustion engine of the Stirling type (Esamco), and a membrane fuel cell (Ansaldo/de Nora) will be realized in Belgium, Germany, Ireland, and Italy in the years to come.[188] A hydrogen bus propelled by an alkaline fuel cell will be demonstrated within an EC Eureka project (Air Products, Ansaldo, Elenco, Saft). For all projects, except for the Irish Stirling engine bus, which will use compressed gaseous hydrogen, liquid hydrogen will be the fuel to be stored on board.

In Denmark, a concept for electricity storage via hydrogen has been investigated in a first-order assessment type of study.[211] The system will consist of an alkaline electrolyzer system, a compressor installation, a salt cavern storage, an expansion installation, and a solid-oxide fuel cell system. The underground hydrogen storage is designed for seasonal storage. The overall system "electricity–electricity" efficiency will reach 59%; the additional thermal efficiency may reach 35%. The storage costs for 1 kWh of electricity are estimated to be on the order of U.S. $0.10.

4. Other Hydrogen-Related Activities in the Rest of the World

In Brazil in 1982 a Commission for the Development of Hydrogen was established for the coordination of all hydrogen activities.[212] The hydrogen policies formulated focus on the research and development, engineering development of prototypes and industrial processes, industrialization, and utilization. The basis for these considerations is the overabundant waterpower resources and the planned development schemes (on the order of 70 to 100 GW).

The production of pig iron and steel in the Amazon region can be achieved by the utilization of hydrogen from hydroelectricity, produced at plants such as Kararao or Tucurui. Huge amounts of rainforest can be saved by the replacement of a part of the charcoal produced from wood originating in the tropical rainforest by hydrogen[213] and by replacement of another part by charcoal from secondary short-rotation forestry.

The hydrogen activities in Saudi Arabia were initiated and are still carried out with the support of and in collaboration with German institutions. The activities are concentrated at the King Abdulaziz University, Jeddah, and the King Abdulaziz City for Science and Technology, Riyadh, and focus on R&D activities and on system component development and testing activities for hydrogen production and application.

In Libya ideal insolation conditions exist and will be used for the large-scale generation of solar electricity, which may then be exported to Europe, either directly via electric transmission lines or via hydrogen, in gaseous or liquid form.[214,215]

Also Morocco, mostly lacking fossil energy resources, considers large-scale solar energy production as an alternative for its future development and also hydrogen as one possible extension of this concept.[147]

In Australia, at the University of Melbourne,[216] and in New Zealand, at the University of Canterbury,[217] some hydrogen research activities focus on the development and testing of complete engines and of engine components such as direct hydrogen injection valves. In the engine tests, dual-fuel operation seems to be of most interest, especially the two-phase combustion process of pilot fuel and of hydrogen.

Problems related to hydrogen combustion in internal combustion engines are also being investigated.[218]

In Romania, the spark-ignition engine of a Dacia car has been modified for dual-fuel operation and tested successfully on the test bed and in the car.[219] In Russia a ZIL gasoline truck was modified for hydrogen admixture.[220] Both the Dacia and ZIL made use of metal hydrides as the storage medium for hydrogen. Similar modification activities on vehicle engines are going on in various other countries including Belgium, Canada, and the United States.

5. Arguments and Concepts for a Hydrogen Energy System and for International Hydrogen Trade

(i) Resources

Many large sources of renewable energy[158,174,176,177,179,181–186] are located in concentrated form in remote areas of the world, far away from where the major world consumers are located. In the case of hydro power potential, this applies, for example, to Zaire, Indonesia, Iceland, Greenland, and some other locations in Africa, Asia, and North and South America. In the case of direct solar radiation, the locations are concentrated in the sun belt—approximately between 35° northern and southern latitude. In this geographical region many large unpopulated or sparsely populated areas exist, for example, northern Africa, northeastern Brazil, Saudi Arabia, southern countries of the Commonwealth of Independent States, Australia, and others.[136,137]

Several long-distance energy transportation technologies are available, such as gas pipelines, high-voltage direct current electricity transmission lines, and liquid gas tanker transportation. For intercontinental (maritime) linkage of production and consumption centers, only large seagoing tankers seem technically feasible in the medium term. At present, experience is limited to containerized small-scale cryogenic transport of liquid hydrogen.

An important reason for the local usage of hydrogen in connection with hydro power could be the conversion of surplus hydro energy in alpine/mountainous regions—typically in Austria and Switzerland—in spring and early summertime and its seasonal storage via a chemical energy carrier, such as methylcyclohexane.[133] This hydro-hydrogen can be used in many local applications, thus improving environmental conditions and also avoiding long-distance transport.

(ii) Consumers

Already today, 50% of the world's population lives in urban agglomerations; in industrialized countries, the figure is 65%. This share will also be reached in developing countries in the not too distant future. Concentrated energy carriers, such as electricity and gas, will be a prerequisite for an efficient energy supply, especially of energy produced from renewable sources. If there is be no major breakthrough in superconductive electricity storage and if the growing constraints on the use of fossil energies are understood and accepted, hydrogen can become increasingly important as an energy storage medium as well as an energy carrier.

In 1991, Japan imported more than 50 billion cubic meters of natural gas in liquid form (LNG), accounting for approximately 70% of the world LNG trade.[210] This amount of gas is about the same as Germany imports via gas pipelines.

(iii) North–South Dialogue

An advanced and simplified hydrogen technology for small-scale stand-alone applications (e.g., photovoltaic/hydrogen or wind/hydrogen systems) might provide a solution also for arid developing countries, where already today the supply of nonconventional fuels (e.g., wood) is not adequate anymore. The accident potential associated with this technology will not be higher than that of liquid gas supply systems, e.g., liquid petroleum gas (LPG) presently in use.

Other countries in regions with high insolation might see an advantage in exporting their solar energy resources via electricity transmission or export of hydrogen, thus participating in future energy trade.[147,214,215]

(iv) Hydrogen As a Facilitating Technology for Renewable Energies

Fluctuating electricity, as generated from renewable sources, can be fed into existing electricity grids at levels between 20 and 30%, at most, of their total capacity.[130] Beyond this percentage, under the conditions of the presently existing electricity-generating infrastructure, electricity storage is required. With modifications in supply and demand management, the percentage of fluctuating energy that can be absorbed by the electric grid without storage might be increased above 30%.

Remote small-scale electricity-generating installations that have to supply energy on daily, weekly, and annual basis and that are not connected to a grid need a storage device. For short-term storage, this can be a battery system. For long-term storage (for longer than a week or so) batteries become very expensive and should be replaced by another electrochemical alternative—electrolysis and fuel cells, and thus hydrogen technology.

For island settings, such as Iceland, Greenland, or Hawaii (see Sections IX.5 and X.2), the close proximity of consumer and resources on the one hand, and the relatively high fuel price level, on the other hand (due to long-distance importation of conventional fuels), may provide more favorable conditions for the introduction of hydrogen technology than exist in many other locations in the world.

For all of the applications or concepts listed above, hydrogen can play a role as a "facilitating" or "enabling" technology; that is, at least in the medium term, all these concepts cannot be realized, or can only be implemented in ways that have large environmental and user-related drawbacks, without the use of hydrogen as a link between supply and demand.

(v) Expected Cost for Renewable Electricity and Hydrogen

For new *optimized* large-scale power plants[137–139,170] based on existing technology and foreseeable long-term technical developments, the following *electricity generating costs* seem obtainable[132,151–155,171]:

Hydro power plant	0.06 DM/kWh (6500 h/a)
Solar thermal parabolic trough plant (California)	0.11 DM/kWh (3600 h/a)
Nuclear power plant	0.13 DM/kWh (7000 h/a)
Wind energy farm	0.17 DM/kWh (2000 h/a)
Photovoltaic plant, North Africa—2000	0.30 DM/kWh (2400 h/a)

Based on the above assumed electricity generating costs and on *mass-produced* pressurized electrolysis (3 MPa), including all system losses, the following *hydrogen costs* (at the site of production) are achievable:

Hydro power plant	0.09 DM/kWh (6500 h/a)
Solar thermal parabolic trough plant (California)	0.15 DM/kWh (3600 h/a)
Nuclear power plant	0.18 DM/kWh (7000 h/a)
Wind energy farm	0.26 DM/kWh (2000 h/a)
Photovoltaic plant, North Africa—2000	0.42 DM/kWh (2400 h/a)

If long-distance transport to Germany of 7000 kilometers is assumed, and two transport modes are considered, namely, high-voltage direct current transmission (HVDC) and a 10-MPa hydrogen pipeline, then transmission costs, including losses, will amount to about 0.05 DM/kWh$_e$ (6500 h/a, hydro) and 0.13 DM/kWh$_e$ (2400 h/a, PV) for HVDC and 0.04 DM/kWh$_{th}$ (6000 h/a, hydro or PV) for hydrogen. Under the given assumptions, this leads to the conclusion that the provision of thermal energy via imported hydrogen is economical for both PV and hydro power. If the imported hydrogen is to be converted into electricity via highly efficient fuel cell technology (4000 h/yr at 65% efficiency), electricity supply costs of more than 0.20 DM/kWh$_e$ are reached. These transmission costs then have to be added to the aforementioned hydrogen costs at the production site.

6. Reasons for Application of Hydrogen in the Transportation Sector

Burning of fossil fuels leads to the emission of greenhouse gases, mainly CO_2. Within city boundaries, environmental problems are caused by various uses of energy, transportation being one of them. In all industrialized countries, awareness of environmental problems has grown significantly during the past years. On the other hand, the desire for increased mobility will still lead to a growing number of vehicles with combustion engines. The congestion of our city centers by automobiles and the omnipresence of the automobile as an emission source has given rise to discussions about additional legal measures, such as the closing down of city centers to polluting vehicles.

These considerations call for improved urban transportation concepts, especially for clean ones. All efforts should aim at the eventual substitution of the presently used fossil fuels by clean energies, not only in the transportation sector, but also in the entire energy system. Clean energies, such as renewable energy and possibly surplus nuclear energy, in conjunction with hydrogen, might open a perspective for an improvement of this situation.

In the long-term perspective, new energy vectors in general should be considered, on the one hand, because of limited availability of hydrocarbon fuels and, on the other hand, with respect to the reduction of energy-related emissions, and accordingly also in relation to possible climatic changes.

Since changes in the existing energy supply and utilization infrastructure, as well as in the transportation sector, have very long lead times and will require huge investments, the development of the necessary technologies has to be initiated very early and pursued continuously. Of high importance are demonstration activities, including the transportation sector, since very valuable practical experience (over the whole process chain down to the user) will be collected and the new energy concepts will be brought to the attention of the public.

Therefore, interest in clean automotive propulsion concepts is growing, and they are increasingly acknowledged as the best solution for vehicle propulsion in metropolitan areas with high air pollution levels. Clean propulsion concepts certainly are not a remedy for our congested cities or our unsuccessful transportation policies. On the other hand, we always will have a need for road vehicles in metropolitan areas, such as city buses, taxis, delivery vehicles, and civic service vehicles as well as a certain number of passenger cars, the degree of penetration with passenger cars depending very much on the accessibility and attractiveness of public transport services.

In many countries around the world, first steps have been taken toward the development of hydrogen-fueled ground vehicles and aircraft. Beginning about 15 years ago, single vehicles, mainly passenger cars, have been converted into hydrogen-operated prototypes, and these have gradually been improved over the years. The best known activities of this kind are those of BMW[221] and Daimler Benz[222] in Germany and of the Mushashi Institute of Technology[223] and Mazda[200] in Japan. More and more companies and institutes worldwide have started basic research work on the conversion of cars, buses, and trucks. Such research is under way, for example, in Australia, Belgium, Canada, Germany, Italy, Ireland, Japan, New Zealand, Russia, Rumania, Sweden, and the United States. In most cases, this work is based on internal combustion engines, but fuel cell systems[209] and external combustion engines of the Stirling type[188] have also been considered. In Russia as well as in Germany[166] and Canada, aircraft jet engines have been modified for hydrogen operation, or parts of them are presently being modified.

If the stringent emission requirements in California, calling for zero-emission vehicles (ZEV) and ultralow-emission vehicles (ULEV), are taken seriously enough at their place of origin and application (and there seems no doubt that they will be[223]) and if a step-by-step introduction of these requirements in Europe also is considered possible in the long run, fuel cell power systems with hydrogen storage presently seem to be the only long-range zero-emission propulsion alternative for cars. For driving ranges within city boundaries, advanced battery systems will also fulfill zero-emission requirements. Near-zero- or almost-zero-emission propulsion concepts can be provided by operating internal combustion engines with lean hydrogen/air mixtures at high lambda values ($\lambda > 2$) and with a catalyst.[223] Conventional emissions from engine lubricants are in the parts per million range and are thus negligible. ULEV standards for heavy-duty vehicles are safely fulfilled with hydrogen- or probably even with hythane- (a hydrogen/methane mixture, typically 20:80) operated internal combustion engines. Demonstration activities, as described in Section IX.2 (EQHHPP[188]), in Section XII.2 (U.S.A.[203,204]), and in Ref. 224, will show the technical feasibility of the vehicle and infrastructure (supply, refueling)[225] concepts.

With respect to operating costs, hydrogen-operated vehicle fleets certainly will require more sophisticated workshops and spare parts as well as more highly qualified personnel, all causing higher operating costs. However, this is already true for natural gas systems, presently being introduced in many metropolitan areas around the world. Presently, for diesel city buses in Europe, between 65 and 80% of the operating costs are for personnel. The additional costs for hydrogen systems[156,188,224] therefore seem to be foreseeable and increasingly justified by the environmental improvements achieved and the benefits gained by switching to clean renewable hydrogen.

XIII. CONCLUSIONS AND RECOMMENDATIONS

The concepts and projects described in this chapter prove the technical feasibility of renewable hydrogen. At present, hydrogen—from fossil or renewable sources—is not competitive with conventional fuels or energy carriers such as coal, oil, natural gas, or electricity in terms of price. If the negative environmental effects of the use of conventional energies are taken into consideration, the economic gap between present fuels and future renewable hydrogen might be closed. Although hydrogen generally has to be produced from a primary source, with the input of energy, its application in end uses is more efficient than that of conventional energies and more environmentally compatible.

In the long-term strategy, renewable hydrogen will be produced on a large scale from direct solar radiation (solar thermal, photovoltaics).[226] As a short-term approach, renewable hydrogen can be produced by the most inexpensive renewable energy source available, namely, hydro power. As a medium-term strategy, wind energy conversion and solar parabolic trough plants may be applicable. Large-scale hydrogen vectorization will be effected either by gaseous hydrogen via pipelines or by liquid hydrogen in superinsulated transport vessels. In decentralized stand-alone applications of renewable energy generation facilities, hydrogen will be produced as a storage medium and for fuel applications.

Owing to greenhouse gas constraints and increasing requirements for the reduction of local emission levels, the switch to fuels with higher hydrogen contents or without any carbon content is mandatory. The cleanest fossil fuel is natural gas, which seems abundantly available for the 21st century[227] and can contribute significantly to a reduction in the amount of CO_2 produced by the energy and transportation system, if properly handled. Natural gas can be the bridging fuel into an electricity/hydrogen energy economy, substantially beginning in the 21st century.

The most important prerequisite for the realization of the concepts as well as for the transfer of the demonstration and pilot projects for electrolysis and solar hydrogen into largescale commercial applications is the declared political will to improve the environmental situation on a world scale, thus helping to introduce these new products and systems (photovoltaics, wind energy conversion, solar thermal electricity production, advanced electrolysis, hydrogen application technologies) into the marketplace. If this approach is not adopted, there will be no real market for these technologies, and therefore the technological concepts described above may never find commercial application, including the necessary cost reductions.

In the view of the authors, the technical concepts of advanced electrolysis and solar hydrogen technology available or under development, the first steps taken in the demonstration and application of these concepts and technologies, and the increasing environmental constraints arising worldwide, lay the basis for and justify the realization of first pilot and then large-scale applications of solar hydrogen technologies and systems.

REFERENCES

[1] G. Sandstede, ed., *Elektrochemische Prozesse*, (Vol. 3 in *Studie Chemische Technik* (*DECHEMA*, 1 (D. Behrens and G. Kreysa, eds.) Frankfurt, 1975.

[2] N. Getoff, K. J. Hadig, G. Kittel, and S. Solar, *Wasserstoff als Energieträger*, Springer-Verlag, Vienna, 1977.

[3] A. P. Fickett and F. R. Kalhammer, in *Hydrogen: Its Technology and Implications* Ed. by K. E. Cox and K. D. Williamson, *Hydrogen Production Technology*, Vol. 1, Chemical Rubber Co., Cleveland, Ohio, 1977, pp. 3–41.

[4] J. O'M. Bockris, in *Encyclopedia of Chemical Technology*, Vol. 12, Ed. by I. Kirk and D. F. Othmer, Wiley-Interscience, 1980, New York, pp. 938–1037.

[5] B. V. Tilak, P. W. T. Lu, J. E. Colman, and S. Strinivasan, in *Comprehensive Treatise of Electrochemistry*, Vol. 2, Ed. by J. O'M. Bockris, B. E. Conway, E. Yeager, and R. E. White, Plenum Press, New York, 1981.

[6] P. Hoffmann, *The Forever Fuel—the Story of Hydrogen*, Westview Press, Boulder, Colorado, 1981.

[7] D. Behrens, ed., *Wasserstofftechnologie—Perspektiven für Forschung und Entwicklung*, DECHEMA, Frankfurt, 1986.

[8] G. Sandstede and G. Collin, eds., *Wasserstoffwirtschaft—Herausforderung für das Chemie-Ingenieurwesen, DECHEMA-Monogr.* **106** (1987).

[9] K. Ledjeff, ed., *Neue Wasserstofftechnologien*, Verlag C.F. Müller, Karlsruhe, 1988.

[10] P. Häussinger, R. Lohmüller and A. M. Watson, in *Ullmann's Encyclopedia of Industrial Chemistry*, Vol. A13, VCH-Verlagsgesallschaft, Weinheim, 1989, pp. 297–442.

[11] C. J. Winter and J. Nitsch, eds., *Wasserstoff als Energieträger*, Springer-Verlag, Berlin, 1989.

[12] M. Fischer, *Chem.-Ing.-Tech.* **61** (1989) 124.

[13] G. Sandstede, *Chem.-Ing.-Tech.* **61** (1989) 349.

[14] G. Sandstede, *Chem.-Ing.-Tech.* **63** (1991) 575.

[15] H. Wendt, ed., *Electrochemical Hydrogen Technologies*, Elsevier, Amsterdam, 1990.

[16] G. Sandstede, *DECHEMA-Monogr.* **125** (1992) 329.

[17] G. Sandstede, *DECHEMA-Monogr.* **128** (1993) 403.

[18] C. Bailleux, ed., H_2 *Perspectives*, Vol. 1, No. 1, Hydrogen Industry Council, Montreal, 1984, p. 2.

[19] J. Poitier and C. Bailleux, in *Hydrogen Energy Progress VI*, Proceedings of the 6th World Hydrogen Energy Conference, Vienna, Austria, Vol. 1, Ed. by T. N. Veziroglu, N. Getoff, and P. Weinzierl, Pergamon Press, New York, 1986, pp. 197–214.

[20] V. Engelhardt, *Die Elektrolyse des Wassers*, Verlag von Wilhelm Knapp, Halle, 1902.

[21] H. Berg and K. Richter, eds., *Entdeckungen zur Elektrochemie, Bioelektrochemie und Photochemie von Johann Wilhelm Ritter*, Akadische Verlagsgesellschaft. Geest und Portig, Leipzig, 1986.

[22] C. Bailleux, H_2 *Perspectives*, Vol. 3, No. 1, Hydrogen Industry Council, Montreal, 1986, p. 2.

[23] D. Ohms, V. Plzak, S. Trasatti, K. Wiesener, and H. Wendt, in *Electrochemical Hydrogen Technologies,* Ed. by H. Wendt, Elsevier, Amsterdam, 1990, pp. 1–135.

[24] J. O'M. Bockris and Z. S. Minevski, *Int. J. Hydrogen Energy* **17** (1992) 423.

[25] A. Damjanovic, in J. O'M. Bockris and B. E. Conway (Eds.), *Modern Aspects of Electrochemistry*, Vol. 5, Plenum Press, New York, 1969, pp. 369–483.

[26] S. Trasatti, in H. Gerischer, C.W. Tobias (Eds.), *Advances in Electrochemical Science and Engineering*, Vol. 2, VCH-Verlag, Weinheim, 1991, pp. 1–85.

[27] K. Kinoshita, *Electrochemical Oxygen Technology*, John Wiley & Sons, New York, 1992.

[29] J. Divisek, J. Mergel, and P. Malinowski, in T. N. Veziroglu *et al.* (Eds.), *Hydrogen Energy Progress IV*, Vol. 1, Pergamon Press, New York, 1982, pp. 437–447.

[29] B. Gupta, F. N. Büchi, and G. G. Scherer, in Annual Report of the Paul-Scherrer-Institut Villigen, 1991, pp. 12–14.

[30] M. Schreiber, G. Lucier, J. A. Ferrante, and R. A. Huggins, *Int. J. Hydrogen Energy* **16** (1991) 373.

[31] H. Iwahara, H. Uchida, and I. Yamasaki, *Int. J. Hydrogen Energy* **12** (1987) 73.

[32] K.-D. Kreuer, *J. Mol. Struct.* **177** (1988) 265.

[33] Th. Dippel and K. D. Kreuer, *Solid State Ionics* **46** (1991) 3.

[34] N. Knudsen, E. Krogh Andersen, I. G. Krogh Andersen, and E. Skou, *Solid State Ionics* **35** (1989) 51.

[35] I. G. Krogh Andersen, E. Krogh Andersen, N. Knudsen, and E. Skou, *Solid State Ionics* **46** (1991) 89.

[36] R. C. T. Slade and N. Sing, *Solid State Ionics* **46** (1991) 111.
[37] R. C. T. Slade and K. E. Young, *Solid State Ionics* **46** (1991) 83.
[38] E. W. Justi and A. W. Winsel, *Kalte Verbrennung—Fuel Cells*, Franz Steiner Verlag, Wiesbaden, 1962.
[39] W. Vielstich, *Fuel Cells*, Wiley-Interscience, London, 1970.
[40] H. Binder, G. Sandstede, A. Köhling, unpublished results.
[41] J. Müller, K. Lohrberg, and H. Wüllenweber, *Chem.-Ing.-Tech.* **52** (1980) 435.
[42] V. Plzak and H. Wendt, *Chem.-Ing.-Tech.* **58** (1986) 415.
[43] H. Wendt, *DECHEMA-Monogr.* **106** (1986) 113.
[44] W. Schnurnberger, M. v. Bradke, and R. Henne, *DECHEMA-Monogr.* **98** (1985) 195.
[45] W. Schnurnberger and J. Divisek, *VDI-Ber.* **602** (1987) 63.
[46] J. Divisek, J. Mergel, and H. Schmitz, *Chem.-Ing.-Tech.* **52** (1980) 465.
[47] J. Divisek, J. Mergel, and P. Malinowski, *DECHEMA-Monogr.* **92** (1982) 349.
[48] J. Divisek, J. Mergel, and H. Schmitz, *DECHEMA-Monogr.* **98** (1985) 389.
[49] J. Balej, *Int. J. Hydrogen Energy* **10** (1985) 89.
[50] J. Divisek, in *Electrochemical Hydrogen Technologies*, Ed. by H. Wendt, Elsevier, Amsterdam, 1990, pp. 137–212.
[51] F. Buteau, P. Demange, C. Moreau, R. Gros-Bonnivard, and J. M. Jud, in Vol. 1, Proceedings of the 9th World Hydrogen Energy Conference, E. by T. N. Veziroglu et al., MCI, Paris, 1992, pp. 345–354.
[52] G. Schiller and V. Borok, *Int. J. Hydrogen Energy* **17** (1992) 261.
[53] G. Schiller, V. Borck, R. Henne, and A. Kayser, in *Hydrogen Energy Progress IX*, Vol. 1 Proceedings of the 9th World Hydrogen Energy Conference, Ed. by T. N. Veziroglu et al., MCI, Paris, 1992, pp. 409–418.
[54] J. Divisek, H. Mergel, and H. Schmitz, *DECHEMA-Monogr.* **123** (1991) 65.
[55] J. Balej, J. Divisek, H. Schmitz, and J. Mergel, *J. Appl. Electrochem.* **22** (1992) 711.
[56] J. Divisek, J. Mergel, and H. Schmik, *Int. J. Hydrogen Energy* **15** (1990) 105.
[57] G. Kreysa and B. Hakansson, *J. Electroanal. Chem.* **201** (1986) 61.
[58] S. Stucki in *Hydrogen Production*, Proceedings of the 2nd IEA Workshop, Ed. by B. D. Struck, KFA, Jülich, Germany, 1991.
[59] H. Wendt and H. Hofmann, in *Hydrogen Energy Progress VII*, Proceedings of the 7th World Hydrogen Energy Conference, Moscow, Ed. by T. N. Venroglu et al., Pergamon Press, New York, 1988, pp. 673–697.
[60] H.-J. Rapp, F. Seifert, and H. Strathmann, in *Energieträger Wasserstoff*, VDI-Verlag, Düsseldorf, 1991, pp. 63–81.
[61] H. Wendt, in M. I. Ismail (Ed.), *Electrochemical Reactors—Their Science and Technology*, Part A, Elsevier, Amsterdam, 1989, pp. 248–274.
[61] H. Wendt and G. Imarisio, *J. Appl. Electrochem.* **18** (1988) 1.
[63] B. D. Struck (Ed.), *Hydrogen Production*, Proceedings of the 2nd IEA Workshop, KFA, Jülich, Germany, 1991.
[64] Lurgi, Wasserstoff aus Wasser—Druckelektrolyse, Lurgi Brochure, Frankfurt.
[65] H. Wüllenweber, From Zero Gap to the New Lurgi Pressure Cell, Paper presented at the 8th World Hydrogen Energy Conference, 1990, Lurgi Brochure, Frankfurt, 1990.
[66] J. Divisek, P. Malinowski, J. Mergel, and H. Schmik, *Int. J. Hydrogen Energy* **13** (1988) 141.
[67] J. Divisek and J. Balej, *DECHEMA-Monogr.* **121** (1990) 267–278.
[68] W. Hug, J. Divisek, J. Mergel, W. M. Seeger and H. Steeb, in T. N. Veziroglu et al. (Eds.), *Hydrogen Energy Progress VIII*, Proceedings of the 8th World Hydrogen Energy Conference, Honolulu, Hawaii, Pergamon Press, New York, 1990, pp. 681–690.
[69] J. Divisek and H. Schmitz, in *Hydrogen Production*, Proceedings of the 2nd IEA Workshop, Ed. by B. D. Struck. KFA, Jülich, Germany, 1991.

[70] O. Führer and Winsel, in D. Bonnet (Ed.), *Solarenergie und Wasserstofftechnik III*, Battelle Congress, Ilmenau, Hessisches Ministerium für Wirtschaft und Technik, Wiesbaden, 1990, pp. 229–239.

[71] O. Führer, St. Rieke, C. Schmitz, B. Willer, and M. Wollay, in *Hydrogen Energy Progress IV*, Vol. 3, MCI, Paris, 1992, pp. 1445–1453.

[72] H. Vandenborre, *DECHEMA-Monogr.* **98** (1985) 313.

[73] Ph. Vermeiren, R. Leysen, H. W. King, G. G. Murphy, and H. Vandenborre, in T. N. Veziroglu, N. Getoff, and P. Weinzierl (Eds.), *Hydrogen Energy Progress VI*, Proceedings of the 6th World Hydrogen Energy Conference, Vienna, Austria, Pergamon Press, New York, 1986, pp. 431–437.

[74] H. Hofinann, V. Plzak, J. Fischer, G. Luft, and H. Wendt, *DECHEMA-Monogr.* **92** (1982) 359.

[75] H. Wendt and H. Hofmann, in T. N. Veziroglu *et al.*, (Eds.), *Hydrogen Energy Progress V*, Proceedings of the 5th World Hydrogen Energy Conference, Pergamon Press, New York, 1984, pp. 893–901.

[76] V. Plzak, H. Hofrnann and H. Wendt, in T. N. Veziroglu, N. Getoff, P. Weinzierl (Eds.), *Hydrogen Energy Progress VI*, Vol. 1, Proceedings of the 6th World Hydrogen Energy Conference, Vienna, Austria, Pergamon Press, New York, 1986, pp. 369–376.

[77] H. Hofmann, R. Brand, and J. Hildebrandt, *DECHEMA-Monogr.* **123** (1991) 3.

[78] H. Hofmann, R. Brand, J. Hildebrandt, and L. Bücher, *DECHEMA-Monogr.* **128** (1993) 425.

[79] E. Hausmann, *Chem.-Ing.-Tech.* **48** (1976) 100.

[80] Krebskosmo, Wasserelektrolyse (company info), Berlin.

[81] A. Szyszka, *Int. J. Hydrogen Energy* **17** (1992) 485.

[82] F. Beck, H. Goldacker, G. Kreysa, H. Vogt, and H. Wendt, in *Ullmann's Encyclopedia of Industrial Chemistry*, Vol. A9, VCH-Verlagsgesellschaft, Weinheim, 1989, pp. 183–249.

[83] A. T. B. Stuart and S. N. Pirani, in *Hydrogen Production*, Proceedings of the 2nd IEA Workshop, Ed. by B. D. Struck, 1991, KFA, Jülich, Germany, 1991.

[84] H. Wendt, in *Wasserstoff-Energietechnik*, VDI-Ber. **602** (1987) 35.

[85] G. Sioli, in *Hydrogen Production*, Proceedings of the 2nd IEA Workshop, Ed. by B. D. Struck, KFA, Jülich, Germany, 1991.

[86] W. Tillmetz, O. Schmid, and G. Dietrich, *DECHEMA-Monogr.* **123** (1992) 31.

[87] K.-H. Tetzlaff and W. Wendel, *DECHEMA-Monogr.* **112** (1988) 325.

[88] H. Matschiner and H. Wenske, *DECHEMA-Monogr.* **128** (1993) 475.

[89] G. Schiller, V. Borok, R. Henne, and W. Hug, DECHEMA-Jahrestagung, Nuremberg, 1993.

[90] S. Stucki, *DECHEMA-Monogr.* **94** (1983) 211.

[91] G. G. Scherer, H. Devantay, R. Oberlin, and S. Stucki, *DECHEMA-Monogr.* **98** (1985) 407.

[92] R. Oberlin, M. Fischer, in T. N. Vezinglu, N. Getoff, and P. Weinzierl (Eds.), *Hydrogen Energy Progress VI*, Vol. 1, Proceedings of the 6th World Hydrogen Energy Conference, Vienna, Austria, Pergamon Press, New York, 1986, pp. 333–340.

[93] G. G. Scherer, in T. N. Veziroglu, N. Getoff, P. Weinzierl (Eds.), *Hydrogen Energy Progress VI*, Vol. 1, Proceedings of the 6th World Hydrogen Energy Conference, Vienna, Austria, Pergamon Press, New York, 1986, pp. 382–389.

[94] N. Römer, K. Hoelmer, and H. Schüle, *VDI-Ber.* **912** (1992) 37.

[95] S. Stucki, H. Buser, and A. Schuler, in Annual Report Paul-Scherrer-Institut Villigen, 1991, pp. 6–8.

[96] K. Ledjeff, J. Ahn, and A. Heinzel, *DECHEMA-Monogr.* **121** (1990) 109.

[97] K. Ledjeff and A. Heinzel, Proceedings of the 26th IECEC, Boston, Massachusetts, 1991, Vol. 3, pp. 538–541.

[98] H. Binder, A. Köhling, H. Krupp, K. Richter, and G. Sandstede, *Electrochim. Acta* **8** (1963) 781.
[99] W. Baukal, H. Döbrich and W. Kuhn, *Chem.-Ing.-Tech.* **48** (1976) 132; **50** (1978) 245.
[100] R. J. Rohr, in *Elektrochemische Energietechnik*, Bundesministerium für Forschung und Technologie, Bonn, 1981, pp. 264–279.
[101] N. J. Maskalick, in *Hydrogen Energy Progress V*, Vol. 2, Ed. by T. N. Veziroglu *et al.*, Pergamon Press, New York, 1984, pp. 801–811.
[102] W. Dönitz, E. Erdle, and R. Streicher, *DECHEMA-Monogr.* **106** (1987) 169.
[103] W. Dönitz, S. Dietrich, E. Erdle, and R. Streicher, *Int. J. Hydrogen Energy* **13** (1988) 283.
[104] A. Kusunoki, H. Matsubara, Y. Kikuoka, C. Yanagi, K. Kugimiya, M. Yoshino, M. Tokura, K. Watanabe, S. Ueda, M. Sumi, H. Miyamoto, and Tokunaga, in T. N. Veziroglu *et al.* (Eds.), *Hydrogen Energy Progress IX*, Vol. 3, MCI, Paris, 1992, pp. 1415–1417.
[105] R. Shimaki, Y. Kikuoka, K. Kugimiya, Suda, M. Okamoto, Ueda, M. Yoshino, C. Yanagi, N. Nakamori and M. Tokura in T. N. Veziroglu *et al.* (Eds.), *Hydrogen Energy Progress IX*, Vol. 3, MCI, Paris, 1992, pp. 1927–1935.
[106] W. Dönitz and E. Erdle, *DECHEMA-Monogr.* **123** (1991) 21.
[107] DECHEMA, A Study for the Generation, Inter-Continental Transport, and Use of Hydrogen as a Source of Clean Energy, on the Basis of Large-Scale and Cheap Hydro-Electricity (Hydrogen Pilot Project—Canada), DECHEMA, Frankfurt, 1987.
[108] J. Gretz, J.P. Baselt, O. Ullmann, and H. Wendt, *Int. J. Hydrogen Energy* **15** (1990) 419.
[109] VEBA, Zukunftsenergien—Fakten und Argumente, VEBA, Düsseldorf, 1989.
[110] C. J. Winter and M. Fuchs, *Int. J. Hydrogen Energy* **16** (1991) 723.
[111] A. Brinner and A. Siegel, Operation of a PV Electrolysis System with Different Coupling Modes, Deutsche Forschungsanstalt für Luft- und Raumfahrt (DLR), Stuttgart, 1989.
[112] A. Brinner, H. Bussmann, W. Hug, and W. Seeger, Test Results of the HYSOLAR 10 kW PV Electrolysis Facility, Deutsche Forschungsanstalt für Luft- und Raumfahrt (DLR), Stuttgart, 1990.
[113] H. Steeb in T. N. Veziroglu *et al.* (Eds.), *Hydrogen Energy Progress VIII*, Proceedings of the 8th World Hydrogen Energy Conference, Honolulu, Hawaii, Pergamon Press, New York, 1990.
[114] R. Brand, J. Hildebrandt, and H. Hofmann, in *Wasserstofftechnik III—Ergebnisse und Optionen*, *VDI-Ber.* **912** (1992) 25.
[115] V. Borok, R. Henne, G. Schiller, and W. Schnumberger, in *Wasserstofftechnik III—Ergebnisse und Optionen*, *VDI-Ber.* **912** (1992) 13.
[116] R. Stucki, *Wasserstofftechnik III—Ergebnisse und Optionen*, *VDI-Ber.* **912** (1992) 7.
[117] J. Divisek, J. Mergel, and H. Schmitz, *Int. J. Hydrogen Energy* **15** (1990) 105.
[118] W. Schnumberger and J. Divisek, *VDI-Ber.* **602** (1987) 63.
[119] Alyzer, Pressurized Hydrogen Generator by Water Electrolysis to Be Powered by Solar Energy (company info), Mendrisio, 1989.
[120] Norsk Hydro Electrolysers, Technical information, 1989.
[121] Electrolyser EI-250 Industrial Plants—Hydrogen and Oxygen Gases by Water Electrolysis, Electrolyser Inc., EI-71705.
[122] H. Vandenborre, R. Leysen, H. Nackaerts, D. van der Eecken, Ph. van Asbroeck, W. Smets, and J. Piepers, *Int. J. Hydrogen Energy* **10** (1985) 719.
[123] H. Wendt, *VDI-Ber.* **602** (1987) 35.
[124] E. Erdle and W. Dönitz, *VDI-Ber.* **602** (1987) 51.
[125] Deutsche Aerospace DASA, GHW Filling Station Electrolyzer (internal company paper), 1993.
[126] W. Hug, J. Divisek, J. Mergel, W. Seeger, and H. Steeb, in T. N. Veziroglu et al. (Eds.), *Hydrogen Energy Progress VIII*, Proceedings of the 8th World Hydrogen Energy Conference, Honolulu, Hawaii, Pergamon Press, New York, 1990.

[127] J. Divisek, in *Electrochemical Hydrogen Technologies*, Ed. by H. Wendt, Elsevier, Amsterdam, 1990, pp. 137–212.
[128] E. Erdle and W. Dönitz, *VDI-Ber.* **602** (1987) 51.
[129] International Water Power & Dam Construction, Handbook, 1992.
[130] J. Nitsch and J. Luther, *Energieversorgung der Zukunft*, Springer-Verlag, Berlin, 1990.
[131] H. Blind, *Energiewirtschaftliche Tagesfragen* **1988**(6) 434.
[132] Enquête Kommission des Deutschen Bundestages "Technikfolgen-Abschätzang und -Bewertung: Aufbaustrategien für eine solare Wasserstoffwirtschaft", Endbericht und Materialienbände I–V, Berlin, Bonn, Cologne, Ottobrunn, Stuttgart, 1990.
[133] Grünenfelder and T. Schucan, *Int. J. Hydrogen Energy* **14** (1989) 579.
[134] H. Breuer, *Energie Akluell* **8**(14) (1984) 590.
[135] L. Bölkow, "Decision Steps for a Long-Term Energy Policy," lecture at the Peutinger-Collegium, reprinted by Messerschmitt-Bölkow, Blohm, Munich, 1982.
[136] C. J. Winter and J. Nitsch, Eds., *Hydrogen as an Energy Carrier—Technologies, Systems, Economy*, Springer-Verlag, Berlin, 1988.
[137] R. Dahlberg, *Int. J. Hydrogen Energy* **7** (1982) 121.
[138] J. O'M. Bockris, *Energy Options*, Taylor and Francis, London, 1980.
[139] J. M. Ogden and R. H. Williams, Hydrogen and the Revolution in Amorphous Silicon Solar Cell Technology, Center for Energy and Environmental Studies, Princeton University, PU/CEES Report No. 231, 1989.
[140] C. F. Kollbrunner and H. Stauber, *Unerschöpfliche saubere Wasser- und Energiequellen in Grönland*, Verlag Leemann, Zurich, 1973.
[141] C. F. Kollbrunner, *Energie und Wasser aus Grönland*, Verlag Leemann, Zurich, 1976.
[142] A. Bartle, ed., *International Water Power & Dam Construction* **41** (1989) 32.
[143] A. Haurie and R. Loulou, Hydrogen Production and Exportation in Québec—a Techno-economic Analysis, Groupe d'Etudes et de Recherche en Analyse des Décisions (G.E.R.A.D.), Work Package No. 242, Euro-Québec Hydro-Hydrogen Pilot Project, 1991.
[144] Solar Energy Research Institute (SERI), The Potential of Renewable Energy—An Interlaboratory White Paper, prepared for the Office of Policy, Planning and Analysis, U.S. Department of Energy, under Contract No. DE-AC02-83CH10093, Golden, Colorado, 1990.
[145] T. Morovic, Macro Analysis, Work Package No. 210, Euro-Québec Hydro-Hydrogen Pilot Project, associated contractor to contract no. 3723-89-05 PC ISP D given by the European Atomic Energy Community, Fraunhofer-Institut für Systemtechnik und Innovationsforschung (FHG-ISI), 1990.
[146] H. Selzer, *Wind Energy—Potential of Wind Energy in the European Community*, D. Reidel, Dordrecht, 1986.
[147] R. Roesler and R. Wurster, Anwendung und Weiterentwicklung umweltverträglicher regenerativer Energietechnologien in Marokko—Konzept für eine Kooperation zwischen dem Königreich Marokko und dem Freistaat Bayem, a short study prepared for and partly funded by the Bavarian Ministries for State Planning and Environmental Affairs and for Economy and Transport, in cooperation with the Centre National de Coordination et de Planification de la Recherche Scientifique et Technique, Rabat, carried out by Ludwig-Bölkow-Systemtechnik GmbH, Munich, 1991.
[148] A. K. Larssen, Paper presented at the IEA Economic Aspects of Wmd Turbine LS-WECS Expert Meeting on Costing, Petten, May 30–31, 1985; data from Vindkraft, Statens Energiverk Stockholm, 1985, 1.
[149] R. Wurster, Various Internal Considerations and Calculations Performed in the Field of Solar Energy Generation and Hydrogen Production, Ludwig-Bölkow-Systemtechnik GmbH, 1982–1991, not published.
[150] K. Hassmann, W. Keller, and D. Stahl, *Breunstoff-Wärme-Kraft (BWK)* **43** (1991) 103.

[151] J. Schindler and D. Strese, Kostendegression Photovoltaik—Stufe 1: Fertigung multikristalliner Solarzellen und im Einsatz im Kraftwerksbereich, research report by Ludwig-Bölkow-Systemtechnik GmbH, prepared for the German Ministry of Research and Technology (BMFT), Ottobrunn, 1988.

[152] D. Strese, Einsatzmöglichkeiten der Photovoltaik zur Entlastung des Energiemarktes, Ludwig-Bölkow-Systemtechnik GmbH, Ottobrunn, 1989.

[153] K. Zweibel and H.S. Ullal, Paper prepared for the 24th Intersociety Energy Conversion Engineering Conference, Washington, D.C., August 6–11, 1989, Solar Energy Research Institute, May 1989.

[154] K. Zweibel, *Harnessing Solar Power—the Photovoltaic Challenge*, Plenum Press, New York, 1990.

[155] Meridian Corporation, Energy System Emissions and Material Requirements, Report prepared for the Deputy Assistant Secretary for Renewable Energy—U.S. Department of Energy, Washington, D.C., Alexandria, Virginia, February 1989.

[156] D. Kluyskens, O. Ullmann, and R. Wurster, Euro-Québec Hydro-Hydrogen Pilot Project Phase II—Executive Summary of the Phase II Feasibility Study, Joint Management Group Hydro-Québec and Ludwig-Bölkow-Stiflung, Montreal and Ottobrunn, March 31, 1991 (European part under CEC contracts no. 354988-12 PD ISP D and 3723-89-05 PC ISP D by the European Atomic Energy Community).

[157] R. Wurster, in T. N. Veziroglu et al. (Eds.), *Hydrogen Energy Progress VIII*, Proceedings of the 8th World Hydrogen Energy Conference, Honolulu, Hawaii, Pergamon Press, New York, 1990.

[158] D. Neill et al., HNEI Wind-Hydrogen Programm, in T. N. Veziroglu et al. (Eds.), *Hydrogen Energy Progress VIII*, Proceedings of the 8th World Hydrogen Energy Conference, Honolulu, Hawaii, Pergamon Press, New York, 1990, 1464–1469.

[159] S. Schulien and M. Steinmetz, Optimierung eines Energieversorgungssystems auf der Basis von Windenergie und Wasserstoff, Proceedings of the 7th International Sonnenforum, Deutsche Gësellschaft für Sonnenenergie (DGS), Frankfurt, October 9–12, 1990, 1464–1469.

[160] A. Haas, J. Luther, and F. Trieb, Hydrogen Energy Storage for an Autonomous Renewable Energy System—Analysis of Experimental Results, Paper by the University of Oldenburg presented at the International Solar Energy Society (ISES) Conference, Denver, Fall 1991.

[161] A. Szyszka, *Elektritzitätswirtschaft* **89**(5) (1990).

[162] Hysolar: Hysolar—Solar Hydrogen Energy/A German–Saudi Arabian Partnership, DLR, KACST, 1989; H. Steeb, personal communication.

[163] W. Stahl and A. Goetzberger, Das Energieautarke Solarhaus, Fhg-ISE Freiburg, in Proceedings of the 7th International Sonnenforum, Deutsche Gësellschaft für Sonnenenergie (DGS), Frankfurt, October 9–12, 1990.

[164] W. Stahl and A. Goetzberger, *Sonnenenergie* **1990**(6).

[165] R. Weber, Europas einziges "Wasserstoff-Wohnhausw"-im Emmental, 1990 (p. 10, press release).

[166] Deutsche Airbus GmbH (MBB Transport Aircraft Group), Pilotprojekt Airbus mit Wasserstoffantrieb "Einsatz von LH_2 in der Luftfahrt," Hamburg, 1989 (company information).

[167] Minister of Supply and Services Canada, Electric Power in Canada 1988, Ottawa, 1989.

[168] Ministère de l'Energie et des Ressources, Energy in Québec, 1988 Edition, Gouvernement du Québec, Québec, April 1988.

[169] Landsvirkjun: Appraisal of the Hydropower Potential in Iceland, Reykjavik, 1986.

[170] W. Häfele, *Energy in a Finite World—a Global Systems Analysis*, Ballinger Publishing, Cambridge, Massachusetts, 1981.

[171] J. Nitsch, H. Klaiß, and J. Meyer, in T. N. Veziroglu et al. (Eds.), *Hydrogen Energy Progress*, Proceedings of the 8th World Hydrogen Energy Conference, Honolulu, Hawaii, Pergamon Press, New York, 1990.

[172] Greenpeace (eds.), Energy Without Oil—The Technical and Economic Feasibility of Phasing Out Global Oil Use, Energy Policy and Research Unit, Greenpeace International, Amsterdam, 1993.

[173] M. Seidet and H. Bradke, Conditions and Possibilities of Hydrogen as Energy Vector, Fraunhofer Institut für Systemtechnik und Innovationsforschung—ISI, EQHHPP Supplementary Task Program work package WP 350 (contract no. 3723-89-05 PC ISP D by the European Atomic Energy Community), Karlsruhe, 1992.

[174] World Energy Conference, 1989 Survey of Energy Resources, Montreal, 1989.

[175] Deutscher Bundestag (eds.), *Enquete-Kommission "Vorsorge zum Schutz der Erdatmosphäre" des Deutschen Bundestages: Energie und Klima, Vol. 3: Erneuerbare Energien*, Economica Verlag, Bonn, and Verlag C. F. Müller, Karlsruhe, 1990.

[176] W. H. Avery, Fuels from the Oceans via Ocean Thermal Energy Conversion (OTEC), Proceedings of Ocean Space Utilization Conference, Tokyo, 1985.

[177] World Resources Institute and International Institute for Environment and Development, Weltressourcen, ecomed, 1991.

[178] J. Raabe, *Hydro Power—The Design, Use and Function of Hydromechanical, Hydraulic and Electrical Equipment*, VDI-Verlag, Düsseldorf, 1985.

[179] A. Strehler, Möglichkeiten und Potentiale zur Nutzung und Produktion von Pflanzen als Träger gespeicherter Sonnenenergie—eine globale Betrachtung, Landtechnik Weihenstephan, in Proceedings of the 7th Internationales Sonnenforum, Deutsche Gesellschaft für Sonnenergie (DGS), Frankfurt, 1990.

[180] R. Wurster (ed.), Bedeutung, Einsatzbereiche und technisch-ökonomische Entwicklungspotential von Wasserstoffnutzungstechniken - Band II: Indirekt energetische Nutzung; Einsatzbereiche und Potential der Nutzungstechniken, ZSW-DLR-LBST, Stuttgart and Ottobrunn, 1992.

[181] A. Strehler, Wärme aus Stroh und Holz, Bayerische Landesanstalt für Landtechnik, Freising, 1980.

[182] T. B. Johansson, H. Kelly, A. K. N. Reddy, and R. H. Williams (eds.), *Renewable Energy—Sources for Fuels and Electricity*, Island Press, Washington, D.C., and Covelo, (California, 1993.

[183] NORWAVE A.S., NORWAVE Wave Power for Island Communities—a Status Report, Oslo, Norway, 1988.

[184] H. Klaiß and F. Staiß (eds.). *Solarthermischen Kraftwerke für den Mittelmeerraum*, DLR Stuttgart, Springer-Verlag, Heidelberg, 1992.

[185] B. Árnason and V. K. Jónsson, Iceland—A Forum for Gradual Introduction of Hydrogen as Marine Fuel, University of Iceland, Paper presented at the Nordic Symposium on Hydrogen and Fuel Cells for Energy Storage, Espoo-Otaniemi, Finland, March 11–12, 1993.

[186] R. Wurster, Renewable Primary Energy Sources—Major Components of a Clean Energy System with Hydrogen, Paper presented at the symposium Hydrogen Energy for Future?, held by the Flemish Community, Brussels, April 17th, 1991.

[187] J. Gtetz, B. Drolet, D. Kluyskens, F. Sandmann, and O. Ullmann, Phase II and Phase III.0 of the 100 MW Euro-Québec Hydro-Hydrogen Pilot Project, in T. N. Veziroglu et al. (Eds.), *Hydrogen Energy Progress IX*, Vol. 3, MCI, Paris, 1992, pp. 1821–1828.

[188] R. Wurster, Status of the Euro-Québec Hydro-Hydrogen Pilot Project [EQHHPP], ARGE EQHHPP LBST/CONOC Munich Office, Ludwig-Bölkow-Systemtechnik GmbH, Paper presented at the 4th Annual Meeting of the U.S. National Hydrogen Association, Washington, D.C., March 25, 1993.

[189] K. Andreassen, U. H. Bünger, N. Henriksen, A. Øyvann, and O. Ullmann, *Int. J. Hydrogen Energy* **18** (1993) 325.

[190] R. Wurster, Überblick über nationale und internationale Pilot- und Demonstrationsprojekte auf dem Solarwasserstoffgebiet, Ludwig-Bölkow-Systemtechnik GmbH, in Proceedings of the 7th International Sonnenforum, Deutsche Gësellschaft für Sonnenenergie (DGS), Frankfurt, 1990.

[191] F. Groß, Entwurf für ein Modellvorhaben: Windkraftanlage zur Heizenergie-, Warmwasser- und Wasserstofferzeugung, Technische Universität Ilmenau, 1993.

[192] H. Blank and A. Syszka, in T. N. Veziroglu *et al.* (Eds.), *Hydrogen Energy Progress IX*, Vol. 2, MCI, Paris, 1992, pp. 677–685.

[193] H. Schubert-Klempnauer, Photovoltaik in der Landwirtschaft—Erfahrungen aus Bau und Betrieb, Ludwig-Bölkow-Systemtechnik GmbH, in Proceedings of the 8th International Sonnenforum, Deutsche Gësellschaft für Sonnenenergie (DGS), Berlin, 1992.

[194] P. S. Kauranen, P. D. Lund, and J. P. Vanhanen, in T. N. Veziroglu et al. (Eds.), *Hydrogen Energy Progress IX*, Vol. 3, MCI, Paris, 1992, pp. 733–742.

[195] A. G. García-Conde and F. Rosa, in T. N. Veziroglu *et al.* (Eds.), *Hydrogen Energy Progress IX*, Vol. 3, MCI, Paris, 1992, 723–731.

[196] H. Barthels, Bau und Betrieb einer Photovoltaik-Wasserstoff-Brennstoffzellen-Demonstrationsanlage, Forschungszentrum Jülich KFA, December 1992.

[197] P. Bernhard and S. Schulien, Optimierungsmessungen an einer Energieversorgungsanlage auf der Basis von solarem Wasserstoff, Fachhochschule-Wiesbaden, information brochure, Spring 1993.

[198] C. Watanabe, Japan's New Sunshine Program and Expectations of the WE-NET Project—International Clean Energy Network Using Hydrogen Conversion, AIST-MITI, Paper presented at the 4th Annual Meeting of the U.S. National Hydrogen Association, Washington, D.C., March 25, 1993.

[199] Agency of Industrial Science and Technology-Ministry of International Trade and Industry (AIST-M1TI), *Sunshine Journal* **1993**(4) 7.

[200] G. Young, Mazda's HR-X Hydrogen Concept Car Makes Canadian Debut at the Montreal International Auto Show, Information Mazda, Scarborough, Ontario, January 13, 1993.

[201] Electrolyser Corporation Ltd., Unicell-Cluster Development—Program Description and Technical Status Report, November 1992.

[202] South Coast Air Quality Management District, Volume II: Project and Technology Status, 1992 Progress Report, Technology Advancement Office, October 1992; S. Romano, The Fuel Cell Powered Bus—The Answer to an Unlimited Range Electric Vehicle, Paper presented at the Hong Kong Electric Vehicle Seminar, 1990.

[203] California Utility Plans Solar Fuel Cell Bus Line, *The Hydrogen Letter*, **VII**(2) (1993) 4.

[204] Ballard PEM Fuel Cell Bus Rolls Out, *The Hydrogen Letter*, **VII**(3) (1993) 1.

[205] P. Patil, J. Ohi, and G.L. Hagey, Energy, Environmental and Economic Benefits Resulting from the Future Commercialization of Fuel Cell Vehicles in Transportation, Fuel Cell Seminar, Tucson, 1992; U.S. Department of Energy, Fuel Cells for Transportation, Info-Brochure, November 1992.

[206] G. A. Crawford and A.F. Hufnagel, *Int. J. Hydrogen Energy* **12** (1987) 297.

[207] F. Buteau, P. Demange, and C. Moreau, in T. N. Veziroglu *et al.* (Eds.), *Hydrogen Energy Progress IX*, Vol. 1, MCI, Paris, 1992, pp. 345–354.

[208] G. Huppmann, Brennstoffzellen großer Leistung bis in den MW-Bereich, DASA internal presentation, Ottobrunn, January 1993; J. Reinecke, G. Huppmann, and W. Drenckhahn, VDI-Ber. **912** (1992) 103.

[209] B. G. Marcenaro, Fuel Cells and Public Transport, Fuel Cell Seminar, Tucson, 1992.

[210] J. V. Koren, Seaborne Transportation of Liquefied Natural Gas, Det Norske Veritas Classification A/S, Paper presented at Cercle Mondial du Consensus (CMDC) World Clean Energy Conference, Geneva, November 6, 1991.

[211] L. H. Nielsen and L. Schleisner, Hydrogen as an Energy Carrier with Focus on Electricity Storage, Riso National Laboratory, Systems Analysis Department, Roskilde, Denmark, 1991.

[212] M. C. Mattos, in *Hydrogen Energy Program VI*, Proceedings of the 6th World Hydrogen Energy Conference, Vienna, 1986.

[213] L. C. de Lima and T. N. Veziroglu, in T. N. Veziroglu *et al.* (Eds.), *Hydrogen Energy Progress IX*, Vol. 3, MCI, Paris, 1992, pp. 1837–1844.

[214] G. S. Eljrushi and M. A. Sharif, in T. N. Veziroglu *et al.* (Eds.), *Hydrogen Energy Progress VIII*, Vol. 1, Proceedings of the 8th World Hydrogen Energy Conference, Honolulu, Hawaii, Pergamon Press, 1990, p. 201.

[215] G.S. Eljrushi and J. Zubia, in T. N. Veziroglu *et al.* (Eds.), *Hydrogen Energy Progress IX*, Vol. 3, MCI, Paris, 1992, p. 1877.

[216] H. C. Watson and S. M. Lambe, in T. N. Veziroglu *et al.* (Eds.), *Hydrogen Energy Progress IX*, Vol. 2, MCI, Paris, 1992, p. 1271.

[217] N. Glasson and R. Green, in T. N. Veziroglu *et al.* (Eds.), *Hydrogen Energy Progress IX*, Vol. 2, MCI, Paris, 1992, p. 1285.

[218] L. M. Das, in T. N. Veziroglu *et al.* (Eds.), *Hydrogen Energy Progress VIII*, Vol. 2, Proceedings of the 8th World Hydrogen Energy Conference, Honolulu, Hawaii, Pergamon Press, New York, 1990, p. 1379.

[219] N. Apostulescu and D. Sfinteanu, in T. N. Veziroglu *et al.* (Eds.), *Hydrogen Energy Progress VIII*, Vol. 2, Proceedings of the 8th World Hydrogen Energy Conference, Honolulu, Hawaii, Pergamon Press, New York, 1990, p. 1323.

[220] A. A. Sheipak and V.N. Kabalkin, in T. N. Veziroglu *et al.* (Eds.), *Hydrogen Energy Progress VIII*, Vol. 2, Proceedings of the 8th World Hydrogen Energy Conference, Honolulu, Hawaii, Pergamon Press, New York, 1990, p. 1355.

[221] D. Reister and W. Strobl, in T. N. Veziroglu *et al.* (Eds.), *Hydrogen Energy Progress IX*, Vol. 2, MCI, Paris, 1992, pp. 1201–1213.

[222] K. Feucht, W. Hurich, N. Komoschinski, and R. Povel, *Hydrogen Energy Progress VI*, Proceedings of the 6th World Hydrogen Energy Conference, Vienna, 1986.

[223] California Air Board May Take Fresh Look at Hydrogen, *The Hydrogen Letter*, **VIII**(8) (1993) 1.

[224] R. Wurster, M. Bracha, J. Braedt, H. Knorr, and W. Stobl, in T. N. Veziroglu *et al.* (Eds.), *Hydrogen Energy Progress IX*, Vol. 2, MCI, Paris, 1992, pp. 1215–1226.

[225] W. Peschka, *VDI-Ber.* **1020** (1992) 279.

[226] Hydrogen Is Flawed Option, Says German Scientist, *The Hydrogen Letter* **VIII**(5) (1993) 4; R. Wurster (reader's comment on *Nachr. Chem. Tech. Lab.* **29** (1991) 1024, *Nachr. Chem. Tech. Lab.* **29** (1991) 503, written by Dr. G. Beckmann).

[227] A. Grubler, *Gewässerschutz, Wasser, Abwasser Gaz-Eaux-Euax Usées* **69** (1989) 763.

Cumulative Author Index for Numbers 1–27

Author	Title	Number
Abruña, H. D.	X Rays as Probes of Electrochemical Interfaces	20
Adžić R.	Reaction Kinetics and Mechanisms on Metal Single Crystal Electrode Surfaces	21
Agarwal, H. P.	Recent Developments in Faradaic Rectification Studies	20
Albella, J. M.	Electric Breakdown in Anodic Oxide Films	23
Allongue, P.	Physics and Applications of Semiconductor Electrodes Covered with Metal Clusters	23
Amokrane, S.	Analysis of the Capacitance of the Metal–Solution Interface. Role of the Metal and the Metal–Solvent Coupling	22
Andersen, H. C.	Improvements upon the Debye-Huckel Theory of Ionic Solutions	11
Andersen, T. N.	Potentials of Zero Charge of Electrodes	5
Appleby, A. J.	Electrocatalysis	9
Arvia, A. J.	Transport Phenomena in Electrochemical Kinetics	6
Augustynski, J.	Application of Auger and Photoelectron Spectroscopy of Electrochemical Problems	13
Badiali, J. P.	Analysis of the Capacitance of the Metal–Solution Interface. Role of the Metal and the Metal–Solvent Coupling	22
Baker, B. G.	Surface Analysis by Electron Spectroscopy	10

Cumulative Author Index

Author	Title	Number
Balsene, L.	Application of Auger and Photoelectron Spectroscopy to Electrochemical Problems	13
Barthel, J.	Temperature Dependence of Conductance of Electrolytes in Nonaqueous Solutions	13
Batchelor, R. A.	Surface States on Semiconductors	22
Bauer, H. H.	Critical Observations on the Measurement of Adsorption at Electrodes	7
Becker, R. O.	Electrochemical Mechanisms and the Control of Biological Growth Processes	10
Beden, B.	Electrocatalytic Oxidation of Oxygenated Aliphatic Organic Compounds at Noble Metal Electrodes	22
Benderskii, V. A.	Phase Transitions in the Double Layer at Electrodes	26
Berg, H.	Bioelectrochemical Field Effects: Electrostimulation of Biological Cells by Low Frequencies	24
Berwick, A.	The Study of Simple Consecutive Processes in Electrochemical Reactions	5
Blank, M.	Electrochemistry in Nerve Excitation	24
Bloom, H.	Models for Molten Salts	9
Bloom, H.	Molten Electrolytes	2
Blyholder, G.	Quantum Chemical Treatment of Adsorbed Species	8
Bockris, J. O'M.	Electrode Kinetics	1
Bockris, J. O'M.	Ionic Solvation	1
Bockris, J. O'M.	The Mechanism of Charge Transfer from Metal Electrodes to Ions in Solution	6
Bockris, J. O'M.	The Mechanism of the ElectrodePosition of Metals	3
Bockris, J. O'M.	Molten Electrolytes	2
Bockris, J. O'M.	Photoelectrochemical Kinetics and Related Devices	14
Boguslavsky, L. I.	Electron Transfer Effects and the Mechanism of the Membrane Potential	18
Breiter, M. W.	Adsorption of Organic Species on Platinum Metal Electrodes	10
Brodskii, A. N.	Phase Transitions in the Double Layer at Electrodes	26

Author	Title	Number
Burke, L. D.	Electrochemistry of Hydrous Oxide Films	18
Burney, H. S.	Membrane Chlor-Alkali Process	24
Charle, K.-P.	Spin-Dependent Kinetics in Dye-Sensitized Charge-Carrier Injection into Organic Crystal Electrodes	19
Cheh, H. Y.	Theory and Applications of Periodic Electrolysis	19
Conway, B. E.	The Behavior of Intermediates in Electrochemical Catalysis	3
Conway, B. E.	Fundamental and Applied Aspects of Anodic Chlorine Production	14
Conway, B. E.	Ionic Solvation	1
Conway, B. E.	Proton Solvation and Proton Transfer Processes in Solution	3
Conway, B. E.	Solvated Electrons in Field- and Photo-assisted Processes at Electrodes	7
Conway, B. E.	The Temperature and Potential Dependence of Electrochemical Reaction Rates, and the Real Form of the Tafel Equation	16
Conway, B. E.	Electroanalytical Methods for Determination of Al_2O_3 in Molten Cryolite	26
Covington, A. K.	NMR Studies of the Structure of Electrolyte Solutions	12
Daikhin, L. I.	Phase Transitions in the Double Layer at Electrodes	26
Damaskin, B. B.	Adsorption of Organic Compounds at Electrodes	3
Damjanovic, A.	The Mechanism of the Electrodeposition of Metals	3
Damjanovic, A.	Mechanistic Analysis of Oxygen Electrode Reactions	5
Desnoyers, J. B.	Hydration Effects and Thermodynamic Properties of Ions	5
Despić, A.	Electrochemistry of Aluminum in Aqueous Solutions and Physics of Its Anodic Oxide	20

Author	Title	Number
Despić, A. R.	Transport-Controlled Deposition and Dissolution of Metals	7
Despić, A. R.	Electrochemical Deposition and Dissolution of Alloys and Metal Components—Fundamental Aspects	27
Djokić, S. S.	Electrodeposition of Nickel–Iron Alloys	22
Djokić, S. S.	Electroanalytical Methods for Determination of Al_2O_3 in Molten Cryolite	26
Drazic, D. M.	Iron and Its Electrochemistry in an Active State	19
Efrima, S.	Surface-Enhanced Raman Scattering (SERS)	16
Eisenberg, H.	Physical Chemistry of Synthetic Polyelectrolytes	1
Elving, P. J.	Critical Observations on the Measurement of Adsorption at Electrodes	7
Enyo, M.	Mechanism of the Hydrogen Electrode Reaction as Studied by Means of Deuterium as a Tracer	11
Erdey-Grúz, T.	Proton Transfer in Solution	12
Fahidy, T. Z.	Recent Advance in the Study of the Dynamics of Electrode Processes	27
Falkenhagen, H.	The Present State of the Theory of Electrolytic Solutions	2
Farges, J.-P.	Charge-Transfer Complexes in Electrochemistry	12
Farges, J.-P.	An Introduction to the Electrochemistry of Charge Transfer Complexes II	13
Findl, E.	Bioelectrochemistry-Electrophysiology-Electrobiology	14
Floyd, W. F.	Electrochemical Properties of Nerve and Muscle	1
Foley, J. K.	Interfacial Infrared Vibrational Spectroscopy	17
Friedman, H. L.	Computed Thermodynamic Properties and Distribution Functions for Simple Models of Ionic Solutions	6

Author	Title	Number
Frumkin, A. A. N.	Adsorption of Organic Compounds at Electrodes	3
Fuller, T. F.	Metal Hydride Electrodes	27
Fuoss, R. M.	Physical Chemistry of Synthetic Polyelectrolytes	1
Galvele, J. R.	Electrochemical Aspects of Stress Corrosion Cracking	27
German, E. D.	The Role of the Electronic Factor in the Kinetics of Charge-Transfer Reactions	24
Gileadi, E.	The Behavior of Intermediates in Electrochemical Catalysis	3
Gileadi, E.	The Mechanism of Oxidation of Organic Fuels	4
Girault, H. H.	Charge Transfer across Liquid–Liquid Interfaces	25
Goddard, E. D.	Electrochemical Aspects of Adsorption on Mineral Solids	13
Goodisman, J.	Theories for the Metal in the Metal–Electrolyte Interface	20
Gores, H.-J.	Temperature Dependence of Conductance of Electrolytes in Nonaqueous Solutions	13
Goruk, W. S.	Anodic and Electronic Currents at High Fields in Oxide Films	4
Grätzel, M.	Interfacial Charge Transfer Reactions in Colloidal Dispersions and Their Application to Water Cleavage by Visible Light	15
Green, M.	Electrochemistry of the Semiconductor–Electrolyte Interface	2
Gregory, D. P.	Electrochemistry and the Hydrogen Economy	10
Gu, Z. H.	Recent Advance in the Study of the Dynamics of Electrode Processes	27
Gurevich, Y. Y.	Electrochemistry of Semiconductors: New Problems and Prospects	16
Gutmann, F.	Charge-Transfer Complexes in Electrochemistry	12
Gutmann, F.	The Electrochemical Splitting of Water	15
Gutmann, F.	An Introduction to the Electrochemistry of Charge Transfer Complexes II	13

Author	Title	Number
Habib, M. A.	Solvent Dipoles at the Electrode-Solution Interface	12
Haering, R. R.	Physical Mechanisms of Intercalation	15
Hamann, S. D.	Electrolyte Solutions at High Pressure	9
Hamelin, A.	Double-Layer Properties at sp and sd Metal Single-Crystal Electrodes	16
Hamnett, A.	Surface States on Semiconductors	22
Hansma, P. K.	Scanning Tunneling Microscopy: A Natural for Electrochemistry	21
Heiland, W.	The Structure of the Metal–Vacuum Interface	11
Herman, P. J.	Critical Observations on the Measurement of Adsorption at Electrodes	7
Hickling, A.	Electrochemical Processes in Glow Discharge at the Gas–Solution Interface	6
Hine, F.	Chemistry and Chemical Engineering in the Chlor-Alkali Industry	18
Hoar, T. P.	The Anodic Behavior of Metals	2
Hopfinger, A. J.	Structural Properties of Membrane Ionomers	14
Humffray, A. A.	Methods and Mechanisms in Electroorganic Chemistry	8
Hunter, R. J.	Electrochemical Aspects of Colloid Chemistry	11
Johnson, C. A.	The Metal–Gas Interface	5
Jolieoeur, C.	Hydration Effects and Thermodynamic Properties of Ions	5
Jović, V. D.	Electrochemical Deposition and Dissolution of Alloys and Metal Components—Fundamental Aspects	27
Kebarle, P.	Gas-Phase Ion Equilibria and Ion Solvation	9
Kelbg, G.	The Present State of the Theory of Electrolytic Solutions	2
Kelly, E. J.	Electrochemical Behavior of Titanium	14
Khan, S. U. M.	Photoelectrochemical Kinetics and Related Devices	14
Kahn, S. U. M.	Some Fundamental Aspects of Electrode Processes	15

Author	Title	Number
Khrushcheva, E. I.	Electroanalytic Properties of Carbon Materials	19
Kinoshita, K.	Preparation and Characterization of Highly Dispersed Electrocatalytic Materials	12
Kinoshita, K.	Small-Particle Effects and Structural Considerations for Electrocatalysis	14
Kita, H.	Theoretical Aspects of Semiconductor Electrochemistry	18
Kitchener, J. A.	Physical Chemistry of Ion Exchange Resins	2
Koch, D. F. A.	Electrochemistry of Sulfide Minerals	10
Kochanova, L. A.	Electric Surface Effects in Solid Plasticity and Strength	24
Kordesch, K. V.	Power Sources for Electric Vehicles	10
Kuhn, A. T.	The Role of Electrochemistry in Environmental Control	8
Kuznetsov, A. M.	Recent Advances in the Theory of Charge Transfer	20
Kuznetsov, A. M.	The Role of the Electronic Factor in the Kinetics of Charge-Transfer Reactions	24
Laidler, K. J.	Theories of Elementary Homogeneous Electron-Transfer Reactions	3
Lamy, C.	Electrocatalytic Oxidation of Oxygenated Aliphatic Organic Compounds at Noble Metal Electrodes	22
Latanision, R. M.	Electrochemistry of Metallic Glasses	21
Léger, J.-M.	Electrocatalytic Oxidation of Oxygenated Aliphatic Organic Compounds at Noble Metal Electrodes	22
Lengyel, S.	Proton Transfer in Solution	12
Lieber, C. M.	Scanning Tunneling Microscopy Investigations of Low-Dimensional Materials: Graphite Intercolation Compounds	12
Lindsay, J. H.	Electrogalvanizing	26
Lipkowski, J.	Ion and Electron Transfer across Monolayers of Organic Surfactants	23
Llopis, J.	Surface Potential at Liquid Interfaces	6
Losev, V. V.	Mechanisms of Stepwise Electrode Processes on Amalgams	7

Author	Title	Number
Lyklema, J.	Interfacial Electrostatics and Electrodynamics in Disperse Systems	17
Lynn, K. G.	The Nickel Oxide Electrode	21
Lyons, M. E. G.	Electrochemistry of Hydrous Oxide Films	18
MacDonald, D. D.	The Electrochemistry of Metals in Aqueous Systems at Elevated Temperatures	11
MacDonald, D. D.	Impedance Measurements in Electrochemical Systems	14
Maksimović, M. D.	Theory of the Effect of Electrodeposition at a Periodically Changing Rate on the Morphology of Metal Deposits	19
Mandel, L. J.	Electrochemical Processes at Biological Interfaces	8
Marchiano, S. L.	Transport Phenomena in Electrochemical Kinetics	6
Marincic, N.	Lithium Batteries with Liquid Depolarizers	15
Markin, V. S.	Thermodynamics of Membrane Energy Transduction in an Oscillating Field	24
Martinez-Duart, J. M.	Electric Breakdown in Anodic Oxide Films	23
Matthews, D. B.	The Mechanism of Charge Transfer from Metal Electrodes to Ions in Solution	6
Mauritz, K. A.	Structural Properties of Membrane Ionomers	14
McBreen, J.	The Nickel Oxide Electrode	21
McKinnon, W. R.	Physical Mechanisms of Intercalation	15
McKubre, M. C. H	Impedance Measurements in Electrochemical Systems	14
Murphy, O. J.	The Electrochemical Splitting of Water	15
Nagarkan, P. V.	Electrochemistry of Metallic Glasses	21
Nagy, Zoltán	DC Electrochemical Techniques for the Measurement of Corrosion Rates	25
Nágy, Z.	DC Relaxation Techniques for the Investigation of Fast Electrode Reactions	21
Newman, J.	Photoelectrochemical Devices for Solar Energy Conversion	18

Author	Title	Number
Newman, J.	Determination of Current Distributions Governed by Laplace's Equation	23
Newman, J.	Metal Hydride Electrodes	27
Newman, K. E.	NMR Studies of the Structure of Electrolyte Solutions	12
Nişanciağlu, K.	Design Techniques in Cathodic Protection Engineering	23
Novak, D. M.	Fundamental and Applied Aspects of Anodic Chlorine Production	14
O'Keefe, T. J.	Electrogalvanizing	26
Orazem, M. E.	Photoelectrochemical Devices for Solar Energy Conversion	18
Oriani, R. A.	The Metal–Gas Interface	5
Padova, J. I.	Ionic Solvation in Nonaqueous and Mixed Solvents	7
Paik, Woon-kie	Ellipsometry in Electrochemistry	25
Parkhutik, V.	Electrochemistry of Aluminum in Aqueous Solutions and Physics of Its Anodic Oxide	20
Parkhutik, V. P.	Electric Breakdown in Anodic Oxide Films	23
Parsons, R.	Equilibrium Properties of Electrified Interphases	1
Pavlovic, M. G.	Electrodeposition of Metal Powders with Controlled Particle Grain Size and Morphology	24
Perkins, R. S.	Potentials of Zero Charge of Electrodes	5
Pesco, A. M.	Theory and Applications of Periodic Electrolysis	19
Piersma, B.	The Mechanism of Oxidation of Organic Fuels	4
Pilla, A. A.	Electrochemical Mechanisms and the Control of Biological Growth Processes	10
Pintauro, P. N.	Transport Models for Ion-Exchange Membranes	19
Pleskov, Y. V.	Electrochemistry of Semiconductors: New Problems and Prospects	16
Plzak, V.	Advanced Electrochemical Hydrogen Technologies: Water Electrolyzers and Fuel Cells	26

Author	Title	Number
Pons, S.	Interfacial Infrared Vibrational Spectroscopy	17
Popov, K. I.	Electrodeposition of Metal Powders with Controlled Particle Grain Size and Morphology	24
Popov, K. I.	Theory of the Effect of Electrodeposition at a Periodically Changing Rate on the Morphology of Metal Deposits	19
Popov, K. I.	Transport-Controlled Deposition and Dissolution of Metals	7
Pound, Bruce G.	Electrochemical Techniques to Study Hydrogen Ingress in Metals	25
Power, G. P.	Metal Displacement Reactions	11
Reeves, R. M.	The Electrical Double Layer: The Current States of Data and Models, with Particular Emphasis on the Solvent	9
Revie, R. W.	Environmental Cracking of Metals: Electrochemical Aspects	26
Ritchie, I. M.	Metal Displacement Reactions	11
Rohland, B.	Advanced Electrochemical Hydrogen Technologies: Water Electrolyzers and Fuel Cells	26
Rusling, J. F.	Electrochemistry and Electrochemical Catalysis in Microemulsions	26
Russell, J.	Interfacial Infrared Vibrational Spectroscopy	17
Rysselberghe, P. Van	Some Aspects of the Thermodynamic Structure of Electrochemistry	4
Sacher, E.	Theories of Elementary Homogeneous Electron-Transfer Reactions	3
Sandstede, G. S.	Water Electrolysis and Solar Hydrogen Demonstration Projects	27
Savenko, V. I.	Electric Surface Effects in Solid Plasticity and Strength	24
Scharifker, B. R.	Microelectrode Techniques in Electrochemistry	22
Schmickler, W.	Electron Transfer Reactions on Oxide-Covered Metal Electrodes	17
Schneir, J.	Scanning Tunneling Microscopy: A Natural for Electrochemistry	21

Author	Title	Number
Schultze, J. W.	Electron Transfer Reactions on Oxide-Covered Metal Electrodes	17
Scott, K.	Reaction Engineering and Digital Simulation in Electrochemical Processes	27
Searson, P. C.	Electrochemistry of Metallic Glasses	21
Seversen, M.	Interfacial Infrared Vibrational Spectroscopy	17
Shchukin, E. D.	Electric Surface Effects in Solid Plasticity and Strength	24
Sides, P. J.	Phenomena and Effects of Electrolytic Gas Evolution	18
Snook, I. K.	Models for Molten Salts	9
Somasundaran, P.	Electrochemical Aspects of Adsorption on Mineral Solids	13
Sonnenfeld, R.	Scanning Tunneling Microscopy: A Natural for Electrochemistry	21
Stonehart, P.	Preparation and Characterization of Highly Dispersed Electrocatalytic Materials	12
Szklarczyk, Marek	Electrical Breakdown of Liquids	25
Taniguchi, I.	Electrochemical and Photoelectrochemical Reduction of Carbon Dioxide	20
Tarasevich, M. R.	Electrocatalytic Properties of Carbon Materials	19
Thirsk, H. R.	The Study of Simple Consecutive Processes in Electrochemical Reactions	5
Tilak, B. V.	Chemistry and Chemical Engineering in the Chlor-Alkali Industry	18
Tilak, B. V.	Fundamental and Applied Aspects of Anodic Chlorine Production	14
Trasatti, S.	Solvent Adsorption and Double-Layer Potential Drop at Electrodes	13
Tributsch, H.	Photoelectrolysis and Photoelectrochemical Catalysis	17
Tsong, T. Y.	Thermodynamics of Membrane Energy Transduction in an Oscillating Field	24
Uosaki, K.	Theoretical Aspects of Semiconductor Electrochemistry	18

Author	Title	Number
Van Leeuwen, H. P.	Interfacial Electrostatics and Electrodynamics in Disperse Systems	17
Velichko, G. I.	Phase Transitions in the Double Layer at Electrodes	26
Verbrugge, M. W.	Transport Models for Ion-Exchange Membranes	19
Vijh, A. K.	Perspectives in Electrochemical Physics	17
Viswanathan, K.	Chemistry and Chemical Engineering in the Chlor-Alkali Industry	18
Von Goldammer, E.	NMR Studies of Electrolyte Solutions	10
Vorotyntsev, M. A.	Modern State of Double Layer Study of Solid Metals	17
Wachter, R.	Temperature Dependence of Conductance of Electrolytes in Nonaqueous Solutions	13
Wendt, H.	Advanced Electrochemical Hydrogen Technologies: Water Electrolyzers and Fuel Cells	26
Wenglowski, G.	An Economic Study of Electrochemical Industry in the United States	4
West, A. C.	Determination of Current Distributions Governed by Laplace's Equation	23
Wieckowski, A.	*In Situ* Surface Electrochemistry: Radioactive Labeling	21
Willig, F.	Spin-Dependent Kinetics in Dye-Sensitized Charge-Carrier Injection into Organic Crystal Electrodes	19
Wojtowicz, J.	Oscillatory Behavior in Electrochemical Systems	8
Wroblowa, H. S.	Batteries for Vehicular Propulsion	16
Wurster, R.	Water Electrolysis and Solar Hydrogen Demonstration Projects	27
Yeager, E. B.	Ultrasonic Vibration Potentials	14
Yeager, H. L.	Structural and Transport Properties of Perfluorinated Ion-Exchange Membranes	16
Yeo, R. S.	Structural and Transport Properties of Perfluorinated Ion-Exchange Membranes	16

Author	Title	Number
Young, L.	Anodic and Electronic Currents at High Fields in Oxide Films	4
Zana, R.	Ultrasonic Vibration Potentials	14
Zobel, F. G. R.	Anodic and Electronic Currents at High Fields in Oxide Films	4

Cumulative Title Index for Numbers 1-27

Title	Author	Number
Adsorption of Organic Compounds at Electrodes	Frumkin, A. A. N. Damaskin, B. B.	3
Adsorption of Organic Species on Platinum Metal Electrodes	Breiter, M. W.	10
Advanced Electrochemical Hydrogen Technologies: Water Electrolyzers and Fuel Cells	Plzak, V. Rohland, B. Wendt, H.	26
Analysis of the Capacitance of the Metal–Solution Interface. Role of the Metal and the Metal-Solvent Coupling	Amokrane, S. Badiali, J. P.	22
The Anodic Behavior of Metals	Hoar, T. P.	2
Anodic and Electronic Currents at High Fields in Oxide Films	Young, L. Goruk, W. S. Zobel, F. G. R.	4
Application of Auger and Photoelectron Spectroscopy to Electrochemical Problems	Augustynski, J. Balsenc, L.	13
Batteries for Vehicular Propulsion	Wroblowa, H. S.	16
The Behavior of Intermediates in Electrochemical Catalysis	Gileadi, E. Conway, B. E.	3
Bioelectrochemical Field Effects: Electrostimulation of Biological Cells by Low Frequencies	Berg, H.	24
Bioelectrochemistry–Electrophysiology–Electrobiology	Findl, E.	14
Charge Transfer across Liquid–Liquid Interfaces	Girault, H. H.	25

Title	Author	Number
Charge-Transfer Complexes in Electrochemistry	Farges, J.-P. Gutmann, F.	12
Chemistry and Chemical Engineering in the Chlor-Alkali Industry	Hine, F. Tilak, B. V. Viswanathan, K.	18
Computed Thermodynamic Properties and Distribution Functions for Simple Models of Ionic Solutions	Friedman, H. L.	6
Critical Observations on the Measurement of Adsorption at Electrodes	Bauer, H. H. Herman, P. J. Elving, P. J.	7
DC Relaxation Techniques for the Investigation of Fast Electrode Reactions	Nagy, Z.	21
DC Electrochemical Techniques for the Measurement of Corrosion Rates	Nagy, Zoltán	25
Design Techniques in Cathodic Protection Engineering	Nişancioğlu, K.	23
Determination of Current Distributions Governed by Laplace's Equation	West, A. C. Neuman, J.	23
Double-Layer Properties at sp and sd Metal Single-Crystal Electrodes	Hamelin, A.	16
An Economic Study of Electrochemical Industry in the United States	Wenglowski, G.	4
Electrical Breakdown of Liquids	Szklarczyk, Marek	25
The Electrical Double Layer: The Current Status of Data and Models, with Particular Emphasis on the Solvent	Reeves, R. M.	9
Electric Breakdown in Anodic Oxide Films	Parkhutik, V. P. Albella, J. M. Martinez-Duart, J. M.	23
Electric Surface Effects in Solid Plasticity and Strength	Shchukin, E. D. Kochanova, L. A. Savenko, V. I.	24

Title	Author	Number
Electroanalytical Methods for Determination of Al_2O_3 in Molten Cryolite	Djokić, S. S. Conway, B. E.	26
Electrocatalysis	Appleby, A. I.	9
Electrocatalytic Oxidation of Oxygenated Aliphatic Organic Compounds at Noble Metal Electrodes	Beden, B. Léger, J.-M. Lamy, C.	22
Electrocatalytic Properties of Carbon Materials	Tarasevich, M. R. Khrushcheva, E. I.	19
Electrochemical Aspects of Adsorption on Mineral Solids	Somasundaran, P. Goddart, E. D.	13
Electrochemical Aspects of Colloid Chemistry	Hunter, R. J.	11
Electrochemical Behavior of Titanium	Kelly, E. J.	14
Electrochemical Mechanisms and the Control of Biological Growth Processes	Becker, R. O. Pilla, A. A.	10
Electrochemical and Photoelectrochemical Reduction of Carbon Dioxide	Taniguchi, I	20
Electrochemical Processes at Biological Interfaces	Mandel, L. J.	8
Electrochemical Processes in Glow Discharge at the Gas–Solution Interface	Hickling, A.	6
Electrochemical Properties of Nerve and Muscle	Floyd, W. F.	1
The Electrochemical Splitting of Water	Gutmann, F. Murphy, O. J.	15
Electrochemical Techniques to Study Hydrogen Ingress in Metals	Pound, Bruce G.	25
Electrochemistry of Aluminum in Aqueous Solutions and Physics of its Anodic Oxide	Despić, A. Parkhutik, V.	20
Electrochemistry and Electrochemical Catalysis in Microemulsions	Rusling, J. F.	26
Electrochemistry and the Hydrogen Economy	Gregory, D. P.	10
Electrochemistry of Hydrous Oxide Films	Burke, L. D. Lyons, M. E. G.	18

Title	Author	Number
Electrochemistry of Metallic Glasses	Searson, P. C. Nagarkan, P. V. Latanision, R. M.	21
The Electrochemistry of Metals in Aqueous Systems at Elevated Temperatures	Macdonald, D. D.	11
Electrochemistry of Nerve Excitation	Blank, M.	24
Electrochemistry of Semiconductors: New Problems and Prospects	Pleskov, Y. V. Gurevich, Y. Y.	16
Electrochemistry of the Semiconductor-Electrolyte Interface	Green, M.	2
Electrochemistry of Sulfide Minerals	Koch, D. F. A.	10
Electrochemical Aspects of Stress Corrosion Cracking	Galvele, J. R.	27
Electrochemical Deposition and Dissolution of Alloys and Metal Components—Fundamental Aspects	Despić, A. R. Jović, V. D.	27
Electrode Kinetics	Bockris, J. O'M.	1
Electrodeposition of Metal Powders with Controlled Particle Grain Size and Morphology	Popov, K. I. Pavlovic, M. G.	24
Electrodeposition of Nickel–Iron Alloys	Djokic, S. S. Maksimovic, M. D.	22
Electrogalvanizing	Lindsay, J. H. O'Keefe, T. J.	26
Electrolyte Solutions at High Pressure	Hamann, S. D.	9
Electron Transfer Effects and the Mechanism of the Membrane Potential	Boguslavsky, L. I.	18
Electron Transfer Reactions on Oxide-Covered Metal Electrodes	Schmickler, W. Schultze, J. W.	17
Ellipsometry in Electrochemistry	Paik, Woon-kie	25
Environmental Cracking of Metals: Electrochemical Aspects	Revie, R. W.	26
Equilibrium Properties of Electrified Interphases	Parsons, R.	1
Fundamental and Applied Aspects of Anodic Chlorine Production	Novak, D. M. Tilak, B. V. Conway, B. E.	14

Title	Author	Number
Gas-Phase Ion Equilibria and Ion Solvation	Kebarle, P.	9
Hydration Effects and Thermodynamic Properties of Ions	Desnoyers, J. B. Jolieoeur, C.	5
Impedance Measurements in Electrochemical Systems	Macdonald, D. D. McKubre, M. C. H.	14
Improvements upon the Debye-Hückel Theory of Ionic Solutions	Andersen, H. C.	11
In Situ Surface Electrochemistry: Radioactive Labeling	Wiekowski, A.	21
Interfacial Charge Transfer Reactions in Colloidal Dispersions and Their Application to Water Cleavage by Visible Light	Grätzel, M.	15
Interfacial Electrostatics and Electrodynamics in Disperse Systems	Van Leeuwen, H. P. Lyklema, J.	17
Interfacial Infrared Vibrational Spectroscopy	Pons, S. Foley, J. K. Russell, J. Seversen, M.	17
An Introduction to the Electrochemistry of Charge Transfer Complexes II	Gutmann, F. Farges, J.-P.	13
Ion and Electron Transfer across Monolayers of Organic Surfactants	Lipkowski, J.	23
Ionic Solvation	Conway, B. E. Bockris, J. O'M.	1
Ionic Solvation in Nonaqueous and Mixed Solvents	Padova, J. I.	7
Iron and Its Electrochemistry in an Active State	Drazic, D. M.	19
Lithium Batteries with Liquid Depolarizers	Marincic, N.	15
The Mechanism of Charge Transfer from Metal Electrodes to Ions in Solution	Matthews, D. B. Bockris, J. O'M.	6

Title	Author	Number
The Mechanism of the Electrodeposition of Metals	Bockris, J. O'M. Damjanovic, A.	3
Mechanism of the Hydrogen Electrode Reaction as Studied by Means of Deuterium as a Tracer	Enyo, M.	11
The Mechanism of Oxidation of Organic Fuels	Gileadi, E. Piersma, B.	4
Mechanisms of Stepwise Electrode Processes on Amalgams	Losev, V. V.	7
Mechanistic Analysis of Oxygen Electrode Reactions	Damjanovic, A.	5
Membrane Chlor-Alkali Process	Burney, H. S.	24
Metal Displacement Reactions	Power, G. P. Ritchie, I. M.	11
The Metal–Gas Interface	Oriani, R. A. Johnson, C. A.	5
Metal Hydride Electrodes	Fuller, T. H. Newman, J.	27
Methods and Mechanisms in Electroorganic Chemistry	Humffray, A. A.	8
Microelectrode Techniques in Electrochemistry	Scharifker, B. R.	22
Models for Molten Salts	Bloom, H. Snook, I. K.	9
Modern State of Double Layer Study of Solid Metals	Vorotyntsev, M. A.	17
Molten Electrolytes	Bloom, H. Bockris, J. O'M.	2
The Nickel Oxide Electrode	McBreen, J. Lynn, K. G.	21
NMR Studies of Electrolyte Solutions	von Goldammer, E.	10
NMR Studies of the Structure of Electrolyte Solutions	Covington, A. K. Newman, K. E.	12
Oscillatory Behavior in Electrochemical Systems	Wojtowicz, J.	8
Perspectives in Electrochemical Physics	Vijh, A. K.	17

Title	Author	Number
Phase Transitions in the Double Layer at Electrodes	Benderskii, V. A. Brodskii, A N. Daikhin, L. I. Velichko, G. I.	26
Phenomena and Effects of Electrolytic Gas Evolution	Sides, P. J.	18
Photoelectrochemical Devices for Solar Energy Conversion	Orazem, M. E. Newman, J.	18
Photoelectrochemical Kinetics and Related Devices	Khan, S. U. M. Bockris, J. O'M.	14
Photoelectrolysis and Photoelectrochemical Catalysis	Tributsch, H.	17
Physical Chemistry of Ion-Exchange Resins	Kitchener, J. A.	2
Physical Chemistry of Synthetic Polyelectrolytes	Eisenberg, H. Fuoss, R. M.	1
Physical Mechanisms of Intercalation	McKinnon, W. R. Haering, R. R.	15
Physics and Applications of Semiconductor Electrodes Covered with Metal Clusters	Allongue, P.	23
Potentials of Zero Charge Electrodes	Perkins, R. S. Andersen, T. N.	5
Power Sources for Electric Vehicles	Kordesch, K. V.	10
Preparation and Characterization of Highly Dispersed Electrocatalytic Materials	Kinoshita, K. Stonehart, P.	12
The Present State of the Theory of Electrolytic Solutions	Falkenhagen, H. Kelbg, G.	2
Proton Solvation and Proton Transfer Processes in Solution	Conway, B. E.	3
Proton Transfer in Solution	Erdey-Grúz, T. Lengyel, S.	12
Quantum Chemical Treatment of Adsorbed Species	Blyholder, G.	8
Reaction Engineering and Digital Simulation in Electrochemical Processes	Scott, K.	27

Title	Author	Number
Reaction Kinetics and Mechanisms on Metal Single Crystal Electrode Surfaces	Adžić, R.	21
Recent Advances in the Study of the Dynamics of Electrode Processes	Fahidy, T. Z. Gu, Z. H.	27
Recent Advances in the Theory of Charge Transfer	Kuznetsov, A. M.	20
Recent Developments in Faradaic Rectification Studies	Agarwal, H. P.	20
The Role of Electrochemistry in Environmental Control	Kuhn, A. T.	8
The Role of the Electronic Factor in the Kinetics of Charge-Transfer Reactions	German, E. D. Kuznetsov, A. M.	24
Scanning Tunneling Microscopy: A Natural for Electrochemistry	Sonnenfeld, R. Schneir, J. Hansma, P. K.	21
Small-Particle Effects and Structural Considerations for Electrocatalysis	Kinoshita, K.	14
Solvated Electrons in Field- and Photo-Assisted Processes at Electrodes	Conway, B. E.	7
Solvent Adsorption and Double-Layer Potential Drop at Electrodes	Trasatti, S.	13
Solvent Dipoles at the Electrode-Solution Interface	Habib, M. A.	12
Some Aspects of the Thermodynamic Structure of Electrochemistry	Rysselberghe, P. van	4
Some Fundamental Aspects of Electrode Processes	Khan, S. U. M.	15
Spin-Dependent Kinetics in Dye-Sensitized Charge-Carrier Injection into Organic Crystal Electrodes	Charle, K.-P. Willig, F.	19
Structural and Transport Properties of Perfluorinated Ion-Exchange Membranes	Yeo, R. S. Yeager, H. L.	16
Structural Properties of Membrane Ionomers	Mauritz, K. A. Hopfinger, A. J.	14
The Structure of the Metal-Vacuum Interface	Heiland, W.	11

Title	Author	Number
The Study of Simple Consecutive Processes in Electrochemical Reactions	Bewick, A. Thirsk, H. R.	5
Surface Analysis by Electron Spectroscopy	Baker, B. G.	10
Surface-Enhanced Raman Scattering (SERS)	Efrima, S.	16
Surface Potential at Liquid Interfaces	Llopis, J.	6
Surface States on Semiconductors	Batchelor, R. A. Hamnett, A.	22
Temperature Dependence of Conductance of Electrolytes in Nonaqueous Solutions	Barthel, J. Wachter, R. Gores, H.-J.	13
The Temperature and Potential Dependence of Electrochemical Reaction Rates, and the Real Form of the Tafel Equation	Conway, B. E.	16
Theoretical Aspects of Semiconductor Electrochemistry	Uosaki, K. Kita, H.	18
Theories for the Metal in the Metal-Electrolyte Interface	Goodisman, J.	20
Theories of Elementary Homogeneous Electron-Transfer Reactions	Sacher, E. Laidler, K. J.	3
Theory and Applications of Periodic Electrolysis	Pesco, A. M. Cheh, H. Y. Popov, K. I.	19
Theory of the Effect of Electrodeposition at a Periodically Changing Rate on the Morphology of Metal Deposits	Maksimovic, M. D.	19
Thermodynamics of Membrane Energy Transduction in an Oscillating Field	Markin, V. S. Tsong, T. Y.	24
Transport-Controlled Deposition and Dissolution of Metals	Despić, A. R. Popov, K. I.	7
Transport Models for Ion-Exchange Membranes	Verbrugge, M. W. Pintauro, P. N.	19
Transport Phenomena in Electrochemical Kinetics	Arvia, A. J. Marchiano, S. L.	6

Title	Author	Number
Ultrasonic Vibration Potentials	Zana, R. Yeager, E. B.	14
Water Electrolysis and Solar Hydrogen Demonstration Projects	Sandstede, G. Wurster, R.	27
X-Rays as Probes of Electrochemical Interfaces	Abruña, H. D.	20

Index

Activation control in codeposition, 166, 169
Activities, very small, 164
Activity, of components in alloy, 155
Adiponitrile, its synthesis, 80, 125
Adsorption
    effect on alloy deposition, 173
    irreversible steps in, 18
    of discharging species, 174
    of metal cations onto inert particles, 198
    rate determining step in, 17
    rate of
        at electrode processes, 14
        generalized, 19
        with various models, 16
        rule of, in electrode processes, 13
        rule of, in electrode processes, 13
Advanced electrolysis system, now undergoing commercialization, 456
Aging, effects on stress corrosion crack velocity, 291
Aging time, effect on hardness, 264
Alkire and Gould
    computational methods, 79
    and electrochemical synthesis, 114
Alkire and Lisius
    digital simulation, 103
    and electrochemical synthesis of propylene oxide, 96
Alkire and Mirarefi, computational methods, 78
Alloy deposition
    and adsorption, 173
    and overpotential, 168

Alloy deposition (*cont.*)
    and solid solutions, 156
    and standard potentials, 165
Alloy phase
    and deposition, 150
    stability in solution, 151
Alloy plating, and first understanding, 144
Alloys,
    containing intermediate phases, 212
    deposition of, 143
    their electrochemical weights, 300
    their optimization, 370
Aluminum alloys, 262
    and stress corrosion cracking, 290
Aluminum cell, simulation of operation, 135
Alvarez, and film induced cleavage, 318
Amatore and Saveant
    building blocks in electrode processes, 70
    and modeling of electrochemical reactions, 52
Amorphous cement theory, 250
Analysis, for nickel-metal hydride cell, 369
Analysis of electrochemical reaction, 74
Anodic dissolution
    of iron, oscillatory, in sodium chloride, 388
    of p-silicon, 392
    and stress corrosion cracking, 283
Anodic films, deleterious, 270
Anodic films, and growth kinetics of Schultze, 310

Anodic processes, 384
Appleby, the solar hydrogen economy concept, 419
Asea Brown Boveri, their water electrolysis systems, 461
Automobiles, hydrogen powered, 471
Autonomous photovoltaic hydrogen supply for library in Julich, 490

Banerjee, an equilibrium phase, in alloy deposition, 147
Barge carrier, of liquid hydrogen, 466
Batch reactor, 89
  and concentration as function of time, 92
  with recirculation, 31
Beck
  and repassivation of fresh metal surfaces, 310
  and time dependence in stressing films, 312
Bennion, and review of reaction engineering, 78
Benzene oxidation, with Fenton's reagent, 108
Bianchi and Galvele, susceptibility of pure copper to stress corrosion cracking, 333
Bipolar arrangements, 451
Bipolar cell for water electrolysis, 425
Blum, and microlayered deposits of copper and nickel in electroplating, 183
Bockris
  and anodic behavior of iron during stressing, 278
  and plastic deformation in crack formation, 279
  and solar-hydrogen concept, 419
Boundary conditions for propylene oxidation, 102
Box–Jenkins approach, to oscillatory behavior, 404
Brass, broken by overloading, 234
Brenner's terms, in codeposition, 171

Bromohydrogenation of olefins, 59
Brown, and solution chemistry inside crack, 252
Brownian motion, and oscillatory behavior, 402
Brownian movement and effect on oscillations, 400

Caban and Chapman, solving current distribution problems, 4
Cadmium alloys and copper deposition, 213
Carranza and Galvele
  on anodic oxide films which stop stress corrosion cracking, 271
  repassivation, 309
Cascade, in plug flow behavior, 105
Categories, of electrochemical rate models, 10
Cathodic overpotential, 167
Cathodic reduction, 106
Cells and incorporation of materials into metal reactors, 202
Cell design, for metal hydrides, 373
Cell stack, for Lurgi pressure electrolyzer, 443
Chao, and migration of oxygen in anion vacancies, 346
Chemical changes, inside cracks, 257
City bus, with hydrogen, 468
Clifton and Savinel
  and batch reactors with Fenton's reagent, 107
  use of Fenton's reagent, 73
Codeposition
  activation control, 166, 169
  of alumina on copper, 204
  at equilibrium, 161
  induced, 182
  of inert particles in metals, 201
  irregular, 160
  of metals and particles, 197
  of metals, in their electrochemical processes, 145
  the various stages, 203

Commercialization, of advanced systems for water splitting, 457
Companies, having hydrogen electrolyzers, 440
Concentration
 as a function of time, in batch reactor, 92
 as a function of time for potentiostatic operations, 94
 during propylene oxidation, 103
 of species inside cracks, 259
Concentration profile, in reaction layers, 133
Conclusions, for solar hydrogen economy, 504
Content of metal, in alloy deposition, 188
Continuous cleavage, and effect on stress corrosion cracking, 315
Continuous flow, stirred tank reactor (CSTR), 34, 35
Conversion equations, for epoxidation of propylene, 100
Copper deposit, and incorporation of alumina, 200
Copper-cadmium, 219
Costs, for renewable electricity and hydrogen, 501
Coupling of solar electric sources, with electrolyzer technology, 455
Crack chemistry measurements, 252
Crack growth, the Doig and Flewitt analysis, 285
Crack, its tip, 241
Crack philosophies, as a function of stress, 287
Crack propagation, 240, 272
 and Doig and Flewitt's theory, 288
 and surface mobility, 347
 as a function of surface compounds, 336
Crack provocation rate versus crack tip stress intensity, 244
Crack tip, and vacancy capture, 323
Crack tip chemistry, 252

Crack velocities, 305
Crack velocity, and current-potential, 306
Cracking
 and acoustic emission, 319
 environmentally induced, 235
 history of, 246
 range of phenomena, 245
 and rate controlling steps, 239
Cracks
 changes in concentration within, 254
 and chemical changes inside them, 257
 and concentration of species in them, 259
Cross sections of electrodes, 445
Current density
 at edges, 280
 increased by straining, 282
 maximum, as a function of strain rate, 279
Current efficiency, for some reactors, 29
Current pulse strain, in alloy deposition, 192
Current pulsing and metal deposition, 190
Current voltage relations for water electrolysis, 426
Cyanide complexes, their distribution as a function of concentration, 163
Cyanide systems, and oscillatory behavior, 385
Cyclic voltammetry, 62

DECHEMA study, of intercontinental hydrogen, 454
Decomposition, of solvent, during reactor function, 27
Deposit composition, and current density, in anomalous codeposition, 181
Deposition of alloys, 143
 and Despic, 143
 and Jovic, 143

Deposition of noble metals, 221
Despic
 and anodic behavior of iron, 278
 and current density at edges, in practical solution, 280
 and deposition of alloys, 143
 and effect of stresses, 283
 and quantative theory of laminar deposition, 183
Despic and Jovic, effect of current pulses in deposition, 184
Despic *et al.*, and effect of elastic strain on stress corrosion cracking, 275
Dewald, and surface reactions, 387
Differential equations, formulation of, in reactor engineering, 85
Diffusion
 basic equations for, 11
 in stress corrosion cracking, 337
Diffusion film, in thin film reactor, 47
Diffusion layer, its growth into the solution, 195
Diffusion limiting currents, and alloy deposition, 178
Digital simulation
 and Alkire, 103
 and electrode processes, 63
 and future developments, 73
 and work of Haines, 59
Dissolution
 of alloys, 206
 anodic, and filmed materials, 297
 of binary alloys, 222, 225
 of cobalt, oscillatory, 389
 of cobalt, oscillatory, in phosphoric acid, 390
 of iron, 387
 of nickel, oscillatory, in various media, 391
 for pure metal, schematic, 294
Distribution, of current density, in electrochemical cells, 2
Divisek, and hydrogen electrolysis on nickel zinc, 437

Doig and Flewitt
 analysis of crack growth 285
 their analysis of stress corrosion cracking, 286
 and plastic deformation at the tip of the crack, 314
Doig and Flewitt's mechanism, and available experimental data, 287
Doig and Flewitt model, in concentration gradients in the crack, 283
Double layers, thickness of, 130
Dual-current pulse laminar deposition, 189
Duffo and Galvele diffusion overpotential in stress corrosion cracking, 338
 in stress corrosion cracking of silver–palladium, 333
Dynamic phenomena, their analysis, 399
Dynamics, of electrode processes, 378

ECE, a response involving it, 38
Edeleanu and Forty, and the mechanism of cracking of brass, 316
Edwards and Newman and model for concentration distribution in parallel plate reactor, 113
Effect of hydrogen in stress corrosion cracking, 341
Effect of yield stress on crack velocity, 288
Efficiency
 for hydrogen plants, 491
 in water electrolysis, 423, 432
Elastic strain, 273
Elastic strain and hydrogen evolution, 276
Elastic stresses, and thermodynamic treatment, 276
Electro-organic synthesis, in flow through porous electrode, 114

Index

Electrocatalysis
  for hydrogen evolution, and Trasatti, 439
  and reaction mechanisms in water electrolysis, 427
  for water dissociation, 422, 437
Electrochemical deposition, of alloys, 143
Electrochemical dynamics, 383
Electrochemical formation, of p-aminophenyl, 90
Electrochemical processes, and codeposition, 145
Electrochemical rate models, categories of, 10
Electrochemical reaction analysis, 74
Electrochemical reaction systems, and rate processes, 6
Electrochemical reactions
  in engineering models, 96
  and mathematical modeling, 81
  and rate processes, 7
Electrochemical reactor
  sieve plate, 134
  three phase, for benzene oxidation, 117
Electrochemical reactors, and dispersion approach, 44
Electrochemical techniques, and the phase structure, 205
Electrochemically generated species, 126
Electrode deposition, of copper and cadmium alloys, 212
Electrode kinetics
  metal hydrides, 365
  and reaction rate modeling, 5
Electrode processes and digital simulation, 63
Electrodeposition
  of alloys, cathodic reactions, 159
  of alloys, in a microstructural analysis, 218
  under current pulses, 185
  and inclusion of nonmetallic alloys, 197
Electrolysis, exhaustive, 71

Electrolytes, used in water electrolysis, 435
Electrolytic cell model, 98
Electrolytic production of hydrogen, why it is increasing, 413
Electrolytic transport, and the metal hydride electrode, 368
Electrolyser, Inc., their electrolysis systems, 460
Electromotive force of silver electrodes, 274
Electron concentration and phase transitions, 157
Embrittlement
  liquid metal, 345
  of metal, 241
Energy, in self sufficient house, diagrammed, 485, 486
Energy sources, for generation of hydrogen, 415
Engineering models
  analytical, 75
  for complex electrochemical reactions, 96
Entropy, of water electrolysis and energy consumption, 424
Epoxidation, of propylene, 100
Equations
  for batch type reactor, 26
  general, for electrochemical processes, 8
  kinetic, for electrochemical reactions, 7
Equilibrium, codeposition, 161
Equilibrium potential, of an alloy in contact with solution, 147
Euro-Quebec hydrogen scheme, 464, 465
Eutectic alloy, 154
Eutectic type of alloys, 206

Face centered cubic lattice, 360
Face centered hydrides, 360
Faraday meeting, 1921, 249
Fast reactions and rotating disk electrode, 58

Fast straining of steel, and current–time curves, 312
Faust, and alloy deposition, 144
Feldburg
    and digital simulation, 64
    and equations for electrochemical reactions, 77
Fenton's reagent, work of Clifton and Savinel, 73
Fick's Law and alloy deposition, 177
Film induced cleavage
    its advantages, 318
    models, 318
Films, in anodic dissolution, 297
Finite difference equations in synthesis of 2-butanone, 122
Fischer, and nonequilibrium phase, in alloy deposition, 147
Fixed bed reactor
    and logic diagram, 117
    Olomann and Riley, 116
Flanagan, and kinetic measurements for hydrides, 365
Fleischmann, and analytical solutions for flow reactor, 76
Floating potential, 153
Flood, and thermodynamic treatment of elastic stresses, 277
Flow models, real and nonideal, 42
Flow pattern, in reactor for chlorination, 131
Flow through porous electrode, and electrochemical synthesis, 114
For and Anderson and slip step dissolution, 307
Forty's model, in stress corrosion cracking, 317
Fractional conversion, and mass transport control conditions, 33
Fractography, in the development of mechanisms of growth processes, 313
Frank–Reed source, and multiplication of dislocations, 250

Fuller, metal hydride electrodes, 359
Future developments in digital simulation, 73

Galvanostatic operation, 90
Galvele, electrochemical aspects of stress corrosion cracking, 233
Gas/liquid, systems, 129
GEC Alsthon group in France, 459
Generation of hydrogen, by means of electrolytic processes, 421
Geometrical techniques, portrayal for oscillatory systems, 399
Gerberich and Chen, and acoustic emission during cracking, 319
Gibbs energy
    for alloys, 149
    of phase formation, 148
    for alloys, 149
Gorbunova and Polukarov, alloy deposition, 144
Grain bodies, in stress corrosion cracking, 265
Grain boundaries, and stress corrosion cracking, 269
Gravano and Galvele, and models of concentrations inside cracks, 260
Greenland, and hydro power, 473
Griffith, his crack, 243
Grovonova and Polukarov
    Tafel lines in alloy deposition, 172
    work on alloy deposition, 172
Growth models, for anodic films, 311
Guglielmi, theory of codeposition, 201

Haines and digital simulation, 59
Hatta's number, 11
Hawaii, wind energy storage test facility, 475
Heyrovksy mechanism, 430
High index planes, and dissolution in stress corrosion cracking, 282
Hines, and theory of stress corrosion cracking, 281

Index

History, for solar hydrogen, 416
Hoar and Galvele, 299
Hoar and Hines, stress corrosion cracking mechanism: anodic dissolution at the tip, 278
Hoar and Scully, currents during straining, 298
Hoar, and the strained rate technique, 298
Hoar and West, continuously straining metals, 298
Horkans, hydroxide suppression mechanism, 180
Hudson, and bifurcation points, 400
Hume-Rotery, and phase transitions as connected with electron concentration, 157
Hy-solar project, Saudi Arabia, 482
Hydro power from Iceland, 472
Hydro power/hydrogen demonstration, 463
Hydrodynamic effects, as seen by Kariapper and Foster, 202
Hydrodynamics, for incorporation of γ alumina into a copper deposit, 200
Hydrogen
   activities, worldwide, outlook, 493
   demonstration programs, listed, 467
   demonstration projects, 411
   effects in stress corrosion cracking, 341
   electrolysis, at high temperatures, 432
   energy systems, in international hydrogen trade, 499
   evolution and stress corrosion cracking, experimental observations, 332
   as a fuel, 414
   fuel for ground vehicles, worldwide, 503
   plant, 470
   related activities, in Europe, 497
   related activities, in rest of world, 498
   related activities, United States and Canada, 495
Hydrogen (*cont.*)
   storage capacity, 365
   in transportation sector, 502
Hydrogen systems, Inc., 458

Iceland, and hydro power, 472
Induction time, and dissolution of cracks, 237
Inert particles, and adsorption of metal cations, 198
Intergranular stress corrosion cracking, 268
Intermediate phases
   and alloy deposition, 157
   and alloys, 212
Intermediates, lifetime less than 10 microseconds, 68
Intermetallic alloys, major classes of, 361
Intermetallic compounds, in alloy deposition, 159
Internal stresses, and provocation of cracking, 247
Irreversible step, in adsorption, 18
Iwakura, and self discharge in metal hydrides, 375

Johnson, a classic case of stress corrosion cracking, 248
Johnson–Mehl equation, for time dependence of kinetics in alloy deposition, 228
Jovic and deposition of alloys, 143
Julich, and water electrolysis, 457
Justi, the solar hydrogen concept, 419

Karp and Meites, and reactions in the diffusion layer, 51
Kelly, and limitation of models and stress corrosion cracking, 348
Krebskosmo, its electrolyzer, 447

Laminar deposition
   and Despic, 183
   theory of, 186

Laminar deposits and pulsing currents, 183
Laminar, electrodeposition of metals, 183
Landolt, and transition from selective to simultaneous dissolution, 334
Lash Miller, mathematical treatment of concentration changes, 254
Lashmore and Dariel, complex pulses, 196
Lave, phases, and hydride electrodes, 371
Lea and Hondros, importance of impurity content in stress corrosion cracking, 269
Leak, and a survey of failure in military aircraft, 248
Ledjeff, and Heinzel, bifunctional electrodes, 448
Levich, and the theory of rotating disk electrodes, 191
Lewis, his monograph on palladium hydrogen systems, 362
Linear sweep voltammetry, and alloy deposition, 179
Liquid metal embrittlement, 345
Liquid–liquid reactions, 124
Localized paths, in stress corrosion, 261
Logic diagram, for numerical analysis for fixed bed reactor, 117
Low alloy steels, their cracking, 309
Lurgi electrodes, planar concepts, 453
Lurgi
  having largest electrolysis plants, 441
  pressure electrolyzer, diagrammed, 442
  rolled plate, 438
  having largest electrolysis plants, 441

Machado and Chapman, and reaction plane problem, 60
Magdowski and Speidel, stress corrosion cracking in steam generators, 289, 341

Maier, and stress corrosion cracking in silver–gold, 321
Markin, and protection of alloys in metal hydrides, 373
Mass balance, and plug flow reactor, 40
Mass balance reaction, 40
Mass transfer, and the reaction rate model, 9
Mass transfer equations in organic reduction reactions, 87
Material
  used in water electrolysis, 433
  in water electrolysis, listing of parts, 434
Mathematical modeling, of electrochemical reactions, 81
Mechanics, and stress corrosion cracking, 251
Mechanism, as a predictive tool, 347
Mechanisms
  and active dissolution, 272
  their limitations in stress corrosion cracking, 293
  of stress corrosion cracking, their limitations, 312
Mediated oxidation of toluenes, 127
Metal hydride electrode, electrolytic transport, 368
Metal hydride electrodes, 359
  and thermal management, 378
Metal hydrides
  and electrode kinetics, 365
  compared with other battery materials, 376
  and thermodynamics, 362
Metal polishing, in phosphoric acid, 386
Metals
  and laminar electrodeposition, 183
  and layered deposits, 183
Methods, steady state, 52
Microstructural analysis of alloys, in electrodeposition, 218
Microstructure, of copper alloys, 263
Military aircraft, and survey of failure, 248

Index

Model calculations for pits, crevices, and cracks, 253
Model equations, for synthesis of 2-butanone, 121
Model
   for rate of adsorption, 15
   of thin film electrochemical reactor, 45
Modeling
   of electrochemical reactors, 22
   mathematical, of multiple reaction, 2
   of reaction rates, 4
Models, and block diagrams, 84
Models of reactor engineering, 74
Molar yield, as function of potential, 91
Molten metal hydrides, in water electrolysis, 435
Mott and Jones, and Brillouin phases, 157

Nernst equation, and deposition of alloys, 146
Neunburg vorm Wald project, 481
Newman and linearizing equations, 66
Newman
   electrochemical reactions in pores, 20
   and metal hydride electrodes, 359, 366
   and model for laminar flow, 110
   and nucleation at the crack tip, 318
Nguyen, and two dimension model in flow reactor, 110
Nickel-zinc as alloys, 179
Nitrobenzene
   and motor yield, in potentiostatic operation, 95
   reduction of, 54
   reduction scheme, 55
Nobe and coworkers, copper dissolution processes, 384
Nonmetallic materials, inclusion of, in deposition, 197
Norwegian hydro energy, for Germany, 469

Notten and Einerhand, exchange current densities for AB5 materials, 366
Nove and Tan, electrode potential response, 275
Numerical methods
   in electrochemical engineering, 76
   listed, 3
Numerical simulation, and product distribution, 67

Okamoto, oxidation of formic acid, 395
Oldenberg, and the solar wind plant for hydrogen, 478
Olefins, their bromohydrination, 59
Olomann and Riley
   and fixed bed reactors, 116, 123
   the oxidation of benzene, 125
Open circuit potential, and metal hydrides, 364
Organic oxidation reactions, from an oscillatory viewpoint, 394
Orowan, and modification of Griffith crack, 243
Oscillations, a summary, 405
Oscillations, and reduction of hydrogen peroxide, 397
Oscillatory anodic dissolution, of copper, 387
Oscillatory behavior
   and the Box–Jenkins approach, 404
   diagrammed, 403
Oscillatory patterns, the solution reactions, 384
Overcharge mechanisms, for metal hydrides, 374
Overpotential
   cathodic, 167
   concept of, in alloy deposition, 168
Ovshinsky, and 400 amp hour per kg current alloys, 372
Oxidation
   of hydrogen, 393
   of silicon, 391
   of silicon in various media, 392

Oxidation systems, numerous, 395
Oxygen bridge complexes, 428

p-silicon, anodic dissolution, 392
Packed bed reactor and Yu, 119
Parallel plate reactor
  diagrammatic, 111
  equations for, 112
  models, 110
Parkins, and anodic dissolution with films, 297
Parkins and Holroyd, and straining electrodes, 298
Parkins
  and slip step dissolution, 307
  and stress corrosion cracking, 270
Passive state, and stress corrosion cracking, 345
Perturbation methods and analysis, 50
Phase composition, and alloy deposition, 214
Phase diagram
  and alloy deposition, 207
  of zinc and nickel systems, 223
Phase diagrams
  copper and nickel, 209
  for copper and cadmium, 215
Phase structure of alloys by electrochemical techniques, 205
Phase transformation kinetics, 226
Phases
  distribution of in electrochemical reactions, 158
  identification of, in thin layer alloys, 205
Photovoltaic hydrogen plant, in Helsinki University, 490
Photovoltaic-wind hydrogen installation, University of Oldenberg, 478
Photovoltaics and water electrolyzers, 454
Pickett, and design of batch flow reactors, 75
Pines, his value for distortion energy, 157

Pit model, 255
Plasma spraying technique, and preparation of electrodes for hydrogen electrolysis, 448
Plastic deformation, and cleavage, 318
Plastic strain
  on film free metal, 278
  on filmed metals, 298
Pletcher, and intermediates in electrochemical reactions, 65
Plot diagram for solar hydrogen project, 494
Plug flow
  behavior, and cascade presentation, 105
  and first order kinetics, 39
  reactor, 32, 36, 37, 39
Polarization, and stress corrosion cracking, 303
Pore, in three dimensional structures, 115
Pores, diffusion in, 21
Porous bed electrodes, 113
Potential response
  in corresponding partial currents, 187
  for current pulses, 189
Potential sweep, 62
  and cyclic voltammetry, 62
Potentiostatic operation
  and current as a function of time, 94
  and nitrobenzene reduction, 95
Potentiostatic step dissolutions, 220
Prentice and Tobias, on numerical methods of current distribution, 3
Processes, near electrode, 12
Product distribution, in oxidation reactions, 116
Propagation, of cracks, 240, 320
Propylene and reaction scheme, 97
Propylene oxide, 96
Proton conducting solids, 436
Proton conductors, 436
Pushin, first measurements of potentials of alloys in solution, 152

Index

Raney, nickel, 436
Rate controlling step, in cracking, 239
Rate equations, for adsorption, 18
Rate, of adsorption, generalized, 19
Rate processes, and electrochemical reaction systems, 6
Raub and Sautter, and phases in silver–thallium alloys, 157
Reaction chemistry for synthesis of 2-butanone, 120
Reaction layer, 69
Reaction layers, in concentration profile, 133
Reaction loop, two phase, 126
Reaction model
  for ECE response, 38
  in electrochemical engineering, 86
Reaction modeling, 49
  and electroanalytic experiments, 49
Reaction models, based on resonance time, 43
Reaction order, of phenyl hydroxylamine reduction, 57
Reaction rate model
  and mass transfer, 9
  and mass transfer coefficient, 30
Reaction rate modeling, and electrode kinetics, 5
Reaction scheme for epoxidation of propylene, 97
Reaction sequence, 83
Reaction zone, for fast reactions, 12
Reactions, liquid–liquid, 124
Reactor
  batch type, 25
  for chlorination in flow pattern, 131
  conservation equations, in cathodic reduction, 106
  continuous flow type, 4
  engineering models, 74
  modeling, 22
  models, 80
  models, for two phase electrochemical processes, 125
  plug flow type, 32

Reactor (*cont.*)
  stirred tank, 34
  well mixed, diagrammatic, 99
Reactors
  batch and continuous, 24
  ideal, 24
  and various configurations in, 23
  view of, 24
Real, and nonideal flow models, 42
Recycling, plugged flow reactor, 37
Reduction of indium, and oscillations, 396
Reduction
  of nitrobenzene, 54
  of tin, under oscillation conditions, 398
  of zinc, and oscillations, 396
Repassivation, 307, 308
Repassivation, and Carranza and Galvele, 309
Research needs, for metal hydrides, 377
Resonance time and reactor models, 43
Rhead, and effect of contaminants on stress corrosion cracking, 327
Rosenhain, and Archbutt, and stress corrosion cracking in 1919, 247
Rotating disk electrode, 52
  and alloy deposition, 199
  applications of, 53
  and fast reactions, 58
  and other applications, 61
  and Levich equation, 191
Rotating disk system, and chemical kinetic measurements, 132
Runge–Kutta procedure, 67, 128

Sakai, and heat treatment for microstructure in metal hydride alloys, 371
Schell, bifurcation, points, 400
Schlapbach, and review of properties in metals and hydrides, 365
Schmidt number, 46
Schultze, and growth kinetics of anodic films, 310

Scott
  and complex reaction scheme, 76
  and mass transport models, 79
Scott and Hayati, an industrial synthesis, 125
Self discharge, for metal hydrides, 375
Shib and Lee, simulation of electrochemical reactions in parallel plate cells, 126
Shuck and Swedlow, transport inside a crack, 261
Side reactions, with super oxide anion, 431
Sieve plate electrochemical reactor, 134
Silver alloys, their stress corrosion cracking, 333
Silver–palladium system, 210
Simulation
  and current yields, 109
  digital, in electrochemical processes, 1
  in propylene formation, 101
  program, 86
Simulations in batch reactors, 93
Slip, step dissolution, 307
Small membrane electrolyzer, 449
Solar house, 488
Solar hydrogen
  and its inventors, 419
  background, 412
  demonstration, 411, 479
  economy, realization of, 505
  its history, 415
  projects, Bavaria, 480
  supply, decentralized, 492
  tests, 479
Solar Hydrogen Economy: conclusions, 504
Solar thermal collectors, 487
Solar village, Riyadh, 484
Solid, distribution of cracks, 236
Solid solution type alloys, 208
Solid solutions, and alloy deposition, 156
Solution chemistry inside crack, 251
Species, electrochemically generated, 126

Specific resistance, of mixed oxides used in electrodes, 452
Speidel
  and crack velocity measurements in chloride containing solutions, 338
  and sharp cracks, 295
  and stress corrosion cracking of aluminum, 256
Spitzer, and understanding of alloy deposition, 144
Stability constant, used in stress corrosion cracking calculations, 258
Stability constants, and alloy deposition, 162
Stability in solution, of alloy phase, 151
Stability, of metal hydride alloys, 372
Staehle, and review of slip step dissolution, 307
Stainless steels, sensitized, 266
Standard potentials, formulated for alloy deposition, 165
Steady state condition, in alloy deposition, 177
Steam boilers, explosions of, 247
Steam turbine rotors, for steel, 343
Steels for steam turbine rotors, 343
Step dissolution, potentiostatic, 220
Stirred tank reactor, 33
Strain rate effects, on stress corrosion cracking, 302
Straining materials, with films, 299
Stress corrosion cracking
  of aluminum alloys, 247
  and bare metal polarization, 303
  and continuous cleavage, 315
  defined, 236
  and diffusion overpotential, 338
  and the early work of Johnson, 248
  and Galvele, 233
  of iron, in water, 340
  and limitation of models, 348
  and limits of mechanisms, 313
  mechanisms based on surface mobility, 321

Index

Stress corrosion cracking (cont.)
  and salt films, 256
  and strain rate effects, 302
  summary of models, 349
  and surface diffusion, 326
  two stage model, 317
Stress intensity
  factor, 242
  and size of anodic front, 292
  and stress corrosion cracking, 296
Stressed steel, diagrammed, 304
Surface compounds, and stress corrosion cracking, 335
Surface diffusion, and stress corrosion cracking, 326
Surface mobility
  and stress corrosion cracking, 322
  in stress corrosion cracking, 321
Swinney, and bifurcation points, 400

Tafel lines
  and partial current densities, 170
  in alloy deposition, Grovonova and Polukarov, 172
Tafel plot
  and alloy deposition, 179
  for nitrobenzene reduction, 56
Tafel reaction, with solvent decomposition, 28
Tafel relationships for anodic and cathodic reactions in cracking, 284
Teledyne Energy Systems, 460
Temkin, and film induced cleavage, 318
Temperature effects, on stress corrosion cracking, 338
Tench and White
  complex pulse application in electrodeposition, 196
  microlayers in metal deposition, 183
Tensile stress, at tip of crack, 324
Theories, modern, of stress corrosion cracking, 249

Theory
  amorphous cement type, 250
  of consecutive reactions involving diffusion, 88
  of pulse deposition, 193
Thermal management, critical metal hydride electrodes, 378
Thermodynamics, of metal hydrides, 362
Thin film, model of thin film reactor, 45
Thin gap models, 48
Thomson and genesis of cracks, 243
Three phase electrochemical reactor, 117
Tip of crack, 241
  tensile stress, 324
  undergoing dissolution, 295
Transition layer thickness, pulse deposition, 194
Transport control, in codeposition, 175
Trasatti, and hydrogen electrolysis, 439
Trickle flow reactor, 118
Tripkovic, electrocatalysis of formic acid solutions, 394
Turnbull
  and solution chemistry in cracks, 253
  his work on mathematical modeling in cracks, 257
Two phase reaction loop, 126
Type of alloy, and Gibbs energy of various types, 153
Tzedakis and Savinel, simulation of indirect oxidation, 128

Vacancies
  and interaction of hydrogen therewith, 331
  sources of, in stress corrosion cracking, 342
Vacancy capture, at crack tip, 323
Vacancy mechanisms for stress corrosion cracking, 342
Vandenborre, and fuel cell commercialization, 444
Verbrugge and Tobias, theoretical treatment of composition of a deposit, 183

Veziroglu, the solar hydrogen economy concept, 419
Vielstich, and Raney, nickel electrodes, 436
Vitanen, and mathematical model for metal hydrides, 369
Volatile systems, and generation of gas, 133
Volta's battery, 417
Voltammograms
   and phase diagrams, 215
   for the solution of alloys, 211

Wagner and Traud, and basic theory of corrosion, 251
Water electrolysis, 411
   and classification thereof, 420
   technical state of the art, 440
   as seen in 1900, 418
   coupled with hydro power, 454
Water electrolyzers, large, 446
Weise, and engineering models for batch reactors, 105
Wendt and Plzak, addition of anodic degenerated radicals, 73
White, and development of models to predict rotating disk behavior, 79

Willens, and the lanthanum nickelide electrode, 367
Wind energy and water electrolyzers, 454
Wind hydrogen installation, in Wiesbaden, Germany, 477
Wind power, and hydrogen demonstration projects, 474
Winter, the solar hydrogen economy concept, 419
Wires, and stress corrosion cracking, 301
Woodridge and oxidation of bromine, 104
Wright and simulation of operation of aluminum cell, 135
Wurster, and water electrolysis, 411

X-ray analysis and phase composition in alloy deposition, 216
X rays and other means of analysis, 217

Yield stress, and crack propagation, 289
Yu, and mathematical modeling of packed bed reactor, 229
Yung and Alkire
   generation of gas, effects on volatile systems, 133
   product distribution, 80
   simulation of electrode reactions, 129